REPRESENTATION THEORY OF ALGEBRAS

PURE AND APPLIED MATHEMATICS

A Program of Monographs, Textbooks, and Lecture Notes

Executive Editors

Earl J. Taft
Rutgers University
New Brunswick, New Jersey

Edwin Hewitt
University of Washington
Seattle, Washington

Chairman of the Editorial Board

S. Kobayashi
University of California, Berkeley
Berkeley, California

Editorial Board

Masanao Aoki
University of California, Los Angeles

Glen E. Bredon
Rutgers University

Sigurdur Helgason
Massachusetts Institute of Technology

G. Leitman
University of California, Berkeley

W.S. Massey
Yale University

Irving Reiner
University of Illinois at Urbana-Champaign

Paul J. Sally, Jr.
University of Chicago

Jane Cronin Scanlon
Rutgers University

Martin Schechter
Yeshiva University

Julius L. Shaneson
Rutgers University

Contributions to *Lecture Notes in Pure and Applied Mathematics* are reproduced by direct photography of the author's typewritten manuscript. Potential authors are advised to submit preliminary manuscripts for review purposes. After acceptance, the author is responsible for preparing the final manuscript in camera-ready form, suitable for direct reproduction. Marcel Dekker, Inc. will furnish instructions to authors and special typing paper. Sample pages are reviewed and returned with our suggestions to assure quality control and the most attractive rendering of your manuscript. The publisher will also be happy to supervise and assist in all stages of the preparation of your camera-ready manuscript.

LECTURE NOTES
IN PURE AND APPLIED MATHEMATICS

1. *N. Jacobson,* Exceptional Lie Algebras
2. *L.-Å. Lindahl and F. Poulsen,* Thin Sets in Harmonic Analysis
3. *I. Satake,* Classification Theory of Semi-Simple Algebraic Groups
4. *F. Hirzebruch, W. D. Newmann, and S. S. Koh,* Differentiable Manifolds and Quadratic Forms
5. *I. Chavel,* Riemannian Symmetric Spaces of Rank One
6. *R. B. Burckel,* Characterization of C(X) Among Its Subalgebras
7. *B. R. McDonald, A. R. Magid, and K. C. Smith,* Ring Theory: Proceedings of the Oklahoma Conference
8. *Y. T. Siu,* Techniques of Extension of Analytic Objects
9. *S. R. Caradus, W. E. Pfaffenberger, and B. Yood,* Calkin Algebras and Algebras of Operators on Banach Spaces
10. *E. O. Roxin, P.-T. Liu, and R. L. Sternberg,* Differential Games and Control Theory
11. *M. Orzech and C. Small,* The Brauer Group of Commutative Rings
12. *S. Thomeier,* Topology and Its Applications
13. *J. M. López and K. A. Ross,* Sidon Sets
14. *W. W. Comfort and S. Negrepontis,* Continuous Pseudometrics
15. *K. McKennon and J. M. Robertson,* Locally Convex Spaces
16. *M. Carmeli and S. Malin,* Representations of the Rotation and Lorentz Groups: An Introduction
17. *G. B. Seligman,* Rational Methods in Lie Algebras
18. *D. G. de Figueiredo,* Functional Analysis: Proceedings of the Brazilian Mathematical Society Symposium
19. *L. Cesari, R. Kannan, and J. D. Schuur,* Nonlinear Functional Analysis and Differential Equations: Proceedings of the Michigan State University Conference
20. *J. J. Schäffer,* Geometry of Spheres in Normed Spaces
21. *K. Yano and M. Kon,* Anti-Invariant Submanifolds
22. *W. V. Vasconcelos,* The Rings of Dimension Two
23. *R. E. Chandler,* Hausdorff Compactifications
24. *S. P. Franklin and B. V. S. Thomas,* Topology: Proceedings of the Memphis State University Conference
25. *S. K. Jain,* Ring Theory: Proceedings of the Ohio University Conference
26. *B. R. McDonald and R. A. Morris,* Ring Theory II: Proceedings of the Second Oklahoma Conference
27. *R. B. Mura and A. Rhemtulla,* Orderable Groups
28. *J. R. Graef,* Stability of Dynamical Systems: Theory and Applications
29. *H.-C. Wang,* Homogeneous Banach Algebras
30. *E. O. Roxin, P.-T. Liu, and R. L. Sternberg,* Differential Games and Control Theory II
31. *R. D. Porter,* Introduction to Fibre Bundles
32. *M. Altman,* Contractors and Contractor Directions Theory and Applications
33. *J. S. Golan,* Decomposition and Dimension in Module Categories
34. *G. Fairweather,* Finite Element Galerkin Methods for Differential Equations
35. *J. D. Sally,* Numbers of Generators of Ideals in Local Rings
36. *S. S. Miller,* Complex Analysis: Proceedings of the S.U.N.Y. Brockport Conference
37. *R. Gordon,* Representation Theory of Algebras

Other Volumes in Preparation

REPRESENTATION THEORY OF ALGEBRAS

Proceedings of the Philadelphia Conference

edited by

Robert Gordon
Department of Mathematics
Temple University
Philadelphia, Pennsylvania

MARCEL DEKKER, INC. New York and Basel

Library of Congress Cataloging in Publication Data

Philadelphia Conference, 1976.
Representation theory of algebras.

(Lecture notes in pure and applied mathematics)
1. Representations of algebras--Congresses.
I. Gordon, Robert, [Date] II. Title.
QA150.P48 1976 512 78-5798
ISBN 0-8247-6714-4

MARCEL DEKKER, INC.

270 Madison Avenue, New York, New York 10016

Current printing (last digit)
10 9 8 7 6 5 4 3 2 1

PRINTED IN THE UNITED STATES OF AMERICA

PREFACE

This collection of papers is the fruit of a National Science Foundation sponsored research conference on representation theory. The conference was held at Temple University's Sugar Loaf facility, May 24-28, 1976. In recognition of his preeminence among American workers in the representation theory of finite dimensional algebras, Maurice Auslander was given the task of delivering approximately half the lectures presented. The reader will find that this is reflected in the text. I wish to thank the contributors for the papers contained therein.

Also I wish to thank Professors G. W. Johnson and M. G. Steisel in their respective capacities as Dean and Chairman of Temple University's College of Liberal Arts and Faculty Senate Lectures and Forums Committee for financial support. The chief donor to the conference was the National Science Foundation, to which institution I am indebted. Finally, I wish to thank my wife, Muriel Gordon, who as secretary to the conference, was instrumental in making things run smoothly.

R. Gordon

CONTENTS

CONTRIBUTORS

MAURICE AUSLANDER
Department of Mathematics, Brandeis University, Waltham, Massachusetts

V. DLAB
Department of Mathematics, Carleton University, Ottowa, Ontario, Canada

K. L. FIELDS
Department of Mathematics, Rider College, Trenton, New Jersey

KENT R. FULLER
Department of Mathematics, University of Iowa, Iowa City, Iowa

EDWARD L. GREEN
Department of Mathematics, Virginia Polytechnic Institute, Blacksburg, Virginia

GERALD J. JANUSZ
Department of Mathematics, University of Illinois, Urbana, Illinois

MARÍA INÉS PLATZECK
Department of Mathematics, University of Illinois, Urbana, Illinois

IRVING REINER
Department of Mathematics, University of Illinois, Urbana, Illinois

C. M. RINGEL
Department of Mathematics, University of Bonn, Bonn, Federal Republic of Germany

MURRAY SCHACHER
Department of Mathematics, University of California at Los Angeles, Los Angeles, California

R. B. WARFIELD, JR.
Department of Mathematics, University of Washington, Seattle, Washington

REPRESENTATION THEORY OF ALGEBRAS

FUNCTORS AND MORPHISMS DETERMINED BY OBJECTS

Maurice Auslander
Brandeis University
Waltham, Massachusetts

Introduction. In the course of our study of the representation theory of artin algebras, I. Reiten and I introduced the notions of right or left almost split morphisms, as well as almost split sequences, of finitely generated modules over artin algebras. In view of the important role these notions have played in elucidating some questions in the representation theory of artin algebras (see [1] for an account of some of the results that have been obtained along these lines), it is natural to wonder if these ideas are special to artin algebras or are of broader interest. This is an especially tempting question since nowhere in the definitions of these notions is the fact that one is dealing with finitely generated modules over artin algebras used. In fact, the formulations given for artin algebras of all these notions make sense in arbitrary abelian categories and some even in arbitrary categories. Moreover, many of the proofs given for artin algebras carry over verbatim to these categorical settings. Thus the problem of generalization boils down basically to the question of when these types of morphisms and exact sequences exist.

This point of view was pursued in [3] where it was shown that right or left almost split morphisms and almost split sequences do indeed exist in more general contexts than artin algebras. That paper dealt with

[1]Written with partial support of NSF Grant MCS 72-04584

various categories of modules over rings Λ which are algebras over complete noetherian local rings R such that Λ is a finitely generated R-module. Amongst other things, it was shown that if R is a complete discrete valuation ring and Λ is an R-order in the classical sense, then the category of Λ-lattices has a theory of right and left almost split morphisms and almost split sequences entirely analogous to that for the category of finitely generated modules over artin algebras.

In this paper we generalize our previous work in a somewhat different direction. Instead of finding various new contexts in which right or left almost split morphisms exist, we introduce a new type of morphisms, those which are either right or left determined by objects. We will presently see that right and left almost split morphisms are special cases of morphisms which are right or left determined by objects. While the notion of a morphism being determined by an object is a general categorical one, as in the case of almost split morphisms, the problem of their existence is not at all clear, and it is with this question that this paper is mainly concerned.

In an effort to make what appears to be a somewhat long and technically involved paper easier to follow, the rest of this introduction is devoted to giving a brief survey of the basic notions and results of the paper. For the most part, the general organization of the survey follows that of the paper, but certain liberties have been taken when this seemed appropriate to aid the exposition of the basic ideas. To simplify the presentation, we assume that all categories are additive, even though much of what is said of a purely categorical nature can be carried out in arbitrary categories.

Let $f: B \longrightarrow C$ be a morphism in the category $\underline{\underline{C}}$. We recall (see [7], [11]) that f is said to be right almost split if a) f is not a splittable epimorphism (f is said to be a splittable epimorphism if there is a morphism $s: C \longrightarrow B$ such that $fs = 1_C$) and b) given a morphism $g: X \longrightarrow C$ in $\underline{\underline{C}}$ which is not a splittable epimorphism, then there is a morphism $h: X \longrightarrow B$ such that $fh = g$. It is easily checked that if $f: B \longrightarrow C$ is right almost split, then a) the endomorphism ring $\text{End}\, C$ of C is a local ring, b) $\text{Im}(\text{Hom}_{\underline{\underline{C}}}(C, B) \xrightarrow{(C, f)} \text{End}\, C)$ is the unique maximal right ideal of $\text{End}\, C$ and c) if $g: Y \longrightarrow C$ is a morphism in $\underline{\underline{C}}$ such that $\text{Im}(\text{Hom}_{\underline{\underline{C}}}(C, Y) \xrightarrow{(C, g)} \text{End}\, C)$ is contained in the unique maximal right ideal of $\text{End}\, C$, then there is a morphism $h: Y \longrightarrow B$ such that $g = fh$. In fact, a little thought suffices to show that a morphism $f: B \longrightarrow C$ in $\underline{\underline{C}}$ is right almost split if and only if it satisfies conditions a), b), and c). This formulation suggests the following definition.

A morphism $f: B \longrightarrow C$ in $\underline{\underline{C}}$ is said to be right determined by an object X in $\underline{\underline{C}}$ (or more simply f is right X-determined) if a morphism $g: Y \longrightarrow C$ has the property $\text{Im}(\text{Hom}_{\underline{\underline{C}}}(X, Y) \xrightarrow{(X, g)} \text{Hom}_{\underline{\underline{C}}}(X, C))$ is contained in $\text{Im}(\text{Hom}_{\underline{\underline{C}}}(X, B) \xrightarrow{(X, f)} \text{Hom}_{\underline{\underline{C}}}(X, C))$ if and only if there is a morphism $h: Y \longrightarrow B$ such that $fh = g$. Before discussing the question of the existence of morphisms which are right determined by objects in $\underline{\underline{C}}$, we take up the question of the uniqueness of such morphisms.

Suppose $f: B \longrightarrow C$ is a morphism in $\underline{\underline{C}}$ and X an object in $\underline{\underline{C}}$. Then it is easily seen that $H = \text{Im}(\text{Hom}_{\underline{\underline{C}}}(X, B) \xrightarrow{(X, f)} \text{Hom}_{\underline{\underline{C}}}(X, C))$ is an $(\text{End}\, X)^{op}$-submodule of the $(\text{End}\, X)^{op}$-module $\text{Hom}_{\underline{\underline{C}}}(X, C)$, where $(\text{End}\, X)^{op}$ denotes the opposite ring of $\text{End}\, X$ and the abelian group $\text{Hom}_{\underline{\underline{C}}}(X, C)$ is considered an $(\text{End}\, X)^{op}$-module (all modules are left modules) by means of the usual operation $t \cdot f$ where $t \cdot f$ is the composition ft for all t in $\text{End}\, X$ and f in $\text{Hom}_{\underline{\underline{C}}}(X, C)$. Now if $f: B \longrightarrow C$ is right X-determined then a morphism $g: Y \longrightarrow C$ has the property that there is a morphism $h\colon Y \longrightarrow B$ such that $g = fh$ if and only if the $(\mathrm{End}\ X)^{op}$-submodule $\mathrm{Im}(\mathrm{Hom}_{\underline{\underline{C}}}(X, Y) \xrightarrow{(X,\ g)} \mathrm{Hom}_{\underline{\underline{C}}}(X, C))$ of $\mathrm{Hom}_{\underline{\underline{C}}}(X, C)$ is contained in the $(\mathrm{End}\ X)^{op}$-submodule $H = \mathrm{Im}(\mathrm{Hom}_{\underline{\underline{C}}}(X, B) \xrightarrow{(X,\ f)} \mathrm{Hom}_{\underline{\underline{C}}}(X, C))$ of $\mathrm{Hom}_{\underline{\underline{C}}}(X, C)$. Thus two right X-determined morphisms $f\colon B \longrightarrow C$ and $f'\colon B' \longrightarrow C$ have the property that the two $(\mathrm{End}\ X)^{op}$-submodules $\mathrm{Im}(X, f)$ and $\mathrm{Im}(X, f')$ of $\mathrm{Hom}_{\underline{\underline{C}}}(X, C)$ are the same, if and only if there is a commutative diagram

$$\begin{array}{ccc} B & \xrightarrow{f} & C \\ h \downarrow & & \| \\ B' & \xrightarrow{f'} & C \\ h' \downarrow & & \| \\ B & \xrightarrow{f} & C \end{array} .$$

This observation easily shows that a right X-determined morphism $f\colon B \longrightarrow C$ need not necessarily be determined even up to isomorphism by the $(\mathrm{End}\ X)^{op}$-submodule $\mathrm{Im}(X, f)$ of $\mathrm{Hom}_{\underline{\underline{C}}}(X, C)$, whereby an isomorphism from $f\colon B \longrightarrow C$ to $f'\colon B' \longrightarrow C$ we mean an isomorphism $u\colon B \longrightarrow B'$ such that $f = f'u$. However, as we now point out, there is a simple condition that can be put on a morphism $f\colon B \longrightarrow C$ which guarantees that if it is right X-determined, then it is uniquely determined, up to isomorphism, by the $(\mathrm{End}\ X)^{op}$-submodule $\mathrm{Im}(X, f)$ of $\mathrm{Hom}_{\underline{\underline{C}}}(X, C)$.

We say that a morphism $f\colon B \longrightarrow C$ is right minimal provided an endomorphism $u\colon B \longrightarrow B$ is an isomorphism whenever $fu = f$. Our previous remarks show that two right X-determined morphisms which are also right minimal $f\colon B \longrightarrow C$ and $f'\colon B' \longrightarrow C$ are isomorphic if and only if $\mathrm{Im}(X, f) = \mathrm{Im}(X, f')$. With these preliminary general remarks in mind we now turn our attention to the question of the existence of right determined morphisms. All of the existence theorems we know are based on the following theorem which is proven in Section 3 of Chapter I.

Let $\underline{\underline{C}}$ be the category Mod Λ of all Λ-modules for some arbitrary ring Λ. Suppose X is a finitely presented Λ-module and C is an arbitrary Λ-module. Let H be an $(\text{End } X)^{op}$-submodule of $\text{Hom}_\Lambda(X, C)$ containing the $(\text{End } X)^{op}$-submodule $P(X, C)$ consisting of all morphisms $h: X \longrightarrow C$ which are a composition $X \longrightarrow P \longrightarrow C$ with P a projective Λ-module. Then there is an epimorphism $f: B \longrightarrow C$ of Λ-modules which is right minimal and right X-determined such that $\text{Im}(X, f) = H$. Furthermore, this right minimal and right X-determined morphism $f: B \longrightarrow C$ has the following property. Suppose $f': B' \longrightarrow C$ is any right X-determined morphism such that $\text{Im}(X, f') = H$, then $f': B' \longrightarrow C$ is isomorphic to $g: B \coprod B'' \longrightarrow C$ where $B \coprod B''$ is the direct sum of B and some other module B'' and $g|B = f$ while $g|B = 0$.

An obvious question to ask concerning the right minimal and right X-determined epimorphism $f: B \longrightarrow C$ is what is Ker f? While a complete answer to this question can be found in Section 3, we briefly indicate the form of the answer.

Let $P_1 \xrightarrow{\nu} P_0 \xrightarrow{\varepsilon} X \longrightarrow 0$ be a projective presentation of X, i.e. $P_1 \xrightarrow{\nu} P_0 \xrightarrow{\varepsilon} X \longrightarrow 0$ is an exact sequence with the P_i finitely generated projective Λ-modules. Then $\text{Hom}_\Lambda(P_i, \Lambda)$ is a finitely generated projective Λ^{op}-module and so
$\text{Tr } X = \text{Coker } (\text{Hom}_\Lambda(P_0, \Lambda) \longrightarrow \text{Hom}_\Lambda(P_1, \Lambda))$ is a finitely presented Λ^{op}-module. Let $\Gamma = \text{End}_{\Lambda^{op}}(\text{Tr } X)$. Then there is an injective Γ-module I such that the abelian group $\text{Hom}_\Gamma(\text{Tr } X, I)$ together with the Λ-module structure induced from the Λ^{op}-module structure of Λ^{op} on $\text{Tr } X$ is isomorphic to the kernel of the right minimal and right

X-determined epimorphism $f: B \longrightarrow C$. The particular injective Γ-module I used in this construction is the injective envelope of $\mathrm{Hom}_\Lambda(X, C)/H$, which is not only an $(\mathrm{End}\ X)^{op}$-module, but also a Γ-module in a natural way, as explained in Section 3.

Before giving some of the applications of this main existence theorem, we point out that no comparable theorem exists for the dual notion of morphisms left determined by an object. More precisely, a morphism $f: A \longrightarrow B$ in a category $\underline{\underline{C}}$ is said to be left determined by Y in $\underline{\underline{C}}$ (or is left Y-determined) if a morphism $g: A \longrightarrow X$ has the property $\mathrm{Im}(\mathrm{Hom}_{\underline{\underline{C}}}(X, Y) \xrightarrow{(g, Y)} \mathrm{Hom}_{\underline{\underline{C}}}(A, Y))$ is contained in $\mathrm{Im}(\mathrm{Hom}_{\underline{\underline{C}}}(B, Y) \xrightarrow{(f, Y)} \mathrm{Hom}_{\underline{\underline{C}}}(A, Y))$ if and only if there is a morphism $h: B \longrightarrow X$ such that $hf = g$. All the known existence theorems for morphisms left determined by objects are derived from existence theorems for morphisms right determined by objects by means of some sort of duality. This and other uses of dualities is part of the reason why most of the theory developed in this paper is for algebras rather than rings.

Rather than getting involved in the technicalities needed to give the applications of the main existence theorem to the derivation of other existence theorems in the full generality given in the latter part of Chapter I, I simply give a few sample results in the classical cases of finite dimensional algebras over a field and classical orders over Dedekind rings.

Let Λ be a finite dimensional algebra over a field k. Suppose $0 \longrightarrow A \xrightarrow{g} B \xrightarrow{f} C \longrightarrow 0$ is an exact sequence of finitely generated Λ-modules. Then f is right determined by $\mathrm{Tr}(\mathrm{Hom}_k(A, k))$ and g is left determined by $\mathrm{Hom}_k(\mathrm{Tr}\ C, k)$. More generally, any morphism $f: B \longrightarrow C$ of finitely generated Λ-modules is right determined by $\mathrm{Tr}(\mathrm{Hom}_k(\mathrm{Ker}\ f, k)) \amalg \Lambda$ and is left determined by

$\text{Hom}_k(\text{Tr}(\text{Coker } f), k) \coprod \text{Hom}_k(\Lambda, k)$. Thus we see that as far as finite dimensional algebras are concerned, the property of morphisms being left and right determined by modules is a universal property of morphisms between finitely generated modules.

Suppose now that R is a Dedekind domain (not a field) and Λ an R-order in the classical sense. Let $L(\Lambda)$ be the category of Λ-lattices, i.e. Λ-modules which are finitely generated projective R-modules. Suppose X is a Λ-lattice. Then we define the Λ^{op}-lattice $\text{Tr}_L X$ as follows. Let $P_1 \xrightarrow{b} P_0 \xrightarrow{\varepsilon} X \longrightarrow 0$ be a projective presentation of X. Then $\text{Tr}_L X = \text{Im}(\text{Hom}_\Lambda(P_0, \Lambda) \longrightarrow \text{Hom}_\Lambda(P_1, \Lambda))$ by definition. Now suppose $0 \longrightarrow A \xrightarrow{g} B \xrightarrow{f} C \longrightarrow 0$ is an exact sequence of Λ-modules each of which is a Λ-lattice. Viewing g and f as morphisms in $L(\Lambda)$, we have that f is right $\text{Tr}_L(\text{Hom}_R(A, R))$-determined in $L(\Lambda)$ and g is left $\text{Hom}_R(\text{Tr}_L C, R)$-determined in $L(\Lambda)$. More generally, a morphism $f: B \longrightarrow C$ in $L(\Lambda)$ is both left and right determined by objects in $L(\Lambda)$.

Chapter II is devoted to applying some of the results of Chapter I to showing that simple functors on suitable categories of modules are finitely presented. In this connection the notion of a subfunctor of a functor being determined by an object in a category is introduced. Suppose $F: \underline{\underline{C}}^{op} \longrightarrow \text{Ab}$ is a functor (all functors are additive) where $\underline{\underline{C}}^{op}$ is the opposite category of $\underline{\underline{C}}$. A subfunctor G of F is said to be determined by an object X in $\underline{\underline{C}}$ if a subfunctor G' of F is contained in G whenever $G'(X) \subseteq G(X)$. The connection between this notion and that of a morphism $f: B \longrightarrow C$ in $\underline{\underline{C}}$ being right determined by X is given by the following easily verified result. A morphism $f: B \longrightarrow C$ is right X-determined if and only if the induced morphism $(\ ,f): (\ ,B) \longrightarrow (\ ,C)$ of representable functors has the property that the subfunctor $\mathrm{Im}(\ ,f)$ of $(\ ,C)$ is determined by X.

Suppose now that G and G' are two X-determined subfunctors of a functor $F: \underline{\underline{C}}^{op} \longrightarrow \mathrm{Ab}$. Then it is obvious that $G = G'$ if and only if $G(X) = G'(X)$. Thus an X-determined subfunctor G of F is uniquely determined by the $(\mathrm{End}\ X)^{op}$-submodule $G(X)$ of $F(X)$. This suggests the question: If H is an $(\mathrm{End}\ X)^{op}$-submodule of $F(X)$ is there an X-determined subfunctor G of F such that $G(X) = H$? It is easy to see that this question has an affirmative answer for each $(\mathrm{End}\ X)^{op}$-submodule H of F. We denote by F_H the unique X-determined subfunctor of F such that $F_H(X) = H$.

In particular, for each C in $\underline{\underline{C}}$ and each right ideal $\underline{a}$ of End C, there is associated the C-determined subfunctor $(\ ,C)_{\underline{a}}$ of $(\ ,C)$. It is not very difficult to see that $\underline{a}$ is a maximal right ideal of End C if and only if $(\ ,C)_{\underline{a}}$ is a maximal subfunctor of $(\ ,C)$. Moreover a subfunctor G of $(\ ,C)$ is maximal if and only if $G(C)$ is a maximal right ideal of End C and $G = (\ ,C)_{G(C)}$. Thus there is a one-to-one correspondence between the maximal right ideals $\underline{m}$ of End C and the maximal subfunctors of $(\ ,C)$, the correspondence being given by $\underline{m} \longmapsto (\ ,C)_{\underline{m}}$ for all maximal right ideals of End C. From this it follows that a simple functor $S: \underline{\underline{C}}^{op} \longrightarrow \mathrm{Ab}$ such that $S(C) \neq 0$ is isomorphic to $(\ ,C)/(\ ,C)_{\underline{m}}$ for some maximal right ideal $\underline{m}$ of End C. Moreover, since each simple functor $S \neq 0$, there is some C in $\underline{\underline{C}}$ such that $S(C) \neq 0$. Thus all simple functors from $\underline{\underline{C}}^{op}$ to Ab are isomorphic to $(\ ,C)/(\ ,C)_{\underline{m}}$ for some C in $\underline{\underline{C}}$ and maximal right ideal $\underline{m}$ of End C.

Next we recall that a functor $F: \underline{\underline{C}}^{op} \longrightarrow \mathrm{Ab}$ is said to be finitely presented if there is an exact sequence of functors $(\ ,A) \longrightarrow (\ ,B) \longrightarrow F \longrightarrow 0$. Therefore the simple functors are finitely presented if and only if for each C in $\underline{\underline{C}}$ and maximal ideal $\underline{m}$ of End C there is a right C-determined morphism f: B ⟶ C such that $\mathrm{Im}((C, B) \xrightarrow{(C, f)} (C, C)) = \underline{m}$.

Chapter II is mainly devoted to showing how the results of Chapter I concerning morphisms which are right determined by objects in certain categories $\underline{\underline{C}}$ of modules over particular rings can be used to prove that the simple functors from $\underline{\underline{C}}^{op}$ to Ab are finitely presented. In particular, if Λ is a finite dimensional algebra over a field k, then the simple functors from $(\mathrm{mod}\ \Lambda)^{op}$ to Ab are finitely presented where mod Λ is the category of all finitely generated Λ-modules. Also if Λ is an order over a Dedekind domain R, then all the simple functors from $(L(\Lambda))^{op}$ to Ab are finitely presented. Similar results for simple functors from a category $\underline{\underline{C}}$ to Ab are also established. These results lead us back to almost split sequences, the starting point of this whole discussion.

While the major portion of this paper is devoted to exploring the notions we just finished discussing, there are sprinkled throughout the paper various applications not directly connected with the development of the general theory. It is hoped that some of these notions will prove useful in studying other such questions as has already happened in the case of finite dimensional algebras over fields as indicated by some of the other papers in this volume (see [1] and [9]).

Finally, I would like to thank Professor Robert Gordon for his skill and diligence in organizing the conference, the NSF and Temple University for their financial backing, and the participants in the conference for their spirit and good humor.

Chapter I

EXISTENCE THEOREMS

This chapter is devoted to proving the basic existence theorems concerning functors and morphisms determined by objects in various categories of modules over particular rings. We begin with the definition and some of the basic properties of functors from a category to abelian groups determined by objects in the category.

§1. *Subfunctors Determined by Objects*. We recall that a category $\underline{\underline{C}}$ is said to be a *preadditive category* if a) for each pair of objects X, Y in $\underline{\underline{C}}$ the set (X, Y) of morphisms from X to Y has an abelian group structure and b) the maps $(X, Y) \times (Y, Z) \longrightarrow (X, Z)$ given by $(f, g) \longmapsto gf$, the composition of f and g in $\underline{\underline{C}}$, is bilinear for all X, Y, Z in $\underline{\underline{C}}$. Also if $\underline{\underline{C}}$ and $\underline{\underline{D}}$ are preadditive categories, then a functor $F: \underline{\underline{C}} \longrightarrow \underline{\underline{D}}$ is said to be *additive* if the morphisms $F: (X, Y) \longrightarrow (F(X), F(Y))$ are group morphisms for all X, Y in $\underline{\underline{C}}$.

From now on we assume, unless stated to the contrary, that all categories are preadditive and all functors are additive. The collection of all functors from the category $\underline{\underline{C}}$ to the category $\underline{\underline{D}}$ will be denoted by $(\underline{\underline{C}}, \underline{\underline{D}})$. Finally, we denote the opposite category of $\underline{\underline{C}}$ by $\underline{\underline{C}}^{op}$. Clearly $\underline{\underline{C}}^{op}$ is a preadditive category. We will usually follow the convention of considering the contravariant functors from $\underline{\underline{C}}$ to $\underline{\underline{D}}$ as the functors (covariant) from $\underline{\underline{C}}^{op}$ to $\underline{\underline{D}}$.

Suppose C is an object in $\underline{\underline{C}}$. Then the abelian group (C, C) together with the multiplication given by the composition of morphisms in $\underline{\underline{C}}$ is a ring called the *endomorphism ring* of C and is denoted by End C. Assume now that $G: \underline{\underline{C}}^{op} \longrightarrow Ab$ is a functor, where Ab is the category of abelian groups. Then for each C in $\underline{\underline{C}}$, the abelian group $G(C)$ is an $(\text{End } C)^{op}$-module, where $(\text{End } C)^{op}$ is the opposite ring of End C, the operation of f on $G(C)$ being given by $f \cdot x = G(f)(x)$ for all f in End C and x in $G(C)$. Clearly if $\alpha: F \longrightarrow G$ is a morphism of functors from $\underline{\underline{C}}^{op}$ to Ab, then for each C in $\underline{\underline{C}}$ the morphism $\alpha_C: F(C) \longrightarrow G(C)$ is a morphism of $(\text{End } C)^{op}$-modules. The morphism $\alpha: F \longrightarrow G$ is said to be an *inclusion morphism* if $\alpha_C: F(C) \longrightarrow G(C)$ is an inclusion of $(\text{End } C)^{op}$-modules for each C in $\underline{\underline{C}}$. A functor F is said to be a *subfunctor* of G, if there is an inclusion morphism $\alpha: F \longrightarrow G$. We now point out a way of describing the subfunctors of a functor.

Lemma 1.1: Suppose G is in $(\underline{\underline{C}}^{op}, Ab)$. If for each X in $\underline{\underline{C}}$ we are given a subgroup A_X of $G(X)$ such that $G(f)(A_Y) \subset A_X$ for all morphisms $f: X \longrightarrow Y$ in $\underline{\underline{C}}$, then there is a unique subfunctor F of G such that $F(X) = A_X$ for all X in $\underline{\underline{C}}$.

Of particular concern to us are the subfunctors of a functor in $(\underline{\underline{C}}^{op}, Ab)$ determined by objects in $\underline{\underline{C}}$, a notion we now define.

Let X be in $\underline{\underline{C}}$. A subfunctor F of a functor $G: \underline{\underline{C}}^{op} \longrightarrow Ab$ is said to be *determined* by X if a subfunctor F' of G is contained in F whenever $F'(X) \subset F(X)$. Obviously if F and F' are two subfunctors of G determined by X, then $F = F'$ if and only if $F(X) = F'(X)$. A natural question to ask at this point is whether for each $(\text{End } X)^{op}$-submodule H of $G(X)$ there is a subfunctor F of G determined by

X such that $F(X) = H$. We now show that his question has an affirmative answer.

Suppose G is in $(\underline{\underline{C}}^{op}, Ab)$, X is an object of $\underline{\underline{C}}$ and H is an $(\text{End } X)^{op}$-submodule of $G(X)$. For each C in $\underline{\underline{C}}$ define A_C to be the subgroup of $G(C)$ consisting of all x in $G(C)$ such that for each $f: X \longrightarrow C$ the element $G(f)(x)$ in $G(X)$ is contained in H. It is easily checked that for each morphism $g: U \longrightarrow V$ in $\underline{\underline{C}}$ we have $G(g)(A_V) \subset A_U$. Thus, by Lemma 1.1, there is a unique subfunctor, which we denote by G_H, of G having the property $G_H(C) = A_C$ for all C in $\underline{\underline{C}}$. Furthermore, the fact that H is an $(\text{End } X)^{op}$-submodule of $G(X)$ implies that $G_H(X) = H$.

It is a straightforward matter to deduce the following facts concerning subfunctors determined by objects from our discussion so far.

<u>Proposition 1.2</u>: Let X be an object in $\underline{\underline{C}}$ and G a functor in $(\underline{\underline{C}}^{op}, Ab)$.

a) If H is an $(\text{End } X)^{op}$-submodule of $G(X)$, then the subfunctor G_H of G is determined by X and has the property $G_H(X) = H$.

b) A subfunctor F of G is determined by X if and only if $F = G_H$ where $H = F(X)$.

c) The map $H \longmapsto G_H$ gives a bijection between the $(\text{End } X)^{op}$-submodules H of $G(X)$ and the subfunctors of G determined by X.

As an immediate consequence of the definition of a subfunctor being determined by an object we have the following.

<u>Proposition 1.3</u>: Suppose X is in $\underline{\underline{C}}$ and $F' \subset F \subset G$ are in $(\underline{\underline{C}}^{op}, Ab)$. Then $F \subset G$ is determined by X if and only if $F/F' \subset G/F'$ is deter-

mined by X.

Since $\underline{\underline{C}} = (\underline{\underline{C}}^{op})^{op}$, we have that $(\underline{\underline{C}}, Ab) = ((\underline{\underline{C}}^{op})^{op}, Ab)$. This observation shows how to give analogous definitions and results for functors in $(\underline{\underline{C}}, Ab)$ to those given for functors in $(\underline{\underline{C}}^{op}, Ab)$. These analogues will be used freely in the rest of the paper.

I would like to thank D. Eisenbud for suggesting the terminology "a subfunctor determined by an object" and the construction of the subfunctors G_H of G.

§2. Morphisms Determined by Objects. This section is mainly devoted to describing when a subfunctor of a functor determined by an object is finitely generated. For simplicity of exposition we assume throughout this section that $\underline{\underline{C}}$ is an additive and not just a preadditive category. We will be using constantly the well known result (Yoneda's lemma) that if G is in $(\underline{\underline{C}}^{op}, Ab)$ and x is in $G(X)$, then there is a unique morphism $\beta\colon (\ ,X) \longrightarrow G$ such that $\beta_X(1_X) = x$.

Recall (see [6] for more details) that a functor F in $(\underline{\underline{C}}^{op}, Ab)$ is said to be finitely generated if there is an epimorphism $(\ ,C) \longrightarrow F$ for some C in $\underline{\underline{C}}$. This definition is based on the fact that the representable functors $(\ ,C)$ are projective objects in $(\underline{\underline{C}}^{op}, Ab)$ which have many of the same categorical properties as finitely generated projective modules. We now develop a criterion for when a subfunctor F of G determined by an object X in $\underline{\underline{C}}$ is finitely generated.

Proposition 2.1: Let B be in $\underline{\underline{C}}$ and G in $(\underline{\underline{C}}^{op}, Ab)$. A morphism $(\ ,B) \xrightarrow{\alpha} G$ has the property that $\operatorname{Im} \alpha$ is determined by an object X in $\underline{\underline{C}}$ if and only if given any morphism $\beta\colon (\ ,L) \longrightarrow G$ with the property $\beta(\ ,L)(X) \subset \alpha(\ ,B)(X)$, then there is a morphism $f\colon L \longrightarrow B$

such that $\beta = \alpha(\ ,f)$.

Proof: Suppose $\operatorname{Im}\alpha$ is a subfunctor of G determined by X. Let $\beta\colon (\ ,L) \longrightarrow G$ be such that $\beta(\ ,L)(X) \subset \alpha(\ ,B)(X)$. Then the subfunctor $\beta(\ ,L)$ of G is contained in $\alpha(\ ,B)$ since $\alpha(\ ,B)$ is determined by X. Therefore there is a morphism $\gamma\colon (\ ,L) \longrightarrow (\ ,B)$ such that $\alpha\gamma = \beta$. But $\gamma\colon (\ ,L) \longrightarrow (\ ,B)$ is of the form $(\ ,f)\colon (\ ,L) \longrightarrow (\ ,B)$ for some $f\colon L \longrightarrow B$. Thus we have shown that if $\operatorname{Im}\alpha$ is determined by X in G, and $\beta\colon (\ ,L) \longrightarrow G$ is such that $\beta(\ ,L)(X) \subset \alpha(\ ,B)(X)$, then there is a morphism $f\colon L \longrightarrow B$ such that $\alpha(\ ,f) = \beta$.

The rest of the proof is trivial.

An easily verified consequence of this result is the following.

Corollary 2.2: Let X be in $\underline{\underline{C}}$ and F a subfunctor of G in $(\underline{\underline{C}}^{op}, Ab)$ determined by X.

a) A morphism $\alpha\colon (\ ,B) \longrightarrow G$ has the property $\operatorname{Im}\alpha = F$ if and only if

i) $\operatorname{Im}\alpha(X) = F(X)$ and

ii) Given $\beta\colon (\ ,L) \longrightarrow G$ such that $\operatorname{Im}\beta(X) \subset F(X)$, then there is a morphism $f\colon L \longrightarrow B$ such that $\alpha(\ ,f) = \beta$.

b) F is finitely generated if and only if there is a morphism $\alpha\colon (\ ,B) \longrightarrow G$ satisfying i) and ii) of a).

We now apply these observations to the case $G = (\ ,C)$ for some C in $\underline{\underline{C}}$. We begin with some terminology which will be used throughout the rest of this paper.

Let $f\colon B \longrightarrow C$ be a morphism in $\underline{\underline{C}}$ and X an object in $\underline{\underline{C}}$.

a) f is said to be <u>right</u> <u>X-determined</u> if the subfunctor $\mathrm{Im}(\ ,f)$ of $(\ ,C)$ is determined by X.

b) f is said to be <u>left</u> <u>X-determined</u> if the subfunctor $\mathrm{Im}(f,\ )$ of $(B,\ )$ is determined by X.

The following characterization of right (left) X-determined morphisms can easily be deduced from Proposition 2.1 and its analogue for covariant functors.

<u>Proposition 2.3</u>: Let $f\colon B \longrightarrow C$ be a morphism in $\underline{\underline{C}}$ and X an object in $\underline{\underline{C}}$.

a) $f\colon B \longrightarrow C$ is right X-determined if and only if given any morphism $g\colon L \longrightarrow C$ in $\underline{\underline{C}}$ such that $\mathrm{Im}(\ ,g)(X) \subset \mathrm{Im}(\ ,f)(X)$, there is an $h\colon L \longrightarrow B$ with the property $fh = g$.

b) $f\colon B \longrightarrow C$ is left X-determined if and only if given any morphism $j\colon B \longrightarrow L$ such that $\mathrm{Im}(j,\ )(X) \subset \mathrm{Im}(f,\ )(X)$, there is a $k\colon C \longrightarrow L$ such that $kf = j$.

The following criteria for when a morphism is right or left determined by an object are very useful in practice.

<u>Proposition 2.4</u>: Let $f\colon B \longrightarrow C$ be a morphism in $\underline{\underline{C}}$ and X an object in $\underline{\underline{C}}$. Suppose H is an $(\mathrm{End}\ X)^{op}$-submodule of (X, C). Then the following are equivalent.

a) A morphism $h\colon L \longrightarrow C$ in $\underline{\underline{C}}$ satisfies $\mathrm{Im}(\ ,h)(X) \subset H$ if and only if there is a morphism $g\colon L \longrightarrow B$ such that $fg = h$.

b) f is a right X-determined morphism with $\mathrm{Im}(\ ,f)(X) = H$.

<u>Proof</u>: a) implies b). It follows from the hypothesis of a) that in order to show that f is right X-determined, it suffices to show that $\mathrm{Im}(\ ,f)(X) = H$. Since for the morphism $f: B \longrightarrow C$ there is a morphism $h: B \longrightarrow B$ satisfying $fh = f$, namely $h = \mathrm{id}_B$, we know by a) that $\mathrm{Im}(\ ,f)(X) \subset H$. We now show that $\mathrm{Im}(\ ,f)(X) \supset H$. For let $h: X \longrightarrow C$ be in H. Then $\mathrm{Im}(\ ,h)(X)$ is simply the $(\mathrm{End}\ X)^{op}$-submodule of (X, C) generated by h and so h is in $\mathrm{Im}(\ ,h)(X) \subset H$. Therefore there is a $g: X \longrightarrow B$ such that $h = fg$. This implies that $\mathrm{Im}(\ ,h)(X) \subset \mathrm{Im}(\ ,f)(X)$ and so h is in $\mathrm{Im}(\ ,f)(X)$. Since this is true for all h in H we have that $\mathrm{Im}(\ ,f)(X) \supset H$. This completes the proof that a) implies b). b) implies a) is trivial.

Of course, we also have the companion result concerning morphisms which are left determined by objects which we now state without proof.

<u>Proposition 2.5</u>: Let $f: B \longrightarrow C$ be a morphism in $\underline{\underline{C}}$ and X an object in $\underline{\underline{C}}$. Suppose H is an End X-submodule of (B, X). Then the following statements are equivalent.

a) A morphism $h: B \longrightarrow L$ in $\underline{\underline{C}}$ satisfies $\mathrm{Im}(h,\ )(X) \subset H$ if and only if there is a morphism $g: C \longrightarrow L$ such that $gf = h$.

b) f is a left X-determined morphism with $\mathrm{Im}(f,\ )(X) = H$.

The following useful result is worth noting.

<u>Proposition 2.6</u>: Let $f: B \longrightarrow C$ be a morphism in $\underline{\underline{C}}$ and suppose $X = X_1 \coprod X_2$.

a) If $f: B \longrightarrow C$ is right X_1-determined then f is right X-determined.

b) If $f: B \longrightarrow C$ is left X_1-determined, then f is left X-determined.

Proof:

a) Suppose $f: B \longrightarrow C$ is right X_1-determined. Let $h: L \longrightarrow C$ be a morphism such that $\mathrm{Im}(\ ,h)(X)$ is contained in $\mathrm{Im}(\ ,f)(X)$. Since $\mathrm{Im}(\ ,h)$ and $\mathrm{Im}(\ ,f)$ are additive subfunctors of $(\ ,C)$, it follows that $\mathrm{Im}(\ ,h)(X_1) \subset \mathrm{Im}(\ ,f)(X_1)$. Thus there is a morphism $t: L \longrightarrow B$ such that $ft = h$ since $f: B \longrightarrow C$ is right X_1-determined.

b) Dual of a).

As a consequence of this result, we have the following.

Proposition 2.7: Suppose $h: A \longrightarrow C$ is the composition $A \xrightarrow{g} B \xrightarrow{f} C$.

a) If f is a monomorphism which is right X_1-determined and g is right X_2-determined, then h is right $X = X_1 \coprod X_2$-determined.

b) If g is an epimorphism which is left X_1-determined and f is left X_2-determined, then h is left $X = X_1 \coprod X_2$-determined.

Proof:

a) Suppose $v: L \longrightarrow C$ is such that $\mathrm{Im}(\ , v)(X) \subset \mathrm{Im}(\ ,h)(X)$. Then $\mathrm{Im}(\ ,v)(X) \subset \mathrm{Im}(\ ,f)(X)$. Since f is right X_1-determined, we know by Proposition 2.6, that f is right X-determined.

Therefore, there is a morphism $t: L \longrightarrow B$ such that $ft = \nu$. Thus $\operatorname{Im}(X, f)(\operatorname{Im}(X, t)(X)) \subset \operatorname{Im}(X, f)(\operatorname{Im}(X, g)(X))$. This implies $\operatorname{Im}(,t)(X) \subset \operatorname{Im}(,g)(X)$ since f and hence (X, f) is a monomorphism. Again by Proposition 2.6, we know that g is right X-determined since $X = X_1 \coprod X_2$ and g is right X_2-determined. Thus there is a morphism $u: L \longrightarrow A$ such that $gu = t$. Therefore $fg\,u = ft = \nu$ and so $u: L \longrightarrow A$ has the property that $hu = \nu$. Consequently, h is right $X_1 \coprod X_2$-determined as claimed.

b) Dual of a).

Usually we are not only interested in when a functor G in $(\underline{\underline{C}}^{op}, Ab)$ is finitely generated, but also in when it has a projective cover which is finitely generated (see [6] for definition and properties of projective covers in $(\underline{\underline{C}}^{op}, Ab)$). The following description of finitely generated projective covers is quite useful in showing when finitely generated functors have projective covers.

Proposition 2.8: Let B be in $\underline{\underline{C}}$ and G in $(\underline{\underline{C}}^{op}, Ab)$.

a) A morphism $\alpha: (,B) \longrightarrow G$ is a projective cover of its image if and only if an endomorphism $f: B \longrightarrow B$ is an isomorphism whenever $\alpha = \alpha(,f)$.

b) In particular, a morphism $g: B \longrightarrow C$ in $\underline{\underline{C}}$ has the property that $(,g): (,B) \longrightarrow (,C)$ is a projective cover of $\operatorname{Im}(,g)$ if and only if an endomorphism $f: B \longrightarrow B$ is an isomorphism whenever $gf = g$.

Proof:

a) Follows easily from the definition of projective covers as can be seen in [6].

b) Trivial consequence of a).

An easily verified consequence of Proposition 2.8 is the following result.

Corollary 2.9: Let $g: B \longrightarrow C$ and $h: A \longrightarrow C$ be morphisms in $\underline{\underline{C}}$ such that $\mathrm{Im}(\,,g) = \mathrm{Im}(\,,h)$. Suppose $(\,,g): (\,,B) \longrightarrow (\,,C)$ is a projective cover of $\mathrm{Im}(\,,g)$.

a) There are morphisms $s : B \longrightarrow A$ and $t: A \longrightarrow B$ satisfying $hs = g$ and $gt = s$ such that $ts = 1_B$.

b) If $(\,,h): (\,,A) \longrightarrow (\,,C)$ is a projective cover of $\mathrm{Im}(\,,h)$ and $t: B \longrightarrow A$ is a morphism such that $ht = g$, then t is an isomorphism.

c) $(\,,h): (\,,A) \longrightarrow (\,,C)$ is a projective cover for $\mathrm{Im}(\,,h)$ if and only if there is an isomorphism $t: B \longrightarrow A$ such that $ht = g$.

The reader should develop the analogues of Proposition 2.6 and Corollary 2.7 for covariant functors. This result suggests the following terminology.

Let $f: B \longrightarrow C$ be a morphism in $\underline{\underline{C}}$.

a) f is said to be <u>right minimal</u> if an endomorphism $h: B \longrightarrow B$ is an isomorphism whenever $fh = f$.

b) f is said to be <u>left minimal</u> if an endomorphism $u: C \longrightarrow C$ is an isomorphism whenever $uf = f$.

c) f is said to be a <u>minimal right X-determined morphism</u> if it is both right X-determined and right minimal.

d) f is said to be a minimal left X-determined morphism if it is both left X-determined and left minimal.

The following uniqueness theorem follows easily from the definitions involved.

Proposition 2.10: Let X be in $\underline{\underline{C}}$.

a) Suppose $f: B \longrightarrow C$ and $f': B' \longrightarrow C$ are two minimal right X-determined morphisms. Then there is an isomorphism $u: B \longrightarrow B'$ such that $f = f'u$ if and only if $\mathrm{Im}(\ ,f)(X) = \mathrm{Im}(\ ,f')(X)$.

b) Suppose $h: A \longrightarrow B$ and $h': A \longrightarrow B'$ are two minimal left X-determined morphisms. Then there is an isomorphism $\upsilon: B \longrightarrow B'$ such that $\upsilon h = h'$ if and only if $\mathrm{Im}(h,\ )(X) = \mathrm{Im}(h',\ )(X)$.

Before giving our next result it is convenient to introduce the following definitions.

Let $f: X \longrightarrow Y$ be a morphism in $\underline{\underline{C}}$. Then f is said to be a splittable monomorphism if there is a morphism $g: Y \longrightarrow X$ such that $gf = 1_X$. Dually, f is said to be a splittable epimorphism if there is an $h: Y \longrightarrow X$ such that $fg = 1_Y$.

Proposition 2.11: Assume idempotents in $\underline{\underline{C}}$ split. Let $f: B \longrightarrow C$ be a right minimal X-determined morphism with $H = \mathrm{Im}(\ ,f)(X)$. Then the following statements are equivalent for a morphism $g: A \longrightarrow C$ with $\mathrm{Im}(\ ,g)(X) \subset H$.

a) If g is the composition $A \xrightarrow{h} L \xrightarrow{j} C$ where j has the property $\mathrm{Im}(\ ,j)(X) \subset H$, then h is a splittable monomorphism.

b) If $h: A \longrightarrow B$ is such that $fh = g$, then h is a splittable monomorphism.

c) There is a splittable monomorphism $h: A \longrightarrow B$ such that $fh = g$.

Proof: a) implies b). Trivial.

b) implies c). Since $f: B \longrightarrow C$ is right X-determined and $\mathrm{Im}(\ ,g)(X) \subset H = \mathrm{Im}(\ ,f)(X)$, it follows that there is an $h: A \longrightarrow B$ such that $fh = g$.

c) implies a). Since idempotents in $\underline{\underline{C}}$ split and $h: A \longrightarrow B$ is a splittable monomorphism, it follows that there is an isomorphism $u: A \amalg A' \longrightarrow B$ such that the composition $A \longrightarrow A \amalg A' \xrightarrow{u} B$ is the morphism h. Suppose g is the composition $A \xrightarrow{h} L \xrightarrow{j} C$ where j has the property $\mathrm{Im}(\ ,j)(X) \subset H$. Then letting $v: L \amalg A' \longrightarrow C$ be the morphism induced by j and the composition $A' \longrightarrow B \xrightarrow{f} C$, we have i) $\mathrm{Im}(\ ,v)(X) \subset H$ and ii) there is a morphism $\omega: B \longrightarrow L \amalg A'$ such that the diagram

$$\begin{array}{ccc} A \amalg A' & \longrightarrow & L \amalg A' \\ \downarrow u & & \| \\ B & \xrightarrow{\omega} & L \amalg A' \end{array},$$

where $A \amalg A' \longrightarrow L \amalg A'$ is induced by $1_A: A' \longrightarrow A'$ and $h: A \longrightarrow L$, commutes. Hence we obtain the commutative diagram

$$\begin{array}{ccc} A & \xrightarrow{h} & L \\ g\downarrow & & \downarrow inc \\ B & \xrightarrow{\omega} & L \coprod A' \\ \downarrow f & & \downarrow \nu \\ C & = & C \end{array}$$

Since $\mathrm{Im}(\ ,\nu)(X) \subset \mathrm{Im}(\ ,f)(X)$ and $f\colon B \longrightarrow C$ is right X-determined, there is a morphism $z\colon L \coprod A' \longrightarrow B$ such that $\nu = fz$. The fact that $f\colon B \longrightarrow C$ is also right minimal implies that $z\omega$ is an isomorphism. Thus we obtain the commutative diagram

$$\begin{array}{ccccc} A & \xrightarrow{h} & L & & \\ g\downarrow & & \downarrow inc & & \\ B & \xrightarrow{\omega} & L \coprod A' & \xrightarrow{(z\omega)^{-1} z} & B \end{array}$$

with $((z\omega)^{-1}z)\omega = 1_B$. Hence if we let $p\colon B \longrightarrow A$ be such that $ph = 1_A$ (remember $h\colon A \longrightarrow B$ is a splittable monomorphism) then the composition $A \xrightarrow{h} L \xrightarrow{inc} L \coprod A' \xrightarrow{(z\omega)^{-1} z} B \xrightarrow{p} A$ is the identity, which shows that $h\colon A \longrightarrow L$ is a splittable monomorphism. This finishes the proof that c) implies a).

As usual we obtain the dual result for left minimal morphisms determined by an object in $\underline{\underline{C}}$ which we state without proof.

Proposition 2.12: Assume idempotents in $\underline{\underline{C}}$ split. Let $f\colon C \longrightarrow B$ be a left minimal X-determined morphism with $H = \mathrm{Im}(f,\ )(X)$. Then the following are equivalent for a morphism $g\colon C \longrightarrow A$ with $\mathrm{Im}(g,\ )(X) \subset H$.

a) If g is the composition $C \xrightarrow{j} L \xrightarrow{h} A$ where j has the property $\mathrm{Im}(j,)(X) \subset H$, then h is a splittable epimorphism.

b) If $h: B \longrightarrow A$ is such that $hf = g$, then h is a splittable epimorphism.

c) There is a splittable epimorphism $h: B \longrightarrow A$ such that $hf = g$.

The rest of this chapter is mainly devoted to developing existence theorems for morphisms which are either left or right determined by objects in certain full subcategories $\underline{\underline{C}}$ of the category of modules of various types of rings. Our usual procedure will be to first establish existence theorems for morphisms right determined by some module and then use duality considerations to derive existence theorems for left determined morphisms. In this connection we point out now how left and right determined morphisms are connected by a duality between categories.

Let $D: \underline{\underline{C}} \longrightarrow \underline{\underline{C}}'$ be a duality between the additive categories $\underline{\underline{C}}$ and $\underline{\underline{C}}'$. Then for each C in $\underline{\underline{C}}$ the duality D induces a ring isomorphism $D: \mathrm{End}\, C \longrightarrow (\mathrm{End}\, D(C))^{op}$ which we will usually consider an identification. With this convention in mind we have the following easliy verified properties of the duality.

Lemma 2.13: Let B and C be in $\underline{\underline{C}}$ and $D: \underline{\underline{C}} \longrightarrow \underline{\underline{C}}'$ a duality.

a) $D: (B, C) \longrightarrow (D(C), D(B))$ is an $(\mathrm{End}\, B)^{op}$-End C bimodule isomorphism.

b) $H \subset (B, C)$ is an End C-submodule of (B, C) if and only if $D(H) \subset (D(C), D(B))$ is an End C-submodule of $(D(C), D(B))$.

c) If $H \subset (B, C)$ is an End C-submodule of (B, C), then $D: (B, C) \longrightarrow (D(C), D(B))$ induces an End C-isomorphism $D: (B, C)/H \longrightarrow (D(C), D(B))/D(H)$.

Applying these observations one easily obtains the following.

Proposition 2.14: Suppose $D: \underline{\underline{C}} \longrightarrow \underline{\underline{C}}'$ is a duality. Let $g: A \longrightarrow B$ be a morphism in $\underline{\underline{C}}$ and X an object in $\underline{\underline{C}}$.

a) If $t: A \longrightarrow L$ is a morphism in $\underline{\underline{C}}$, then $D(\mathrm{Im}(t, \ )(X)) = \mathrm{Im}(\ ,D(t))(D(X))$.

b) $D(\mathrm{Im}(g, \ )(X)) = \mathrm{Im}(\ ,D(g))(D(X))$.

c) g is left X-determined in $\underline{\underline{C}}$ if and only if $D(g): D(B) \longrightarrow D(A)$ is right $D(X)$-determined in $\underline{\underline{C}}'$.

d) g is left minimal if and only if $D(g): D(B) \longrightarrow D(A)$ is right minimal.

e) g is minimal left X-determined in $\underline{\underline{C}}$ if and only if $D(g): D(B) \longrightarrow D(A)$ is minimal right $D(X)$-determined in $\underline{\underline{C}}'$.

§3. General Existence Theorems.

Throughout this section Λ is a fixed ring. We denote the category of all Λ-modules by $\mathrm{Mod}\ \Lambda$ and the category of all finitely presented Λ-modules by $\mathrm{mod}\ \Lambda$. Our primary purpose in this section is to investigate when an epimorphism $f: B \longrightarrow C$ in $\mathrm{Mod}\ \Lambda$ is right X-determined with X in $\mathrm{mod}\ \Lambda$. The results of this section form the basis of all existence theorems that we know for morphisms determined by modules. In order to state more explicitly the main result of this section we need the following definitions and notations.

We designate the opposite ring of Λ by Λ^{op}. Following the convention of considering right Λ-modules as left Λ^{op}-modules, we denote the category of right Λ-modules by $\mathrm{Mod}\ \Lambda^{op}$ and the category of right finitely presented Λ-modules by $\mathrm{mod}\ \Lambda^{op}$. In accordance with our previous notational practice in this paper, we will often denote $\mathrm{Hom}_\Lambda(A, B)$, the group of Λ-morphisms from A to B in Mod Λ, by (A, B). As usual we consider (A, B) an $(\text{End } A)^{op}$-End B bimodule by means of the obvious operations of End A on A and End B on B. Finally for each pair A, B in Mod Λ we denote by P(A, B) the abelian subgroup of (A, B) consisting of all morphisms $f: A \longrightarrow B$ which can be factored as $A \xrightarrow{g} P \xrightarrow{h} B$ with P projective. It is easily checked that P(A, B) is an $(\text{End } A)^{op}$-End B submodule of (A, B). The importance of this submodule is indicated by the following result.

Proposition 3.1: Suppose $f: B \longrightarrow C$ is a morphism in Mod Λ and X is in Mod Λ.

a) If f is an epimorphism, then the $(\text{End } X)^{op}$-submodule Im(,f)(X) of (X, C) contains P(X, C).

b) If f is right X-determined, then f is an epimorphism if and only if Im(,f)(X) contains P(X, C).

Proof:

a) Suppose $f: B \longrightarrow C$ is an epimorphism. Let $t: X \longrightarrow C$ be in P(X, C). Then t can be written as a composition $X \xrightarrow{u} P \xrightarrow{v} C$ with P projective. Since $f: B \longrightarrow C$ is an epimorphism, there is a morphism $\omega: P \longrightarrow B$ such that $f\omega = v$. Therefore $t: X \longrightarrow C$ can be written as the composition $f\omega u$ and so t is in Im(,f)(X). Therefore $P(X, C) \subset \text{Im}(,f)(X)$.

b) Since we know by a) that if $f: B \longrightarrow C$ is an epimorphism, then $\text{Im}(,f)(X) \supset P(X, C)$ we only have to show that if $f: B \longrightarrow C$ is right X-determined and $\text{Im}(,f)(X) \supset P(X, C)$, then $f: B \longrightarrow C$ is an epimorphism. Let $g: P \longrightarrow C \longrightarrow 0$

be exact with P projective. Then

$\text{Im}(\ ,g)(X) \subset P(X, C) \subset \text{Im}(\ ,f)(X)$. Since by hypothesis $f: B \longrightarrow C$ is right X-determined, it follows that there is a morphism $t: P \longrightarrow B$ such that $g = ft$. The fact that g is an epimorphism implies that f is an epimorphism, our desired result.

We can now state the main result of this section.

Let X be in $\text{mod}\ \Lambda$ and C in $\text{Mod}\ \Lambda$. Suppose H is an $(\text{End}\ X)^{op}$-submodule of (X, C) containing $P(X, C)$. Then there is a minimal right X-determined epimorphism $f: B \longrightarrow C$ such that $(\text{Im}(\ ,f))(X) = H$.

Before proceeding to the proof of this theorem, we recall some definitions and results we will need in the course of the proof (see [12] for more details).

We begin by recalling some easily verified properties of the subgroups $P(A, B)$ of (A, B).

a) If $f \in P(A, B)$ and $g \in (B, C)$, then $gf \in P(A, C)$.

b) If $f \in (A, B)$ and $g \in P(B, C)$, then $gf \in P(A, C)$.

c) If $f: P \longrightarrow C$ is an epimorphism with P projective, then $\text{Im}((X, P) \xrightarrow{(X,\ f)} (X, C)) = P(X, C)$ for all X in $\text{Mod}\ \Lambda$.

Next we recall the definition of the additive category $\underline{\text{Mod}}\ \Lambda$, the category of Λ-modules modulo projectives. The objects of $\underline{\text{Mod}}\ \Lambda$ are the same as those of $\text{Mod}\ \Lambda$. For each pair of objects A, B in $\underline{\text{Mod}}\ \Lambda$, the group of morphisms $\underline{\text{Hom}}_\Lambda(A, B)$ from A to B is the group $\text{Hom}_\Lambda(A, B)/P(A, B)$. For each f in $\text{Hom}_\Lambda(A, B)$ we denote by $\underline{f}$ the image of f in $\text{Hom}_\Lambda(A, B)/P(A, B)$. Further the composition of morphisms in $\underline{\text{Mod}}\ \Lambda$ is defined by the formula $\underline{g}\,\underline{f} = \underline{gf}$ for $f: A \longrightarrow B$ and $g: B \longrightarrow C$. The fact that this composition is well defined follows from properties a) and b) cited above.

We denote by $\underline{\text{mod}}\ \Lambda$ the full subcategory of $\underline{\text{Mod}}\ \Lambda$ whose objects are in $\text{mod}\ \Lambda$, the category of finitely presented Λ-modules. There is a well known duality (see [8] and [10]) $\text{Tr}: \underline{\text{mod}}\ \Lambda \longrightarrow \underline{\text{mod}}\ \Lambda^{op}$, called the <u>transpose</u>, which plays a central role throughout this paper and is now briefly described.

Let $\underline{\underline{p}}(\Lambda)$ and $\underline{\underline{p}}(\Lambda^{op})$ denote the categories of finitely generated projective Λ and Λ^{op}-modules respectively. It is well known that if P is in $\underline{\underline{p}}(\Lambda)$, then the Λ^{op}-module $\text{Hom}_\Lambda(P, \Lambda)$ is in $\underline{\underline{p}}(\Lambda^{op})$ and that the functor $\text{Hom}_\Lambda(\ ,\Lambda): \underline{\underline{p}}(\Lambda) \longrightarrow \underline{\underline{p}}(\Lambda^{op})$ is a duality. While this duality cannot be extended to a duality $\text{mod}\ \Lambda \longrightarrow \text{mod}\ \Lambda^{op}$, it can be extended to a duality $\text{Tr}: \underline{\text{mod}}\ \Lambda \longrightarrow \underline{\text{mod}}\ \Lambda^{op}$ as follows.

For each A in $\text{mod}\ \Lambda$ choose a fixed projective presentation $P_1(A) \xrightarrow{u} P_0(A) \longrightarrow A \longrightarrow 0$ with the $P_i(A)$ in $\underline{\underline{p}}(\Lambda)$. Then $\text{Hom}_\Lambda(u, \Lambda): \text{Hom}_\Lambda(P_0(A), \Lambda) \longrightarrow \text{Hom}_\Lambda(P_1(A), \Lambda)$ is a morphism in $\underline{\underline{p}}(\Lambda^{op})$ and so $\text{Coker}\ \text{Hom}_\Lambda(u, \Lambda)$ is in $\text{mod}\ \Lambda^{op}$. Define $\text{Tr}A = \text{Coker}(\text{Hom}_\Lambda(u, \Lambda))$. Suppose $f: A \longrightarrow B$ is a morphism in $\text{mod}\ \Lambda$. There is then an exact commutative diagram

$$\begin{array}{ccccccc} P_1(A) & \longrightarrow & P_0(A) & \longrightarrow & A & \longrightarrow & 0 \\ \downarrow f_1 & & \downarrow f_0 & & \downarrow f & & \\ P_1(B) & \longrightarrow & P_0(B) & \longrightarrow & B & \longrightarrow & 0 \end{array}$$

which induces an exact commutative diagram

$$\begin{array}{ccccccc}
\mathrm{Hom}_\Lambda(P_0(B), \Lambda) & \longrightarrow & \mathrm{Hom}_\Lambda(P_1(B), \Lambda) & \longrightarrow & \mathrm{Tr}B & \longrightarrow & 0 \\
\downarrow \mathrm{Hom}_\Lambda(f_0, \Lambda) & & \downarrow \mathrm{Hom}_\Lambda(f_1, \Lambda) & & \downarrow h & & \\
\mathrm{Hom}_\Lambda(P_0(A), \Lambda) & \longrightarrow & \mathrm{Hom}_\Lambda(P_1(A), \Lambda) & \longrightarrow & \mathrm{Tr}A & \longrightarrow & 0
\end{array}$$

While the morphism h depends on the choice of f_0 and f_1, the image $\underline{h}$ of h in $\underline{\mathrm{Hom}}_\Lambda(\mathrm{Tr}B, \mathrm{Tr}A)$ does not depend on these choices and so we define $\mathrm{Tr}(f) = \underline{h}$. This then gives rise to a contravariant functor $\mathrm{Tr}\colon \underline{\mathrm{mod}}\ \Lambda \longrightarrow \underline{\mathrm{mod}}\ \Lambda^{op}$ which can be seen to be a duality using the fact that $\mathrm{Hom}_\Lambda(\ ,\Lambda)\colon \underline{\underline{p}}(\Lambda) \longrightarrow \underline{\underline{p}}(\Lambda^{op})$ is a duality.

Obviously the construction of the duality Tr depends on the initial choices of the projective presentations $P_1(A) \longrightarrow P_0(A) \longrightarrow A \longrightarrow 0$. However, standard homological arguments show that changing these presentations changes the associated dualities $\underline{\mathrm{mod}}\ \Lambda \longrightarrow \underline{\mathrm{mod}}\ \Lambda^{op}$ by prescribed isomorphisms and so the duality $\mathrm{Tr}\colon \underline{\mathrm{mod}}\ \Lambda \longrightarrow \underline{\mathrm{mod}}\ \Lambda^{op}$ is essentially unique.

Before going on to show the role the dualities $\mathrm{Tr}\colon \underline{\mathrm{mod}}\ \Lambda \longrightarrow \underline{\mathrm{mod}}\ \Lambda^{op}$ play in the study of epimorphisms in Mod Λ which are right determined by modules in mod Λ, it is important to make a few observations concerning these dualities.

First it is important to note that if Tr_1 and Tr_2 are two dualities gotten by using different projective presentations for the modules in mod Λ, then $\mathrm{Tr}_1(X)$ and $\mathrm{Tr}_2(X)$ need not be isomorphic in mod Λ even though they are isomorphic in $\underline{\mathrm{mod}}\ \Lambda$. What can be shown is that there are projective modules P and Q in mod Λ such that $\mathrm{Tr}_1(X) \coprod P \approx \mathrm{Tr}_2(X) \coprod Q$ in mod Λ.

Second, it is not difficult to show that given two dualities $\mathrm{Tr}_1\colon \underline{\mathrm{mod}}\ \Lambda \longrightarrow \underline{\mathrm{mod}}\ \Lambda^{op}$ and $\mathrm{Tr}_2\colon \underline{\mathrm{mod}}\ \Lambda^{op} \longrightarrow \underline{\mathrm{mod}}\ \Lambda$, then $\mathrm{Tr}_2\,\mathrm{Tr}_1 \approx 1_{\underline{\mathrm{mod}}\ \Lambda}$ and $\mathrm{Tr}_1\,\mathrm{Tr}_2 \cong 1_{\underline{\mathrm{mod}}\ \Lambda}$. However, it is not necessarily the case thet $Tr_2\ Tr_1(X) \approx X$ in mod Λ for all X in mod Λ. However, it can be shown without a great deal of difficulty that given Tr_1· mod $\Lambda \longrightarrow$ mod Λ^{op} there is a Tr_2: mod $\Lambda^{op} \longrightarrow$ mod Λ such that $Tr_2\ Tr_1(X) \approx X$ in mod Λ for all X in mod Λ. The reader should be cautioned that it is not necessarily the case that just because the dualities Tr_1: mod $\Lambda \longrightarrow$ mod Λ^{op} and Tr_2: mod $\Lambda^{op} \longrightarrow$ mod Λ have the property that $Tr_2\ Tr_1(X) \approx X$ in mod Λ for all nonprojective X in mod Λ, then $Tr_1\ Tr_2(Y) \approx Y$ in mod Λ^{op} for all nonprojective Y in mod Λ^{op}. In order to simplify notation, we assume from now on that we have chosen two dualities Tr_1: mod $\Lambda \longrightarrow$ mod Λ^{op} and Tr_2: mod $\Lambda^{op} \longrightarrow$ mod Λ with the property $Tr_2\ Tr_1(X) \approx X$ for all X in mod Λ both of which we denote simply by Tr.

With this convention in mind we turn our attention to showing how the duality Tr: mod $\Lambda \longrightarrow$ mod Λ^{op} is connected with epimorphisms which are right determined by modules in mod Λ.

Our use of the transpose in this section is based on the following.

Proposition 3.2: For each X in mod Λ the functors $\underline{Hom}_\Lambda(X,\ )$: Mod $\Lambda \longrightarrow$ Ab and $Tor_1^\Lambda(TrX,\ )$: Mod $\Lambda \longrightarrow$ Ab are isomorphic in a way which is functional in X. Specifically, given $f: X \longrightarrow Y$ in mod Λ the diagram

$$\begin{array}{ccc} \underline{Hom}_\Lambda(Y,\ ) & \xrightarrow{\alpha_Y} & Tor_1^\Lambda(TrY,\ ) \\ \downarrow(\underline{f},\ ) & & \downarrow Tor_1^\Lambda(Trf,\ ) \\ \underline{Hom}_\Lambda(X,\ ) & \xrightarrow{\alpha_X} & Tor_1^\Lambda(TrX,\ ) \end{array}$$

commutes.

Proof: The proof is an obvious generalization of the one given in [10, Proposition 2.2] for artin algebras Λ.

With these preliminaries in mind, we turn our attention to proving the result cited in the beginning of this section. We first prove a somewhat weaker result, which is very useful in its own right, from which we deduce our desired result.

Let X be in Mod Λ. Then X is also an End X-module. Since the operations of Λ and End X on X commute, X is a Λ-End X bimodule. More generally given any ring morphism $\Sigma \longrightarrow \text{End } X$, the induced operation of Σ on X gives rise to a $\Lambda - \Sigma$ bimodule structure on X. All bimodule structures $\Lambda - \Sigma$ on X are obtained in this way.

Let X be in mod Λ. Then $\text{Tr}X$ is in mod Λ^{op}. Suppose $\Gamma = \text{End Tr}X$ and $\Sigma \longrightarrow \Gamma$ is a ring morphism so that $\text{Tr}X$ is a Λ^{op}- Σ bimodule. Then the operation of Λ^{op} on $\text{Tr}X$ makes $\text{Hom}_\Sigma(\text{Tr}X, Y)$ a Λ-module for each Σ-module Y. Moreover, it is important to observe that the usual isomorphism $\text{Hom}_\Gamma(\text{Tr}X, \text{Hom}_\Sigma(\Gamma, Y)) \longrightarrow \text{Hom}_\Sigma(\text{Tr}X, Y)$ is an isomorphism of Λ-modules for each Σ-module Y. We recall the following well known result which plays a critical role in much of this paper.

Proposition 3.3: Let X and Σ be as above. Suppose C is in Mod Λ and Y is in Mod Σ. Then there are morphisms

$$\text{Ext}^1_\Lambda(C, \text{Hom}_\Sigma(\text{Tr}X, Y)) \longrightarrow \text{Hom}_\Sigma(\text{Tor}^\Lambda_1(\text{Tr}X, C), Y)$$

which are functorial in C and Y and which are isomorphisms whenever Y is an injective Σ-module. Moreover the diagram of isomorphisms

$$\begin{array}{ccc} \mathrm{Ext}^1_\Lambda(C, \mathrm{Hom}_\Sigma(\mathrm{Tr}X, Y)) & \longrightarrow & \mathrm{Hom}_\Sigma(\mathrm{Tor}^\Lambda_1(\mathrm{Tr}X, C), Y) \\ \downarrow & & \downarrow \\ \mathrm{Ext}^1_\Lambda(C, \mathrm{Hom}_\Gamma(\mathrm{Tr}X, \mathrm{Hom}_\Sigma(\Gamma, Y))) & \longrightarrow & \mathrm{Hom}^1_\Gamma(\mathrm{Tor}^\Lambda_1(\mathrm{Tr}X, C), \mathrm{Hom}_\Sigma(\Gamma, Y)) \end{array}$$

commutes.

Proof: See [14, VI, Proposition 5.1].

Our use of Proposition 3.3 depends on the following observation. The duality Tr: $\underline{\mathrm{mod}}\ \Lambda^{op} \longrightarrow \underline{\mathrm{mod}}\ \Lambda$ gives rise to an isomorphism $\underline{\mathrm{End}}\ \mathrm{Tr}X \longrightarrow (\underline{\mathrm{End}}\ X)^{op}$ where $\underline{\mathrm{End}}\ \mathrm{Tr}X = \mathrm{End}\ X/P(X, X)$. Thus for each C in Mod Λ the $(\underline{\mathrm{End}}\ X)^{op}$-module $\underline{\mathrm{Hom}}_\Lambda(X, C)$ is also an $\underline{\mathrm{End}}$ TrX-module and thus an End TrX-module. Hence if TrX is $\Lambda^{op} - \Sigma$ bimodule, then $\underline{\mathrm{Hom}}_\Lambda(X, C)$ is a Σ-module for all C in Mod Λ by means of the morphism $\Sigma \longrightarrow \mathrm{End}\ \mathrm{Tr}X$. Moreover the functorial isomorphisms $\mathrm{Tor}^\Lambda_1(\mathrm{Tr}X, C) \longrightarrow \underline{\mathrm{Hom}}_\Lambda(X, C)$ referred to in Proposition 3.2 are Σ-isomorphisms. Thus we obtain the following version of Proposition 3.3.

Proposition 3.4: Let C and X be Λ-modules with X in mod Λ. Suppose TrX is a $\Lambda^{op} - \Sigma$ bimodule and I is an injective Σ-module. Then there are isomorphisms

$$\mathrm{Ext}^1_\Lambda(C, \mathrm{Hom}_\Sigma(\mathrm{Tr}X, I)) \longrightarrow \mathrm{Hom}_\Sigma(\underline{\mathrm{Hom}}_\Lambda(X, C), I)$$

functorial in C and I.

Moreover the diagram of isomorphisms

$$\begin{array}{ccc} \operatorname{Ext}^1_\Lambda(C, \operatorname{Hom}_\Sigma(\operatorname{Tr}X, I)) & \longrightarrow & \operatorname{Hom}_\Sigma(\underline{\operatorname{Hom}}_\Lambda(X, C), I) \\ \downarrow & & \downarrow \\ \operatorname{Ext}^1_\Lambda(C, \operatorname{Hom}_\Gamma(\operatorname{Tr}X, \operatorname{Hom}_\Sigma(\Gamma, I))) & \longrightarrow & \operatorname{Hom}_\Gamma(\underline{\operatorname{Hom}}_\Lambda(X, C), \operatorname{Hom}_\Sigma(\Gamma, I)) \end{array}$$

commutes.

Proof: Immediate once one observes that since I is an injective Σ-module, $\operatorname{Hom}_\Sigma(\Gamma, I)$ is an injective Γ-module.

We now investigate some of the connections between an exact sequence $x\colon 0 \longrightarrow \operatorname{Hom}_\Sigma(\operatorname{Tr}X, I) \xrightarrow{g} B \xrightarrow{f} C \longrightarrow 0$ in $\operatorname{Ext}^1_\Lambda(C, \operatorname{Hom}_\Sigma(\operatorname{Tr}X, I))$ and the Σ-morphism $\upsilon\colon \underline{\operatorname{Hom}}_\Lambda(X, C) \longrightarrow I$ in $\operatorname{Hom}_\Sigma(\underline{\operatorname{Hom}}_\Lambda(X, C), I)$ corresponding to x under the isomorphism $\operatorname{Ext}^1_\Lambda(C, \operatorname{Hom}_\Sigma(\operatorname{Tr}X, I)) \longrightarrow \operatorname{Hom}_\Sigma(\underline{\operatorname{Hom}}_\Lambda(X, C), I)$. Under these circumstances we have the following result.

Theorem 3.5:

a) A morphism $h\colon L \longrightarrow C$ in $\operatorname{Mod}\Lambda$ can be written as $ft = h$ for for some morphism $t\colon L \longrightarrow B$ if and only if $\operatorname{Im}(\ ,h)(X) \subset H$, where H is the preimage of $\operatorname{Ker}\upsilon$ under the epimorphism $\operatorname{Hom}_\Lambda(X, C) \longrightarrow \underline{\operatorname{Hom}}_\Lambda(X, C)$.

b) f is right X-determined with $\operatorname{Im}(\ ,f)(X)$ contained in H.

c) $\operatorname{Im}(\ ,f)(X)$ is the maximal $(\operatorname{End}X)^{op}$-submodule of H, i.e. $\operatorname{Im}(\ ,f)(X)$ is the $(\operatorname{End}X)^{op}$-submodule of H generated by all $(\operatorname{End}X)^{op}$-submodules.

d) $\operatorname{Im}(\ ,f)(X) = H$ if and only if H is an $(\operatorname{End}X)^{op}$-submodule of $\operatorname{Hom}_\Lambda(X, C)$.

Proof:

a) The assertion that there is a $t: L \longrightarrow B$ such that $ft = h$ is equivalent to the top line splitting in the pull back diagram

$$\begin{array}{ccccccccc} 0 & \longrightarrow & \mathrm{Hom}_{\Sigma}(\mathrm{Tr}X, I) & \longrightarrow & V & \longrightarrow & L & \longrightarrow & 0 \\ & & \downarrow & & \downarrow & & \downarrow h & & \\ 0 & \longrightarrow & \mathrm{Hom}_{\Sigma}(\mathrm{Tr}X, I) & \xrightarrow{g} & B & \xrightarrow{f} & C & \longrightarrow & 0 \end{array},$$

or equivalently, that $\mathrm{Ext}^1_{\Lambda}(h, \mathrm{Hom}_{\Sigma}(\mathrm{Tr}X, I))(x) = 0$ where x is the element of $\mathrm{Ext}^1(C, \mathrm{Hom}_{\Sigma}(\mathrm{Tr}X, I))$ represented by $0 \longrightarrow \mathrm{Hom}_{\Sigma}(\mathrm{Tr}X, I) \xrightarrow{g} B \xrightarrow{f} C \longrightarrow 0$.

From the commutative diagram

$$\begin{array}{ccc} \mathrm{Ext}^1_{\Lambda}(C, \mathrm{Hom}_{\Sigma}(\mathrm{Tr}X, I)) & \approx & \mathrm{Hom}_{\Sigma}(\underline{\mathrm{Hom}}_{\Lambda}(X, C), I) \\ {\scriptstyle \mathrm{Ext}^1_{\Lambda}(h, \mathrm{Hom}_{\Sigma}(\mathrm{Tr}X, I))}\downarrow & & \downarrow{\scriptstyle (\underline{\mathrm{Hom}}_{\Lambda}(X, h), I)} \\ \mathrm{Ext}^1_{\Lambda}(L, \mathrm{Hom}_{\Sigma}(\mathrm{Tr}X, I) & \approx & \mathrm{Hom}_{\Sigma}(\underline{\mathrm{Hom}}_{\Lambda}(X, L), I) \end{array}$$

we deduce that $\mathrm{Ext}^1_{\Lambda}(h, \mathrm{Hom}_{\Sigma}(\mathrm{Tr}X, I))(x) = 0$ if and only if $(\underline{\mathrm{Hom}}_{\Lambda}(X, h), I)(\upsilon) = 0$. Since $(\underline{\mathrm{Hom}}_{\Lambda}(X, h), I)(\upsilon)$ is the composition

$$\underline{\mathrm{Hom}}_{\Lambda}(X, L) \xrightarrow{\underline{\mathrm{Hom}}_{\Lambda}(X, h)} \underline{\mathrm{Hom}}_{\Lambda}(X, C) \xrightarrow{\upsilon} I,$$

we have that $(\underline{\mathrm{Hom}}_{\Lambda}(X, h), I)(\upsilon) = 0$ if and only if $\mathrm{Im}(\underline{\mathrm{Hom}}_{\Lambda}(X, h)) \subset \mathrm{Ker}\ \upsilon$. From this it follows that $(\underline{\mathrm{Hom}}_{\Lambda}(X, h), I)(\upsilon) = 0$ if and only if

$\mathrm{Im}(\mathrm{Hom}_\Lambda(X, h)) \subset H$, the preimage of $\mathrm{Ker}\ \nu$ under the epimorphism $(X, C) \longrightarrow \underline{\mathrm{Hom}}_\Lambda(X, C)$. Since $\mathrm{Im}(\mathrm{Hom}_\Lambda(X, h)) = \mathrm{Im}(\ ,h)(X)$, the proof of a) is complete.

b) Since $f1_B = f$, it follows from a), that $\mathrm{Im}(\ ,f)(X) \subset H$. Now suppose $h: L \longrightarrow C$ is such that $\mathrm{Im}(\ ,h)(X) \subset \mathrm{Im}(\ ,f)(X)$. Then $\mathrm{Im}(\ ,h)(X) \subset H$ and so by part a) there is $t: L \longrightarrow B$ such that $h = ft$. Therefore f is right X-determined.

c) Suppose J is the maximal $(\mathrm{End}\ X)^{op}$-submodule of H. Then for each $h: L \longrightarrow C$ such that $\mathrm{Im}(\ ,h)(X) \subset H$ we have that $\mathrm{Im}(\ ,h)(X) \subset J$. Thus J is an $(\mathrm{End}\ X)^{op}$-submodule of $\mathrm{Hom}_\Lambda(X, C)$ with the property that a morphism $h: L \longrightarrow C$ has the property there is a morphism $t: L \longrightarrow B$ such that $ft = h$ if and only if $\mathrm{Im}(\ ,h)(X)$ is contained in J. It then follows from Proposition 2.4 that $\mathrm{Im}(\ ,f)(X) = J$.

d) Trivial consequence of c).

Before going on to give various applications of this result, we point out that the proof of Theorem 3.5 can be used to establish a somewhat more general result which we state now for future reference.

Theorem 3.6: Let $\underline{\underline{C}}$ be a full additive subcategory of $\mathrm{Mod}\ \Lambda$. Suppose X is a $\Lambda - \Sigma^{op}$ bimodule with X in $\underline{\underline{C}}$ and Y is a Σ-module such that there exists an isomorphism of functors from $\underline{\underline{C}}$ to Ab

$$\mathrm{Ext}^1_\Lambda(\ ,A) \cong \mathrm{Hom}_\Sigma(\underline{\mathrm{Hom}}_\Lambda(X,\ ), Y)$$

for some A in $\mathrm{Mod}\ \Lambda$. Finally suppose

$x: 0 \longrightarrow A \xrightarrow{g} B \xrightarrow{f} C \longrightarrow 0$ is in $\operatorname{Ext}^1_\Lambda(C, A)$ with C in $\underline{\underline{C}}$ and $\nu: \underline{\operatorname{Hom}}_\Lambda(X, C) \longrightarrow Y$ the element of $\operatorname{Hom}_\Sigma(\underline{\operatorname{Hom}}_\Lambda(X, C), Y)$ corresponding to x under the isomorphism $\operatorname{Ext}^1_\Lambda(C, A) \xrightarrow{\sim} \operatorname{Hom}_\Sigma(\underline{\operatorname{Hom}}_\Lambda(X, C), Y)$. Then

a) A morphism $h: L \longrightarrow C$ in $\underline{\underline{C}}$ can be written as $ft = h$ for some $t: L \longrightarrow B$ if and only if $\operatorname{Im}(\ ,h)(X) \subset H$ where H is the preimage of $\operatorname{Ker} \nu$ under the epimorphism $\operatorname{Hom}_\Lambda(X, C) \longrightarrow \underline{\operatorname{Hom}}_\Lambda(X, C)$.

Further, suppose A is in $\underline{\underline{C}}$ and $\underline{\underline{C}}$ is closed under extensions. Then we have the following.

b) f is right X-determined in $\underline{\underline{C}}$ with $\operatorname{Im}(\ ,f)(X) \subset H$.

c) $\operatorname{Im}(\ ,f)(X)$ is the maximal $(\operatorname{End} X)^{op}$-submodule of H.

d) $\operatorname{Im}(\ ,f)(X) = H$ if and only if H is an $(\operatorname{End} X)^{op}$-submodule of (X, C).

We now apply Theorem 3.5 to obtain part of the existence theorem for right X-determined morphisms given at the beginning of this section.

Theorem 3.7: Let X be in $\operatorname{mod} \Lambda$, Σ a ring such that $\operatorname{Tr}X$ is a Λ^{op}-Σ bimodule. Suppose C is in $\operatorname{Mod} \Lambda$ and H is an $(\operatorname{End} X)^{op}$-submodule of (X, C) containing $P(X, C)$. Finally, let I be an injective Σ-module such that there is a Σ-morphism $\nu: \underline{\operatorname{Hom}}_\Lambda(X, C) \longrightarrow I$ with $\operatorname{Ker} \nu = H/P(X, C)$. Then there is an exact sequence of Λ-modules $x: 0 \longrightarrow \operatorname{Hom}_\Sigma(\operatorname{Tr}X, I) \xrightarrow{g} B \xrightarrow{f} C \longrightarrow 0$ satisfying

a) $\operatorname{Im}(\ ,f)(X) = H$ and

b) f is right X-determined.

Furthermore the element x in $\operatorname{Ext}^1_\Lambda(C, \operatorname{Hom}_\Sigma(\operatorname{Tr}X, I))$ is the element corresponding to $\nu: \underline{\operatorname{Hom}}(X, C) \longrightarrow I$ under the isomorphism $\operatorname{Ext}^1_\Lambda(C, \operatorname{Hom}_\Sigma(\operatorname{Tr}X, I)) \longrightarrow \operatorname{Hom}_\Sigma(\underline{\operatorname{Hom}}_\Lambda(X, C), I)$.

Proof: Let $x: 0 \longrightarrow \operatorname{Hom}_\Sigma(\operatorname{Tr}X, I) \longrightarrow B \xrightarrow{f} C \longrightarrow 0$ in $\operatorname{Ext}^1_\Lambda(C, \operatorname{Hom}_\Sigma(\operatorname{Tr}X, I))$ be the element corresponding to $\nu: \underline{\operatorname{Hom}}_\Lambda(X, C) \longrightarrow I$ under the isomorphism $\operatorname{Ext}^1_\Lambda(C, \operatorname{Hom}_\Sigma(\operatorname{Tr}X, I)) \longrightarrow \operatorname{Hom}_\Sigma(\underline{\operatorname{Hom}}_\Lambda(X, C), I)$. It then follows trivially from Theorem 3.5 that $0 \longrightarrow \operatorname{Hom}_\Sigma(\operatorname{Tr}X, I) \longrightarrow B \xrightarrow{f} C \longrightarrow 0$ has our desired properties.

Having shown that for X in mod Λ and C in Mod Λ, and H an $(\operatorname{End} X)^{op}$-submodule of (X, C) containing $P(X, C)$, there is an epimorphism $f: B \longrightarrow C$ which is right X-determined with $(\operatorname{Im}(\ ,f))(X) = H$, we now show that there is a minimal right X-determined epimorphism $f: B \longrightarrow C$ such that $\operatorname{Im}(\ ,f)(X) = H$. To do this we will need the following property of finitely presented modules which will be proven in the appendix of this chapter.

Proposition 3.8: Let Y be a finitely presented Λ^{op}-module and let $\Gamma = \operatorname{End} Y$. Further, let $\underline{\underline{I}}$ be the full subcategory of Mod Γ whose objects are the injective Γ-modules. Then the functor $\operatorname{Hom}_\Gamma(Y, \ ): \underline{\underline{I}} \longrightarrow \operatorname{Mod} \Lambda$, given by $I \longmapsto \operatorname{Hom}_\Gamma(Y, I)$ for each injective Γ-module I in $\underline{\underline{I}}$, is fully faithful.

We now apply this result to obtain the following consequence of Theorem 3.5.

Theorem 3.9: Let X be in mod Λ and $\Gamma = \operatorname{End} \operatorname{Tr}X$. Suppose C is in Mod Λ and H is an $(\operatorname{End} X)^{op}$-submodule of (X, C) containing $P(X, C)$. Further, let $\nu: \underline{\operatorname{Hom}}_\Lambda(X, C) \longrightarrow I$ be a Γ-morphism with I an injective Γ-module such that $\operatorname{Ker} \nu = H/P(X, C)$ and the induced morphism

$\underline{\mathrm{Hom}}_\Lambda(X, C)/H/P(X, C) \longrightarrow I$ is an injective envelope. Then the short exact sequence $x: 0 \longrightarrow \mathrm{Hom}_\Gamma(\mathrm{Tr}X, I) \xrightarrow{g} B \xrightarrow{f} C \longrightarrow 0$ in $\mathrm{Ext}^1_\Lambda(C, \mathrm{Hom}_\Gamma(\mathrm{Tr}X, I))$ corresponding to the morphism $\nu: \underline{\mathrm{Hom}}_\Lambda(X, C) \longrightarrow I$ under the isomorphism $\mathrm{Ext}^1_\Lambda(C, \mathrm{Hom}_\Gamma(\mathrm{Tr}X, I)) \longrightarrow \mathrm{Hom}_\Gamma(\underline{\mathrm{Hom}}_\Lambda(X, C), I)$ has the property that $f: B \longrightarrow C$ is a minimal right X-determined morphism such that $(\mathrm{Im}(\;,f))(X) = H$.

Proof: Letting $\Sigma = \Gamma$ in Theorem 3.5, we know that $f: B \longrightarrow C$ is a right X-determined morphism such that $(\mathrm{Im}(\;,f))(X) = H$. Therefore we only have to show that $f: B \longrightarrow C$ is right minimal.

Let $h: B \longrightarrow B$ be such that $fh = f$. We want to show that h is an isomorphism. Since $0 \longrightarrow \mathrm{Hom}_\Gamma(\mathrm{Tr}X, I) \xrightarrow{g} B \xrightarrow{f} C \longrightarrow 0$ is exact, there is a commutative diagram

$$\begin{array}{ccccccccc} 0 & \longrightarrow & \mathrm{Hom}_\Gamma(\mathrm{Tr}X, I) & \xrightarrow{g} & B & \xrightarrow{f} & C & \longrightarrow & 0 \\ & & j\downarrow & & h\downarrow & & \| & & \\ 0 & \longrightarrow & \mathrm{Hom}_\Gamma(\mathrm{Tr}X, I) & \xrightarrow{g} & B & \xrightarrow{f} & C & \longrightarrow & 0 \end{array} .$$

Therefore to show that h is an isomorphism, it suffices to show that j is an isomorphism. By Proposition 3.8 we know that there is a Γ-morphism $t: I \longrightarrow I$ such that $\mathrm{Hom}_\Gamma(\mathrm{Tr}X, t): \mathrm{Hom}_\Gamma(\mathrm{Tr}X, I) \longrightarrow \mathrm{Hom}_\Gamma(\mathrm{Tr}X, I)$ is the morphism $j: \mathrm{Hom}_\Gamma(\mathrm{Tr}X, I) \longrightarrow \mathrm{Hom}_\Gamma(\mathrm{Tr}X, I)$. We now show that j is an isomorphism by showing that $t: I \longrightarrow I$ is an isomorphism, thus finishing the proof of the theorem.

By Proposition 3.4 we have the commutative diagram

$$\begin{array}{ccc}
\operatorname{Ext}^1_\Lambda(C, \operatorname{Hom}_\Gamma(\operatorname{Tr}X, I)) & \longrightarrow & \operatorname{Hom}_\Gamma(\underline{\operatorname{Hom}}_\Lambda(X, C), I) \\
\operatorname{Ext}^1_\Lambda(C, j)\downarrow & & \downarrow \operatorname{Hom}_\Gamma(\underline{\operatorname{Hom}}_\Lambda(X, C), t) \\
\operatorname{Ext}^1_\Lambda(C, \operatorname{Hom}_\Gamma(\operatorname{Tr}X, I)) & \longrightarrow & \operatorname{Hom}_\Gamma(\underline{\operatorname{Hom}}_\Lambda(X, C), I)
\end{array}$$

Since $\operatorname{Ext}^1_\Lambda(C, j)(x) = x$, we have that $\operatorname{Hom}_\Gamma(\underline{\operatorname{Hom}}_\Lambda(X, C), t)(\upsilon) = \upsilon$. Hence

$$\begin{array}{ccc}
\underline{\operatorname{Hom}}_\Lambda(X, C) & \xrightarrow{\upsilon} & I \\
\parallel & & \downarrow t \\
\underline{\operatorname{Hom}}_\Lambda(X, C) & \xrightarrow{\upsilon} & I
\end{array}$$

commutes. This gives rise to the exact commutative diagram

$$(*) \qquad \begin{array}{ccccc}
0 & \longrightarrow & \underline{\operatorname{Hom}}_\Lambda(X, C)/\operatorname{Ker} \upsilon & \longrightarrow & I \\
 & & \parallel & & \downarrow t \\
0 & \longrightarrow & \underline{\operatorname{Hom}}_\Lambda(X, C)/\operatorname{Ker} \upsilon & \longrightarrow & I
\end{array} .$$

The fact that $0 \longrightarrow \underline{\operatorname{Hom}}_\Lambda(X, C)/\operatorname{Ker} \upsilon \longrightarrow I$ is an injective envelope implies that $t\colon I \longrightarrow I$ is an isomorphsim.

To see this observe that $\operatorname{Ker} t = 0$ since from the commutative diagram (*) it follows that $\operatorname{Ker} t \cap \underline{\operatorname{Hom}}_\Lambda(X, C)/\operatorname{Ker} \upsilon = (0)$. Hence $\operatorname{Im} t$ is an injective submodule of I containing $\underline{\operatorname{Hom}}_\Lambda(X, C)/\operatorname{Ker} \upsilon$. Therefore $\operatorname{Im} t = I$ since $0 \longrightarrow \underline{\operatorname{Hom}}_\Lambda(X, C)/\operatorname{Ker} \upsilon \longrightarrow I$ is an injective envelope. Consequently $t\colon I \longrightarrow I$ is an isomorphism and the proof of the theorem is complete.

Before going on with our discussion of epimorphisms which are right determined by a module, we point out the following well known and easily verified facts about extensions which we will use freely throughout the rest of the paper.

<u>Proposition 3.10</u>: Suppose $A = A_1 \coprod A_2$ and $x: 0 \longrightarrow A \xrightarrow{g} B \xrightarrow{f} C \longrightarrow 0$ is an exact sequence. Let $x_i: 0 \longrightarrow A_i \xrightarrow{g_i} B_i \xrightarrow{f_i} C \longrightarrow 0$ for $i = 1, 2$ be the components of x under the natural isomorphism $\text{Ext}^1_\Lambda(C, A_1 \coprod A_2) \longrightarrow \text{Ext}^1_\Lambda(C, A_1) \coprod \text{Ext}^1_\Lambda(C, A_2)$. Then the following are equivalent:

a) $x_1 = 0$.

b) there exists a commutative diagram

$$\begin{array}{ccccccccc}
0 & \longrightarrow & A_2 & \xrightarrow{g_2} & B_2 & \xrightarrow{f_2} & C & \longrightarrow & 0 \\
 & & \downarrow{\scriptstyle inc} & & \downarrow{\scriptstyle u} & & \| & & \\
0 & \longrightarrow & A & \xrightarrow{g} & B & \xrightarrow{f} & C & & \\
 & & \downarrow{\scriptstyle p} & & \downarrow{\scriptstyle \nu} & & \| & & \\
0 & \longrightarrow & A_2 & \longrightarrow & B_2 & \longrightarrow & C & &
\end{array}$$

with $inc: A_2 \longrightarrow A_1 \coprod A_2$ the natural inclusion, $p: A_1 \coprod A_2 \longrightarrow A_2$ the natural projection and νu an isomorphism.

c) $g(A_1)$ has a complement in B containing $g(A_2)$.

As a consequence of Theorem 3.9 we have the following.

<u>Proposition 3.11</u>: Suppose X is in $\text{mod}\,\Lambda$ and $x: 0 \longrightarrow A \xrightarrow{g} B \xrightarrow{f} C \longrightarrow 0$ is an exact sequence in $\text{Mod}\,\Lambda$ with f right X-determined. Let $0 \longrightarrow A' \xrightarrow{g'} B' \xrightarrow{f'} C \longrightarrow 0$ be an exact sequence in $\text{Mod}\,\Lambda$ satisfying $\text{Im}(\ ,f')(X) = \text{Im}(\ ,f)(X)$ and f' is a minimal right X-determined morphism.

a) There exists a commutative diagram

$$\begin{array}{ccccccccc}
0 & \longrightarrow & A' & \xrightarrow{g'} & B' & \xrightarrow{f'} & C & \longrightarrow & 0 \\
 & & u\downarrow & & \nu\downarrow & & \| & & \\
0 & \longrightarrow & A & \xrightarrow{g} & B & \xrightarrow{f} & C & \longrightarrow & 0 \\
 & & \omega\downarrow & & z\downarrow & & \| & & \\
0 & \longrightarrow & A' & \xrightarrow{g'} & B' & \xrightarrow{f'} & C & &
\end{array}$$

such that $z\nu = 1_{B'}$ and $\omega u = 1_{A'}$.

b) x splits if and only if $\mathrm{Im}(\ ,f)(X) = (X, C)$.

c) Suppose x does not split. Then f is a right minimal morphism if and only if the element x in $\mathrm{Ext}^1_\Lambda(C, A)$ has the property that if $A \cong A_1 \coprod A_2$ with A_1 and A_2 not zero, then the components x_i in $\mathrm{Ext}^1_\Lambda(C, A_i)$ for $i = 1, 2$ of x under the canonical isomorphism $\mathrm{Ext}^1_\Lambda(C, A) \longrightarrow \prod_{i\varepsilon 1}^{2} \mathrm{Ext}^1_\Lambda(C, A_i)$ are both not zero.

Proof:

a) Since f and f' are right X-determined morphisms with $\mathrm{Im}(\ ,f)(X) = \mathrm{Im}(\ ,f')(X)$, there is a commutative exact diagram

$$\begin{array}{ccccccccc}
0 & \longrightarrow & A' & \xrightarrow{g'} & B' & \xrightarrow{f'} & C & \longrightarrow & 0 \\
 & & \downarrow u & & \downarrow \nu & & \| & & \\
0 & \longrightarrow & A & \xrightarrow{g} & B & \xrightarrow{f} & C & & \\
 & & \downarrow u' & & \downarrow \nu' & & \| & & \\
0 & \longrightarrow & A' & \xrightarrow{g'} & B' & \xrightarrow{f'} & C & &
\end{array}$$

Since $f': B' \longrightarrow C$ is right minimal, the fact that $f'\nu'\nu = f'$, implies that $\nu'\nu$ is an isomorphism. Letting $z = (\nu'\nu)^{-1}\nu'$ and ω the morphism such that $g'\omega = zg$, we have that $z\nu = 1_{B'}$ and $\omega u = 1_{A'}$.

b) Clearly if x splits, then $\operatorname{Im}(\ ,f)(X) = (X, C)$. Suppose $\operatorname{Im}(\ ,f)(X) = (X, C)$. Then $1_C : C \longrightarrow C$ is a minimal right X-determined morphism such that $\operatorname{Im}(\ ,1_C)(X) = (X, C) = \operatorname{Im}(\ ,f)(X)$. It then follows from a) that $A' = 0$. Applying a) again we have that x splits.

c) Since x does not split, $\operatorname{Im}(\ ,f)(X) \neq (X, C)$ and so $\operatorname{Im}(\ ,f')(X) \neq (X, C)$. Hence $A' \neq 0$ (see part a)). Suppose f is not right minimal. Then the splittable monomorphism $u: A' \longrightarrow A$ is not an isomorphism. Hence letting $\operatorname{Coker} u = A''$, we have that $A \cong A' \coprod A''$ with $A' \neq (0) \neq A''$. It then follows from a) that we have a commutative exact diagram

$$\begin{array}{ccccccccc}
 & & A'' & = & A'' & & & & \\
 & & \coprod & & \coprod & & & & \\
0 & \longrightarrow & A' & \xrightarrow{g'} & B' & \xrightarrow{f'} & C & \longrightarrow & 0 \\
 & & \wr\wr & & \wr\wr & & \| & & \\
0 & \longrightarrow & A & \longrightarrow & B & \xrightarrow{f} & C & \longrightarrow & 0
\end{array}$$

From this it follows that the component of x in $\operatorname{Ext}^1_\Lambda(C, A'')$ is zero under the isomorphism $\operatorname{Ext}^1_\Lambda(C, A) \approx (\operatorname{Ext}^1_\Lambda(C, A') \coprod \operatorname{Ext}^1_\Lambda(C, A''))$. Thus f is a minimal right morphism if given any representation of $A \approx A_1 \coprod A_2$ with $A_i \neq 0$ for $i = 1, 2$, both of the components of x in $\operatorname{Ext}^1_\Lambda(C, A_i)$ are nonzero.

The rest of c) is easily established.

The following uniqueness theorem is worth noting.

Proposition 3.12: Let X be in $\text{mod}\,\Lambda$ and suppose $0 \longrightarrow A \xrightarrow{g} B \xrightarrow{f} C \longrightarrow 0$ and $0 \longrightarrow A' \xrightarrow{g'} B' \xrightarrow{f'} C \longrightarrow 0$ are two exact sequences with f and f' minimal right X-determined morphisms.

a) If $\text{Im}(\ ,f)(X) = \text{Im}(\ ,f')(X)$, then there is a commutative exact diagram

$$\begin{array}{ccccccccc} 0 & \longrightarrow & A & \xrightarrow{g} & B & \xrightarrow{f} & C & \longrightarrow & 0 \\ & & \downarrow u & & \downarrow v & & \| & & \\ 0 & \longrightarrow & A' & \xrightarrow{g'} & B' & \xrightarrow{f'} & C & \longrightarrow & \end{array}$$

and any such commutative diagram has the property that v, and hence u, is an isomorphism.

b) $\text{Im}(\ ,f)(X) = \text{Im}(\ ,f')(X)$ if and only if there is a commutative diagram

$$\begin{array}{ccccccccc} 0 & \longrightarrow & A & \xrightarrow{g} & B & \xrightarrow{f} & C & \longrightarrow & 0 \\ & & \downarrow u & & \downarrow v & & \| & & \\ 0 & \longrightarrow & A' & \longrightarrow & B' & \xrightarrow{f'} & C & \longrightarrow & 0 \end{array}$$

with v, and hence u, is an isomorphism.

Proof: Is a trivial consequence of Proposition 3.10. See also Proposition 2.8.

Let $\underline{\underline{U}}$ be the full subcategory of $\text{Mod}\,\Lambda$ consisting of finite sums of Λ-modules isomorphic to modules of the form $\text{Hom}_\Sigma(X, I)$, where X is a Λ^{op}- Σ bimodule which is a finitely presented Λ^{op}-module and I is an injective Σ-module. The rest of this section is devoted to showing that if $0 \longrightarrow A \xrightarrow{g} B \xrightarrow{f} C$ is an exact sequence of Λ-modules with A in $\underline{\underline{U}}$, then f is right X-determined for some X in $\mathrm{mod}\ \Lambda$. To this end it is convenient to have the following description of when an exact sequence $0 \longrightarrow A \xrightarrow{g} B \xrightarrow{f} C \longrightarrow 0$ of Λ-modules has the property f is right X-determined for some X in $\mathrm{Mod}\ \Lambda$.

Lemma 3.13: Let $x\colon 0 \longrightarrow A \xrightarrow{g} B \xrightarrow{f} C \longrightarrow 0$ be an exact sequence in $\mathrm{Mod}\ \Lambda$ and X a Λ-module. Then f is right X-determined if and only if the element x in $\mathrm{Ext}^1_\Lambda(C, A)$ has the following property.

Let $h\colon L \longrightarrow C$ be a morphism in $\mathrm{Mod}\ \Lambda$. Then $\mathrm{Ext}^1_\Lambda(h, A)(x) = 0$ whenever for each morphism $j\colon X \longrightarrow L$ we have that $\mathrm{Ext}^1_\Lambda(hj, A)(x) = 0$.

Proof: Suppose $t\colon Y \longrightarrow C$ is an arbitrary morphism in $\mathrm{Mod}\ \Lambda$. It is well known that $\mathrm{Ext}^1_\Lambda(t, A)(x) = 0$ if and only if there exists $u\colon Y \longrightarrow B$ such that $fu = t$. In view of this observation it is easy to show that a morphism $h\colon L \longrightarrow C$ has the property that $\mathrm{Ext}^1_\Lambda(hj, A)(x) = 0$ for all j in (X, L) if and only if $\mathrm{Im}(\ ,h)(X) \subset \mathrm{Im}(\ ,f)(X)$. Our desired result now follows trivially.

As a consequence of this description of when an epimorphism is determined by a module, we have the following general result.

Proposition 3.14: Let $\{A_i\}_{i\in I}$ be a family of Λ-modules and $x\colon 0 \longrightarrow \prod_{i\in I} A_i \xrightarrow{g} B \xrightarrow{f} C \longrightarrow 0$ an exact sequence in $\mathrm{Mod}\ \Lambda$. Let x_i in $\mathrm{Ext}^1_\Lambda(C, A_i)$ be the ith-component of x under the canonical isomorphism $\mathrm{Ext}^1_\Lambda(C, \prod_{i\in I} A_i) \longrightarrow \prod \mathrm{Ext}^1_\Lambda(C, A_i)$ and let $x_i\colon 0 \longrightarrow A_i \xrightarrow{g_i} B_i \xrightarrow{f_i} C \longrightarrow 0$ be the exact sequence corresponding to x_i. Then a Λ-module X has the property that $f\colon B \longrightarrow C$ is right X-determined if each $f_i\colon B_i \longrightarrow C$ is right X-determined.

Proof: Recall that the projections $\prod_{i\in I} A_i \longrightarrow A_j$ induce an isomorphism of functors $\text{Ext}^1_\Lambda(\ ,\prod_{i\in I} A_i) \longrightarrow \prod_{i\in I} \text{Ext}^1_\Lambda(\ ,A_i)$. In particular, given any morphism $t: Y \longrightarrow C$ we have that $\text{Ext}^1_\Lambda(t, C)(x) = (\text{Ext}^1_\Lambda(t, C)(x_i))_{i\in I}$. Thus we have that if $h: L \longrightarrow C$ is a morphism, then a morphism $j: X \longrightarrow L$ has the property $\text{Ext}^1_\Lambda(jh, \prod_{i\in I} A_i)(x_i)_{i\in I}$ if and only if $\text{Ext}^1_\Lambda(jh, A_i)(x_i) = 0$ for each i in I.

Assume that each f_i is right X-determined and let $h: L \longrightarrow C$ be a morphism such that for each $j: X \longrightarrow L$ we have $\text{Ext}^1_\Lambda(hj, \prod_{i\in I} A_i)(x_i) = 0$. Then by our preliminary remarks we have that $\text{Ext}^1_\Lambda(hj, A_i)(x_i) = 0$ for all i in I. Hence $\text{Ext}^1_\Lambda(h, A_i)(x) = 0$ for all i in I by Lemma 3.12 since each f_i is right X-determined. Consequently $\text{Ext}^1_\Lambda(h, \prod A_i)(x) = 0$. Thus we have by Lemma 3.13 that f is right X-determined.

As a consequence of this proposition we have the following result.

Corollary 3.15: Let $x: 0 \longrightarrow \prod_{i\in I} A_i \xrightarrow{g} B \xrightarrow{f} C \longrightarrow 0$ be in $\text{Ext}^1_\Lambda(C, \prod A_i)$ and let $x_i: 0 \longrightarrow A_i \xrightarrow{g_i} B \xrightarrow{f_i} C \longrightarrow 0$ in $\text{Ext}^1_\Lambda(C, A_i)$ be the ith projection of x. If each f_i is right X_i-determined, then f is right $X = \coprod_{i\in I} X_i$-determined.

Proof: The fact that each f_i is right X_i-determined implies that each f_i is right X-determined since X contains each X_i as a summand (see Proposition 2.6). Thus f is right X-determined by Proposition 3.14.

The following is a trivial consequence of this corollary.

Proposition 3.16: Let $\{X_j\}_{j\in J}$ be a family of Λ^{op}-Σ_j bimodules with each

X_j a finitely presented Λ^{op}-module and let I_j be an injective Σ_j-module. Finally let A_j be the Λ-module $\mathrm{Hom}_{\Sigma_j}(X_j, I_j)$ for each j in A_j. If $x: 0 \longrightarrow \prod_{j\varepsilon J} A_j \xrightarrow{g} B \xrightarrow{f} C \longrightarrow 0$ is exact, then f is right $\coprod_{j\varepsilon J} \mathrm{TrX}_j$-determined.

Proof: Let $x_j: 0 \longrightarrow A_j \xrightarrow{g_j} B_j \xrightarrow{f_j} C \longrightarrow 0$ be the jth projection of x. Since $\mathrm{Tr}(\mathrm{TrX}_j) \approx X_j$, and so $A_j = \mathrm{Hom}_{\Sigma_j}(\mathrm{Tr}(\mathrm{TrX}_j), I_j)$ it follows from Theorem 3.5 that f_j is right TrX_j-determined. Since this is true for each j in J, it follows from Corollary 3.15 that f is right $\coprod \mathrm{TrX}_j$-determined.

We are now in a position to prove the result cited earlier.

Theorem 3.17: Let $0 \longrightarrow A \xrightarrow{g} B \xrightarrow{f} C$ be an exact sequence of Λ-modules. Suppose $A = \prod_{j\varepsilon J} \mathrm{Hom}_{\Sigma_j}(X_j, I_j)$ where each X_j is a Λ^{op}-Σ_j bimodule which is a finitely presented Λ^{op}-module and each I_j is Σ_j-injective. Let $X = \coprod_{j\varepsilon J} \mathrm{TrX}_j$. Then

a) If f is an epimorphism, then f is right X-determined.

b) f is right $X \coprod \Lambda$-determined in general.

c) If J is finite, then $X \coprod \Lambda$ is in $\mathrm{mod}\ \Lambda$.

Proof:

a) This is simply Proposition 3.15.

b) By a) we know that the induced epimorphism $h: B \longrightarrow \mathrm{Imf}$ is right X-determined. Further it is trivial to show that the inclusion $\mathrm{inc}: \mathrm{Imf} \longrightarrow C$ is right Λ-determined. Thus the composition $B \xrightarrow{h} \mathrm{Imf} \xrightarrow{\mathrm{inc}} C$ is $X \coprod \Lambda$-determined by Proposition 2.7.

c) Obvious.

Added in proof:

The main existence theorem of this section shows that if X is a finitely presented Λ-module and C is an arbitrary Λ-module, then for each $(\text{End } X)^{op}$-submodule H of (X, C) containing $P(X, C)$ there is a minimal right X-determined morphism $f\colon B \longrightarrow C$ such that $\text{Im}(\ ,f)(X) = H$. In this addendum we show that the same result is valid for arbitrary $(\text{End } X)^{op}$-submodules H of (X, C), not only for those containing $P(X, C)$. The method of proof is to reduce the general case to the special one we have studied in this section. This reduction is based on the following general observations. Throughout this discussion we consider the canonical isomorphism $(\Lambda, C) \longrightarrow C$ given by $f \longmapsto f(1)$ an identification for all Λ-modules C.

Proposition 3.18: Suppose F is a subfunctor of $(\ ,C)$ for some Λ-module C.

a) Let L be a Λ-module. A morphism $f\colon L \longrightarrow C$ has the property $\text{Im}(\ ,f) \subset F$ if and only if the element f in (L, C) is contained in $F(L)$.

b) Suppose $F(\Lambda) = (\Lambda, C)$. Then $F(X) \supset P(X, C)$ for all finitely generated Λ-modules X.

c) If C' is a submodule of C, then $(\ ,C')$ is a Λ-determined subfunctor of $(\ , C)$.

d) Letting C' be the submodule $F(\Lambda)$ of C, we have that $F \subset (\ ,C')$ and $F(X) \supset P(X, C')$ for all finitely generated Λ-modules X.

Proof:

a) Trivial.

b) Since $F(\Lambda) = (\Lambda, C)$, we have that $F(\Lambda^n) = (\Lambda^n, C)$ for all positive integers n. Hence by a), we know that if f is in (Λ^n, C) then $\mathrm{Im}(\ ,f)(\ ,\Lambda^n) \subset F$ for all positive integers n. Suppose $h: \coprod_{i\varepsilon I} \Lambda_i \longrightarrow C$ is an epimorphism with each $\Lambda_i = \Lambda$. Then for each finite subset J of I, the restriction $h_J: \coprod_{i\varepsilon J} \Lambda_i \longrightarrow C$ of h to $\coprod_{i\varepsilon J} \Lambda_i$ has the property $\mathrm{Im}(\ ,h_J) \subset F$.

Let f be in $P(X, C)$ where X is a finitely generated Λ-module. Then there is a morphism $g: X \longrightarrow \coprod_{i\varepsilon I} \Lambda_i$ such that $f = hg$. Since X is finitely generated, there is a finite subset J of I such that $\mathrm{Im}\, g \subset \coprod_{i\varepsilon J} \Lambda_i$. Thus $f = h_J g$. Since $\mathrm{Im}(\ ,h_J) \subset F$, it follows that $\mathrm{Im}(\ ,f) \subset F$. This means by a), that f is in $F(X)$. Hence $F(X) \supset P(X, C)$.

c) Suppose C' is a subfunctor of C. In order to show that $(\ ,C')$ is a Λ-determined subfunctor of $(\ ,C)$ we have to show that a subfunctor F of $(\ ,C)$ is contained in $(\ ,C')$ if $F(\Lambda) \subset (\Lambda, C')$. Suppose $F(\Lambda) \subset (\Lambda, C')$. Let L be a Λ-module. Then $F(L) \subset (L, C)$. In order to show that $F(L) \subset (L, C')$, we have to show that each morphism $f: L \longrightarrow C$ in $F(L)$ has the property $\mathrm{Im}\, f \subset C'$, or, equivalently, $\mathrm{Im}((\Lambda, L) \xrightarrow{(\Lambda, f)} (\Lambda, C)) \subset (\Lambda, C')$. Since $f: L \longrightarrow C$ is in $F(L)$, we know by a) that $\mathrm{Im}(\ ,f) \subset F$. In particular, $\mathrm{Im}((\Lambda, L) \xrightarrow{(\Lambda, f)} (\Lambda, C)) \subset F(\Lambda) \subset C'$, which finishes the proof of c).

d) Trivial consequence of c) and d).

We now apply these observations to obtain our promised result.

Theorem 3.19: Let X and C be Λ-modules with X finitely presented. Suppose H is an $(\text{End } X)^{op}$-submodule of (X, C). Then there is an exact sequence $0 \longrightarrow A \xrightarrow{g} B \xrightarrow{f} C$ of Λ-modules with f a minimal right X-determined morphism such that Im(,f)(X) = H. Moreover, if we let C' = Im f and let $0 \longrightarrow A \xrightarrow{g} B \xrightarrow{f'} C' \longrightarrow 0$ be the induced exact sequence, then

a) H is an $(\text{End } X)^{op}$-submodule of (X, C') containing P(X, C').

b) f' is minimal right X-determined with Im(,f')(X) = H.

Proof: Let H be an $(\text{End } X)^{op}$-submodule of (X, C) and suppose $(,C)_H$ is the X-determined subfunctor of (,C) such that $(,C)_H(X) = H$. Let C' be the submodule $(,C)_H(\Lambda)$ of C. Then by Proposition 3.18 we know that $(,C)_H \subset (,C')$ and $H = (,C)_H(X)$ is an $(\text{End } X)^{op}$-submodule of (X, C') containing P(X, C'). Therefore we know there is an exact sequence $0 \longrightarrow A \xrightarrow{g} B \xrightarrow{f'} C' \longrightarrow 0$ with f' a minimal right X-determined morphism such that Im(,f')(X) = H. Thus Im(,f') is an X-determined subfunctor of (,C') with Im(,f')(X) = H. But the fact that $(,C)_H$ is an X-determined subfunctor of (,C) implies that $(,C)_H$ is an X-determined subfunctor of (,C'). Hence $\text{Im}(,f') = (,C)_H$ since they are both X-determined subfunctors of (,C') such that $\text{Im}(,f')(X) = (,C)_H(X)$. Letting $f: B \longrightarrow C$ be the composition $B \xrightarrow{f'} C' \xrightarrow{inc} C$, we conclude from the above remarks that $\text{Im}(,f) = (,C)_H$. Hence f is a right X-determined morphism with Im(,f)(X) = H. The fact that f is right minimal is an immediate consequence of the fact that f' is right minimal. Hence the proof of the theorem is complete.

§4. Noetherian R-algebras. This section is devoted to specializing the results of Section 3 to the case Λ is a noetherian algebra, a notion we now define.

By an R-algebra Λ we mean a commutative ring R together with a ring morphism $f: R \longrightarrow \Lambda$ with $f(R)$ contained in the center of Λ. If $f: R \longrightarrow \Lambda$ is an R-algebra, then the morphism $g: R \longrightarrow \Lambda^{op}$ given by $g(r) = f(r)$ for all r in R is also an R-algebra which we denote by Λ^{op}. We say that an R-algebra Λ is a noetherian R-algebra if R is a noetherian ring and Λ is a finitely generated R-module. Clearly an R-algebra Λ is a noetherian R-algebra if and only if the R-algebra Λ^{op} is a noetherian R-algebra. Thus if Λ is a noetherian R-algebra, then Λ is a noetherian ring, i.e. Λ is left and right noetherian.

It is not difficult to see that a ring Λ has the property that Λ is a noetherian R-algebra for some R if and only if the center of Λ is a noetherian ring and Λ is a finitely generated module over its center. We say that a ring is a noetherian algebra if its center is noetherian and it is a finitely generated module over its center. Throughout the rest of this section we assume that R is a commutative noetherian ring (unless stated explicitly to the contrary).

Before applying the results of Section 3 to noetherian algebras we remind the reader of some well known facts concerning modules over noetherian algebras.

Let R be a commutative noetherian ring. We denote by $I(M)$ an injective envelope of an R-module M. The injective R-module $\coprod_{\underline{m}} I(R/\underline{m})$ where $\underline{m}$ runs over all maximal ideals of R, which we denote by I_R, plays an important role in our considerations of noetherian R-algebras primarily because of the following properties which we state without proof. The reader is referred to [13], [16], and [17] for the basic facts concerning commutative noetherian rings we need.

Proposition 4.1: Let R be a commutative noetherian ring. Then the R-module $I_R = \coprod_{\underline{m}} I(R/\underline{m})$, where $\underline{m}$ ranges over all maximal ideals of R, has the following properties.

a) I_R is an injective cogenerator of Mod R.

b) Let A be a submodule of I_R. Then A is of finite length if and only if it is finitely generated.

c) A finitely generated R-module is of finite length if and only if there is a monomorphism $A \longrightarrow I_R^n$, where I_R^n means the sum of n copies of I_R with n a nonnegative integer ($I^0 = (0)$ by convention).

d) An R-module A has finite length if and only if $\text{Hom}_R(A, I_R)$ has finite length.

e) Denoting the full subcategory of Mod Λ whose objects are the R-modules of finite length by $f. \ell. R$, the functor $\text{Hom}_R(\ ,I_R): f. \ell. R \longrightarrow f. \ell. R$ is a duality with inverse $\text{Hom}_R(\ ,I_R)$. The isomorphism $1_{f. \ell. R} \longrightarrow \text{Hom}_R(\text{Hom}_R(\ ,I_R), I_R)$ is given by the functorial isomorphism $\varphi: A \longrightarrow \text{Hom}_R(\text{Hom}_R(A, I_R), I_R)$ where $\varphi_M(a)(f) = f(a)$ for all a in A and f in $\text{Hom}_R(M, I_R)$.

Suppose now that Λ is a noetherian R-algebra. Then every Λ-module A is also an R-module. Moreover for each pair of Λ-modules A and B, $\text{Hom}_\Lambda(A, B)$ is also an R-submodule of $\text{Hom}_R(A, B)$. In particular, $\text{End}_\Lambda(A)$ is an R-subalgebra of $\text{End}_R(A)$. We now list some well known connections between viewing a Λ-module as a Λ-module and as an R-module.

Proposition 4.2: Let Λ be a noetherian R-algebra.

a) A Λ-module A is a finitely generated Λ-module if and only if A is a finitely generated R-module.

b) If A and B are finitely generated Λ-modules, then $\mathrm{Hom}_\Lambda(A, B)$ is a finitely generated R-module.

c) If A is a finitely generated Λ-module, then $\mathrm{End}_\Lambda A$ is a noetherian R-algebra.

d) The following statements are equivalent for a Λ-module A.

i) A is of finite length over Λ.

ii) A is of finite length over R.

iii) $\mathrm{Hom}_R(A, I_R)$ is a Λ^{op}-module of finite length.

iv) $A_{\underline{p}} = 0$ for all nonmaximal prime ideals $\underline{p}$ of R.

e) The functor $\mathrm{Hom}_R(\ ,I_R) : \mathrm{Mod}\ \Lambda \longrightarrow \mathrm{Mod}\ \Lambda^{op}$ induces a duality $\mathrm{Hom}_R(\ ,I_R): \mathrm{f.}\ \ell.\ \Lambda \longrightarrow \mathrm{f.}\ \ell.\ \Lambda^{op}$ with inverse $\mathrm{Hom}_R(\ ,I_R): \mathrm{f.}\ \ell.\ \Lambda^{op} \longrightarrow \mathrm{f.}\ \ell.\ \Lambda$ and the functorial isomorphisms $\varphi_A : A \longrightarrow \mathrm{Hom}_R(\mathrm{Hom}_R(A, I_R), I_R)$ the usual ones.

f) If A and B are Λ-modules of finite length, then $\mathrm{Hom}_\Lambda(A, B)$ is an R-module of finite length.

As a consequence of these observations we have the following.

Proposition 4.3: Let Λ be a noetherian R-algebra.

a) $\mathrm{Hom}_R(\Lambda^{op}, I_R)$ is an injective cogenerator for $\mathrm{Mod}\ \Lambda$.

b) A finitely generated Λ-module A is of finite length if and only if there is a monomorphism $A \longrightarrow \mathrm{Hom}_R(\Lambda^{op}, I_R)^n$ for some nonnegative integer n.

c) Suppose A is a Λ-module of finite length. Then $\mathrm{Soc}\ A$ the socle of A, is isomorphic to $\coprod_{i=1}^{k} S_i^{k_i}$ with the S_i a finite set of nonisomorphic simple Λ-modules. Thus there is a monomorphism $A \longrightarrow \mathrm{Hom}_R(\Lambda^{op}, I_R)^n$ where $n = \max$ of $n_1, \ldots, n_k$.

Proof:

a) Since Λ^{op} is a projective Λ^{op}-module and I_R is an injective R-module, it is well known that $\mathrm{Hom}_R(\Lambda^{op}, I_R)$ is an injective Λ-module. Therefore to finish showing that $\mathrm{Hom}_R(\Lambda^{op}, I_R)$ is an injective cogenerator for Mod Λ, we have to show that every simple Λ-module is isomorphic to a submodule of $\mathrm{Hom}_R(\Lambda^{op}, I_R)$.

Let S be a simple Λ-module. The fact that $\mathrm{Hom}_R(\ ,I_R)$: f. ℓ. $\Lambda \longrightarrow$ f. ℓ. Λ^{op} is a duality shows that $T = \mathrm{Hom}_R(S, I_R)$ is a simple Λ^{op}-module. Thus there is an epimorphism $\Lambda^{op} \longrightarrow T$ which induces a monomorphism $\mathrm{Hom}_R(T, I_R) \longrightarrow \mathrm{Hom}_R(\Lambda^{op}, I_R)$. Since $S \approx \mathrm{Hom}_R(T, I_R)$, we are done.

The rest of the proof of the proposition follows in a straightforward way from a) and some of the previous observations.

As our final preliminary remark we point out the following.

Proposition 4.4: Let X be a Λ-module. Then the standard isomorphisms

$$\mathrm{Hom}_\Lambda(X, \mathrm{Hom}_R(\Lambda^{op}, I_R)) \cong \mathrm{Hom}_R(X, I_R),$$

which are functorial in X, give an isomorphism of functors $\mathrm{Hom}_\Lambda(\ , \mathrm{Hom}_R(\Lambda^{op}, I_R)) \cong \mathrm{Hom}_R(\ ,I_R)$ which we will sometimes view as an identification.

With these preliminary remarks in mind, we turn to applying some of the results of Section 3 to noetherian algebras. Throughout the rest of this paper we will denote the full subcategory of Mod Λ whose objects are the finitely generated Λ-modules by noeth Λ and the full subcategory of Mod Λ whose objects are the artinian Λ-modules by art Λ.

We begin with the following analogue of Theorem 3.5. However before stating this result, it is convenient to introduce the following terminology.

Let $0 \longrightarrow A \xrightarrow{g} B \xrightarrow{f} C \longrightarrow 0$ be an exact sequence. An exact sequence $0 \longrightarrow A' \xrightarrow{g'} B' \xrightarrow{f'} C \longrightarrow 0$ is said to be a <u>summand over C</u> of $0 \longrightarrow A \xrightarrow{g} B \xrightarrow{f} C \longrightarrow 0$ if there is a commutative diagram

$$\begin{array}{ccccccccc}
0 & \longrightarrow & A' & \xrightarrow{g'} & B' & \xrightarrow{f'} & C & \longrightarrow & 0 \\
 & & \downarrow u & & \downarrow \upsilon & & \| & & \\
0 & \longrightarrow & A & \xrightarrow{g} & B & \xrightarrow{f} & C & \longrightarrow & 0 \\
 & & \downarrow u' & & \downarrow \upsilon' & & \| & & \\
0 & \longrightarrow & A' & \xrightarrow{g'} & B' & \xrightarrow{f'} & C & &
\end{array}$$

with $\upsilon'\upsilon = 1_{B'}$ and $u'u = 1_{A'}$.

An exact sequence $0 \longrightarrow A \xrightarrow{g'} B' \longrightarrow C' \longrightarrow 0$ is said to be a <u>summand over</u> A of $0 \longrightarrow A \xrightarrow{g} B \xrightarrow{f} C \longrightarrow 0$ if there is a commutative diagram

$$\begin{array}{ccccccccc}
0 & \longrightarrow & A & \xrightarrow{g'} & B' & \xrightarrow{f'} & C' & \longrightarrow & 0 \\
 & & \| & & \downarrow u & & \downarrow \upsilon & & \\
0 & \longrightarrow & A & \xrightarrow{g} & B & \xrightarrow{f} & C & \longrightarrow & 0 \\
 & & \| & & \downarrow u' & & \downarrow \upsilon' & & \\
0 & \longrightarrow & A & \xrightarrow{g'} & B' & \xrightarrow{f'} & C' & \longrightarrow & 0
\end{array}$$

with $u'u = 1_{B'}$ and $\upsilon'\upsilon = 1_{C'}$.

Theorem 4.5: Let Λ be a noetherian R-algebra, X and C in noeth Λ and A the Λ-module $\mathrm{Hom}_R(\mathrm{Tr}X, I_R^n)$ with n a positive integer. Further let $\Gamma = \mathrm{End}\,\mathrm{Tr}X$. Suppose $0 \longrightarrow A \xrightarrow{g} B \xrightarrow{f} C \longrightarrow 0$ is an exact sequence in Mod Λ.

a) f is right X-determined.

b) $(X, C)/\mathrm{Im}(\ ,f)(X)$ is a Γ-module of finite length which is isomorphic to a submodule of $\mathrm{Hom}_R(\Gamma^{op}, I)^n$.

c) There is an exact sequence $0 \longrightarrow A' \xrightarrow{g'} B' \xrightarrow{f'} C \longrightarrow 0$ which is a summand over C of $0 \longrightarrow A \xrightarrow{g} B \xrightarrow{f} C \longrightarrow 0$ with f' a minimal right X-determined morphism such that $\mathrm{Im}(\ ,f)(X) = \mathrm{Im}(\ ,f)(X)$.

d) The following are equivalent for an exact sequence $0 \longrightarrow A'' \xrightarrow{g''} B'' \xrightarrow{f''} C \longrightarrow 0$ with f" right X-determined which is a summand of C over $0 \longrightarrow A \xrightarrow{g} B \xrightarrow{f} C \longrightarrow 0$:

i) f" is minimal right X-determined.

ii) If A"' is a submodule of A" such that g"(A"') is a summand of B", then A"' = 0.

Proof:

a) An immediate consequence of Theorem 3.5.

b) Using the identification $\mathrm{Hom}_R(\mathrm{Tr}X, I_R^n) = \mathrm{Hom}_\Gamma(\mathrm{Tr}X, \mathrm{Hom}_R(\Gamma, I_R^n))$, we have by Theorem 3.5 that the Γ-morphism $\upsilon\colon (X, C)/P(X, C) \longrightarrow \mathrm{Hom}_R(\Gamma^{op}, I_R^n)$ corresponding to the extension $0 \longrightarrow A \xrightarrow{g} B \xrightarrow{f} C \longrightarrow 0$ under the isomorphism $\mathrm{Ext}^1_\Lambda(C, \mathrm{Hom}_\Gamma(\mathrm{Tr}X, \mathrm{Hom}_R(\Gamma, I_R^n)) \longrightarrow \mathrm{Hom}_\Gamma(\underline{\mathrm{Hom}}_\Lambda(X, C), \mathrm{Hom}_R(\Gamma^{op}, I_R^n))$ has the property $\mathrm{Ker}\,\upsilon = \mathrm{Im}(\ ,f)(X)/P(X, C)$. Hence the Γ-module

$(X, C)/\mathrm{Im}(\ ,f)(X)$ is isomorphic to a submodule of $\mathrm{Hom}_R(\Gamma^{op}, I_R^n)$. Since X and C are noetherian Λ-modules, (X, C), and hence $(X, C)/\mathrm{Im}(\ ,f)(X)$, is a finitely generated R-module. Therefore $(X, C)/\mathrm{Im}(\ ,f)(X)$ is a finitely generated Γ-submodule of $\mathrm{Hom}_R(\Gamma^{op}, I_R^n)$ and is thus of finite length over Γ.

c) An immediate consequence of Proposition 3.11.

d) The fact that i) and ii) are equivalent follows from Proposition 3.10 and 3.11.

Proposition 4.5 suggests the following existence theorem for morphisms which are right determined by objects in noeth Λ.

Theorem 4.6: Let X and C be in noeth Λ. Suppose H is an $(\mathrm{End}\ X)^{op}$-submodule of (X, C) containing $P(X, C)$ such that $(X, C)/H$ is an $(\mathrm{End}\ X)^{op}$-module of finite length. Then $\mathrm{Soc}((X, C)/H) = \coprod_{i=1}^{k} S_i^{n_i}$ where $S_1,\dots,S_k$ is a finite number of nonisomorphic simple $(\mathrm{End}\ X)^{op}$-modules. Letting $n = \max$ of $n_1,\dots,n_k$, we have that there is an exact sequence of Λ-modules

$$0 \longrightarrow \mathrm{Hom}_R(\mathrm{Tr}X, I_R)^n \longrightarrow B \xrightarrow{f} C \longrightarrow 0$$

such that f is right X-determined and $\mathrm{Im}(\ ,f)(X) = H$.

Proof: Since H contains $P(X, C)$ we have that $(X, C)/H$ is an $(\mathrm{End}\ X)^{op}$-module. Let $\Gamma = \mathrm{End}\ \mathrm{Tr}X$. Since the duality Tr: noeth $\Lambda^{op} \longrightarrow$ noeth Λ gives an R-algebra isomorphism $\mathrm{End}\ \mathrm{Tr}X \longrightarrow (\mathrm{End}\ X)^{op}$, $(X, C)/H$ can be viewed as Γ-module. Moreover the simple $(\mathrm{End}\ X)^{op}$-modules $S_1,\dots,S_k$ are $(\mathrm{End}\ X)^{op}$-modules and thus simple Γ-modules with the property that $\mathrm{Soc}((X, C)/H) = \coprod_{i=1}^{k} S_i^{n_i}$. By our preliminary remarks, we know that $\mathrm{Hom}_R(\Gamma^{op}, I_R)$ is an injective Γ-module which is a cogenerator. Thus each $S_i \subset \mathrm{Hom}_R(\Gamma^{op}, I_R)^n$. Hence there is a Γ-morphism ν: $(X, C)/P(X, C) \longrightarrow \mathrm{Hom}_R(\Gamma^{op}, I_R)^n$ with $\mathrm{Ker}\ \nu = H/P(X, C)$.

By Theorem 3.5 we know that the extension $0 \longrightarrow \mathrm{Hom}_\Gamma(\mathrm{Tr}X, \mathrm{Hom}_R(\Gamma^{op}, I_R)^n) \longrightarrow B \xrightarrow{f} C \longrightarrow 0$ in $\mathrm{Ext}^1_\Lambda(C, \mathrm{Hom}_\Gamma(C, \mathrm{Hom}_\Gamma(\mathrm{Tr}X, \mathrm{Hom}_R(\Gamma^{op}, I_R)^n)$ corresponding to the morphism ν: $(X, C)/P(X, C) \longrightarrow \mathrm{Hom}_R(\Gamma^{op}, I_R)^n$ has the property that f is a right X-determined morphism such that $\mathrm{Im}(\ ,f)(X) = H$. Since $\mathrm{Hom}_\Gamma(\mathrm{Tr}X, \mathrm{Hom}_R(\Gamma^{op}, R))$ is isomorphic to $\mathrm{Hom}_R(\mathrm{Tr}X, I_R)$, see Proposition 4.4, we have that $\mathrm{Hom}_\Gamma(\mathrm{Tr}X, \mathrm{Hom}_R(\Gamma^{op}, I_R)^n) \cong \mathrm{Hom}_\Gamma(\mathrm{Tr}X, \mathrm{Hom}_R(\Gamma^{op}, I_R))^n \cong \mathrm{Hom}_R(\mathrm{Tr}X, I_R)^n$. This finishes the proof of the theorem.

Added in proof:

Using arguments similar to those given in the proof of Theorem 3.19, we can generalize Theorem 4.6 as follows.

Theorem 4.7: Let X and C be in noeth Λ. Suppose H is an $(\mathrm{End}\ X)^{op}$-submodule of (X, C) such that $(X, C)/H$ is an $(\mathrm{End}\ X)^{op}$-module of finite length. Then there is an exact sequence $0 \longrightarrow \mathrm{Hom}_R(\mathrm{Tr}X, I_R)^n \xrightarrow{g} B \xrightarrow{f} C$ such that f is right X-determined and $\mathrm{Im}(\ ,f)(X) = H$.

Proof: Let $(\ ,C)_H$ be the X-determined subfunctor of $(\ ,C)$ such that $(\ ,C)_H(X) = H$. Letting C' be the submodule $(\ ,C)_H(\Lambda)$ of C, we know by Proposition 3.18, that $(\ ,C)_H \subset (\ ,C')$ and $H \supset P(X, C')$. Since $(X, C')/H \subset (X, C)/H$ and $(X, C)/H$ is of finite length over $(\mathrm{End}\ X)^{op}$, we know that $(X, C'/H$ is of finite length over $(\mathrm{End}\ X)^{op}$. Therefore by Theorem 4.6, we know there is an exact sequence

$0 \longrightarrow \mathrm{Hom}_R(\mathrm{Tr}X, I)^n \xrightarrow{g} B \xrightarrow{f'} C' \longrightarrow 0$ such that f' is right X-determined with $\mathrm{Im}(\ ,f')(X) = H$. Using the same arguments as those given in the proof of Theorem 3.19, we see that the exact sequence $0 \longrightarrow \mathrm{Hom}_R(\mathrm{Tr}X, I)^n \xrightarrow{g} B \xrightarrow{f} C$, where f is the composition $B \xrightarrow{f'} C' \xrightarrow{inc} C$, has our desired properties.

§5. Noetherian Algebras Over Complete Local Rings. Our purpose in this section is to show how some of the special properties of the category of Λ-modules in the case Λ is a noetherian R-algebra with R a complete local ring are reflected when our previous results are considered in this particular context. We assume throughout this section, unless stated to the contrary, that R is a complete, noetherian local ring and Λ is a noetherian R-algebra. Our discussion begins with a review of the properties of Mod Λ pertinent to our purposes.

Every noetherian R-algebra Λ is semiperfect, i.e. every finitely generated Λ-module has a projective cover or, equivalently, $\Lambda/$ rad Λ is semi-simple and every idempotent in Λ/rad Λ is the image of an idempotent in Λ. In particular if A is in noeth Λ, then End A is a noetherian R-algebra by means of the morphism $g: R \longrightarrow \mathrm{End}\, A$ given by $g(r)(a) = ra$ for all a in A (remember A is also an R-module). Hence End A is semiperfect. From this the following properties of the category noeth Λ can be deduced.

Proposition 5.1:

a) A in noeth Λ is indecomposable if and only if End A is a local ring.

b) Each nonzero A in mod Λ can be written as a finite sum

$\coprod_{i=1}^{n} A_i$ of indecomposable Λ-modules A_i and this decomposition is unique up to isomorphism (i.e. if $\coprod_{i=1}^{n} A_i = A = \coprod_{j=1}^{m} B_j$ with the A_i and B_j indecomposable modules, then $m = n$ and, after suitable relabeling, $A_i \approx B_i$ for all i).

Since R is a local ring, rad R is the unique maximal ideal in R and $I_R = I(R/\text{rad } R)$. Because R is also complete the following properties of the functors $D\colon \text{Mod } \Lambda \longrightarrow \text{Mod } \Lambda^{op}$ and $D\colon \text{Mod } \Lambda^{op} \longrightarrow \text{Mod } \Lambda$ defined by $D(A) = \text{Hom}_R(A, R)$ can be deduced from the corresponding well known properties of the functor $D\colon \text{Mod } R \longrightarrow \text{Mod } R$. For each A in Mod Λ or in Mod Λ^{op} we define $\varphi_A\colon A \longrightarrow D^2A$ by $\varphi_A(a)(f) = f(a)$ for all a in A. Then we have the following.

a) The φ_A are monomorphisms which are functorial in A.

b) If A is in noeth Λ, then $D(A)$ is in art Λ^{op}, the category of artinian Λ-modules, and φ_A is an isomorphism.

c) If A is in art Λ, then $D(A)$ is in noeth Λ and φ_A is an isomorphism.

d) The functors $D\colon \text{Mod } \Lambda \longrightarrow \text{Mod } \Lambda^{op}$ and $D\colon \text{Mod } \Lambda^{op} \longrightarrow \text{Mod } \Lambda$ induce inverse dualities $D\colon \text{noeth } \Lambda \longrightarrow \text{art } \Lambda^{op}$ and $D\colon \text{art } \Lambda^{op} \longrightarrow \text{noeth } \Lambda$ with the isomorphisms $A \longrightarrow D^2A$ being the morphisms φ_A for all A in noeth Λ or in art Λ^{op}.

The duality $D\colon \text{noeth } \Lambda \longrightarrow \text{art } \Lambda^{op}$ shows that the category art Λ has the following properties analogous to those given for noeth Λ in Proposition 5.1.

Proposition 5.2:

a) End A is a noetherian R-algebra and so is semi-perfect for all A in art Λ.

b) If A is in art Λ then A is indecomposable if and only if End A is local.

c) Every A in art Λ can be written as a finite sum $\coprod_{i=1}^{n} A_i$ of indecomposable Λ-modules A_i and this representation is unique up to isomorphism.

d) If $A \longrightarrow I(A)$ is an injective envelope of A in art Λ, then I(A) is in art Λ.

Not only does the functor $\mathrm{Hom}_R(\ ,I_R)$ have special properties when R is a complete local ring, but also the duality Tr: $\underline{\text{noeth}}\ \Lambda \longrightarrow \underline{\text{noeth}}\ \Lambda^{op}$, as we now point out. This is due to the fact that the modules in noeth Λ have minimal projective presentations.

The duality Tr: $\underline{\text{mod}}\ \Lambda \longrightarrow \underline{\text{mod}}\ \Lambda^{op}$ described in Section 3 for arbitrary rings Λ has the property that for each M in mod Λ the object TrM in mod Λ depends on the particular projective presentation $P_1 \longrightarrow P_0 \longrightarrow M \longrightarrow 0$ used in the construction of the functor Tr. It is only when TrM is viewed as an object in $\underline{\text{mod}}\ \Lambda$ that it is independent (up to a canonical isomorphism) of the particular projective presentation for M used in constructing Tr.

Suppose now that each M in mod Λ has a minimal projective presentation $P_1 \xrightarrow{f} P_0 \longrightarrow M \longrightarrow 0$. Then the exact sequence $(P_0, \Lambda) \longrightarrow (P_1, \Lambda) \longrightarrow \mathrm{Coker}(f, \Lambda) \longrightarrow 0$ is easily seen to be a minimal projective presentation for the Λ^{op}-module Coker(f, Λ). From this it follows that each Λ^{op}-module in mod Λ^{op} has a minimal projective presentation. Hence in this case, it is natural to use only minimal projective presentations to define the functors Tr: $\underline{\text{mod}}\ \Lambda \longrightarrow \underline{\text{mod}}\ \Lambda^{op}$ and

Tr: $\underline{\text{mod}}\ \Lambda^{op} \longrightarrow \underline{\text{mod}}\ \Lambda$. This convention has the advantage that since minimal projective presentations are unique up to isomorphism, TrM is uniquely determined (up to a canonical isomorphism) not only in $\underline{\text{mod}}\ \Lambda$ but in mod Λ as well. So from now on whenever the ring Λ has the property that each M in mod Λ has a minimal projective presentation, we will use only minimal projective presentations whenever projective presentations are needed. In particular, we will only use minimal projective presentations in constructing the duality Tr: $\underline{\text{mod}}\ \Lambda \longrightarrow \underline{\text{mod}}\ \Lambda^{op}$.

Assume now that not only does each M in mod Λ and mod Λ^{op} have a minimal projective presentation, but also that End M is semiperfect for all M in mod Λ and mod Λ^{op}. Let $\text{mod}_P\Lambda$ denote the full subcategory of mod Λ consisting of those M with no nonzero projective summands. Then it is not difficult to verify that the map
Tr: $\text{Ob}(\text{mod}\ \Lambda) \longrightarrow \text{Ob}(\text{mod}\ \Lambda^{op})$ has the following properties.

a) TrM = 0 if and only if M is projective.

b) TrM is in $\text{mod}_P\Lambda^{op}$ for all M in mod Λ. Hence Tr induces Tr: $\text{mod}_P\Lambda \longrightarrow \text{mod}_P\Lambda^{op}$.

c) $M \approx \text{Tr}(\text{TrM}) \amalg P$ with P a projective Λ-module.

d) $M \approx \text{Tr}(\text{TrM})$ if and only if M is in $\text{mod}_P\Lambda$.

e) If M_1 and M_2 are in $\text{mod}_P\Lambda$, then $M_1 \approx M_2$ if and only if $\text{Tr}(M_1) \approx \text{TrM}_2$.

f) If $M = M_1 \amalg M_2$ in mod Λ, then $\text{Tr}(M) \approx \text{TrM}_1 \amalg \text{TrM}_2$.

g) If M is in $\text{mod}_P\Lambda$, then M is indecomposable if and only if TrM is indecomposable.

h) Let M be in $\text{mod}_P\Lambda$, then $M \approx M_1 \amalg \cdots \amalg M_n$ is a decomposition of M into indecomposable modules if and only if $\text{Tr}(M) \approx \text{Tr}(M_1) \amalg \cdots \amalg \text{Tr}(M_n)$ is a decomposition of TrM into indecomposable Λ^{op}-modules.

In connection with these observations we recall the notion of a representation equivalence. Let $\underline{\underline{C}}$ and $\underline{\underline{D}}$ be additive categories. A map $F\colon \text{Ob}\underline{\underline{C}} \longrightarrow \text{Ob}\underline{\underline{D}}$ is said to be a <u>representation equivalence</u> (see [2] for a discussion of this notion when F is given by a functor from $\underline{\underline{C}}$ to $\underline{\underline{D}}$) if it has the following properties:

a) C_1 and C_2 in $\underline{\underline{C}}$ are isomorphic if and only if $F(C_1) \approx F(C_2)$ in $\underline{\underline{D}}$;

b) If $C \approx C_1 \coprod C_2$, then $F(C) \approx F(C_1) \coprod F(C_2)$;

c) Given D in $\underline{\underline{D}}$, there is a C in $\underline{\underline{C}}$ such that $F(C) \approx D$.

It is not difficult to see that a representation equivalence $F\colon \text{Ob}\underline{\underline{C}} \longrightarrow \text{Ob}\underline{\underline{D}}$ has the additional properties:

d) C in $\underline{\underline{C}}$ is indecomposable if and only if $F(C)$ is indecomposable in $\underline{\underline{D}}$ and

e) If $C \approx C_1 \coprod \dots \coprod C_n$ with the C_i indecomposable, then $F(C) \approx F(C_1) \coprod \dots \coprod F(C_n)$ with the $F(C_i)$ indecomposable.

Suppose the objects in mod Λ and mod Λ^{op} have minimal projective resolution and that End M is semiprimary for all M in mod Λ and mod Λ^{op}. Suppose now that the duality $\text{Tr}\colon \underline{\text{mod}}\ \Lambda \longrightarrow \underline{\text{mod}}\ \Lambda^{op}$ is defined using minimal projective presentations. Then our previous remarks show that the induced duality $\text{Tr}\colon \underline{\text{mod}}_P\Lambda \longrightarrow \underline{\text{mod}}_P\Lambda^{op}$ has the property that the associated map $\text{Tr}\colon \text{mod}_P\Lambda \longrightarrow \text{mod}_P\Lambda^{op}$ is a representation equivalence. We will see other examples of representation equivalences as we go along.

In particular, if Λ is a noetherian R-algebra with R a complete local ring, then the above remarks apply to noeth Λ = mod Λ since the necessary hypothesis is satisfied by noeth Λ and noeth Λ^{op}. It should also be observed that the inclusion $\underline{\text{noeth}_P\Lambda} \longrightarrow \underline{\text{noeth}\ \Lambda}$ is an equivalence of categories where $\text{noeth}_P\Lambda$ is the full subcategory of noeth Λ whose objects are those of $\text{noeth}_P\Lambda$. This equivalence $\underline{\text{noeth}_P\Lambda} \rightarrow \underline{\text{noeth}\ \Lambda}$ will often be considered an identification.

Suppose now that Λ is a noetherian R-algebra with R a complete local ring. Let D: noeth $\Lambda \longrightarrow$ art Λ^{op} be the usual duality. From the fact that each module in noeth Λ has a projective cover in noeth Λ, it follows that each module in art Λ has an injective envelope in art Λ^{op}. As we have already seen, each X in art Λ^{op} can be written uniquely (up to isomorphism) as $X_1 \coprod \cdots \coprod X_n$ with the X_i indecomposable in X. Thus each X in art Λ^{op} can be written uniquely up to isomorphism as $U \coprod I$ where U has no nonzero injective summands and I is injective.

Let $\text{art}_I\Lambda^{op}$ denote the full subcategory of art Λ^{op} whose objects are those with no nonzero injective summands. Then the duality D: noeth $\Lambda \longrightarrow$ art Λ^{op} induces a duality D: $\underline{\text{noeth}_P\Lambda} \longrightarrow \overline{\text{art}_I\Lambda^{op}}$. We thus obtain maps DTr: Objects($\text{noeth}_P\Lambda$) $\longrightarrow$ Objects($\text{art}_I\Lambda$) and TrD: Objects($\text{art}_I\Lambda$) $\longrightarrow$ Objects($\text{noeth}_P\Lambda$) such that $\text{TrD(DTr(X))} \approx X$ for all X in $\text{noeth}_P\Lambda$ and $\text{DTr(TrD(Y))} \approx Y$ for all Y in $\text{art}_I\Lambda$. In addition the maps DTr: Ob($\text{noeth}_P\Lambda$) $\longrightarrow$ Ob($\text{art}_I\Lambda$) and TrD: Ob($\text{art}_I\Lambda$) $\longrightarrow$ Ob($\text{noeth}_P\Lambda$) are representation equivalences. This follows from the fact that Tr: Ob($\text{noeth}_P\Lambda$) $\longrightarrow$ Ob($\text{noeth}_P\Lambda^{op}$) and D: Ob($\text{noeth}_P\Lambda$) $\longrightarrow$ Ob($\text{art}_I\Lambda^{op}$) are representation equivalences.

In order to give a categorical interpretation of these maps on the Ob($\text{noeth}_P\Lambda$) and Ob($\text{art}_I\Lambda$) we need to introduce the following general definitions and notations.

In analogy with the definition of $\underline{\text{Mod}}\ \Lambda$, we define for an arbitrary ring Λ the category Mod Λ modulo injectives, which we denote by $\overline{\text{Mod}}\ \Lambda$, in the following way. The objects of $\overline{\text{Mod}}\ \Lambda$ are those of Mod Λ and the morphisms $\overline{\text{Hom}}_\Lambda(X, Y)$ in $\overline{\text{Mod}}\ \Lambda$ is defined to be $\text{Hom}_\Lambda(X, Y)/I(X, Y)$ where $I(X, Y)$ is the subgroup consisting of all those morphisms $f: X \longrightarrow Y$ which can be written as $X \longrightarrow I \longrightarrow Y$ with I injective. For each f in $\text{Hom}_\Lambda(X, Y)$, denote its image in $\overline{\text{Hom}}_\Lambda(X, Y)$ by $\overline{f}$. Then the composition of morphisms $\overline{f}, \overline{g}$ in $\overline{\text{Mod}}\ \Lambda$ is given by $\overline{f}\,\overline{g} = (\overline{fg})$. It is worth noting that if $X \longrightarrow I$ is a monomorphism with I injective, then the sequence $(I, Y) \longrightarrow (X, Y) \longrightarrow \overline{\text{Hom}}_\Lambda(X, Y) \longrightarrow 0$ is exact.

Returning to the case where Λ is a noetherian R-algebra with R a complete local ring, we denote by $\overline{\text{art}}\ \Lambda$ the full subcategory of $\overline{\text{Mod}}\ \Lambda$ whose objects are the objects of art Λ. Denoting by $\overline{\text{art}}_I\Lambda$ the full subcategory of $\overline{\text{art}}\ \Lambda$ whose objects are in $\text{art}_I\Lambda$, we see that the inclusion $\overline{\text{art}}_I\Lambda \longrightarrow \overline{\text{art}}\ \Lambda$ is an equivalence of categories which we often consider an identification. Also it is not difficult to see that the duality D: noeth $\Lambda \longrightarrow$ art Λ induces a duality D: $\underline{\text{noeth}}\ \Lambda \longrightarrow \overline{\text{art}}\ \Lambda$. Combining this duality with the dualities Tr: $\underline{\text{noeth}}\ \Lambda \longrightarrow \underline{\text{noeth}}\ \Lambda^{op}$ and Tr: $\underline{\text{noeth}}\ \Lambda^{op} \longrightarrow \underline{\text{noeth}}\ \Lambda$, we obtain the equivalences DTr: $\underline{\text{noeth}}\ \Lambda \longrightarrow \overline{\text{art}}\ \Lambda^{op}$ and TrD: $\overline{\text{art}}\ \Lambda^{op} \longrightarrow \underline{\text{noeth}}\ \Lambda$ which are inverses of each other. These play a fundamental role in much of what follows.

With these preliminaries in mind, we now begin our discussion of morphisms determined by modules for Λ a noetherian R-algebra with R a complete local ring. We start with the following analogue of the first part of Theorem 4.5.

<u>Proposition 5.3</u>: Suppose Λ is a noetherian R-algebra with R a

complete local ring. Let $0 \longrightarrow A \xrightarrow{g} B \xrightarrow{f} C \longrightarrow 0$ be an exact sequence in Mod Λ with A in $\text{art}_I \Lambda$ and C in noeth Λ and let X be the Λ-module TrD A in noeth Λ. Finally, let $\Gamma = \text{End } D(A)$.

a) f is right X-determined.

b) $(X, C)/\text{Im}(\ ,f)(X)$ is a Γ-module of finite length isomorphic to a submodule of $\text{Hom}_R(\Gamma^{op}, I_R)$.

c) There is a summand $0 \longrightarrow A' \xrightarrow{g'} B' \xrightarrow{f'} C \longrightarrow 0$ over C of $0 \longrightarrow A \xrightarrow{g} B \xrightarrow{f} C \longrightarrow 0$ with f' a minimal right X-determined morphism such that $\text{Im}(\ ,f')(X) = \text{Im}(\ ,f)(X)$.

<u>Proof</u>: This result is essentially a restatement of a) - c) of Theorem 4.5 with $n = 1$ once one observes that since A is in $\text{art}_I \Lambda$, $A = \text{DTr}X$ where $X = \text{TrD } A$. This fact is a trivial consequence of our preliminary comments.

In order to simplify the statements of our next result which is a strengthened version of the remainder of Theorem 4.5, we introduce the following terminology.

Suppose $0 \longrightarrow A \xrightarrow{g} B \xrightarrow{f} C \longrightarrow 0$ is an exact sequence. An exact sequence $0 \longrightarrow A' \xrightarrow{g'} B' \xrightarrow{f'} C \longrightarrow 0$ is said to be <u>isomorphic over</u> C to $0 \longrightarrow A \xrightarrow{g} B \xrightarrow{f} C \longrightarrow 0$ if there is a commutative diagram

$$\begin{array}{ccccccccc} 0 & \longrightarrow & A' & \xrightarrow{g'} & B' & \xrightarrow{f'} & C & \longrightarrow & 0 \\ & & \downarrow u & & \downarrow v & & \| & & \\ 0 & \longrightarrow & A & \xrightarrow{g} & B & \longrightarrow & C & \longrightarrow & 0 \end{array}$$

with u and v isomorphisms. An exact sequence

$0 \longrightarrow A \xrightarrow{g'} B' \xrightarrow{f'} C' \longrightarrow 0$ is said to be <u>isomorphic</u> <u>over</u> A to $0 \longrightarrow A \xrightarrow{g} B \xrightarrow{f} C \longrightarrow 0$ if there is a commutative diagram

$$\begin{array}{ccccccccc} 0 & \longrightarrow & A & \xrightarrow{g'} & B' & \xrightarrow{f'} & C' & \longrightarrow & 0 \\ & & \| & & \downarrow u & & \downarrow \nu & & \\ 0 & \longrightarrow & A & \xrightarrow{g} & B & \xrightarrow{f} & C & \longrightarrow & 0 \end{array}$$

with u and ν isomorphisms.

<u>Proposition 5.4</u>: Suppose $0 \longrightarrow A \xrightarrow{g} B \xrightarrow{f} C \longrightarrow 0$ is an exact sequence in Mod Λ with A in $\text{art}_I\Lambda$ and C in noeth Λ.

a) f is right minimal if and only if the zero module is the only submodule A' of A such that g(A') is a summand of B.

b) There are summands A' of A maximal with respect to g(A') being a summand of B. Suppose A_1 and A_2 are are two such summands of A.

i) The induced exact sequences $0 \longrightarrow A/A_i \xrightarrow{g_i} B/g(A_i) \xrightarrow{f_i} C \longrightarrow 0$ have the property that f_i is minimal right TrD A-determined and $\text{Im}(\ ,f_i)(X) = \text{Im}(\ ,f)(X)$ for $i = 1, 2$.

ii) The exact sequences $0 \longrightarrow A/A_i \xrightarrow{g_i} B/g(A_i) \xrightarrow{f_i} C \longrightarrow 0$ are isomorphic over C.

iii) A_1 and A_2 are isomorphic summands of A and A/A_1 and A/A_2 are isomorphic summands of A.

Proof: Since f is right TrD A-determined by Proposition 5.3, a) follows from a) of Theorem 4.5.

b) Since A is artinian, the set of summands of A has both the ascending and descending chain conditions. Thus there are summands A' of A maximal with respect to $g(A')$ being a summand of B.

i) First we observe that because of the way the A_i were chosen, it is not hard to show that (0) is the only submodule A' of A/A_i such that $g_i(A')$ is a summand of $B/g(A_i)$ for $i = 1, 2$. Hence i) follows immediately from a).

ii) This follows from the fact that the f_i are minimal right TrD A-determined morphisms such that $\text{Im}(\ ,f_1)(X) = \text{Im}(\ ,f)(X) = \text{Im}(\ ,f_2)(X)$ (see Proposition 3.12).

iii) That A_1 and A_2 are isomorphic summands of A is a consequence of ii). That $A/A_1 \approx A/A_2$ follows from the fact that art Λ is a Krull-Schmidt category.

As our next result we have the following restatement of Theorem 4.6.

Theorem 5.5: Let X and C be in $\text{noeth}_P\Lambda$. Suppose H is an $(\text{End } X)^{op}$-submodule of (X, C) containing $P(X, C)$ such that $(X, C)/H$ is an $(\text{End } X)^{op}$-module of finite length. Then $\text{Soc}((X, C)/H) = \coprod_{i=1}^{k} S_i^{n_i}$ where $S_1,\ldots,S_k$ is a finite number of nonisomorphic simple $(\text{End } X)^{op}$-modules. Letting $n = \max$ of $n_1,\ldots,n_k$, we have that there is an exact sequence of Λ-modules

$$0 \longrightarrow (\mathrm{DTr}X)^n \longrightarrow B \xrightarrow{f} C \longrightarrow 0$$

such that f is right X-determined and $\mathrm{Im}(\ ,f)(X) = H$.

We now establish some results which are not for the most part merely rephrasing of general results for arbitrary noetherian algebras, but are particular to noetherian R-algebras where R is a complete local ring. We begin with the following uniqueness theorem.

<u>Proposition 5.6</u>: Let $0 \longrightarrow A \xrightarrow{g} B \xrightarrow{f} C \longrightarrow 0$ be an exact sequence in $\mathrm{Mod}\,\Lambda$ with A in $\mathrm{art}_I\Lambda$ and C in $\mathrm{noeth}\,\Lambda$ such that f is right minimal. Suppose $A = \coprod_{i=1}^{k} V_i^{n_i}$ with the $n_i > 0$ and the $V_1,\ldots,V_k$ non-isomorphic indecomposable Λ-modules. Letting $V = \coprod_{i=1}^{k} V_i$, we have the following.

a) f is right $\mathrm{TrD}\,V$-determined.

b) If X is in $\mathrm{noeth}\,\Lambda$ such that f is right X-determined but not right X'-determined for any proper summand X' of X, then $X \approx \mathrm{TrD}\,V$.

<u>Proof</u>: a) Syppose $Y = \coprod_{j=1}^{m} Y_j$ where some of the Y_j may be isomorphic. Let $h\colon L \longrightarrow C$ be a morphism in $\mathrm{Mod}\,\Lambda$. If for every morphism $t_j\colon Y_j \longrightarrow L$ there is a morphism $s_j\colon Y_j \longrightarrow B$ such that $ht_j = fs_j$, then clearly for each morphism $t\colon \coprod_{j=1}^{m} Y_j \longrightarrow L$ there is a morphism $s\colon \coprod_{j=1}^{m} Y_j \longrightarrow B$ such that $ht = fs$. On the other hand, suppose for every

$t\colon \coprod_{j=1}^{m} Y_j \longrightarrow L$ there is an $s\colon \coprod_{j=1}^{m} Y_j \longrightarrow B$ such that $ht = fs$. Then given $t_k\colon Y_k \longrightarrow L$ for some $k = 1,\ldots,m$, there is a morphism $t\colon \coprod_{j=1}^{m} Y_j \longrightarrow L$ such that t_k is the composition $Y_k \xrightarrow{i_k} \coprod_{j=1}^{m} Y_j \xrightarrow{t} L$, where i_k is the natural inclusion $Y_k \longrightarrow \coprod_{j=1}^{m} Y_j$.

Let $s: \coprod Y_j \longrightarrow B$ be such that $ht = fs$. Then $h(t_k) = h(t\, i_k) = f(s\, i_k)$ and so the morphism $s_k = s\, i_k: Y_k \longrightarrow B$ has the property $ht_k = f(s_k)$. Hence we have shown that for each morphism $t: \coprod_{j=1}^{m} Y_j \longrightarrow L$ there is a morphism $s: \coprod_{j=1}^{m} Y_j \longrightarrow B$ such that $ht = fs$ if and only if for each $t_j: Y_j \longrightarrow L$ there is an $s_j: Y \longrightarrow B$ such that $ht_j = fs_j$.

Now suppose that the indexing is such that $Y_1, \ldots, Y_n$ are all the nonisomorphic Y_j where $j = 1, \ldots, m$. It now follows from what we have shown that a morphism $h: L \longrightarrow C$ has the property that for each $t: \coprod_{i=1}^{m} Y_i \longrightarrow L$ there is a morphism $s: \coprod_{i=1}^{m} Y_i \longrightarrow B$ such that $ht = fs$ if and only if for each $u: \coprod_{i=1}^{n} Y_i \longrightarrow L$ there is a $v: \coprod_{i=1}^{n} Y_i \longrightarrow B$ such that $hu = fv$. From this it follows that f is right $\coprod_{j=1}^{m} Y_j$-determined if and only if f is right $\coprod_{i=1}^{n} Y_i$-determined.

Now we know by Proposition 5.3 that f is right TrD A-determined, hence by what we have just shown we know that f is right TrD V-determined. This finishes the proof of a).

b) We first show that X is in $\text{noeth}_P \Lambda$. For suppose $\coprod_{i=1}^{m} X_i$ is a representation of X as a sum of indecomposable modules and one X_i projective, say X_1. Since $f: B \longrightarrow C$ is an epimorphism, given any morphism $h: L \longrightarrow C$ in $\text{Mod}\,\Lambda$ and any morphism $t: X_1 \longrightarrow L$ there is an $s: X_1 \longrightarrow B$ such that $ht = fs$. Thus from the proof of a) it follows that if $h: L \longrightarrow C$ is a morphism in $\text{Mod}\,\Lambda$ then for each $t: \coprod_{i=1}^{m} X_i \longrightarrow L$ there is an $s: \coprod_{i=1}^{m} X_i \longrightarrow B$ such that $ht = fs$ if and only if for each

$u: \coprod_{i=2}^{m} X_i \longrightarrow L$ there is a morphism $v: \coprod_{i=2}^{m} X_i \longrightarrow B$ such that $hu = fv$. Therefore the fact that f is right X-determined implies that f is right $\coprod_{i=2}^{m} X_i$-determined. But this is impossible since $\coprod_{i=2}^{m} X_i$ is a proper summand of X. Hence X is in $\text{noeth}_P\Lambda$.

Suppose f is right X-determined. Then there is by Theorem 5.5 an exact sequence $0 \longrightarrow (\text{DTr}X)^n \longrightarrow B' \xrightarrow{f'} C \longrightarrow 0$ such that f' is right X-determined and $\text{Im}(\ ,f')(X) = \text{Im}(\ ,f)(X)$. The fact that f is right minimal means that f is a minimal right X-determined morphism. Hence $0 \longrightarrow A \longrightarrow B \longrightarrow C \longrightarrow 0$ is a summand over C of $0 \longrightarrow (\text{DTr}X)^n \longrightarrow B' \xrightarrow{f'} C \longrightarrow 0$. Therefore A is isomorphic to a summand of $(\text{DTr}X)^n$ which implies that V is isomorphic to a summand of DTrX. From this it follows that TrD V is isomorphic to a summand of TrD(DTr X) which is isomorphic to X since X is in $\text{noeth}_P\Lambda$. But by a) we know that f is right TrD V-determined. Therefore $\text{TrD } V \approx X$ since no proper summand X' of X has the property f is right X'-determined. This finishes the proof of part b) as well as the entire proposition.

In our next result we apply our methods to derive a general theorem concerning extensions of noetherian by artinian modules.

Theorem 5.7: Let A and C be in $\text{art}_I\Lambda$ and noeth Λ respectively. Suppose $(X, C)/P(X, C)$ is an $(\text{End } X)^{op}$-module of finite length where $X = \text{TrD } A$. Let $S_1,\dots,S_d$ be a complete set of nonisomorphic simple $(\text{End } X)^{op}$-modules and for each $(\text{End } X)^{op}$-submodule H of (X, C) containing P(X, C) let $n_1(H),\dots,n_d(H)$ be the uniquely determined nonnegative integers such that the $(\text{End } X)^{op}$-socle of $(X, C)/H$ is isomorphic to $\coprod_{i=1}^{d} S_i^{n_i(H)}$. Finally let $n = \max\{n_i(H)\}$ as i runs through $1,\dots,d$ and H runs through all $(\text{End } X)^{op}$-submodules of (X, C) containing $P(X, C)$.

a) n is finite.

b) If $k > n$ and $0 \longrightarrow A^k \xrightarrow{g} B \xrightarrow{f} C \longrightarrow 0$ is exact, then there is a submodule A' of A^k isomorphic to A^{k-n} such that $g(A')$ is a summand of B.

Proof:

a) Since each $n_i(H) \leq t$, the length of $(X, C)/P(X, C)$, it follows that $n \leq t$ and is thus finite.

b) Let $\Gamma = \text{End TrX}$. Since $X = \text{TrD } A$ is in $\text{noeth}_P\Lambda$ we know that $P(\text{TrX}, \text{TrX})$ is contained in the radical of Γ and so $\underline{\Gamma} = \underline{\text{End}}\ \text{TrX}$ has the same simple modules as Γ. Similarly the simple $(\underline{\text{End}}\ X)^{op}$-modules are the same as the simple $(\text{End } X)^{op}$-modules. Thus $S_1, \ldots, S_d$ is a complete set of nonisomorphic $\underline{\Gamma}$-modules by means of the isomorphism $\text{Tr}: (\underline{\text{End}}\ X)^{op} \longrightarrow \underline{\text{End}}\ \text{TrX}$. Finally, letting P_i be a Γ^{op} projective cover of the simple Γ^{op}-module $D(S_i)$ we have that the induced morphisms $S_i \longrightarrow D(P_i)$ are injective envelopes for the S_i and $\Gamma^{op} = \coprod_{i=1}^{d} P_i^{m_i}$ for some positive integers m_i.

Let $k > n$ and let $x: 0 \longrightarrow A^k \xrightarrow{g} B \xrightarrow{f} C \longrightarrow 0$ be an exact sequence. Then since A is in $\text{art}_I\Lambda$ we know that $A = \text{DTrX} = \text{Hom}_R(\text{TrX}, I_R) = \text{Hom}_\Gamma(\text{TrX}, \text{Hom}_R(\Gamma^{op}, I_R))$, thus $A^k = \text{Hom}_\Gamma(\text{TrX}, D(\Gamma^{op})^k)$. Let $v: (X, C)/P(X, C) \longrightarrow D(\Gamma^{op})^k$ be the Γ-morphism corresponding to the element x in $\text{Ext}^1_\Lambda(C, \text{Hom}_\Gamma(\text{TrX}, D(\Gamma^{op})^k))$. Denoting by H the submodule of (X, C) containing $P(X, C)$ such that $H/P(X, C) = \text{Ker } v$, we have that the socle of $(X, C)/H$ is $\coprod_{i=1}^{d} S_i^{n_i(H)}$ and therefore $I = D(\coprod_{i=1}^{d} P_i^{n_i(H)})$ is an injective envelope of $(X, C)/H$. Since $D(\Gamma^{op})^k = \coprod D(P_i^{k m_i})$ we can write $D(\Gamma^{op})^k$ as a sum

$$I \coprod \coprod_{i=1}^{d} D(P_i^{k m_i - n_i(H)}) = I \coprod \coprod_{i=1}^{d} D(P_i^{n m_i - n_i(H)}) \coprod D(\Gamma^{op})^{k-n}$$

since $n m_i \geq n_i(H)$ for all $i = 1,\ldots,d$. This gives a decomposition $U \coprod D(\Gamma^{op})^{k-n}$ of $D(\Gamma^{op})^k$ such that $\operatorname{Im} \nu \subset U$ and $U \approx D(\Gamma^{op})^n$. Therefore the projection $p\colon U \coprod D(\Gamma^{op})^{k-n} \longrightarrow D(\Gamma^{op})^{k-n}$ has the property $p\,\nu = 0$. Since the diagram

$$\begin{array}{ccc} \operatorname{Ext}^1_\Lambda(C, \operatorname{Hom}_\Gamma(\operatorname{Tr}X, D(\Gamma^{op})^k)) & \xrightarrow{\approx} & \operatorname{Hom}_\Gamma((X, C)/P(X, C), D(\Gamma^{op})^k) \\ \downarrow (\operatorname{Ext}^1(C, \operatorname{Hom}_\Gamma(\operatorname{Tr}X, p)) & & \downarrow \operatorname{Hom}_\Gamma((X, C)/P(X, C), p) \\ \operatorname{Ext}^1_\Lambda(C, \operatorname{Hom}_\Gamma(\operatorname{Tr}X, D(\Gamma^{op})^{k-n}) & \xrightarrow{\approx} & \operatorname{Hom}_\Gamma((X, C)/P(X, C), D(\Gamma^{op})^{k-n}) \end{array}$$

commutes, it follows that $\operatorname{Ext}^1_\Lambda(C, \operatorname{Hom}_\Gamma(\operatorname{Tr}X, p))(x) = 0$. Hence there is a commutative diagram

$$\begin{array}{l} 0 \longrightarrow \operatorname{Hom}_\Gamma(\operatorname{Tr}X, D(\Gamma^{op})^k) \xrightarrow{g} B \longrightarrow C \longrightarrow 0 \\ \operatorname{Hom}_\Gamma(\operatorname{Tr}X, p) \downarrow \qquad \swarrow \\ \quad \operatorname{Hom}_\Gamma(\operatorname{Tr}X, D(\Gamma^{op})^{k-n}) \end{array}$$

Therefore the submodule $A' = \operatorname{Hom}_\Gamma(\operatorname{Tr}X, D(\Gamma^{op})^k) = A^k$ has the property that $g(A')$ is a summand of B. Since $A' = \operatorname{Hom}_\Gamma(\operatorname{Tr}X, D(\Gamma^{op})^{k-n})$ is isomorphic to $D(\operatorname{Tr}X)^{k-n} \cong A^{k-n}$, the proposition has been established.

Added in proof:

Using arguments similar to those already used in the addenda to

Sections 3 and 4, we can generalize Theorem 5.5 as follows.

Theorem 5.8: Let X and C be in $\text{noeth}_P\Lambda$. Suppose H is an $(\text{End } X)^{op}$-submodule of (X, C) such that (X, C)/H is an $(\text{End } X)^{op}$-module of finite length. Then there is an exact sequence of Λ-modules

$$0 \longrightarrow (\text{DTr}X)^n \xrightarrow{g} B \xrightarrow{f} C$$

such that f is right X-determined and $\text{Im}(\ ,f)(X) = H$. Moreover there is a submodule A of B such that g(A) is a summand of B such that the exact sequence $0 \longrightarrow (\text{DTr}X)^n/A \xrightarrow{g'} B/g(A) \xrightarrow{f'} C$ has the property that f' is minimal right X-determined with $\text{Im}(\ ,f')(X) = H$.

§6. Morphisms Left Determined by Modules. As in Section 5, we assume throughout this section, unless stated explicitly to the contrary, that Λ is a noetherian R-algebra with R a complete local ring. We showed in Section 5 that if $0 \longrightarrow A \xrightarrow{g} B \xrightarrow{f} C \longrightarrow 0$ is an exact sequence with A in $\text{art}_I\Lambda$ and C in noeth Λ, then f is right TrD A-determined. In view of this result it is natural to ask if g is left X-determined for some module X. Eventually (see Section 10) we will show that g is left DTrC-determined. However, we now only show that g is left DTrC-determined relative to a certain full subcategory arno Λ of Λ, which we now describe.

Let $0 \longrightarrow A \longrightarrow B \longrightarrow C \longrightarrow 0$ be an exact sequence of Λ-modules with A in art Λ and C in noeth Λ. In general the module B is neither artinian nor noetherian. This suggests considering the full subcategory arno Λ of Mod Λ whose objects consist of those Λ-modules B such that there is an exact sequence $0 \longrightarrow B' \longrightarrow B \longrightarrow B'' \longrightarrow 0$

with B' in art Λ and B" in noeth Λ. The following result concerning the category arno Λ is easily verified.

Proposition 6.1: The full subcategory arno Λ of Mod Λ has the following properties.

a) Let $0 \longrightarrow A' \longrightarrow A \longrightarrow A'' \longrightarrow 0$ be an exact sequence in Mod Λ. Then A is in arno Λ if and only if A' and A" are in arno Λ.

b) A Λ-module A is in arno Λ if and only if D(A) is in arno Λ^{op}.

c) If A is in arno Λ, then $\varphi_A : A \longrightarrow D^2A$ is an isomorphism.

d) The functors D: Mod $\Lambda \longrightarrow$ Mod Λ^{op} and D: Mod $\Lambda^{op} \longrightarrow$ Mod Λ induce inverse dualities D: arno $\Lambda \longrightarrow$ arno Λ^{op} and D: arno $\Lambda^{op} \longrightarrow$ arno Λ.

Thus we see that arno Λ is a full abelian subcategory of Mod Λ which is closed under extensions such that the inclusion arno $\Lambda \longrightarrow$ Mod Λ is exact. Viewing arno Λ as an additive category in its own right we can ask if a morphism $f: B \longrightarrow C$ in arno Λ is right X-determined for some X in arno Λ, i.e. if f and X have the property that for a morphism $h: L \longrightarrow C$ in arno Λ there is an $s: L \longrightarrow B$ such that $h = sf$ whenever $\mathrm{Im}(,f)(X) \supset \mathrm{Im}(,h)(X)$. Clearly if f is right X-determined in Mod Λ, then f is right X-determined in arno Λ. However the converse is not so clear. Analogous definitions and comments can be made for morphisms being left determined in arno Λ. The rest of this section is devoted to showing how the results in Section 5 concerning right determined morphisms in Mod Λ can be translated to results concerning left determined morphisms in arno Λ using the duality

$D\colon \text{arno}\ \Lambda \longrightarrow \text{arno}\ \Lambda^{op}$. These results are based on the observations made at the end of Section 2 concerning dual additive categories which will be used freely.

For each X in $\text{art}\ \Lambda$ the duality $D\colon \text{art}\ \Lambda \longrightarrow \text{noeth}\ \Lambda^{op}$ induces a ring isomorphism $D\colon \text{End}\ X \longrightarrow (\text{End}\ D(X))^{op}$ which we will often view as an identification. With this convention in mind we have the following easily verified properties of the duality D.

<u>Proposition 6.2</u>: Let X and Y be in $\text{art}\ \Lambda$ and $D\colon \text{art}\ \Lambda \longrightarrow \text{noeth}\Lambda^{op}$ the usual duality.

a) $D\colon (X, Y) \longrightarrow (D(Y), D(X))$ is an $(\text{End}\ X)^{op}$-End Y bimodule isomorphism.

b) If $H \subset (X, Y)$ is an End Y-submodule of (X, Y), then $D(H) \subset (D(Y), D(X))$ is an $(\text{End}\ D(Y))^{op}$-submodule of $(D(Y), D(X))$.

c) Suppose H is an End Y-submodule of (X, Y).

i) $H = I(X, Y)$ if and only if $D(H) = P(D(Y), D(X))$.

ii) $H \supset I(X, Y)$ if and only if $D(H) \supset P(D(Y), D(X))$.

iii) $D\colon (X, Y) \longrightarrow (D(Y), D(X))$ induces an End Y-isomorphism $D\colon (X, Y)/H \longrightarrow (D(Y), D(X))/D(H)$.

<u>Proof</u>: Left to the reader.

We say that a sequence $0 \longrightarrow A \xrightarrow{g} B \xrightarrow{f} C \longrightarrow 0$ of Λ-modules is exact in $\text{arno}\ \Lambda$ if it is an exact sequence of Λ-modules and A, B, and C are in $\text{arno}\ \Lambda$.

It is the following easily verified result which enables us to transfer what has been shown in Section 5 to morphisms which are left determined in arno Λ.

Proposition 6.3: Let $0 \longrightarrow A \xrightarrow{g} B \xrightarrow{f} C \longrightarrow 0$ be an exact sequence of Λ-modules with A in art Λ and C in noeth Λ. Let X be in art Λ and let H be the End X-submodule $\mathrm{Im}(g, \;)(X)$ of (A, X).

a) $H \supset I(A, X)$.

b) The exact sequence of Λ^{op}-modules $0 \longrightarrow D(C) \xrightarrow{D(f)} D(B) \xrightarrow{D(g)} D(A) \longrightarrow 0$ has the property that $D(H)$ is the End $D(X)^{op}$-submodule $\mathrm{Im}(\;, D(g))(D(X))$ of $(D(X), D(A))$.

c) More generally, if $t: A \longrightarrow L$ is a morphism in arno Λ, then $D(\mathrm{Im}(t, \;)(X)) = \mathrm{Im}(\;, D(t))(D(X))$.

d) g is left X-determined in arno Λ if and only if $D(g)$ is right $D(X)$-determined in arno Λ.

e) g is minimal left X-determined in arno Λ if and only if $D(g)$ is minimal right $D(X)$-determined in arno Λ.

A straightforward application of Proposition 6.3 to Proposition 5.3 yields the following result.

Proposition 6.4: Let $0 \longrightarrow A \xrightarrow{g} B \xrightarrow{f} C \longrightarrow 0$ be an exact sequence in arno Λ with A in art Λ and C in $\mathrm{noeth}_P\Lambda$ and let $X = D\mathrm{Tr}C$.

a) g is left X-determined in arno Λ.

b) $(A, X)/\mathrm{Im}(g, \;)(X)$ is an End X-module of finite length which is isomorphic to a submodule of $\mathrm{Hom}_R(\mathrm{End}\,(X)^{op}, I_R)$.

c) There is a summand $0 \longrightarrow A \xrightarrow{g'} B' \xrightarrow{f'} C' \longrightarrow 0$ over A of $0 \longrightarrow A \xrightarrow{g} B \xrightarrow{f} C \longrightarrow 0$ with g' a minimal left X-determined morphism such that

$\text{Im}(g', \ )(X) = \text{Im}(g, \ )(X)$.

As a consequence of Proposition 6.4, or of applying Proposition 6.3 to Proposition 5.4, we obtain the following analogue of Proposition 5.4.

Proposition 6.5: Suppose $0 \longrightarrow A \xrightarrow{g} B \xrightarrow{f} C \longrightarrow 0$ is an exact sequence in arno Λ with A in art Λ and C in noeth Λ.

a) g is left minimal if and only if there are no proper submodules C' of C such that the composition $B \longrightarrow C \longrightarrow C/C'$ is a splittable epimorphism.

b) Obviously there are submodules C' minimal with respect to the composition $B \longrightarrow C \longrightarrow C/C'$ being a splittable epimorphism. Suppose C_1 and C_2 are two such submodules.

i) C_1 and C_2 are summands of C and there are summands B_1 and B_2 of B such that $f(B_i) = C_i$ and $f|B_i : B_i \longrightarrow C_i$ are isomorphisms for $i = 1, 2$. Moreover the exact sequences $0 \longrightarrow A \xrightarrow{g_i} B/B_i \xrightarrow{f_i} C/C_i \longrightarrow 0$ have the property that g_i is minimal left X-determined and $\text{Im}(g_i, \ )(X) = \text{Im}(g, \ )(X)$ for $i = 1, 2$.

ii) The exact sequences $0 \longrightarrow A \longrightarrow B/B_i \longrightarrow C/C_i \longrightarrow 0$ are isomorphic over A.

iii) C/C_1 and C/C_2 are isomorphic summands of C and C_1 and C_2 are isomorphic summands of C.

We now prove the following existence theorem which is an analogue of Theorem 5.5.

Theorem 6.6: Let A and X be in art Λ. Let H be an End X-submodule of (A, X) such that $H \supset I(A, X)$ and the End X-module $(A, X)/H$ is of finite length with socle $\coprod_{i=1}^{k} S_i^{n_i}$ where $S_1, \ldots, S_k$ is a complete set of nonisomorphic simple End X-modules. Finally let $n = \max\{n_1, \ldots, n_k\}$. Then

a) There exists an exact sequence $0 \longrightarrow A \xrightarrow{g} B \xrightarrow{f} \mathrm{TrD}(X)^n \rightarrow 0$ with g a left X-determined morphism in arno Λ such that $(\mathrm{Im}(g, \ ))(X) = H$.

b) Letting $C = (\mathrm{TrD}(X))^n$, the exact sequence $0 \longrightarrow A \xrightarrow{g} B \xrightarrow{f} C \longrightarrow 0$ can be written as a sum

$$\begin{array}{ccccccccc} 0 & \longrightarrow & A & \xrightarrow{g'} & B' & \xrightarrow{f'} & C_1 & \longrightarrow & 0 \\ & & & & \coprod & & \coprod & & \\ & & & & C_2 & = & C_2 & & \end{array}$$

where g' is a minimal left X-determined morphism in arno Λ such that $(\mathrm{Im}(g', \ ))(X) = H$.

Proof:

a) Since A and X are in art Λ, we know that $D(A)$ and $D(X)$ are in noeth Λ^{op}. Also we know by Proposition 6.2, that the End $D(X)^{op}$-submodule $D(H)$ of $(D(X), D(A))$ contains $P(D(X), D(A))$ and that the $(\mathrm{End}\ D(X))^{op}$-module $(D(X), D(A))/D(H)$ is an essential extension of its socle which is of the form $\coprod_{i=1}^{k} T_i^{n_i}$ where $T_1, \ldots, T_k$ is a complete set of nonisomorphic simple $\mathrm{End}(D(X))^{op}$-modules. Thus by Theorem 5.5 we know there is an exact sequence $0 \longrightarrow (\mathrm{DTrD}(X))^n \xrightarrow{h} B' \xrightarrow{j} D(A) \longrightarrow 0$ of Λ^{op}-modules with j a right $D(X)$-determined morphism such that $(\operatorname{Im}(\ ,j))(D(X)) = D(H)$. Applying the functor D to this exact sequence we obtain the exact sequence $0 \longrightarrow A \xrightarrow{D(j)} D(B') \longrightarrow (\operatorname{TrD}(X))^n \longrightarrow 0$. By Proposition 6.3 we know that $D(j)$ is a left X-determined morphism with $(\operatorname{Im}(D(j),\ ))(X) = H$. This completes the proof of a).

b) This follows readily from Theorem 5.3.

The following translation of Proposition 5.6 can be obtained by applying Proposition 6.3 to Proposition 5.6.

Proposition 6.7: Let $0 \longrightarrow A \xrightarrow{g} B \xrightarrow{f} C \longrightarrow 0$ be an exact sequence in arno Λ with A in art Λ and C in $\text{noeth}_P\Lambda$ such that g is left minimal. Suppose $C = \coprod_{i=1}^{k} W_i^{n_i}$ with the $n_i > 0$ and the W_i nonisomorphic indecomposable Λ-modules. Letting $W = \coprod_{i=1}^{k} W_i$, we have the following.

a) g is left DTrW-determined.

b) If X is in art Λ such that g is left X-determined but not left X'-determined for any proper summand X' of X, then $X \approx \text{DTrW}$.

Our next result is the following analogue of Proposition 5.7.

Proposition 6.8: Let A and C be in art Λ and $\text{noeth}_P\Lambda$ respectively. Suppose $(A, X)/I(A, X)$ is an End X-module of finite length t where $X = \text{DTrC}$. Let $S_1,\dots,S_d$ be a complete set of nonisomorphic simple End X-modules and for each End X-submodule H of (A, X) containing $I(A, X)$, let $n_1(H),\dots,n_d(H)$ be the uniquely determined nonnegative integers such that the (End X)-socle of $(A, X)/H$ is isomorphic to

$\coprod_{i=1}^{d} S_i^{n_i(H)}$. Finally let $n = \max\{n_i(H)\}$ as i runs through $1, \ldots, d$ and H runs through all End X-submodules of (A, X) containing $I(A, X)$.

a) n is finite.

b) If $k > n$ and $0 \longrightarrow A \longrightarrow B \longrightarrow C^k \longrightarrow 0$ is exact, then there is an exact sequence $0 \longrightarrow A \longrightarrow B \longrightarrow C^n \longrightarrow 0$ which is a summand over A of $0 \longrightarrow A \longrightarrow B \longrightarrow C^k \longrightarrow 0$.

Using the duality D: noeth $\Lambda \longrightarrow$ art Λ^{op}, the following result can be derived which gives a different starting point for the results of this section. It is included here without proof for the sake of completeness.

Proposition 6.9: Let C be in noeth Λ and A in art Λ. Further, let Γ = End DTrC. Then there is an isomorphism of R-modules

$$\mathrm{Ext}^1_\Lambda(C, A) \longrightarrow \mathrm{Hom}_\Gamma(\overline{\mathrm{Hom}}_\Lambda(A, DTrC), \mathrm{Hom}_R(\Gamma^{op}, I_R))$$

which is functorial in A in art Λ. Moreover if $0 \longrightarrow A \xrightarrow{g} B \xrightarrow{f} C \longrightarrow 0$ in $\mathrm{Ext}^1_\Lambda(C, A)$ and the Γ-morphism $\nu: \overline{\mathrm{Hom}}_\Lambda(A, DTrC) \longrightarrow \mathrm{Hom}_R(\Gamma^{op}, I_R)$ correspond under this isomorphism, then

a) g is left DTrC-determined. Im(g,) is a Γ-submodule of $(\Lambda, DTrC)$ such that

b) Im(g,)(DTrC) $\supset$ I(A, DTrC) and Im(g,)(DTrC)/I(A, DTrC) = Ker ν

Added in proof:

Applying the duality D: noeth $\Lambda \longrightarrow$ art Λ^{op} to Theorem 5.8 we obtain the following generalization of Theorem 6.6.

Theorem 6.10: Let A and X be in art Λ. Let H be an End X-submodule of (A, X) such that $(A, X)/H$ is of finite length over End X. Then there is an exact sequence of Λ-modules

$$A \xrightarrow{g} B \xrightarrow{f} \mathrm{TrD}(X)^n \longrightarrow 0$$

with g left X-determined morphism in arno Λ such that $\mathrm{Im}(g, \ )(X) = H$. Moreover, letting $C = (\mathrm{TrD}(X))^n$, the exact sequence $A \xrightarrow{g'} B \xrightarrow{f} C \longrightarrow 0$ can be written as a sum

$$\begin{array}{ccccccc} A & \xrightarrow{g'} & B' & \xrightarrow{f'} & C' & \longrightarrow & 0 \\ & & \amalg & & \amalg & & \\ & & C_1 & \cong & C_2 & & \end{array}$$

where g' is a minimal left X-determined morphism in arno Λ such that $(\mathrm{Im}(g', \ ))(X) = H$.

§7. Orders and Lattices. The two classical situations where the representation theory of rings have been studied most extensively are a) finitely generated modules over a finite dimensional algebra over a field and b) lattices over orders over Dedekind domains. One of the aims of the rest of this chapter is to see what our earlier results tell us about representation theory in these classical situations. Rather than dealing directly with these two seemingly different situations, we develop in this section a more general notion of orders and lattices which includes, amongst a great many other examples, these classical situations as special cases. We will then interpret our previous results for these generalized orders and lattices.

We say that a commutative noetherian ring R is equidimensional if $\dim R_{\underline{m}} = \dim R$ for all maximal ideals $\underline{m}$ of R. In particular if R is equidimensional, then dim R is finite. It is easily seen that if R is equidimensional then R/(x) is equidimensional for each nonzero divisor x in R.

We recall that a local ring R is said to be Gorenstein if inj. dim. $R < \infty$ (see [13] and [16] for basic properties of Gorenstein rings). If R is a local Gorenstein ring of dimension d, then inj. dim. R = d and a minimal injective resolution $0 \longrightarrow R \longrightarrow I_0(R) \longrightarrow \ldots \longrightarrow I_d(R) \longrightarrow 0$ of R over R has the following properties: a) $I_d(R) = I_R$ the injective envelope of R|rad R and b) $\text{Hom}_R(M, I_j(R)) = 0$ for all R-modules M of finite length and $j < d$. An arbitrary commutative noetherian ring R is said to be a <u>Gorenstein</u> ring if $R_{\underline{p}}$ is <u>Gorenstein</u> for all prime ideals $\underline{p}$ of R. The following properties of equidimensional rings are well known.

<u>Proposition 7.1</u>: Let R be an equidimensional ring of dimension d.

a) R is Gorenstein if and only if inj. dim. $R < \infty$.

b) If R is Gorenstein, then inj. dim. R = d.

c) If R is Gorenstein and $0 \longrightarrow R \longrightarrow I_0(R) \longrightarrow \ldots \longrightarrow I_d(R) \longrightarrow 0$ is a minimal injective resolution of R, then $I_d(R) \cong I_R$ and a module M in noeth R is of finite length if and only if $\text{Hom}_R(M, I_j(R)) = 0$ for all $j < d$.

d) If R is Gorenstein and x in R is regular (i.e. x is a nonunit which is not a zero divisor in R), then R/(x) is an equidimensional Gorenstein ring.

e) If R is Gorenstein, then every localization of R is a Gorenstein ring.

Before discussing equidimensional Gorenstein rings further we point out some examples: a) fields, b) Dedekind domains, c) equidimensional regular rings such as affine rings of nonsingular affine varieties, d) 0-dimensional local rings with a simple socle, e) rings derived from examples a) through c) by taking factor rings by ideals generated by a regular element.

Let M be a nonzero finitely generated R-module with R a local ring. A sequence $x_1,\ldots,x_r$ of elements in rad R is said to be a regular M-sequence if x_j is $M/(x_1,\ldots,x_{j-1})M$-regular for all $j = 1,\ldots,d$, i.e. x_j is not a zero divisor in $M/(x_1,\ldots,x_{j-1})M$ where $(x_1,\ldots,x_{j-1})$ is the ideal generated by $x_1,\ldots,x_{j-1}$. A regular M-sequence $x_1,\ldots,x_r$ is said to be maximal if every element of rad R is not regular in $M/(x_1,\ldots,x_r)M$. It is known that every regular M-sequence can be extended to a maximal regular M-sequence and that all maximal regular M-sequences have the same length (i.e. same number of elements). The length of the maximal regular M-sequences is denoted by depth M. In general depth $M \leq \dim R$. We recall that depth M is the smallest integer i such that $\mathrm{Ext}^i_R(R/\operatorname{rad} R, M) \neq 0$. A module M over R is said to be a <u>Cohen-Macauley module</u> if depth M = dim R.

Suppose R is an arbitrary commutative noetherian ring. A finitely generated R-module M is called a <u>Cohen-Macauley module</u> if $M_{\underline{p}}$ is a Cohen-Macauley $R_{\underline{p}}$-module for all prime ideals $\underline{p}$ of R. Moreover R is said to be a <u>Cohen-Macauley ring</u> if it is a Cohen-Macauley module over itself. Of particular importance to us is the fact that if R is a Gorenstein ring then R is a Cohen-Macauley ring. Thus all our previous examples of Gorenstein rings also give examples of Cohen-Macauley rings. The following are examples of Cohen-Macauley modules: a) if R is a field, then all nonzero finite dimensional vector spaces are Cohen-Macauley; b) if R is an artin ring, then all finitely generated nonzero modules are Cohen-Macauley modules; c) if R is a Dedekind domain, then a nonzero finitely generated module M is Cohen-Macauley if and only if M is projective, or equivalently, if and only if M is torsion free; d) if R is a regular ring, then a finitely generated R-module M is Cohen-Macauley if and only if M is projective; e) if R is a two-dimensional integrally closed domain, then a nonzero module M is Cohen-Macauley if and only if M is reflexive, i.e. the usual morphism $M \longrightarrow \mathrm{Hom}_R(\mathrm{Hom}(M, R), R)$ is an isomorphism. In particular R is Cohen-Macauley as are all the nonzero finitely generated projective R-modules.

We now summarize the main properties of Cohen-Macauley modules over a Gorenstein ring R which we will need. All of these properties are obvious in case R is a field or a Dedekind ring.

Proposition 7.2: Assume that R is a Gorenstein ring of dimension d (not necessarily equidimensional).

a) Suppose $0 \longrightarrow M_1 \longrightarrow M_2 \longrightarrow M_3 \longrightarrow 0$ is an exact sequence in noeth R.

   i) If M_1, M_3 are Cohen-Macauley then M_2 is Cohen-Macauley.

   ii) If M_2 and M_3 are Cohen-Macauley then M_1 is Cohen-Macauley.

   iii) If the sequence splits, then M_2 is Cohen-Macauley if and only if M_1 and M_3 are Cohen-Macauley.

b) If $0 \longrightarrow M_d \xrightarrow{f_d} M_{d-1} \longrightarrow \cdots \xrightarrow{f_1} M_o$ is an exact sequence in noeth R with the M_i Cohen-Macauley for $i = 0,\dots,d-1$, then depth Im $f_i \geq i$ for all $i \geq 1$ and so M_d is Cohen-Macauley.

c) If M is Cohen-Macauley, then $\mathrm{Ext}_R^i(M, R) = 0$ for all $i > 0$.

d) If M is Cohen-Macauley, then $\mathrm{Hom}_R(M, R) = M^*$ is Cohen-Macauley.

e) Each Cohen-Macauley module is reflexive, i.e. the natural morphism $M \longrightarrow M^{**}$ is an isomorphism.

f) Let $\underline{\underline{CM}}(R)$ denote the full subcategory of noeth R whose objects are the Cohen-Macauley modules. The functor $\mathrm{Hom}_R(\ ,R): \underline{\underline{CM}}(R) \longrightarrow \underline{\underline{CM}}(R)$ is a duality.

g) If $0 \longrightarrow M_1 \longrightarrow M_2 \longrightarrow M_3 \longrightarrow 0$ is exact with the M_i in $\underline{\underline{CM}}(R)$, then $0 \longrightarrow M_3^* \longrightarrow M_2^* \longrightarrow M_1^* \longrightarrow 0$ is exact with the M_i^* in $\underline{\underline{CM}}(R)$.

h) If M is in $\underline{\underline{CM}}(R)$, then $M_{\underline{p}}$ is in $\underline{\underline{CM}}(R_p)$ for all prime ideals $\underline{p}$ in R.

Let Λ be a noetherian R-algebra with R an equidimensional Gorenstein ring. We say that a Λ-module is a Cohen-Macauley module if it is finitely generated and is Cohen-Macauley when viewed as an R-module. We say that a Λ-module M is a <u>lattice</u> if it is a Cohen-Macauley Λ-module and $M_{\underline{p}}$ and $\mathrm{Hom}_R(M, R)_{\underline{p}}$ are $\Lambda_{\underline{p}}$-projective and $\Lambda_{\underline{p}}^{op}$-projective modules respectively for all nonmaximal prime ideals $\underline{p}$ in R. We denote the full subcategory of noeth Λ whose objects are the Λ-lattices by $L(\Lambda)$. Finally we say a noetherian R-algebra Λ is an <u>R-order</u> if R is an equidimensional Gorenstein ring and Λ is in $L(\Lambda)$. Before discussing lattices over R-orders we point out some examples.

a) If R is a field or, more generally, a 0-dimensional Gorenstein ring, then every noetherian R-algebra Λ is an R-order and $L(\Lambda) = \text{noeth } \Lambda = \text{art } \Lambda$.

b) Suppose Λ is a noetherian R-algebra with R a Dedekind ring with field of quotients K. Then Λ is an R-order if and only if Λ is a projective R-module and $\Sigma = K \otimes_R \Lambda$ is a selfinjective ring. If Λ is an R-order, then $L(\Lambda)$ consists of the Λ-modules M which are projective R-modules such that $K \otimes_R M$ is a projective Σ-module. It should be noted in particular that if Λ is a projective R-module such that $\Sigma = K \otimes \Lambda$ is semisimple, then Λ is an R-order and the Λ-lattices are precisely the Λ-modules which are projective R-modules. Thus we see that the classical orders and lattices fit into this more general scheme.

c) Suppose R is an equidimensional Gorenstein ring. Then R is an R-order and the R-lattices consist of those Cohen-Macauley R-modules M such that $M_{\underline{p}}$ is $R_{\underline{p}}$-free for all nonmaximal primes $\underline{p}$ of R.

d) Suppose R is a two-dimensional integrally closed Gorenstein ring. Then R is an R-order and the R-lattices are precisely the reflexive R-modules.

We now give some of the basic properties of Λ-lattices over an R-order Λ. These follow readily from the definitions and Proposition 7.2. Moreover they are obvious for R a field or a Dedekind domain.

<u>Proposition 7.3</u>: Let Λ be an R-order with $d = \dim R$.

a) Λ^{op} is an R-order.

b) For M in noeth Λ the following are equivalent:

i) M is in $L(\Lambda)$.

ii) M is Cohen-Macauley and $M_{\underline{p}}$ is $\Lambda_{\underline{p}}$-projective for all nonmaximal primes $\underline{p}$ in R.

iii) M is Cohen-Macauley and $M_{\underline{m}}$ is $\Lambda_{\underline{m}}$-projective for all but a finite number of maximal ideals $\underline{m}$ of R.

iv) M is Cohen-Macauley and $\mathrm{Hom}_R(M, R)$ is in $L(\Lambda^{op})$.

c) The functor $\mathrm{Hom}_R(\ ,R): L(\Lambda) \longrightarrow L(\Lambda^{op})$ is a duality.

d) Suppose $0 \longrightarrow M_1 \longrightarrow M_2 \longrightarrow M_3 \longrightarrow 0$ is an exact sequence in noeth Λ.

i) If the sequence splits, then M_2 is in $L(\Lambda)$ if and only if M_1 and M_3 are in $L(\Lambda)$.

ii) If M_1 and M_3 are in $L(\Lambda)$, then M_2 is in $L(\Lambda)$.

iii) If M_2 and M_3 are in $L(\Lambda)$, then M_1 is in $L(\Lambda)$.

iv) If the M_i are in $L(\Lambda)$, then $0 \longrightarrow M_3^* \longrightarrow M_2^* \longrightarrow M_1^* \longrightarrow 0$ is an exact sequence with the M_i^* in $L(\Lambda^{op})$ where $M_i = \mathrm{Hom}\ (M_i, R)$.

e) If $\cdots \longrightarrow P_{i+1} \xrightarrow{d_{i+1}} P_1 \longrightarrow \cdots \longrightarrow P_1 \xrightarrow{d_1} P_0 \xrightarrow{d_0} M \longrightarrow 0$ is a projective resolution of M and M has the property $M_{\underline{p}}$ is $\Lambda_{\underline{p}}$ projective for all nonmaximal prime ideals $\underline{p}$, then Imd_i is a lattice for all $i \geq \dim R$.

We now use these basic facts concerning lattices to define various functors which will enable us to translate most of our results in earlier sections into results concerning lattices.

First we recall the functor Ω: $\underline{\text{noeth}}\ \Lambda \longrightarrow \underline{\text{noeth}}\ \Lambda$ defined as follows. For each M in noeth Λ choose an epimorphism $P(M) \longrightarrow M \longrightarrow 0$ with $P(M)$ projective and define $\Omega M = \text{Ker}(P(M) \longrightarrow M)$. As usual if M has a projective cover we choose $P(M) \longrightarrow M \longrightarrow 0$ to be a projective cover. Clearly if $f: M \longrightarrow N$ is a morphism in noeth Λ, then there exists a commutative exact diagram

$$\begin{array}{ccccccccc} 0 & \longrightarrow & \Omega M & \longrightarrow & P(M) & \longrightarrow & M & \longrightarrow & 0 \\ & & \downarrow f_1 & & \downarrow f_0 & & \downarrow f & & \\ 0 & \longrightarrow & \Omega N & \longrightarrow & P(N) & \longrightarrow & N & \longrightarrow & 0 \end{array} .$$

While the morphism $f_1: \Omega(M) \longrightarrow \Omega(N)$ depends on the particular choice of the morphism $f_0: P(M) \longrightarrow P(N)$ which makes the diagram commute, different choices of the f_0 result in morphisms $\Omega M \longrightarrow \Omega N$ which differ by a morphism factoring through a projective module. Thus if we define $\Omega(f): \Omega M \longrightarrow \Omega N$ to be $\underline{f}_1$ in $\underline{\text{Hom}}_\Lambda(\Omega M, \Omega N)$ where $f_1: \Omega M \longrightarrow \Omega N$ is any morphism for which there is a commutative diagram

$$\begin{array}{ccccccccc} 0 & \longrightarrow & \Omega M & \longrightarrow & P(M) & \longrightarrow & M & \longrightarrow & 0 \\ & & \downarrow f_1 & & \downarrow f_0 & & \downarrow & & \\ 0 & \longrightarrow & \Omega N & \longrightarrow & P(N) & \longrightarrow & N & \longrightarrow & 0 \end{array} ,$$

we obtain the functor Ω: $\underline{\text{noeth}}\ \Lambda \longrightarrow \underline{\text{noeth}}\ \Lambda$.

More generally, we define Ω^i: $\underline{\text{noeth}}\ \Lambda \longrightarrow \underline{\text{noeth}}\ \Lambda$ for all nonnegative integers by induction on i as follows: Ω^0 is the identity and $\Omega^{i+1} = \Omega(\Omega^i)$ for all $i \geq 0$.

Next we observe that if $f_1, f_2: M \longrightarrow N$ are morphisms in noeth Λ, then the morphisms $\text{Ext}^1_\Lambda(f_1, \ )$, $\text{Ext}^1_\Lambda(f_2, \ )$: $\text{Ext}^1_\Lambda(N, \ ) \longrightarrow \text{Ext}^1_\Lambda(M, \ )$ are the same if and only if $\underline{f}_1$ and $\underline{f}_2$ are equal in $\underline{\text{Hom}}_\Lambda(M, N)$ (see [15]). Thus for each $\underline{f}$ in $\underline{\text{Hom}}_\Lambda(M, N)$ we can define $\text{Ext}^1_\Lambda(\underline{f}, \ )$: $\text{Ext}^1_\Lambda(N, \ ) \longrightarrow \text{Ext}^1_\Lambda(M, \ )$ to be $\text{Ext}^1_\Lambda(f, \ )$: $\text{Ext}^1_\Lambda(N, \ ) \longrightarrow \text{Ext}^1_\Lambda(M, \ )$ without introducing ambiguities. We now list some of the properties of the functor Ω: $\underline{\text{noeth}}\ \Lambda \longrightarrow \underline{\text{noeth}}\ \Lambda$ which are well known and easily checked.

Proposition 7.4: Let Λ be an arbitrary noetherian ring. Then the functors Ω^i: $\underline{\text{noeth}}\ \Lambda \longrightarrow \underline{\text{noeth}}\ \Lambda$ have the following properties for each $i > 0$.

a) For each M in noeth Λ there is an isomorphism φ_M: $\text{Ext}^1_\Lambda(\Omega^i M, \ ) \longrightarrow \text{Ext}^{i+1}(M, \ )$ which is functorial in M, i.e. given a morphism $\underline{f}: M \longrightarrow N$ in $\underline{\text{noeth}}\ \Lambda$, the diagram

$$\begin{array}{ccc} \text{Ext}^1_\Lambda(N, \ ) & \xrightarrow{\varphi_N} & \text{Ext}^{i+1}_\Lambda(\Omega^i(N), \ ) \\ \downarrow \text{Ext}^1_\Lambda(\underline{f}, \ ) & & \downarrow \text{Ext}^1_\Lambda(\Omega^i(\underline{f}), \ ) \\ \text{Ext}^1_\Lambda(M, \ ) & \xrightarrow[\varphi_M]{} & \text{Ext}^{i+1}_\Lambda(\Omega^i(N), \ ) \end{array}$$

commutes.

b) If $\text{Ext}^j_\Lambda(M, \Lambda) = 0$ for $j = 1, \ldots, i$, then Ω^i: $\underline{\text{Hom}}_\Lambda(\ , M) \longrightarrow \underline{\text{Hom}}_\Lambda(\ , \Omega^i M)$ is an isomorphism.

Proof:

a) Since $0 \longrightarrow \Omega^i(M) \longrightarrow P_{i-1} \longrightarrow \dots \longrightarrow P_o \longrightarrow M \longrightarrow 0$ is exact with the P_j projective, we have the usual dimension shift isomorphism $\varphi_M: \mathrm{Ext}^1_\Lambda(\Omega^i M, \ ) \longrightarrow \mathrm{Ext}^{i+1}_\Lambda(M, \ )$. It is not difficult to check that the φ_M have desired properties.

b) Follows from [8, Proposition 1.39].

Suppose now that Λ is an R-order with $d = \dim R$. We denote by $J(\Lambda)$ the full subcategory of noeth Λ consisting of all M in noeth Λ satisfying both of the following conditions: a) $M_{\underline{p}}$ is $\Lambda_{\underline{p}}$-projective for all nonmaximal prime ideals $\underline{p}$ of R and b) $\mathrm{Ext}^i_\Lambda(M, \Lambda) = 0$ for all $i = 1,\dots,d$. Let $\underline{J}(\Lambda)$ be the full subcategory of $\underline{\text{noeth}}\ \Lambda$ whose objects are in $J(\Lambda)$. By Proposition 7.3, condition a) implies that if M is in $J(\Lambda)$, then $\Omega^d M$ is in $L(\Lambda)$. Thus Ω^d induces a functor $\Omega^d: \underline{J}(\Lambda) \longrightarrow \underline{L}(\Lambda)$. Condition b) together with Proposition 7.4 shows that $\Omega^d: \underline{J}(\Lambda) \longrightarrow \underline{L}(\Lambda)$ is fully faithful. It is our aim now to show that Ω^d is dense and hence an equivalence of categories. That is, we want to show that if X is a nonprojective lattice, then there is an M in $J(\Lambda)$ such that $\Omega^d(M) \approx X$ in $\underline{L}(\Lambda)$. The proof of this fact depends on the following result.

Proposition 7.5: Suppose Λ is an R-order with $\dim R = d$. For each M in $L(\Lambda)$ we have that $\mathrm{Tr}M$ is in $J(\Lambda^{op})$.

Proof: The proof takes several steps. We begin with the following lemmas.

Lemma 7.6: Let Λ be an arbitrary noetherian R-algebra (R an arbitrary commutative noetherian ring) and let M be in noeth Λ.

a) M is of finite length over Λ if and only if $M_{\underline{p}} = 0$ for all nonmaximal prime ideals $\underline{p}$ of R.

b) The natural morphism $\operatorname{Ext}^i_\Lambda(M, N)_{\underline{p}} \longrightarrow \operatorname{Ext}^i_{\Lambda_{\underline{p}}}(M_{\underline{p}}, N_{\underline{p}})$ is an isomorphism for all N in Mod Λ and all $i \geq 0$.

c) $\underline{\operatorname{Hom}}_\Lambda(M, N)_{\underline{p}} \cong \underline{\operatorname{Hom}}_{\Lambda_{\underline{p}}}(M_{\underline{p}}, N_{\underline{p}})$.

d) The following are equivalent:

   i) $M_{\underline{p}}$ is $\Lambda_{\underline{p}}$-projective for all nonmaximal prime ideals $\underline{p}$ of R.

   ii) $\underline{\operatorname{Hom}}_\Lambda(M, N)$ and $\underline{\operatorname{Hom}}_\Lambda(N, M)$ are of finite length over R for all N in noeth Λ.

   iii) $\underline{\operatorname{End}}(M)$ is of finite length over R.

e) $M_{\underline{p}}$ is $\Lambda_{\underline{p}}$-projective for all nonmaximal prime ideals $\underline{p}$ of R if and only if $(\operatorname{Tr}M)_{\underline{p}}$ is $\Lambda_{\underline{p}}$-projective for all nonmaximal prime ideals $\underline{p}$ of R.

<u>Proof</u>:

a) and b) Standard facts in commutative ring theory.

c) Let $P \longrightarrow N \longrightarrow 0$ be exact with P projective. Then $\operatorname{Hom}_\Lambda(M, P) \longrightarrow \operatorname{Hom}_\Lambda(M, N) \longrightarrow \underline{\operatorname{Hom}}_\Lambda(M, N) \longrightarrow 0$ is exact. Hence for each prime ideal $\underline{p}$ or R we have the commutative exact diagram

$$\begin{array}{ccccccc}
\operatorname{Hom}_\Lambda(M, P)_{\underline{p}} & \longrightarrow & \operatorname{Hom}_\Lambda(M, N)_{\underline{p}} & \longrightarrow & \underline{\operatorname{Hom}}_\Lambda(M, N)_{\underline{p}} & \longrightarrow & 0 \\
\| & & \| & & \downarrow & & \\
\operatorname{Hom}_{\Lambda_{\underline{p}}}(M_{\underline{p}}, P_{\underline{p}}) & \longrightarrow & \operatorname{Hom}_{\Lambda_{\underline{p}}}(M_{\underline{p}}, N_{\underline{p}}) & \longrightarrow & \underline{\operatorname{Hom}}_{\Lambda_{\underline{p}}}(M_{\underline{p}}, N_{\underline{p}}) & \longrightarrow & 0
\end{array}.$$

From this it follows that $\underline{\operatorname{Hom}}_\Lambda(M, N)_{\underline{p}} \longrightarrow \underline{\operatorname{Hom}}_{\Lambda_{\underline{p}}}(M_{\underline{p}}, N_{\underline{p}})$ is an isomorphism.

d) i) implies ii). Since $M_{\underline{p}}$ is $\Lambda_{\underline{p}}$-projective for all nonmaximal prime ideals, we have that $\underline{\mathrm{Hom}}_{\Lambda_{\underline{p}}}(N_{\underline{p}}, M_{\underline{p}}) = 0 = \underline{\mathrm{Hom}}_{\Lambda_{\underline{p}}}(M_{\underline{p}}, N_{\underline{p}})$ for all N in noeth Λ and all nonmaximal prime ideals $\underline{p}$ of R. The fact that $\underline{\mathrm{Hom}}_{\Lambda_{\underline{p}}}(N_{\underline{p}}, M_{\underline{p}}) = (\underline{\mathrm{Hom}}_{\Lambda}(N, M))_{\underline{p}}$, implies that $\underline{\mathrm{Hom}}_{\Lambda}(N, M)_{\underline{p}} = 0 = \underline{\mathrm{Hom}}_{\Lambda}(M, N)_{\underline{p}}$ for all nonmaximal prime ideals $\underline{p}$ of R. Hence $\underline{\mathrm{Hom}}_{\Lambda}(N, M)$ and $\underline{\mathrm{Hom}}_{\Lambda}(M, N)$ are of finite length over R by part a).

ii) implies iii). Trivial.

iii) implies i). By a) we know that $\underline{\mathrm{End}}_{\Lambda}(M)$ is of finite length if and only if $\underline{\mathrm{End}}_{\Lambda}(M)_{\underline{p}} = 0$ for all nonmaximal primes $\underline{p}$ of R. Then by c) we have that $\underline{\mathrm{End}}_{\Lambda}(M)$ is of finite length if and only if $\underline{\mathrm{End}}_{\Lambda_{\underline{p}}}(M_{\underline{p}}) = 0$ for all nonmaximal prime ideals $\underline{p}$ of R. Since $\underline{\mathrm{End}}_{\Lambda_{\underline{p}}}(M_{\underline{p}}) = 0$ if and only if $M_{\underline{p}}$ is $\Lambda_{\underline{p}}$-projective, we have our desired result.

e) The duality Tr: $\underline{\text{noeth}}\ \Lambda \longrightarrow \underline{\text{noeth}}\ \Lambda^{op}$ gives an isomorphism Tr: $\underline{\mathrm{End}}_{\Lambda}(M) \longrightarrow \underline{\mathrm{End}}_{\Lambda^{op}}(\mathrm{Tr}M)$ of R-modules. Thus $\underline{\mathrm{End}}_{\Lambda}(M)$ has finite length if and only if $\underline{\mathrm{End}}_{\Lambda^{op}}(\mathrm{Tr}M)$ is of finite length.

Remark: The connection between this lemma and Proposition 7.5 is that the lemma shows that if M is in $L(\Lambda)$, then $(\mathrm{Tr}M)_{\underline{p}}$ is $\Lambda_{\underline{p}}$-projective for all nonmaximal prime ideals of $\underline{p}$, a necessary condition for TrM to be in $J(\Lambda^{op})$.

The proof of the next lemma can be found in [8, Proposition 2.6].

Lemma 7.7: Let Λ be a noetherian ring and M in noeth Λ. There is an exact sequence

$$0 \longrightarrow \mathrm{Ext}^1_\Lambda(\mathrm{Tr}M, \Lambda) \longrightarrow M \longrightarrow \mathrm{Hom}_\Lambda(\mathrm{Hom}_\Lambda(M, \Lambda), \Lambda) \longrightarrow \mathrm{Ext}^2(\mathrm{Tr}M, \Lambda) \to 0$$

where $M \longrightarrow \mathrm{Hom}_\Lambda(\mathrm{Hom}_\Lambda(M, \Lambda), \Lambda)$ is the usual morphism.

We now apply this to prove the following which establishes Proposition 7.5 for $\dim R = 0, 1, 2$.

Lemma 7.8: Let Λ be an R-order with $\dim R = d$ and M a module in $L(\Lambda)$.

a) If $d \geq 1$, then $\mathrm{Ext}^1_\Lambda(\mathrm{Tr}M, \Lambda) = 0$ or, equivalently, $0 \longrightarrow M \longrightarrow \mathrm{Hom}_\Lambda(\mathrm{Hom}_\Lambda(M, \Lambda), \Lambda)$ is exact.

b) If $d \geq 2$, then $\mathrm{Ext}^i_\Lambda(\mathrm{Tr}M, \Lambda) = 0$ for $i = 1, 2$ or, equivalently $M \longrightarrow \mathrm{Hom}_\Lambda(\mathrm{Hom}_\Lambda(M, \Lambda), \Lambda)$ is an isomorphism.

Proof: The usual localization arguments show that it suffices to prove the lemma when R is a local ring. So we assume R is a local ring.

a) Since $(\mathrm{Tr}M)_{\underline{p}}$ is $\Lambda_{\underline{p}}$-projective for all nonmaximal prime ideals $\underline{p}$ of R, we have by Lemma 7.6 that $\mathrm{Ext}^1_\Lambda(\mathrm{Tr}M, \Lambda)_{\underline{p}} = \mathrm{Ext}^1_{\Lambda_{\underline{p}}}((\mathrm{Tr}M)_{\underline{p}}, \Lambda_{\underline{p}}) = (0)$ for all nonmaximal prime ideals $\underline{p}$ of R. Hence $\mathrm{Ext}^1_\Lambda(\mathrm{Tr}M, \Lambda)$ has finite length. By Lemma 7.7 we have that $0 \longrightarrow \mathrm{Ext}^1_\Lambda(\mathrm{Tr}M, \Lambda) \longrightarrow M \longrightarrow \mathrm{Hom}_\Lambda(\mathrm{Hom}_\Lambda(M, \Lambda), \Lambda)$ is exact which means that $\mathrm{Ext}^1_\Lambda(\mathrm{Tr}M, \Lambda) = 0$ since depth $M \geq 1$.

b) Suppose $d \geq 2$. Then by a) and Lemma 7.7 we have the exact sequence

$$(*) \quad 0 \longrightarrow M \longrightarrow \mathrm{Hom}_\Lambda(\mathrm{Hom}_\Lambda(M, \Lambda), \Lambda) \longrightarrow \mathrm{Ext}^2_\Lambda(\mathrm{Tr}M, \Lambda) \longrightarrow 0 .$$

Now the fact that $d \geq 2$ implies that depth M and depth Λ is at least 2. From the fact that depth $\Lambda \geq 2$ it follows trivially that depth $\mathrm{Hom}_\Lambda(\mathrm{Hom}_\Lambda(M, \Lambda), \Lambda) \geq 1$. But $\mathrm{Ext}^2_\Lambda(\mathrm{Tr}M, \Lambda)$ has finite length since $(\mathrm{Tr}M)_{\underline{p}}$ is $\Lambda_{\underline{p}}$-projective for all nonmaximal prime ideals $\underline{p}$ of R. Hence if $\mathrm{Ext}^2_\Lambda(\mathrm{Tr}M, \Lambda) \neq (0)$, then depth $\mathrm{Ext}^2_\Lambda(\mathrm{Tr}M, \Lambda) = 0$. Since depth X is the smallest integer i such that $\mathrm{Ext}^i_R(R/\mathrm{rad}\ R, X) \neq 0$, it follows from the exact sequence (*) that depth M = 1. This contradicts the fact that depth M = dim R ≥ 2. Therefore if dim R ≥ 2 and M is a lattice, then $\mathrm{Ext}^i_\Lambda(\mathrm{Tr}M, \Lambda) = 0$ for $i = 1, 2$.

We now return to the proof of Proposition 7.5.

If $d = 0$ there is nothing to prove. The cases $d = 1, 2$ have been established in the last lemma. We now show that if dim R ≥ 3, then $\mathrm{Ext}^i_\Lambda(\mathrm{Tr}M, \Lambda) = 0$ for $i = 1, 2, 3$.

Let $P_1 \longrightarrow P_0 \longrightarrow M \longrightarrow 0$ be a projective presentation for M in $L(\Lambda)$. Then we have the exact sequence

$$0 \longrightarrow \mathrm{Hom}_\Lambda(M, \Lambda) \longrightarrow \mathrm{Hom}_\Lambda(P_0, \Lambda) \longrightarrow \mathrm{Hom}_\Lambda(P_1, \Lambda) \longrightarrow \mathrm{Tr}M \longrightarrow 0$$

so $\mathrm{Ext}^{i+2}_{\Lambda^{op}}(\mathrm{Tr}M, \Lambda^{op}) \approx \mathrm{Ext}^i_{\Lambda^{op}}(\mathrm{Hom}_\Lambda(M, \Lambda), \Lambda^{op})$ for all $i \geq 1$. Since dim R ≥ 3, we know that $\mathrm{Ext}^i_\Lambda(\mathrm{Tr}M, \Lambda) = 0$ for $i = 1, 2$ (see Lemma 7.8). Hence in order to show that $\mathrm{Ext}^i_\Lambda(\mathrm{Tr}M, \Lambda) = 0$ for $i = 3$, it suffices to prove that $\mathrm{Ext}^1_{\Lambda^{op}}(\mathrm{Hom}_\Lambda(M, \Lambda), \Lambda^{op}) = 0$.

The fact that depth R = depth Λ = depth M ≥ 1 implies that there is an element x in rad R which is simultaneously R, Λ, and M regular.

From the exact sequence $0 \longrightarrow \Lambda^{op} \xrightarrow{x} \Lambda^{op} \longrightarrow \Lambda^{op}/x\Lambda^{op} \longrightarrow 0$ we deduce the exact sequence

$$0 \longrightarrow \mathrm{Hom}_{\Lambda^{op}}(\mathrm{Hom}_\Lambda(M, \Lambda), \Lambda^{op}) \xrightarrow{x} \mathrm{Hom}_{\Lambda^{op}}(\mathrm{Hom}_\Lambda(M, \Lambda), \Lambda^{op}) \longrightarrow$$
$$\mathrm{Hom}_{\Lambda^{op}}(\mathrm{Hom}_\Lambda(M, \Lambda), \Lambda^{op}/x\Lambda^{op}) \longrightarrow \mathrm{Ext}^1_{\Lambda^{op}}(\mathrm{Hom}_\Lambda(M, \Lambda), \Lambda^{op})$$
$$\xrightarrow{x} \mathrm{Ext}^1_{\Lambda^{op}}(\mathrm{Hom}_\Lambda(M, \Lambda), \Lambda^{op}).$$

Since $M_{\underline{p}}$ is $\Lambda_{\underline{p}}$-projective for $\underline{p}$ nonmaximal, we see that $\operatorname{Hom}_\Lambda(M, \Lambda)_{\underline{p}}$ is $\Lambda_{\underline{p}}^{op}$-projective for all $\underline{p}$ nonmaximal. Hence $\operatorname{Ext}^1_{\Lambda^{op}}(\operatorname{Hom}_\Lambda(M, \Lambda), \Lambda^{op})$ is of finite length. Suppose it is not zero. Then multiplication by x has a non-trivial kernel K of finite length. Also since $\dim R \geq 3$, we know that $M = \operatorname{Hom}_{\Lambda^{op}}(\operatorname{Hom}_\Lambda(M, \Lambda), \Lambda^{op})$. So we deduce the exact sequence

$$0 \longrightarrow M/xM \longrightarrow \operatorname{Hom}_{\Lambda^{op}}(\operatorname{Hom}_\Lambda(M, \Lambda), \Lambda^{op}/x\Lambda^{op}) \longrightarrow K \longrightarrow 0$$

whose first 2 terms have depth ≥ 2 while depth $K = 0$. Since this is impossible (use the fact that depth X is the smallest integer i such that $\operatorname{Ext}^i_R(R/\operatorname{rad} R, X) \neq 0$), this shows that $K = 0 = \operatorname{Ext}^1_{\Lambda^{op}}(\operatorname{Hom}_\Lambda(M, \Lambda), \Lambda)$ if $\dim R \geq 3$. Hence we know that Proposition 7.5 is true for $\dim R \leq 3$. We now finish the proof of Proposition 7.5 by induction on $\dim R = d$.

Suppose Proposition 7.5 is true for $\dim R = k \geq 2$ and let $\dim R = k + 1$. Since $\operatorname{Ext}^{i+2}_{\Lambda^{op}}(\operatorname{Tr}M, \Lambda^{op}) \cong \operatorname{Ext}^i_\Lambda(\operatorname{Hom}_\Lambda(M, \Lambda), \Lambda^{op})$ for all $i \geq 1$ it suffices to show that $\operatorname{Ext}^i_\Lambda(\operatorname{Hom}_\Lambda(M, \Lambda), \Lambda^{op}) = 0$ for $i = 1,\ldots,k - 1$. As before we can find an x in $\operatorname{rad} R$ such that x is R, Λ, and M regular at the same time. Hence M/xM is a lattice over the R/xR-order $\Lambda/x\Lambda$ with $\dim R/xR = k$. Thus $\operatorname{Ext}^i_{\Lambda/x\Lambda}(\operatorname{Hom}_{\Lambda/x\Lambda}(M/xM, \Lambda/x\Lambda))$ is zero for $i = 1,\ldots,k - 2$ by the induction hypothesis. Since $k \geq 3$, we know that $\operatorname{Ext}^1_{\Lambda^{op}}(\operatorname{Hom}_\Lambda(M, \Lambda), \Lambda^{op}) = 0$ so $\operatorname{Hom}_\Lambda(M, \Lambda)/x\operatorname{Hom}_\Lambda(M, \Lambda) \approx \operatorname{Hom}_\Lambda(M, \Lambda/x\Lambda) \cong \operatorname{Hom}_{\Lambda/x\Lambda}(M/xM, \Lambda/x\Lambda)$. From this it follows that

$\operatorname{Ext}^i_{\Lambda^{op}/x\Lambda^{op}}(\operatorname{Hom}_{\Lambda/x\Lambda}(M/x\Lambda, \Lambda/x\Lambda), \Lambda^{op}/x\Lambda^{op}) \cong \operatorname{Ext}^i_{\Lambda^{op}}(\operatorname{Hom}_\Lambda(M, \Lambda), \Lambda^{op}/x\Lambda^{op})$ for all i. Thus we have that $\operatorname{Ext}^i_{\Lambda^{op}}(\operatorname{Hom}_\Lambda(M, \Lambda), \Lambda^{op}/x\Lambda^{op}) = 0$ for $i = 1,\ldots,k - 2$.

From the exact sequence $0 \longrightarrow \Lambda^{op} \xrightarrow{x} \Lambda^{op} \longrightarrow \Lambda^{op}/x\Lambda^{op}$ we deduce the long exact sequence

$$\mathrm{Ext}^1_{\Lambda^{op}}(\mathrm{Hom}_\Lambda(M, \Lambda), \Lambda^{op}/x\Lambda^{op}) \longrightarrow \mathrm{Ext}^2_{\Lambda^{op}}(\mathrm{Hom}_\Lambda(M, \Lambda), \Lambda^{op}) \xrightarrow{x}$$
$$\mathrm{Ext}^2_{\Lambda^{op}}(\mathrm{Hom}_\Lambda(M, \Lambda), \Lambda^{op}) \longrightarrow \cdots \longrightarrow \mathrm{Ext}^{k-2}_{\Lambda^{op}}(\mathrm{Hom}_\Lambda(M, \Lambda), \Lambda^{op}/x\Lambda^{op})$$
$$\longrightarrow \mathrm{Ext}^{k-1}_{\Lambda^{op}}(\mathrm{Hom}_\Lambda(M, \Lambda), \Lambda^{op}) \xrightarrow{x} \mathrm{Ext}^{k-1}_{\Lambda^{op}}(\mathrm{Hom}_\Lambda(M, \Lambda), \Lambda^{op}).$$

Since $\mathrm{Ext}^i_{\Lambda^{op}}(\mathrm{Hom}_\Lambda(M, \Lambda), \Lambda^{op}/x\Lambda^{op}) = 0$ for $i = 1,\ldots,k - 2$, it follows that multiplication by x in rad R is a monomorphism on $\mathrm{Ext}^i_{\Lambda^{op}}(\mathrm{Hom}_\Lambda(M, \Lambda), \Lambda^{op})$ for $i = 2,\ldots,k - 1$. Therefore $\mathrm{Ext}^i_{\Lambda^{op}}(\mathrm{Hom}_\Lambda(M, \Lambda), \Lambda^{op}) = 0$ for $i = 2,\ldots,k - 1$ since each $\mathrm{Ext}^i_{\Lambda^{op}}(\mathrm{Hom}_\Lambda(M, \Lambda), \Lambda^{op})$ is finite length for $i \geq 1$ (remember $\mathrm{Hom}_\Lambda(M, \Lambda)_{\underline{p}}$ is $\Lambda_{\underline{p}}$-projective for all nonmaximal $\underline{p}$). Because $\dim R = k + 1 \geq$ with $k \geq 3$, we also know that $\mathrm{Ext}^1_{\Lambda^{op}}(\mathrm{Hom}_\Lambda(M, \Lambda, \Lambda^{op}) = 0$. Thus we have shown that if $\dim R = k + 1$ with $k \geq 3$, then $\mathrm{Ext}^i_{\Lambda^{op}}(\mathrm{Hom}_\Lambda(M, \Lambda), \Lambda^{op}) = 0$ for $i = 1,\ldots,(k + 1) - 2$. This finishes the proof that $\mathrm{Ext}^i_{\Lambda^{op}}(\mathrm{Tr}M, \Lambda^{op}) = 0$ for $i = 1,\ldots,\dim R$ for M in $L(\Lambda)$. Hence we have established Proposition 7.5, that $\mathrm{Tr}M$ is in $J(\Lambda^{op})$ if M is in $L(\Lambda)$.

We now point out several consequences of Proposition 7.5 of particular interest to us in this paper.

Theorem 7.8: Let Λ be an R-order with $\dim R = d$. The functor $\Omega^d\colon \underline{J}(\Lambda) \longrightarrow \underline{L}(\Lambda)$ is an equivalence of categories.

Proof: As noted before, we already know that $\Omega^d\colon \underline{J}(\Lambda) \longrightarrow \underline{L}(\Lambda)$ is fully faithful. Therefore we only have to show that Ω^d is dense.

Let M be in $L(\Lambda)$. Since $\mathrm{Tr}M$ is in $J(\Lambda^{op})$ by Proposition 7.5, we know that $\mathrm{Ext}^{i}_{\Lambda^{op}}(\mathrm{Tr}M, \Lambda^{op}) = 0$ for $i = 1,\ldots,d$. Hence in the terminology of [8, page 59], we know that M is d-torsion free. Thus by [8, Theorem 2.17], there is an exact sequence of Λ-modules

$$(*) \qquad 0 \longrightarrow M \longrightarrow P_1 \longrightarrow \ldots \longrightarrow P_d \longrightarrow U \longrightarrow 0$$

with the P_i projective Λ-modules such that

(**)

$$0 \to \mathrm{Hom}_{\Lambda}(U, \Lambda) \to \mathrm{Hom}_{\Lambda}(P_d, \Lambda) \to \ldots \to \mathrm{Hom}_{\Lambda}(P_1, \Lambda) \to \mathrm{Hom}_{\Lambda}(M, \Lambda) \to 0$$

is exact. We now show that U is in $\underline{J}(\Lambda)$.

Since for each nonmaximal prime ideal $\underline{p}$ in R, we know that $M_{\underline{p}}$ is $\Lambda_{\underline{p}}$-projective, it follows from (*) that $\mathrm{pd}_{\Lambda_{\underline{p}}} U_{\underline{p}} \leq d$ for all nonmaximal prime ideals $\underline{p}$ in R. On the other hand we also have by (**) that $\mathrm{Ext}^{i}_{\Lambda}(U, \Lambda) = 0$ for $i = 1,\ldots,d$ so that $\mathrm{Ext}^{i}_{\Lambda_{\underline{p}}}(U_{\underline{p}}, \Lambda_{\underline{p}}) = 0$ for $i = 1,\ldots,d$ for all nonmaximal $\underline{p}$ in R. Since $\Lambda_{\underline{p}}$ is noetherian, the fact that $\mathrm{pd}_{\Lambda_{\underline{p}}} U_{\underline{p}} \leq d$ and $\mathrm{Ext}^{i}_{\Lambda_{\underline{p}}}(U_{\underline{p}}, \Lambda_{\underline{p}}) = 0$ for $i = 1,\ldots,d$, for all nonmaximal prime ideals $\underline{p}$ of R implies that $U_{\underline{p}}$ is $\Lambda_{\underline{p}}$-projective for all nonmaximal prime ideals $\underline{p}$ of R. Combining this with the fact that $\mathrm{Ext}^{i}_{\Lambda}(U, \Lambda) = 0$ for $i = 1,\ldots,d$ we have that U is in $\underline{J}(\Lambda)$.

Since $0 \longrightarrow M \longrightarrow P_1 \longrightarrow \ldots \longrightarrow P_d \longrightarrow U \longrightarrow 0$ is exact with the P_i projective Λ-modules for $i = 1,\ldots,d$, we know that $\Omega^{d}(U) \approx M$ in $\underline{L}(\Lambda)$. Therefore we have extablished that $M \approx \Omega^{d}(U)$ in $\underline{L}(\Lambda)$ with U in $\underline{J}(\Lambda)$. This completes the proof that $\Omega^{d}: \underline{J}(\Lambda) \longrightarrow \underline{L}(\Lambda)$ is an equivalence of categories.

Next we strengthen Proposition 7.5 as follows.

Theorem 7.9: The duality Tr: $\underline{\text{noeth}}\ \Lambda \longrightarrow \underline{\text{noeth}}\ \Lambda^{op}$ induces a duality $\mathrm{Tr}: \underline{L}(\Lambda) \longrightarrow \underline{J}(\Lambda^{op})$.

Proof: By Proposition 7.5, we know that if M is in $L(\Lambda)$, then TrM is in $J(\Lambda^{op})$. Hence the duality Tr: $\underline{\text{noeth}}\ \Lambda \longrightarrow \underline{\text{noeth}}\ \Lambda^{op}$ induces a contravariant functor $Tr: \underline{L}(\Lambda) \longrightarrow \underline{J}(\Lambda^{op})$ which is fully faithful. Hence to show that this fully faithful functor $Tr: \underline{L}(\Lambda) \longrightarrow \underline{J}(\Lambda^{op})$ is a duality it suffices to show that it is dense. Since $TrTr$: $\underline{\text{noeth}}\ \Lambda \longrightarrow \underline{\text{noeth}}\ \Lambda$ is isomorphic to the identity, this amounts to showing that if M is in $J(\Lambda^{op})$, then TrM is in $L(\Lambda)$.

Suppose M is in $J(\Lambda^{op})$. Then $M_{\underline{p}}$ is $\Lambda_{\underline{p}}^{op}$-projective for all non-maximal prime ideals $\underline{p}$ of R, or equivalently, $\underline{\text{End}}\ M$ is an artin ring. Since Tr: $\underline{\text{End}}\ M \longrightarrow (\underline{\text{End}}\ TrM)^{op}$ is a ring isomorphism, we have that $\underline{\text{End}}\ TrM$ is an artin ring and so $(TrM)_{\underline{p}}$ is $\Lambda_{\underline{p}}$-projective for all non-maximal prime ideals $\underline{p}$ of R. Therefore to show that TrM is in $L(\Lambda)$, it suffices to show that TrM is Cohen-Macauley.

Let $d = \dim R$ and let

$\cdots \longrightarrow P_{d+1} \longrightarrow P_d \longrightarrow \cdots \longrightarrow P_0 \longrightarrow M \longrightarrow 0$ be a projective Λ^{op}-resolution of M. Since M is in $J(\Lambda^{op})$ we know that $\text{Ext}^i_{\Lambda^{op}}(M, \Lambda^{op}) = 0$ for $i = 1,\ldots,d$. Therefore

$$0 \to \text{Hom}_{\Lambda^{op}}(M, \Lambda^{op}) \to \text{Hom}_{\Lambda^{op}}(P_0, \Lambda^{op}) \to \cdots \to \text{Hom}_{\Lambda^{op}}(P_{d+1}, \Lambda^{op})$$

os an exact sequence of Λ-modules with the $\text{Hom}_{\Lambda^{op}}(P_i, \Lambda^{op})$ projective-modules. Thus we have the exact sequence

$$0 \longrightarrow TrM \longrightarrow \text{Hom}_{\Lambda^{op}}(P_2, \Lambda^{op}) \longrightarrow \cdots \longrightarrow \text{Hom}_{\Lambda^{op}}(P_{d+1}, \Lambda^{op})$$

Since the $\text{Hom}_{\Lambda^{op}}(P_i, \Lambda^{op})$ are Cohen-Macauley modules for $i = 2,\ldots,d + 1$ because they are projective Λ-modules, it follows that TrM is a Cohen-Macauley module (see Proposition 7.2).

As a consequence of Theorems 7.8 and 7.9 we have that the composition of functors $\underline{L}(\Lambda) \xrightarrow{Tr} \underline{J}(\Lambda^{op}) \xrightarrow{\Omega^d} \underline{L}(\Lambda^{op})$ is a duality since Tr is a duality and Ω^d is an equivalence of categories. We will denote this duality by $Tr_L: \underline{L}(\Lambda) \longrightarrow \underline{L}(\Lambda^{op})$. As we shall see, the duality Tr_L plays the role for lattices that the duality Tr plays for noetherian modules.

Using the results of this section, it is a matter of carrying out obvious calculations to verify the following.

<u>Proposition 7.10</u>: The contravariant functors $Tr_L: \underline{L}(\Lambda) \longrightarrow \underline{L}(\Lambda^{op})$ and $Tr_L: \underline{L}(\Lambda^{op}) \longrightarrow \underline{L}(\Lambda)$ are inverse dualities.

As in the case of the ordinary transpose, the functors Tr_L depend on choices of projective resolutions of modules in noeth Λ. So while we always have that $Tr_L(Tr_L X) \simeq X$ in $\underline{L}(\Lambda)$ for each X in $L(\Lambda)$, it need not be the case that $Tr_L Tr_L X \approx X$ in $L(\Lambda)$ for all X in $L(\Lambda)$ for arbitrary choices of transpose functors $\underline{L}(\Lambda) \longrightarrow \underline{L}(\Lambda^{op})$ and $\underline{L}(\Lambda^{op}) \longrightarrow \underline{L}(\Lambda)$. However there are transpose functors $Tr_L': \underline{L}(\Lambda) \longrightarrow \underline{L}(\Lambda^{op})$ and $Tr_L'': \underline{L}(\Lambda^{op}) \longrightarrow \underline{L}(\Lambda)$ such that $Tr_L''Tr_L'(X) \approx X$ for all X in $L(\Lambda)$. Throughout this section we denote by $Tr_L: \underline{L}(\Lambda) \longrightarrow \underline{L}(\Lambda^{op})$ and $Tr_L: \underline{L}(\Lambda^{op}) \longrightarrow \underline{L}(\Lambda)$ a fixed pair of transpose functors with the property $Tr_L Tr_L(X) \cong X$ for all X in $L(\Lambda)$.

We end this section by pointing out the following easily verified facts concerning the categories $L(\Lambda)$, $J(\Lambda)$ and the duality $Tr_L: \underline{L}(\Lambda) \longrightarrow \underline{L}(\Lambda^{op})$ for low values of $d = \dim R$.

Case 1: $d = 0$. Then noeth $\Lambda = L(\Lambda) = J(\Lambda)$ and

Tr_L: $\underline{L}(\Lambda) \longrightarrow \underline{L}(\Lambda^{op})$ is the ordinary transpose Tr: noeth $\Lambda \rightarrow$ noeth Λ^{op}.

Case 2: $d = 1$. Let M be in $L(\Lambda)$ and

$P_1 \longrightarrow P_0 \longrightarrow M \longrightarrow 0$ is a projective presentation of M. If

$0 \longrightarrow K \longrightarrow \mathrm{Hom}_\Lambda(P_1, \Lambda) \longrightarrow \mathrm{Tr}M \longrightarrow 0$ is exact, then

$K \approx \mathrm{Tr}_L(M)$ in $\underline{L}(\Lambda^{op})$.

Case 3: $d = 2$. $\mathrm{Tr}_L(M) \cong \mathrm{Hom}_\Lambda(M, \Lambda)$ in $\underline{L}(\Lambda^{op})$ for all M in $L(\Lambda)$.

§8. Lattice Determined Morphisms of Lattices.

Throughout this section we assume that Λ is an R-order with $d = \dim R$. Our aim is to study morphisms in the additive category $L(\Lambda)$ of Λ-lattices. In particular, we derive results for $L(\Lambda)$ analogous to those derived for Mod Λ in Section 3 and noeth Λ in Section 6. In order to accomplish this we need some preliminary definitions and results. We begin with some analogues of definitions and results which are well known for classical orders.

We say that a sequence $\longrightarrow M_1 \xrightarrow{f_1} M_2 \xrightarrow{f_2} M_3 \longrightarrow M_4 \longrightarrow \cdots$ of morphisms in $L(\Lambda)$ is exact if it is exact as a sequence of Λ-modules and each nonzero $\mathrm{Im}\, f_i$ is a lattice. We say that a lattice C is projective in $L(\Lambda)$ if every exact sequence $0 \longrightarrow A \longrightarrow B \longrightarrow C \longrightarrow 0$ in $L(\Lambda)$ splits. We say that a lattice A is injective in $L(\Lambda)$ if every exact sequence $0 \longrightarrow A \longrightarrow B \longrightarrow C \longrightarrow 0$ in $L(\Lambda)$ splits.

Proposition 8.1: Let X be in $L(\Lambda)$.

a) X is projective in $L(\Lambda)$ if and only if X is a projective Λ-module.

b) X is injective in $L(\Lambda)$ if and only if there is a projective lattice P in $L(\Lambda^{op})$ such that $X \approx P^*$ where

$P^* = \mathrm{Hom}_R(P, R)$.

Proof:

a) Suppose X is projective in $L(\Lambda)$. Then there is an exact sequence $0 \longrightarrow K \longrightarrow P \longrightarrow X \longrightarrow 0$ in noeth Λ with P projective. Since X and P are in $L(\Lambda)$, it follows that K is in $L(\Lambda)$ (see Proposition 7.3). Thus $0 \longrightarrow K \longrightarrow P \longrightarrow X \longrightarrow 0$ is an exact sequence in $L(\Lambda)$ and therefore splits, which means that X is a projective Λ-module. The fact that X being a projective Λ-module implies that X is projective in $L(\Lambda)$ is trivial.

b) Suppose X is injective in $L(\Lambda)$. Let $0 \longrightarrow A \longrightarrow B \longrightarrow X^* \longrightarrow 0$ be an exact sequence in $L(\Lambda^{op})$. Then $0 \longrightarrow X^{**} \longrightarrow B^* \longrightarrow A^* \longrightarrow 0$ is exact in $L(\Lambda)$ (see Proposition 7.3). Since, by Proposition 7.3, we know that $X \approx X^{**}$, the fact that X is injective in $L(\Lambda)$ implies that the sequence $0 \longrightarrow X^{**} \longrightarrow B^* \longrightarrow A^* \longrightarrow 0$ splits. Using the fact that the functor $\mathrm{Hom}_R(\ ,R)\colon L(\Lambda) \longrightarrow L(\Lambda^{op})$ is a duality, (see Proposition 7.3) it follows that the sequence $0 \longrightarrow A \longrightarrow B \longrightarrow X^* \longrightarrow 0$ splits, which implies that X^* is a projective lattice. Thus $X \approx (X^*)^*$ with X^* a projective Λ^{op}-lattice, which gives our desired result. A similar argument shows that if $X \approx P^*$ with P a projective Λ^{op}-lattice, then X is injective in $L(\Lambda)$. Thus b) is established.

As a consequence of this result we have the following.

Proposition 8.2: Let X be in $L(\Lambda)$.

a) There is an injective resolution of X in $L(\Lambda)$, i.e. there is an exact sequence in $L(\Lambda)$

$$0 \longrightarrow X \longrightarrow I_0 \longrightarrow I_1 \longrightarrow \dots \longrightarrow I_n \longrightarrow \dots$$

with the I_j injective lattices for all $j \geq 0$.

b) If A is a lattice and

$0 \longrightarrow X \longrightarrow I_0 \longrightarrow I_1 \longrightarrow \dots \longrightarrow I_n \longrightarrow \dots$ is an injective resolution of X in $L(\Lambda)$, then

$H^i(\mathrm{Hom}_\Lambda(A, I)) \cong \mathrm{Ext}^i_\Lambda(A, X)$ for all $i \geq 0$ where I is the complex $0 \longrightarrow I_0 \longrightarrow I_1 \longrightarrow \dots \longrightarrow I_n \longrightarrow \dots$

Proof:

a) Let $\dots \longrightarrow P_n \longrightarrow \dots \longrightarrow P_1 \longrightarrow P_0 \longrightarrow X^* \longrightarrow 0$ be a projective resolution for X^* in $L(\Lambda^{op})$. Then

$\dots \longrightarrow P_n \longrightarrow \dots \longrightarrow P_1 \longrightarrow P_0 \longrightarrow X^* \longrightarrow 0$ is an exact sequence in $L(\Lambda^{op})$ and so

$0 \longrightarrow X^{**} \longrightarrow P_0^* \longrightarrow P_1^* \longrightarrow \dots \longrightarrow P_n^* \longrightarrow \dots$ is an exact sequence in $L(\Lambda)$ (see Proposition 7.3). Since, by Proposition 8.1, each of the P_i^* is an injective Λ-lattice, part a) has been established.

b) We first observe that if A and C are in $L(\Lambda)$ and

$0 \longrightarrow A \longrightarrow B \longrightarrow C \longrightarrow 0$ is an exact sequence in noeth Λ, then B is in $L(\Lambda)$ and so the sequence

$0 \longrightarrow A \longrightarrow B \longrightarrow C \longrightarrow 0$ is exact in $L(\Lambda)$. Part b) follows easily from this observation using standard arguments from homological algebra.

Let $I(L(\Lambda))$ be the full subcategory of $L(\Lambda)$ whose objects are the injective Λ-lattices. For each pair of Λ-lattices A and C define

$I(A, C)$ to be the R-submodule of $\mathrm{Hom}_\Lambda(A, C)$ consisting of all morphisms $A \longrightarrow C$ which can be written as a composition $A \longrightarrow I \longrightarrow C$ with I in $I(L(\Lambda))$. Using the duality $\mathrm{Hom}_R(\ ,R)\colon L(\Lambda) \longrightarrow L(\Lambda^{op})$ it is not difficult to see that if $0 \longrightarrow A \longrightarrow I \longrightarrow L \longrightarrow 0$ is exact in $L(\Lambda)$ with I an injective Λ-lattice, then

$\mathrm{Im}(\mathrm{Hom}_\Lambda(I, C) \longrightarrow \mathrm{Hom}_\Lambda(A, C)) = I(A, C)$.

Define the category $\overline{L}(\Lambda)$ to be the preadditive category whose objects are the Λ-lattices and whose morphisms are

$\overline{\mathrm{Hom}}_\Lambda(A, C) = \mathrm{Hom}_\Lambda(A, C)/I(A, C)$ for all A and C in $L(\Lambda)$. As an immediate consequence of the definitions involved we have the following.

<u>Proposition 8.3</u>:

a) The duality $\mathrm{Hom}_R(\ ,R)\colon L(\Lambda) \longrightarrow L(\Lambda^{op})$ induces a duality $\underline{L}(\Lambda) \longrightarrow \overline{L}(\Lambda^{op})$.

b) the composition $\underline{L}(\Lambda) \xrightarrow{Tr_L} \underline{L}(\Lambda^{op}) \xrightarrow{\mathrm{Hom}_R(\ ,R)} \overline{L}(\Lambda)$ is an equivalence of categories with inverse the composition $\overline{L}(\Lambda) \xrightarrow{\mathrm{Hom}_R(\ ,R)} \underline{L}(\Lambda^{op}) \xrightarrow{Tr_L} \underline{L}(\Lambda)$.

The following is another useful preliminary result.

<u>Lemma 8.4</u>: Let $0 \longrightarrow R \longrightarrow I_0 \longrightarrow \dots \longrightarrow I_d \longrightarrow 0$ be a minimal injective resolution of R over R and let M be in $L(\Lambda)$.

a) $\mathrm{Hom}_R(M, I_j)$ is an injective Λ^{op}-module for $j = 0,\dots,d - 1$.

b) $0 \longrightarrow \mathrm{Hom}_R(M, R) \longrightarrow \mathrm{Hom}_R(M, I_0) \longrightarrow \dots \longrightarrow \mathrm{Hom}_R(M, I_d) \to 0$ is an exact sequence of Λ^{op}-modules with the $\mathrm{Hom}_R(M, I_j)$ injective for $j = 0,\dots,d - 1$.

c) There is an isomorphism of functors $\mathrm{Ext}^1_{\Lambda^{op}}(\ ,\mathrm{Hom}_R(M, I_d)) \cong \mathrm{Ext}^{d+1}_{\Lambda^{op}}(\ ,\mathrm{Hom}_R(M, R))$ which is functorial in M in $L(\Lambda)$.

Proof:

a) Suppose $0 \leq j \leq d - 1$. If $d = 0$, there is nothing to prove so we may as well assume that $d > 0$. Since I_j is an injective R-module, we know that for each Λ^{op}-module A we have an isomorphism $\operatorname{Ext}^1_{\Lambda^{op}}(A, \operatorname{Hom}_R(M, I_j)) \cong \operatorname{Hom}_R(\operatorname{Tor}_1^{\Lambda}(A, M), I_j)$. Because M is in $L(\Lambda)$, we know that $M_{\underline{p}}$ is $\Lambda_{\underline{p}}$-projective for all nonmaximal prime ideals $\underline{p}$ in R. Hence $\operatorname{Tor}^{\Lambda}_{\Lambda}(A, M)_{\underline{p}} = \operatorname{Tor}_1^{\Lambda_{\underline{p}}}(A_{\underline{p}}, M_{\underline{p}}) = 0$ for all nonmaximal prime ideals $\underline{p}$ in R. Thus $\operatorname{Tor}_1^{\Lambda}(A, M)$ is an R-module which is locally of finite length, i.e. every finitely generated submodule of $\operatorname{Tor}_1^{\Lambda}(A, M)$ is of finite length. Since R is a Gorenstein ring and $0 \leq j \leq d - 1$, we know that $\operatorname{Hom}_R(X, I_j) = 0$ for all X of finite length. Therefore $\operatorname{Hom}_R(\operatorname{Tor}_1^{\Lambda}(A, M), I_j) = 0 = \operatorname{Ext}^1_{\Lambda^{op}}(A, \operatorname{Hom}_R(M, I_j))$. Since this is true for all Λ^{op}-modules A, we have that $\operatorname{Hom}_R(M, I_j)$ is Λ^{op}-injective for $0 \leq j < d - 1$.

b) The fact that M is in $L(\Lambda)$ implies that M is a Cohen-Macauley module over R. Therefore we know that $\operatorname{Ext}^i_R(M, R) = 0$ for $i > 0$ since R is a Gorenstein ring. Hence the sequence of Λ^{op}-modules $0 \longrightarrow \operatorname{Hom}_R(M, R) \longrightarrow \operatorname{Hom}_R(M, I_0) \longrightarrow \dots \longrightarrow \operatorname{Hom}_R(M, I_d) \rightarrow 0$ is exact. The fact that $\operatorname{Hom}_R(M, I_j)$ is Λ^{op}-injective for $j = 0,\dots,d - 1$ was shown in a).

c) Suppose M and N are in $L(\Lambda)$ and $f\colon M \longrightarrow N$ is a morphism. We then have the commutative diagram of Λ^{op}-modules

$$\begin{array}{ccccccccc} 0 \longrightarrow & \mathrm{Hom}_R(N, R) & \longrightarrow & \mathrm{Hom}_R(N, I_0) & \longrightarrow & \cdots & \longrightarrow & \mathrm{Hom}_R(N, I_d) & \longrightarrow 0 \\ & \downarrow \mathrm{Hom}_R(f, R) & & \downarrow \mathrm{Hom}_R(f, I_0) & & & & \downarrow \mathrm{Hom}_R(f, I_d) & \\ 0 \longrightarrow & \mathrm{Hom}_R(M, R) & \longrightarrow & \mathrm{Hom}_R(M, I_0) & \longrightarrow & \cdots & \longrightarrow & \mathrm{Hom}_R(M, I_d) & \longrightarrow 0. \end{array}$$

By b) we know that this diagram is exact and the $\mathrm{Hom}_R(M, I_j)$ are injective Λ^{op}-modules for $j = 0,\ldots,d - 1$. Therefore by standard dimension shift arguments we obtain the following exact commutative diagram

$$\begin{array}{ccccc} 0 \longrightarrow & \mathrm{Ext}^1_{\Lambda^{op}}(\ , \mathrm{Hom}_R(N, I_d)) & \longrightarrow & \mathrm{Ext}^{d+1}_{\Lambda^{op}}(\ , \mathrm{Hom}_R(N, R)) & \longrightarrow 0 \\ & \downarrow \mathrm{Hom}_R(f, I_d) & & \downarrow \mathrm{Hom}_R(f, R) & \\ 0 \longrightarrow & \mathrm{Ext}^1_{\Lambda^{op}}(\ , \mathrm{Hom}_R(M, I_d)) & \longrightarrow & \mathrm{Ext}^{d+1}_{\Lambda^{op}}(\ , \mathrm{Hom}_R(M, R)) & \longrightarrow 0 \end{array}$$

which is our desired result.

Next we recall from Section 7 that the functor Ω^d: $\underline{\text{noeth}}\ \Lambda \longrightarrow \underline{\text{noeth}}\ \Lambda$ induces an equivalence of categories $\Omega^d: \underline{J}(\Lambda) \longrightarrow \underline{L}(\Lambda)$ where $J(\Lambda)$ is the full subcategory of noeth Λ consisting of those M in noeth Λ satisfying $M_{\underline{p}}$ is $\Lambda_{\underline{p}}$-projective for nonmaximal prime ideals $\underline{p}$ in R and $\mathrm{Ext}^i_\Lambda(M, \Lambda) = 0$ for $i = 1,\ldots,d$. We also agreed to denote the inverse of $\Omega^d: \underline{J}(\Lambda) \longrightarrow \underline{L}(\Lambda)$ by $\Omega^{-d}: \underline{L}(\Lambda) \longrightarrow \underline{J}(\Lambda)$.

Let C be in $\underline{L}(\Lambda)$. Since $\Omega(\Omega^{-d}\ C) \cong C$ in $\underline{L}(\Lambda)$, we have isomorphisms $\mathrm{Ext}^1_\Lambda(C,\ ) \cong \mathrm{Ext}^{d+1}(\Omega^{-d}\ C,\ )$ which are functorial in C in $\underline{L}(\Lambda)$. Further let X be in $L(\Lambda)$. Then combining the various isomorphisms given above we have

$$\mathrm{Ext}^1_\Lambda(C, X) \cong \mathrm{Ext}^1_\Lambda(C, \mathrm{Hom}_R(X^*, R)) \cong$$
$$\mathrm{Ext}^{d+1}_\Lambda(C, \mathrm{Hom}_R(X^*, I_d)) \cong \mathrm{Ext}^1_\Lambda(\Omega^{-d}C, \mathrm{Hom}_R(X^*, I_d))$$

which are functorial in C and X in $\underline{L}(\Lambda)$. By Proposition 3.3 we have the isomorphism $\mathrm{Ext}^1_\Lambda(\Omega^{-d}C, \mathrm{Hom}_R(X^*, I_d)) \cong \mathrm{Hom}_R(\underline{\mathrm{Hom}}_\Lambda(\mathrm{Tr}(X), \Omega^{-d}C), I_d)$ which is functorial in C and X in $L(\Lambda)$. Since X^* is in $L(\Lambda^{op})$, we have that $\mathrm{Tr}(X^*)$ is in $J(\Lambda)$ (see Proposition 7.5) as is $\Omega^{-d}C$. Hence the equivalence of categories $\Omega^d\colon \underline{J}(\Lambda) \longrightarrow \underline{L}(\Lambda)$ gives an isomorphism $\Omega^d\colon \underline{\mathrm{Hom}}_\Lambda(\mathrm{Tr}X^*, \Omega^{-d}C) \longrightarrow \underline{\mathrm{Hom}}_\Lambda(\Omega^d\mathrm{Tr}(X^*), C)$ functorial in X and C in $L(\Lambda)$. But by definition $\Omega^d\mathrm{Tr}(X^*) = \mathrm{Tr}_L(X^*)$. Therefore we have the isomorphism $\Omega^d\colon \underline{\mathrm{Hom}}_\Lambda(\mathrm{Tr}X^*, \Omega^{-d}C) \longrightarrow \underline{\mathrm{Hom}}_\Lambda(\mathrm{Tr}_LX^*, C)$ which is functorial in C and X in $L(\Lambda)$. Hence we have the isomorphism

$$\mathrm{Ext}^1_\Lambda(\Omega^{-d}C, \mathrm{Hom}_R(X^*, I_d)) \cong \mathrm{Hom}_R(\underline{\mathrm{Hom}}_\Lambda(\mathrm{Tr}_LX^*, C), I_d)$$

which is functorial in C and X in $L(\Lambda)$. Therefore combining these isomorphisms we obtain the following result.

<u>Proposition 8.7</u>: Let C and X be in $L(\Lambda)$. Then we have an isomorphism

$$\mathrm{Ext}^1_\Lambda(C, X) \cong \mathrm{Hom}_R(\underline{\mathrm{Hom}}_\Lambda(\mathrm{Tr}_LX^*, C), I_d)\ ,$$

functorial in C and X in $L(\Lambda)$.

Using the equivalence of categories $\mathrm{Hom}_R(\ ,R)\mathrm{Tr}_L\colon \underline{L}(\Lambda) \longrightarrow \overline{L}(\Lambda)$, this result can be restated as follows.

<u>Proposition 8.8</u>: Let C and X be in $L(\Lambda)$. Then there is an isomorphism $\mathrm{Ext}^1_\Lambda(C, \mathrm{Hom}_R(\mathrm{Tr}_LX, R)) \cong \mathrm{Hom}_R(\underline{\mathrm{Hom}}_\Lambda(X, C), I_d)$ which is functorial in C and X in $L(\Lambda)$.

The following result is an easy consequence of Proposition 8.8.

Proposition 8.9: Let C and X be in $L(\Lambda)$ and n a positive integer. Then there is an isomorphism

$$\operatorname{Ext}^1_\Lambda(C, \operatorname{Hom}_R(\operatorname{Tr}_L X, R^n)) \cong \operatorname{Hom}_R(\underline{\operatorname{Hom}}_\Lambda(X, C), I^n_d)$$

which is functorial in C and X in $L(\Lambda)$.

We now investigate some of the connections between an exact sequence $x\colon 0 \longrightarrow \operatorname{Hom}_R(\operatorname{Tr}_L X, R^n) \xrightarrow{g} B \xrightarrow{f} C \longrightarrow 0$ in $L(\Lambda)$ and the R-morphism $\upsilon\colon \underline{\operatorname{Hom}}_\Lambda(X, C) \longrightarrow I^n_d$ corresponding to x under the isomorphism

$$\operatorname{Ext}^1_\Lambda(C, \operatorname{Hom}_R(\operatorname{Tr}_L X, R^n)) \cong \operatorname{Hom}_R(\underline{\operatorname{Hom}}_\Lambda(X, C), I^n_d).$$

Under this hypothesis we have the following consequence of Theorem 3.6.

Theorem 8.10: Let H be the R-submodule of (X, C) containing $P(X, C)$ such that $H/P(X, C) = \operatorname{Ker} \upsilon$.

a) A morphism $h\colon L \longrightarrow C$ in $L(\Lambda)$ can be written as $ft = h$ for some morphism $t\colon L \longrightarrow B$ if and only if $\operatorname{Im}(\ ,h)(X) \subset H$.

b) f is right X-determined in $L(\Lambda)$ and $\operatorname{Im}(\ ,f)(X)$ is the maximal $(\operatorname{End} X)^{op}$-submodule of H.

c) $\operatorname{Im}(\ ,f)(X) = H$ if and only if H is an $(\operatorname{End} X)^{op}$-submodule of (X, C).

Proof: All one has to do is check that the hypothesis of Theorem 3.6 is satisfied in this case.

We now restate Proposition 8.9 and Theorem 8.10 in a form particularly useful for our immediate purposes.

Theorem 8.11: Let X and C be in $L(\Lambda)$ and n a positive integer and let $\Sigma = (\text{End } X)^{op}$.

a) There is an isomorphism

$\text{Ext}^1_\Lambda(C, \text{Hom}_R(\text{Tr}_L X, R^n)) \cong \text{Hom}_\Sigma((X, C), \text{Hom}_R(\Sigma^{op}, I^n_d))$ which is functorial in C in $L(\Lambda)$.

b) Let $\upsilon: \underline{\text{Hom}}_\Lambda(X, C) \longrightarrow \text{Hom}_R(\Sigma^{op}, I^n_d)$ be the Σ-morphism corresponding to an extension $0 \longrightarrow \text{Hom}_R(\text{Tr}_L X, R^n) \xrightarrow{g} B \xrightarrow{f} C \rightarrow 0$ under the isomorphism given in a) and let H be the Σ-submodule of (X, C) containing $P(X, C)$ such that $H/P(X, C) = \text{Ker } \upsilon$.

i) f is right X-determined.

ii) $\text{Im}(\ ,f)(X) = H$.

Proof:

a) We know that there is an isomorphism

$\text{Hom}_R(\underline{\text{Hom}}_\Lambda(X, C), I^n_d) \approx \text{Hom}_\Sigma(\underline{\text{Hom}}_\Lambda(X, C), \text{Hom}_R(\Sigma^{op}, I^n_d)$ which is functorial in C in $L(\Lambda)$. Part a) now follows trivially from Proposition 8.9.

b) Follows from a) and Theorem 3.6.

As an easy consequence of Theorem 8.11 we have the following.

Proposition 8.12: Let $0 \longrightarrow A \xrightarrow{g} B \xrightarrow{f} C \longrightarrow 0$ be an exact sequence in $L(\Lambda)$. Then f is right $\text{Tr}_L A^*$-determined in $L(\Lambda)$.

Proof: We know that $A \approx \text{Hom}_R(\text{Tr}_L(\text{Tr}_L A^*), R)$. Using this observation, the proposition is a trivial consequence of Theorem 8.11.

In order to give further applications of Theorem 8.11 it is convenient to have the following facts.

Lemma 8.13: Let X and C be in $L(\Lambda)$ and let $\Sigma = (\mathrm{End}\ X)^{op}$.

a) $\underline{\mathrm{Hom}}_\Lambda(X, C)$ is a Σ-module of finite length since it is an R-module of finite length.

b) Suppose H is a Σ-submodule of (X, C) containing $P(X, C)$. Then $(X, C)/H$ is a Σ-module of finite length. Hence the socle of $(X, C)/H$ is of the form $\coprod_{i=1}^{k} S_i^{n_i}$ with the S_i nonisomorphic Σ-modules and the n_i positive integers. Let $n = \max\{n_1, \ldots, n_k\}$. Then there is a Σ-morphism $\upsilon\colon \underline{\mathrm{Hom}}_\Lambda(X, C) \longrightarrow \mathrm{Hom}_R(\Sigma^{op}, I_d^n)$ with $H/P(X, C) = \mathrm{Ker}\ \upsilon$.

Proof:

a) Since X and C are in noeth Λ and $X_{\underline{p}}$ is $\Lambda_{\underline{p}}$-projective for all nonmaximal prime ideals $\underline{p}$ we have by Lemma 7.6 that $\underline{\mathrm{Hom}}_\Lambda(X, C)$ is an R-module of finite length and so a Σ-module of finite length.

b) We know that $\mathrm{Hom}_R(\Sigma^{op}, I_d)$ is an injective cogenerator for Σ and so each simple Σ-module is contained in $\mathrm{Hom}_R(\Sigma^{op}, I_d)$. From this observation it follows trivially that $\coprod S_i^{n_i}$ is contained in $\mathrm{Hom}_R(\Sigma^{op}, I_d^n) = \mathrm{Hom}_R(\Sigma^{op}, I_d)^n$. Thus there is a Σ-monomorphism $(X, C)/H \longrightarrow \mathrm{Hom}_R(\Sigma^{op}, I_d^n)$ and hence the composition $\underline{\mathrm{Hom}}_\Lambda(X, C) \longrightarrow (X, C)/H \longrightarrow \mathrm{Hom}_R(\Sigma^{op}, I_d^n)$ is our desired Σ-morphism.

As a consequence of Theorem 8.11 and Lemma 8.13 we obtain the following analogue of Theorem 3.7.

Theorem 8.14: Let C and X be in $L(\Lambda)$ and let $\Sigma = (\text{End } X)^{op}$. Suppose H is a Σ-submodule of (X, C) containing P(X, C) and suppose $\coprod_{i=1}^{k} S_i^{n_i}$ is isomorphic to the socle of (X, C)/H where the S_i are non-isomorphic simple Σ-modules and the n_i are positive integers. Let $n = \max\{n_1, \ldots, n_k\}$. Then there is an exact sequence $x: 0 \longrightarrow \text{Hom}_R(\text{Tr}_L X, R^n) \xrightarrow{g} B \xrightarrow{f} C \longrightarrow 0$ satisfying:

a) Im(,f)(X) = H.

b) f is right X-determined in $L(\Lambda)$.

So far we have been concentrating on how epimorphisms in $L(\Lambda)$ are right determined in $L(\Lambda)$. The duality $\text{Hom}_R(\, ,R): L(\Lambda) \longrightarrow L(\Lambda^{op})$ enables us to translate these results to results about how monomorphisms in $L(\Lambda)$ are left determined in $L(\Lambda)$. The proofs, which are not given, follow easily from the remarks concerning dual categories at the end of Section 2. The reader should also consult the analogous results in Section 6 for noetherian algebras over complete local rings.

Proposition 8.15: Let $0 \longrightarrow A \xrightarrow{g} B \xrightarrow{f} C \longrightarrow 0$ be an exact sequence in $L(\Lambda)$. Then g is left $(\text{Tr}_L C)^*$-determined.

Proof: Dual of Proposition 8.13.

As the dual of Theorem 8.14 we obtain the following.

Theorem 8.16: Let A and X be in $L(\Lambda)$. Let $\Sigma = \text{End } X$ and suppose H is a Σ-submodule of (A, X) containing I(A, X). Then (A, X)/I is a Σ-module of finite length whose socle is of the form $\coprod_{i=1}^{k} S_i^{n_i}$ with the S_i nonisomorphic Σ-modules and the n_i positive integers. Let

$n = \max\{n_1, \ldots, n_k\}$. Then there is an exact sequence $0 \longrightarrow A \xrightarrow{g} B \xrightarrow{f} \mathrm{Tr}_L(\mathrm{Hom}_R(X, R^n)) \longrightarrow 0$ in $L(\Lambda)$ such that

a) g is left X-determined.

b) $\mathrm{Im}(g, \ )(X) = H$.

Finally we point out the following version of Proposition 8.7 which could have been used as the starting point for deriving the last two results.

Proposition 8.17: Let X and A be in $L(\Lambda)$. Then there is an isomorphism $\mathrm{Ext}^1_\Lambda(X, A) \cong \mathrm{Hom}_R(\overline{\mathrm{Hom}}_\Lambda(A, (\mathrm{Tr}_L X)^*), I_d)$ which is functorial in X and A.

Proof: The equivalence of categories $\underline{L}(\Lambda) \longrightarrow \overline{L}(\Lambda)$ given by $X \longmapsto \mathrm{Hom}_R(\mathrm{Tr}_L X, R)$ gives isomorphisms $\underline{\mathrm{Hom}}_\Lambda(U, V) \longrightarrow \overline{\mathrm{Hom}}_\Lambda(\mathrm{Hom}_R(\mathrm{Tr}_L U, R), \mathrm{Hom}_R(\mathrm{Tr}_L V, R))$ functorial in U and V. Letting $U = \mathrm{Tr}_L A^*$ and $V = X$ we obtain isomorphisms $\underline{\mathrm{Hom}}_\Lambda(\mathrm{Tr}_L A^*, X) \longrightarrow \overline{\mathrm{Hom}}_\Lambda(A, (\mathrm{Tr}_L X)^*)$ functorial in A and X. Thus the functorial isomorphism given in Proposition 8.7

$$\mathrm{Ext}^1_\Lambda(X, A) \longrightarrow \mathrm{Hom}_R(\underline{\mathrm{Hom}}_\Lambda(\mathrm{Tr}_L A^*, X), I_d)$$

becomes the functorial isomorphism

$$\mathrm{Ext}^1_\Lambda(X, A) \longrightarrow \mathrm{Hom}_R(\overline{\mathrm{Hom}}_\Lambda(A, (\mathrm{Tr}_L X)^*), I_d),$$

which is our desired result.

§9. Orders Over Complete Local Rings. Throughout this section we assume, unless stated to the contrary, that Λ is an R-order where R is a complete local ring of dimension d. Our purpose is to explore what some of the special features of the category noeth Λ already explored in Section 5 enable us to say about Λ-lattices. The results here parallel those of Sections 5 and 6.

Let $L(\Lambda)$ be the category of lattices. Then we know that $L(\Lambda)$ is a Krull-Schmidt category, i.e. every lattice M has a unique (up to isomorphism) representation as a sum $M_1 \coprod \cdots \coprod M_n$ with the M_i indecomposable lattices. Thus each lattice M has a unique (up to isomorphism) representation as a sum $M_1 \coprod P$ where M_1 has no nonzero projective summands and P is projective. Similarly every lattice M has a unique (up to isomorphism) representation as $N \coprod I$ where N has no nonzero injective lattice summands and I is an injective lattice. We denote by $L_P(\Lambda)$ the full subcategory of $L(\Lambda)$ whose objects are those lattices with no nonzero projective summands. Similarly we denote by $L_I(\Lambda)$ the full subcategory of $L(\Lambda)$ whose objects are those lattices with no nonzero summands which are injective lattices. Obviously the duality $D\colon L(\Lambda) \longrightarrow L(\Lambda^{op})$ given by $D(M) = \mathrm{Hom}_R(M, R)$ induces a duality $D\colon L_P(\Lambda) \longrightarrow L_I(\Lambda^{op})$. Also if we denote by $\underline{L}_P(\Lambda)$ the full subcategory of $\underline{L}(\Lambda)$ whose objects are in $L_P(\Lambda)$, then the inclusion $\underline{L}_P(\Lambda) \longrightarrow \underline{L}(\Lambda)$ is an equivalence of categories. Similarly, letting $\overline{L}_I(\Lambda)$ be the full subcategory of $\overline{L}(\Lambda)$ whose objects are in $L_I(\Lambda)$, then the inclusion $\overline{L}_I(\Lambda) \longrightarrow \overline{L}(\Lambda)$ is an equivalence of categories. Finally the duality $D\colon L_P(\Lambda) \longrightarrow L_I(\Lambda^{op})$ induces a duality $D\colon \underline{L}_P(\Lambda) \longrightarrow \overline{L}_I(\Lambda^{op})$.

Following our usual convention for noetherian algebras over complete local rings when talking about a projective resolution of a lattice M we will mean, unless stated to the contrary, that the resolution is minimal. The fact that every object in $L(\Lambda)$ has a minimal projective resolution implies that every object in $L(\Lambda)$ has a minimal lattice injective resolution. For let M be in $L(\Lambda)$. Then $D(M)$ is in $L(\Lambda^{op})$ and so has a minimal projective resolution

$\dots \longrightarrow P_n \longrightarrow \dots \longrightarrow P_0 \longrightarrow D(M) \longrightarrow 0$ in $L(\Lambda)$. Then

$M = D^2(M) \longrightarrow D(P_0) \longrightarrow \dots \longrightarrow D(P_n) \longrightarrow \dots$ is a minimal lattice injective resolution of M.

We have already seen in Section 5 that the transpose $Tr: \underline{\text{noeth}}\ \Lambda \longrightarrow \underline{\text{noeth}}\ \Lambda^{op}$ has the property that TrM is in $\text{noeth}_P\Lambda^{op}$. Thus if M is in $L(\Lambda)$, then TrM is in $J_P(\Lambda^{op})$ where $J_P(\Lambda^{op})$ is the full subcategory of $J(\Lambda^{op})$ whose objects have no nonzero projective summands. Clearly the inclusion $\underline{J}_P(\Lambda^{op}) \longrightarrow \underline{J}(\Lambda^{op})$ is an equivalence of categories. Thus the duality $Tr: \underline{\text{noeth}}\ \Lambda \longrightarrow \underline{\text{noeth}}\ \Lambda^{op}$ induces the duality $Tr: \underline{L}_P(\Lambda) \longrightarrow \underline{J}_P(\Lambda^{op})$. Also the fact that $Tr: \text{Ob noeth}_P\Lambda \longrightarrow \text{Ob noeth}_P\Lambda^{op}$ is a representation equivalence implies that $Tr: \text{Ob } L_P(\Lambda) \longrightarrow \text{Ob } J_P(\Lambda^{op})$ is also a representation equivalence.

We now want to show that the duality $Tr_L: \underline{L}(\Lambda) \longrightarrow \underline{L}(\Lambda^{op})$ has the property that if M is in $L_P(\Lambda)$, then $Tr_L(M)$ is in $L_P(\Lambda^{op})$ and the map $Tr_L: \text{Ob } L_P(\Lambda) \longrightarrow \text{Ob } L_P(\Lambda^{op})$ is a representation equivalence. We begin with the following.

Lemma 9.1: Let $\underline{\underline{V}}$ be the full subcategory of $\text{noeth}_P\Lambda$ consisting of all objects M such that $\text{Ext}^1_\Lambda(M, \Lambda) = 0$.

a) If M is in $\underline{\underline{V}}$, then $\Omega^1(M)$ is in $\text{noeth}_P\Lambda$.

b) $\Omega^1: \underline{\text{Hom}}_\Lambda(M, N) \longrightarrow \underline{\text{Hom}}_\Lambda(\Omega^1(M), \Omega^1(N))$ is an isomorphism for all M and N in $\underline{\underline{V}}$.

c) If M and N are in $\underline{\underline{V}}$, then $M \approx N$ if and only if $\Omega^1(M) \approx \Omega^1(N)$.

d) If $M \approx M_1 \coprod M_2$ in $\underline{\underline{V}}$, then $\Omega^1(M) \approx \Omega^1(M_1) \coprod \Omega^1(M_2)$.

e) M in $\underline{\underline{V}}$ is indecomposable if and only if $\Omega^1(M)$ is indecomposable.

f) If $M \cong M_1 \amalg \dots \amalg M_k$ with the M_i indecomposable, then $\Omega^1(M) \approx \Omega^1(M_1) \amalg \dots \amalg \Omega^1(M_k)$ with the $\Omega^1(M_i)$ indecomposable modules in noeth Λ_P.

Proof:

a) Let $P \longrightarrow M \longrightarrow 0$ be a projective cover. Then $0 \longrightarrow \Omega^1(M) \xrightarrow{f} P \xrightarrow{g} M \longrightarrow 0$ is exact. Suppose Q is a projective summand of $\Omega^1(M)$, i.e. there is a morphism $u\colon Q \longrightarrow \Omega^1(M)$ which is a splittable monomorphism. Since $\mathrm{Ext}^1_\Lambda(M, \Lambda) = 0$, we have the following exact commutative diagram.

$$\begin{array}{ccccccccc} 0 \longrightarrow & \mathrm{Hom}_\Lambda(M, \Lambda) & \longrightarrow & \mathrm{Hom}_\Lambda(P, \Lambda) & \xrightarrow{\mathrm{Hom}(f, \Lambda)} & \mathrm{Hom}_\Lambda(\Omega^1(M), \Lambda) & \longrightarrow 0 \\ & & & & & \downarrow \mathrm{Hom}_\Lambda(u, \Lambda) & \\ & & & & & \mathrm{Hom}_\Lambda(Q, \Lambda) & \\ & & & & & \downarrow & \\ & & & & & 0 & \end{array}$$

Because $\mathrm{Hom}_\Lambda(Q, \Lambda)$ is Λ^{op}-projective and the composition $\mathrm{Hom}_\Lambda(fu, \Lambda)\colon \mathrm{Hom}_\Lambda(P, \Lambda) \longrightarrow \mathrm{Hom}_\Lambda(Q, \Lambda)$ is an epimorphism, there is a morphism $s\colon \mathrm{Hom}_\Lambda(Q, \Lambda) \longrightarrow \mathrm{Hom}_\Lambda(P, \Lambda)$ such that $\mathrm{Hom}_\Lambda(fu, \Lambda)\, s = 1_{\mathrm{Hom}_\Lambda(Q, \Lambda)}$. Hence $1_{\mathrm{Hom}_\Lambda(\mathrm{Hom}_\Lambda(Q, \Lambda), \Lambda)} = \mathrm{Hom}_\Lambda(s, \Lambda)\, \mathrm{Hom}_\Lambda(\mathrm{Hom}_\Lambda(fu, \Lambda), \Lambda)$. From this it follows that the composition

$$Q \xrightarrow{fu} P \longrightarrow \mathrm{Hom}_\Lambda(\mathrm{Hom}_\Lambda(P, \Lambda), \Lambda) \xrightarrow{\mathrm{Hom}_\Lambda(s, \Lambda)} \mathrm{Hom}_\Lambda(\mathrm{Hom}_\Lambda(Q, \Lambda), \Lambda) \longrightarrow Q$$

is the identity, where $P \longrightarrow \mathrm{Hom}_\Lambda(\mathrm{Hom}_\Lambda(P, \Lambda), \Lambda)$ and $\mathrm{Hom}_\Lambda(\mathrm{Hom}_\Lambda(Q, \Lambda), \Lambda) \longrightarrow Q$ are the standard isomorphisms. Thus

$Q \xrightarrow{fu} P$ is a splittable monomorphism. Since $gfu = 0$, a complement P' of $\mathrm{Im} fu$ has the property $g|P' : P' \longrightarrow M$ is an epimorphism. The fact that $g: P \longrightarrow M \longrightarrow 0$ is a projective cover implies that $P' = P$ or equivalently $Q = 0$. Hence $\Omega^1(M)$ is in $\mathrm{noeth}_P \Lambda$.

b) Follows from the fact that $\mathrm{Ext}^1_\Lambda(N, \Lambda) = 0$ (see Proposition 7.4).

c) It is obvious that if $M \approx N$, then $\Omega^1(M) \approx \Omega^1(N)$. Suppose $f: \Omega^1(M) \longrightarrow \Omega^1(N)$ and $g: \Omega^1(N) \longrightarrow \Omega^1(M)$ have the property $gf = 1_{\Omega^1(M)}$ and $fg = 1_{\Omega^1 N}$. Then by b) there are morphisms $u: M \longrightarrow N$ such that $\Omega^1(\underline{u}) = \underline{f}$ and $\nu: N \longrightarrow M$ such that $\Omega^1(\underline{\nu}) = \underline{g}$. Since $\Omega^1(\underline{\nu}\,\underline{u}) = \Omega^1(\underline{\nu})\Omega^1(\underline{u}) = \underline{g}\,\underline{f} = \underline{gf} = \underline{1}_{\Omega^1(M)}$, it follows from b) that $\underline{\nu}\,\underline{u} = \underline{1}_M$. Because R is complete and M is in $\mathrm{noeth}_P \Lambda$, the fact that $\underline{\nu u} = \underline{1}_M$ implies that $\nu u: M \longrightarrow M$ is an isomorphism. A similar argument shows that $u\nu: N \longrightarrow N$ is an isomorphism. Hence M and N are isomorphic.

d) Follows from the fact that the projective cover of a sum of two modules is the sum of the projective covers of each module.

e) Since R is complete, we know that X in $\mathrm{noeth}\,\Lambda$ is indecomposable if and only if $\mathrm{End}\, X$ is local. Further, if X is in $\mathrm{noeth}_P \Lambda$, we have that $P(X, X) \subset \mathrm{rad}\, \mathrm{End}\, X$ and so X is indecomposable if and only if $\underline{\mathrm{End}}\, X$ is local. But by b) $\Omega^1: \underline{\mathrm{End}}\, M \longrightarrow \underline{\mathrm{End}}\, \Omega^1(M)$ is an isomorphism. Hence $\underline{\mathrm{End}}\, M$ is local if and only if $\underline{\mathrm{End}}\, \Omega^1(M)$ is local. Since M and $\Omega^1(M)$ are both in $\mathrm{noeth}_P \Lambda$, it follows that M is indecomposable if and only if $\Omega^1(M)$ is indecomposable.

f) Follows from previous results.

As a consequence of this lemma we have

Proposition 9.2: Let $d = \dim R$. Then the equivalence of categories $\Omega^d: \underline{J}(\Lambda) \longrightarrow \underline{L}(\Lambda)$ has the following properties.

a) If M is in $J_P(\Lambda)$, then $\Omega^d(M)$ is in $L_P(\Lambda)$.

b) The induced map $\Omega^d: \mathrm{Ob}(\underline{J}_P(\Lambda)) \longrightarrow \mathrm{Ob}(\underline{L}_P(\Lambda))$ is a representation equivalence.

Proof: If $d = 0$, then $J(\Lambda) = L(\Lambda)$ and $\Omega^0: J(\Lambda) \longrightarrow L(\Lambda)$ is the identity so there is noething to prove. Hence we can assume that $d > 0$.

a) If M is in $J(\Lambda)$, then we know that $\mathrm{Ext}^i_\Lambda(M, \Lambda) = 0$ for $i = 1,\ldots,d$. Suppose M is in $J_P(\Lambda)$. Then arguing by induction on i, we have by Lemma 9.1 that $\Omega^i(M)$ is in $\mathrm{noeth}_P\Lambda$ for $i = 1,\ldots,d$. Thus a) has been established.

b) Suppose M and N are in $J_P(\Lambda)$. Then arguing by induction on i, we have by Lemma 9.1 that $M \approx N$ if and only if $\Omega^i(M) \approx \Omega^i(N)$ for $i = 1,\ldots,d$. Thus $M \approx N$ if and only if $\Omega^d(M) \approx \Omega^d(N)$. A similar argument shows that if M is in $J_P(\Lambda)$ and $M \approx M_1 \coprod M_2$, then $\Omega^d(M) \approx \Omega^d(M_1) \coprod \Omega^d(M_2)$. Finally, the fact that $\Omega^d: \underline{J}(\Lambda) \longrightarrow \underline{L}(\Lambda)$ is an equivalence of categories implies that the induced functor $\Omega^d: \underline{J}_P(\Lambda) \longrightarrow \underline{L}_P(\Lambda)$ is an equivalence of categories. Hence given N in $L_P(\Lambda)$, there is an M in $J_P(\Lambda)$ such that $\Omega^d(M) \approx N$ in $\underline{L}_P(\Lambda)$. Since R is complete and $\Omega^d(M)$ and N are in $L_P(\Lambda)$, the fact that $\Omega^d(M) \approx N$ in $\underline{L}_P(\Lambda)$ implies that $\Omega^d(M) \approx N$ in $L_P(\Lambda)$. This completes the proof that $\Omega^d: \mathrm{Ob}(\underline{J}_P(\Lambda)) \longrightarrow \mathrm{Ob}(\underline{L}_P(\Lambda))$ is a representation equivalence.

Combining these results we obtain the following.

Corollary 9.3: Let $d = \dim R$. Then we have the following.

a) The duality $Tr\colon \underline{L}(\Lambda) \longrightarrow \underline{J}(\Lambda^{op})$ induces a representation equivalence $Tr\colon Ob(\underline{L}_P(\Lambda)) \longrightarrow Ob(\underline{J}_P(\Lambda^{op}))$.

b) The equivalence $\Omega^d\colon \underline{J}(\Lambda) \longrightarrow \underline{L}(\Lambda)$ of categories induces a representation equivalence $\Omega^d\colon Ob(\underline{J}_P(\Lambda)) \longrightarrow Ob(\underline{L}_P(\Lambda))$.

c) The duality $Tr_L\colon \underline{L}(\Lambda) \longrightarrow \underline{L}(\Lambda^{op})$ induces a representation equivalence $Tr_L\colon Ob(\underline{L}_P(\Lambda)) \longrightarrow Ob(\underline{L}_P(\Lambda^{op}))$ with the property $Tr_L Tr_L(X) \approx X$ for all X in $\underline{L}_P(\Lambda)$ and all X in $\underline{L}_P(\Lambda^{op})$.

d) The equivalence of categories $DTr_L\colon \underline{L}(\Lambda) \longrightarrow \overline{L}(\Lambda^{op})$ induces a representation equivalence $DTr_L\colon Ob(\underline{L}_P(\Lambda)) \longrightarrow Ob(\overline{L}_I(\Lambda^{op}))$ with the property $Tr_L D(DTr_L(X)) \approx X$ in $L_P(\Lambda)$ for all X in $\underline{L}_P(\Lambda)$ and $DTr_L(Tr_L D(Y)) \approx Y$ in $L_I(\Lambda)$ for all Y in $\overline{L}_I(\Lambda)$.

The rest of this section is devoted to giving various analogues for lattices over an R-order Λ with R a complete local ring of results already given in Sections 5 and 6 for arbitrary noetherian R-algebras with R a complete local ring. In fact, the results concerning lattices are deduced from the earlier results using the following basic facts.

Lemma 9.4: Let $d = \dim R$.

a) The functor $L(\Lambda) \longrightarrow \text{art}\,\Lambda$ given by $X \longmapsto \text{Hom}_R(D(X), I_d)$ is fully faithful.

b) For C and X in $L(\Lambda)$, there are isomorphisms

$$\text{Ext}^1_\Lambda(C, X) \cong \text{Ext}^1_\Lambda(\Omega^{-d}C, \text{Hom}_R(D(X), I_d))$$

which are functorial in C and X in $L(\Lambda)$.

Proof:

a) The functor $L(\Lambda) \longrightarrow \text{art } \Lambda$ is the composition of dualities $D: L(\Lambda) \longrightarrow L(\Lambda^{op})$ given by $D(X) = \text{Hom}_R(X, R)$ and $\text{Hom}_R(\ ,I_d)$: noeth $\Lambda^{op} \longrightarrow \text{art } \Lambda$ and is therefore fully faithful.

b) Proven in Section 8 using Lemma 8.4 (see argument preceding Proposition 8.7).

As our first application we prove the following.

Proposition 9.5: Let $0 \longrightarrow A \xrightarrow{g} B \xrightarrow{f} C \longrightarrow 0$ be an exact sequence of lattices and let $X = \text{Tr}_L D(A)$. Then

a) f is right X-determined in $L(\Lambda)$.

b) There is a summand $0 \longrightarrow A' \xrightarrow{g'} B' \xrightarrow{f'} C \longrightarrow 0$ over C of $0 \longrightarrow A \xrightarrow{g} B \xrightarrow{f} C \longrightarrow 0$ with f' a minimal right X-determined morphism in $L(\Lambda)$ such that $\text{Im}(\ ,f')(X) = \text{Im}(\ ,f)(X)$.

c) f is right minimal if and only if the zero module is the only summand A' of A such that $g(A')$ is a summand of B.

d) There are summands A' of A maximal with respect to $g(A')$ being a summand of B. Suppose A_1 and A_2 are two such summands of A.

   i) The induced exact sequences $0 \longrightarrow A/A_i \xrightarrow{g_i} B/g(A_i) \xrightarrow{f_i} C \longrightarrow 0$ in $L(\Lambda)$ have the property that f_i is minimal right $\text{Tr}_L DA$-determined in $L(\Lambda)$ and $\text{Im}(\ ,f_i)(X) = \text{Im}(\ ,f)(X)$ for $i = 1, 2$.

   ii) The exact sequences $0 \longrightarrow A/A_i \xrightarrow{g_i} B/g(A_i) \xrightarrow{f_i} C \longrightarrow 0$ are isomorphic over C.

iii) A_1 and A_2 are isomorphic summands of A and A/A_1 and A/A_2 are isomorphic summands of A.

<u>Proof</u>:

a) Proven in Section 7.

b) Let y be the element of $\text{Ext}^1_\Lambda(\Omega^{-d}C, \text{Hom}_R(D(A), I_d))$ corresponding to $x: 0 \longrightarrow A \xrightarrow{g} B \xrightarrow{f} C \longrightarrow 0$ in $\text{Ext}^1_\Lambda(C, A)$ under the functorial isomorphism $\text{Ext}^1_\Lambda(C, A) \xrightarrow{\simeq} \text{Ext}^1_\Lambda(\Omega^{-d}C, \text{Hom}_R(D(A), I_d))$. Then by Proposition 5.3 we know that $\text{Hom}_R(D(A), I_d)$ can be written as $U_1 \coprod U_2$ such that the components y_i in $\text{Ext}^1_\Lambda(\Omega^{-d}C, U_i)$ of y have the following properties: a) $y_2 = 0$ and b) $y_1: 0 \longrightarrow U_1 \xrightarrow{j} E \xrightarrow{h} C \longrightarrow 0$ has the property h is right minimal or equivalently, if $t: U_1 \longrightarrow U_1$ is such that $\text{Ext}^1_\Lambda(\Omega^{-d}C, t)(y_1) = y_1$, then t is an isomorphism.

Since the functor $F: L(\Lambda) \longrightarrow \text{art}\ \Lambda$ given by $F(X) = \text{Hom}_R(D(X), I_d)$ is fully faithful, there is a decomposition $A \cong A_1 \coprod A_2$ such that $F(A_i) \cong U_i$. Thus if we let x_i in $\text{Ext}^1_\Lambda(C, A_i)$ be the components of x, then x_i corresponds to y_i under the isomorphisms $\text{Ext}^1_\Lambda(C, A_i) \simeq \text{Ext}^1_\Lambda(C, F(A_i))$. Hence $x_2 = 0$ and x_1 has the property that if $h: A_i \longrightarrow A_i$ is such that $\text{Ext}^1_\Lambda(C, u)(x_1) = x_1$, then u is an isomorphism. From this it follows that the exact sequence $x_1: A_1 \xrightarrow{g_1} B_1 \xrightarrow{f_1} C \longrightarrow 0$ is right minimal. The fact that $x_2 = 0$, implies that f_1 is right X-determined with $\text{Im}(\ ,f_1)(X) = \text{Im}(\ ,f)(X)$. Thus the exact sequence $0 \longrightarrow A_1 \xrightarrow{g_1} B_1 \xrightarrow{f_1} C \longrightarrow 0$ is our desired exact sequence

$$0 \longrightarrow A' \xrightarrow{g'} B' \xrightarrow{f'} C \longrightarrow 0.$$

c) and d). Proven in the same way as analogues in Proposition 5.4.

For the convenience of the reader we state, without proof, the dual version of Proposition 9.5 which can easily be deduced from that proposition using the duality $D: L(\Lambda) \longrightarrow L(\Lambda^{op})$.

Proposition 9.6: Let $0 \longrightarrow A \xrightarrow{g} B \longrightarrow C \longrightarrow 0$ be an exact sequence in $L(\Lambda)$ and let $X = DTr_L C$.

a) g is left X-determined in $L(\Lambda)$.

b) There is a summand $0 \longrightarrow A \xrightarrow{g'} B' \xrightarrow{f'} C' \longrightarrow 0$ over A of $0 \longrightarrow A \xrightarrow{g} B \xrightarrow{f} C \longrightarrow 0$ with g' a minimal left X-determined morphism in $L(\Lambda)$ such that $\operatorname{Im}(g', \ )(X) = \operatorname{Im}(g, \ )(X)$.

c) g is left minimal if andonly if the zero module is the only summand C' of C such that the composition $B \longrightarrow C \longrightarrow C/C'$ is a splittable epimorphism.

d) There are summands C' of C minimal with respect to the composition $B \longrightarrow C \longrightarrow C/C'$ being a splittable epimorphism. Suppose C_1 and C_2 are two such submodules.

i) C_1 and C_2 are summands of C and there are summands B_1 and B_2 of B such that $f(B_i) = C_i$ and $f|B_i: B_i \longrightarrow C_i$ are isomorphisms for $i = 1, 2$. Moreover the exact sequences $0 \longrightarrow A \xrightarrow{g_i} B/B_i \xrightarrow{f_i} C/C_i \longrightarrow 0$ have the property that the g_i are minimal left X-determined and $\operatorname{Im}(g_i, \ )(X) = \operatorname{Im}(g, \ )(X)$ for $i = 1, 2$.

ii) The exact sequences $0 \longrightarrow A \longrightarrow B/B_i \longrightarrow C/C_i \longrightarrow 0$

are isomorphic over A.

iii) C/C_1 and C/C_2 are isomorphic summands of C and C_1 and C_2 are isomorphic summands of C.

The following is the analogue for lattices of Proposition 5.6. It is stated without proof since the proof of Proposition 5.6 tranlates verbatim to the lattice situation.

Proposition 9.7: Let $0 \longrightarrow A \xrightarrow{g} B \xrightarrow{f} C \longrightarrow 0$ be exact in $L(\Lambda)$ such that f is right minimal. Suppose $A = \coprod_{i=1}^{k} V_i^{n_i}$ with the $n_i > 0$ and the $V_1,\dots,V_k$ nonisomorphic indecomposable Λ-lattices. Letting $V = \coprod_{i=1}^{k} V_i$, we have the following.

a) f is right $Tr_L DV$-determined in $L(\Lambda)$.

b) If X is in $L(\Lambda)$ such that f is right X-determined but not right X'-determined for any proper summand X' of X, then $X \approx Tr_L DV$.

Again by duality we have the following dual of Proposition 9.7 which is the analogue of Proposition 6.7.

Proposition 9.8: Let $0 \longrightarrow A \xrightarrow{g} B \xrightarrow{f} C \longrightarrow 0$ be an exact sequence in $L(\Lambda)$ with g left minimal. Suppose $C = \coprod_{i=1}^{k} W_i^{n_i}$ with the $n_i > 0$ and the W_i nonisomorphic indecomposable Λ-lattices. Letting $W = \coprod_{i=1}^{k} W_i$ we have the following.

a) g is left $DTr_L W$-determined.

b) If X is in $L(\Lambda)$ such that g is left X-determined but not left X'-determined for any proper summand X' of X, then $X \approx DTr_L W$.

As our final results of this section we have the following analogue of Theorem 5.7 and Theorem 6.6. Since the proofs are essentially the same as those for Theorems 5.7 and 6.6, we omit the proofs.

Proposition 9.9: Let A and C be in $L(\Lambda)$ and let $X = Tr_L DA$. Then $(X, C)/P(X, C)$ is an $(\text{End } X)^{op}$-module of finite length. Let $S_1, \ldots, S_d$ be a complete set of nonisomorphic simple $(\text{End } X)^{op}$-modules and for each $(\text{End } X)^{op}$-submodule H of (X, C) containing $P(X, C)$, let $n_1(H), \ldots, n_d(H)$ be the uniquely determined nonnegative integers such that the $(\text{End } X)^{op}$-socle of $(X, C)/H$ is isomorphic to $\coprod_{i=1}^{d} S_i^{n_i(H)}$. Finally let $n = \max\{n_i(H)\}$ as i runs through all $(\text{End } X)^{op}$-submodules of (X, C) containing $P(X, C)$.

a) n is finite.

b) If $k > n$ and $0 \longrightarrow A^k \xrightarrow{g} B \xrightarrow{f} C \longrightarrow 0$ is exact, then there is a submodule A' of A^k isomorphic to A^{k-n} such that $g(A')$ is a summand of B.

Finally, dualizing Proposition 9.9 we obtain the following.

Proposition 9.10: Let A and C be in $L(\Lambda)$. Then $(A, X)/I(A, X)$ is an End X-module of finite length where $X = DTr_L C$. Let $S_1, \ldots, S_d$ be a complete set of nonisomorphic simple End X-modules and for each End X-sumbodule H of (A, X) containing $I(A, X)$, let $n_1(H), \ldots, n_d(H)$ be the uniquely determined nonnegative integers such that (End X)-socle of $(A, X)/H$ is isomorphic to $\coprod_{i=1}^{d} S_i^{n_i(H)}$. Finally, let $n = \max\{n_i(H)\}$ as i runs through $1, \ldots, d$ and H runs through all End X-submodules H of (X, C) containing $I(A, X)$.

a) n is finite.

b) If $k > n$ and $0 \longrightarrow A \longrightarrow B \longrightarrow C^k \longrightarrow 0$ is an exact sequence of lattices, then there is an exact sequence $0 \longrightarrow A \longrightarrow B \longrightarrow C^{k-n} \longrightarrow 0$ which is a summand over A of $0 \longrightarrow A \longrightarrow B \longrightarrow C^k \longrightarrow 0$.

§10. A Generalization. Combining the results in Sections 3 and 6, we have the following theorem.

Let Λ be a noetherian R-algebra with R a complete local ring. Suppose $0 \longrightarrow A \xrightarrow{g} B \xrightarrow{f} C \longrightarrow 0$ is an exact sequence of Λ-modules with A in art Λ and C in noeth Λ. Then f is right TrDA-determined in Mod Λ and g is left DTrC-determined in arno Λ.

Our main aim in this section is to strengthen this result by showing that g is left DTrC-determined in Mod Λ, not just in arno Λ. To accomplish this we need some preliminary results. We start by recalling some of the well known basic facts and definitions of pure exact sequences, pure injectives, etc.

An exact sequence of Λ-modules $0 \longrightarrow A \longrightarrow B \longrightarrow C \longrightarrow 0$ is said to be pure exact if the sequence of functors $0 \longrightarrow \otimes A \longrightarrow \otimes B \longrightarrow \otimes C \longrightarrow 0$ is exact. A Λ-module A is said to be pure injective if every pure exact sequence $0 \longrightarrow A \longrightarrow B \longrightarrow C \longrightarrow 0$ splits. We now give some characterizations of when a Λ-module is pure injective. To this end it is convenient to make a few observations.

Let R be a commutative ring. An R-algebra consists of a ring morphism $R \longrightarrow \Lambda$. Obviously every ring Λ can be considered as an R-algebra for some commutative ring R, for instance $R = Z$, the integers.

Thus we lose no generality be considering R-algebras Λ rather than just rings Λ.

Suppose Λ is an R-algebra. Also suppose that I is an R-module which is an injective cogenerator for Mod R. Clearly for each M in Mod Λ, $\mathrm{Hom}_R(M, I)$ is a Λ^{op}-module by means of the operations of Λ on M. Thus we obtain a functor $\mathrm{Hom}_R(\ ,I)$: Mod $\Lambda \longrightarrow$ Mod Λ^{op} which we denote by D: Mod $\Lambda \longrightarrow$ Mod Λ^{op}. Since I is an injective cogenerator, D is a faithful exact functor. Moreover for the same reason, a sequence $A \longrightarrow B \longrightarrow C$ of Λ-modules is exact if and only if the sequence of R-modules $D(C) \longrightarrow D(B) \longrightarrow D(A)$ is exact. Returning to characterizing the Λ-modules which are pure injectives, we have the following.

<u>Proposition 10.1</u>: The following statements are equivalent for a Λ-module A.

a) A is pure injective.

b) The Λ-morphism $\varphi: A \longrightarrow D^2(A)$ given by $\varphi(a)(f) = f(a)$ for all a in A and f in D(A), is a splittable monomorphism.

c) There is a Λ^{op}-Σ bimodule X for some ring Σ such that A is isomorphic to a summand of the Λ-module $\mathrm{Hom}_\Sigma(X, E)$ for some injective Σ-module E.

d) If $\{C_k\}_{k\varepsilon K}$ is a direct limit system of Λ-modules over a directed set K, then the natural morphism $\mathrm{Ext}^1_\Lambda(\varinjlim_{k\varepsilon K} C_k, A) \longrightarrow \varprojlim_{k\varepsilon K} \mathrm{Ext}^1_\Lambda(C_k, A)$ is an isomorphism.

e) An exact sequence $0 \longrightarrow A \longrightarrow B \longrightarrow C \longrightarrow 0$ of Λ-modules splits if for each morphism $h: X \longrightarrow C$ with X a finitely presented Λ-module the pull back exact sequence $0 \longrightarrow A \longrightarrow B \times_C X \longrightarrow X \longrightarrow 0$ splits.

f) An exact sequence of Λ-modules $0 \longrightarrow A \longrightarrow B \longrightarrow C \longrightarrow 0$

splits whenever $(X, B) \longrightarrow (X, C) \longrightarrow 0$ is exact for all X in $\text{mod}\ \Lambda$.

Proof:

a) implies b) Since I is an injective cogenerator, $A \longrightarrow D^2(A)$ is a monomorphism. Let $0 \longrightarrow A \longrightarrow D^2(A) \longrightarrow T \longrightarrow 0$ be exact. Since we are assuming by a) that A is pure injective, to show that this exact sequence $0 \longrightarrow A \longrightarrow D^2A \longrightarrow T \longrightarrow 0$ splits, it suffices to show that it is pure exact.

From the exact sequence $0 \longrightarrow A \longrightarrow D^2(A) \longrightarrow T \longrightarrow 0$ we obtain the exact sequence of Λ^{op}-modules $0 \longrightarrow D(T) \longrightarrow D^3(A) \longrightarrow D(A) \longrightarrow 0$ (remember that D is exact). It is well known that the composition $D(A) \longrightarrow D^3(A) \longrightarrow D(A)$ is the identity where $D(A) \longrightarrow D^3(A)$ is the natural morphism $D(A) \longrightarrow D^2(D(A))$. Thus the exact sequence $0 \longrightarrow D(T) \longrightarrow D^3(A) \longrightarrow D(A) \longrightarrow 0$ splits. Hence for each X in $\text{Mod}\ \Lambda^{op}$ we have that $0 \longrightarrow (X, D(T)) \longrightarrow (X, D^3(A)) \longrightarrow (X, D(A))$ is exact. Applying a well known associative law we obtain that $0 \longrightarrow D(X \otimes_\Lambda T) \longrightarrow D(X \otimes_\Lambda D^2(A)) \longrightarrow D(X \otimes_\Lambda D(A)) \longrightarrow 0$ is exact. Therefore $0 \longrightarrow X \otimes_\Lambda D(A) \longrightarrow X \otimes_\Lambda D^2(A) \longrightarrow X \otimes_\Lambda T \longrightarrow 0$ is exact since D is fully faithful and exact on $\text{Mod}\ R$. Since this is true for all X in $\text{Mod}\ \Lambda^{op}$, the exact sequence $0 \longrightarrow A \longrightarrow D^2(A) \longrightarrow T \longrightarrow 0$ is pure exact. Therefore the fact that A is pure injective implies the sequence splits, showing that a) implies b).

b) implies c) Let $\Sigma = R$ and $E = I$.

c) implies d) We have the well known functorial isomorphism

$\operatorname{Ext}^1_\Lambda(C, \operatorname{Hom}_\Sigma(X, E)) \longrightarrow \operatorname{Hom}_\Sigma(\operatorname{Tor}^\Lambda_1(X, C), E)$ (see Proposition 6.2). Suppose $\{C_k\}_{k\varepsilon K}$ is a direct limit system over a directed set K. Then $\operatorname{Tor}^\Lambda_1(X, \varinjlim_{k\varepsilon K} C_k) = \varinjlim_{k\varepsilon K} \operatorname{Tor}^\Lambda_1(X, C_k)$ since Tor commutes with direct limits. Hence

$\operatorname{Hom}_\Sigma(\operatorname{Tor}^\Lambda_1(X, \varinjlim_{k\varepsilon K} C_k), E) = \varprojlim_{k\varepsilon K} \operatorname{Hom}_\Sigma(\operatorname{Tor}^\Lambda_1(X, C_k), E)$. This shows that the natural morphism

$\operatorname{Ext}^1_\Lambda(\varinjlim_{k\varepsilon K} C_k, \operatorname{Hom}_\Sigma(X, E)) \longrightarrow \varprojlim \operatorname{Ext}^1_\Lambda(C_k, \operatorname{Hom}_\Sigma(X, E))$ is an isomorphism. From the fact that A is a summand of $\operatorname{Hom}_\Sigma(X, E)$, it now follows that $\operatorname{Ext}^1_\Lambda(\varinjlim_{k\varepsilon K} C_k, A) = \varprojlim \operatorname{Ext}^1_\Lambda(C_k, A)$.

d) implies e) Let x be the element $0 \longrightarrow A \longrightarrow B \longrightarrow C \longrightarrow 0$ in $\operatorname{Ext}^1_\Lambda(C, A)$. The hypothesis of e) is equivalent to given any morphism $f: X \longrightarrow C$ with X finitely presented $\operatorname{Ext}^1_\Lambda(f, A)(x)$ in $\operatorname{Ext}^1_\Lambda(X, A)$ is zero. We know that $C = \varinjlim_{k\varepsilon K} C_k$ with the C_k in mod Λ. Thus the hypothesis of e) tells us that x goes to zero under each of the morphisms $\operatorname{Ext}^1_\Lambda(C, A) \longrightarrow \operatorname{Ext}^1_\Lambda(C_k, A)$. Since $\operatorname{Ext}^1_\Lambda(C, A) = \varprojlim \operatorname{Ext}^1_\Lambda(C_k, A)$, this means that $x = 0$, our desired result.

e) implies f) Suppose $x: 0 \longrightarrow A \longrightarrow B \longrightarrow C \longrightarrow 0$ is our exact sequence such that $(X, B) \longrightarrow (X, C) \longrightarrow 0$ is exact for all X in mod Λ. Let $f: X \longrightarrow C$. Then $\operatorname{Ext}^1(f, A)(x)$ in $\operatorname{Ext}^1_\Lambda(X, A)$ is the same as the image of f in $\operatorname{Ext}^1(X, A)$ given by the connecting morphism $(X, C) \longrightarrow \operatorname{Ext}^1_\Lambda(X, A)$ induced by the exact sequence $0 \longrightarrow A \longrightarrow B \longrightarrow C \longrightarrow 0$. Since $(X, B) \longrightarrow (X, C) \longrightarrow 0$ is exact, $(X, C) \longrightarrow \operatorname{Ext}^1_\Lambda(X, A)$ is zero and so $\operatorname{Ext}^1_\Lambda(f, A)(x) = 0$. The fact that this is true for all morphisms $f: X \longrightarrow C$ with X in mod Λ tells us by the hypothesis of e) that $0 \longrightarrow A \longrightarrow B \longrightarrow C \longrightarrow 0$ splits.

f) implies a) To show that A is pure injective we have to show that an exact sequence $0 \longrightarrow A \longrightarrow B \longrightarrow C \longrightarrow 0$ splits if it is pure exact. Suppose $0 \longrightarrow A \longrightarrow B \longrightarrow C \longrightarrow 0$ is pure exact and X is in $\operatorname{Mod} \Lambda^{op}$. Then

$0 \longrightarrow X \otimes_{\Lambda} A \longrightarrow X \otimes_{\Lambda} B \longrightarrow X \otimes_{\Lambda} C \longrightarrow 0$ is exact and so

$0 \longrightarrow D(X \otimes_{\Lambda} C) \longrightarrow D(X \otimes_{\Lambda} B) \longrightarrow D(X \otimes_{\Lambda} A) \longrightarrow 0$ is exact.

Applying a well known associative law we obtain that the sequence

$0 \longrightarrow (X, D(C)) \longrightarrow (X, D(B)) \longrightarrow (X, D(A)) \longrightarrow 0$ is exact.

Since this is true for all X in $\operatorname{Mod} \Lambda^{op}$, we have that the exact sequence of Λ^{op}-modules

$0 \longrightarrow D(C) \longrightarrow D(B) \longrightarrow D(A) \longrightarrow 0$ splits. Hence for X in $\operatorname{mod} \Lambda$ we have that

$0 \longrightarrow D(C) \otimes_{\Lambda} X \longrightarrow D(B) \otimes_{\Lambda} X \longrightarrow D(A) \otimes_{\Lambda} X \longrightarrow 0$ is exact.

Since X is a finitely presented Λ-module and I is R injective, it follows that there are isomorphisms

$\operatorname{Hom}_R(Y, I) \otimes X \cong \operatorname{Hom}_R(\operatorname{Hom}(X, Y), I)$ functorial in Y (see 14, VI Proposition 5.2]). Thus the exact sequence

$0 \longrightarrow A \longrightarrow B \longrightarrow C \longrightarrow 0$ has the property

$0 \longrightarrow (X, A) \longrightarrow (X, B) \longrightarrow (X, C) \longrightarrow 0$ is exact for all X in $\operatorname{mod} \Lambda$ since

$0 \longrightarrow D((X, C)) \longrightarrow D((X, B)) \longrightarrow D((X, A)) \longrightarrow 0$ is exact

with D exact and faithful. Thus by the hypothesis of f) the sequence $0 \longrightarrow A \longrightarrow B \longrightarrow C \longrightarrow 0$ splits. Hence f) implies a), which finishes the proof of the proposition.

Suppose $f\colon B \longrightarrow C$ is a morphism in $\operatorname{Mod} \Lambda$ and X is in $\operatorname{Mod} \Lambda$. In order to check whether f is right X-determined it is necessary to show that if $h\colon L \longrightarrow C$ is any morphism such that

$(\mathrm{Im}(\ ,f))(X) \supset (\mathrm{Im}(\ ,h))(X)$, then there is a morphism $g\colon L \longrightarrow B$ such that $h = fg$. We now show that under certain circumstances one only has to check if morphisms $h\colon L \longrightarrow C$ with L finitely presented have this property.

Proposition 10.2: Suppose $0 \longrightarrow A \xrightarrow{g} B \xrightarrow{f} C \longrightarrow 0$ is an exact sequence of Λ-modules with A pure injective and let X be in $\mathrm{Mod}\,\Lambda$. Then $f\colon B \longrightarrow C$ is right X-determined if given any morphism $h\colon L \longrightarrow C$ with L finitely presented such that $(\mathrm{Im}(\ ,h))(X) \subset (\mathrm{Im}(\ ,f))(X)$, there is a morphism $g\colon L \longrightarrow B$ such that $h = fg$.

Proof: Let $u\colon U \longrightarrow C$ be an arbitrary morphism such that $(\mathrm{Im}(\ ,u))(X) \subset (\mathrm{Im}(\ ,f))(X)$. We want to show this implies that there is a $v\colon U \longrightarrow B$ such that $u = fv$. Or equivalently, the pull back exact sequence $0 \longrightarrow A \longrightarrow U \times_C B \longrightarrow U \longrightarrow 0$ splits.

Since A is pure injective, to show that $0 \longrightarrow A \longrightarrow U \times_C B \longrightarrow U \longrightarrow 0$ splits it suffices to show that given any $g\colon L \longrightarrow U$ with L finitely presented the pull back exact sequence $0 \longrightarrow A \longrightarrow (U \times_C B) \times_U L \longrightarrow 0$ splits. From the exact commutative diagram

$$\begin{array}{ccccccccc}
0 & \longrightarrow & A & \longrightarrow & (U \times_C B) \times_U L & \longrightarrow & L & \longrightarrow & 0 \\
 & & \| & & \downarrow & & \downarrow g & & \\
0 & \longrightarrow & A & \longrightarrow & U \times_C B & \longrightarrow & U & \longrightarrow & 0 \\
 & & \| & & \downarrow & & \downarrow u & & \\
0 & \longrightarrow & A & \longrightarrow & B & \xrightarrow{f} & C & \longrightarrow & 0
\end{array}$$

we deduce the following:

a) $0 \longrightarrow A \longrightarrow (U \times_C B) \times_U L \longrightarrow L \longrightarrow 0$ is the pull back exact sequence for $ug\colon L \longrightarrow C$.

b) $(\mathrm{Im}(\ ,ug))(X) \subset (\mathrm{Im}(\ ,u))(X) \subset (\mathrm{Im}(\ ,f))(X)$.

Since L is finitely presented, it follows from a) and b) that the exact sequence $0 \longrightarrow A \longrightarrow (U \times_C B) \times_U L \longrightarrow L \longrightarrow 0$ splits. Because this is true for all morphisms $L \longrightarrow U$ with L finitely presented, it follows from Proposition 10.1 that the exact sequence $0 \longrightarrow A \longrightarrow B \times_C U \longrightarrow U \longrightarrow 0$ splits. This finishes the proof that $0 \longrightarrow A \longrightarrow B \xrightarrow{f} C \longrightarrow 0$ is right X-determined.

With these preliminary results in mind, we turn our attention to establishing the results cited in the beginning of this section. So we assume throughout the rest of this section that Λ is a noetherian R-algebra with R a complete local ring. We begin with the following general observations.

Proposition 10.3: Let $0 \longrightarrow X \longrightarrow Y \xrightarrow{h} C \longrightarrow 0$ be a nonsplit exact sequence in Mod Λ with C in noeth Λ.

a) There is an exact commutative diagram

$$\begin{array}{ccccccccc} 0 & \longrightarrow & X & \longrightarrow & Y & \longrightarrow & C & \longrightarrow & 0 \\ & & \downarrow & & \downarrow & & \| & & \\ 0 & \longrightarrow & DTrC & \longrightarrow & B & \longrightarrow & C & \longrightarrow & 0 \end{array}$$

where $0 \longrightarrow DTrC \longrightarrow B \longrightarrow C \longrightarrow 0$ does not split.

b) There is an exact commutative diagram

$$\begin{array}{ccccccccc} 0 & \to & X & \to & Y & \to & C & \to & 0 \\ & & \downarrow & & \downarrow & & \| & & \\ 0 & \to & X'' & \to & Y'' & \to & C & \to & 0 \\ & & \downarrow & & \downarrow & & & & \\ & & 0 & & 0 & & & & \end{array}$$

such that $0 \to X'' \to Y'' \to C \to 0$ does not split with X'' an indecomposable module in $\text{art}_I\Lambda$.

Proof:

a) Since $h: Y \to C$ is an epimorphism we know that the right ideal $J = (\text{Im}(\ ,h))(C)$ of End C contains $P(C, C)$. Further, J is a proper right ideal of End C since h is not a splittable epimorphism. Let H be a maximal right ideal of End C containing J. Since $(C, C)/H$ is a simple $(\text{End } C)^{op}$-module we have by Theorem 5.5 that there is an exact sequence $0 \to DTrC \to B \xrightarrow{f} C \to 0$ with f a right C-determined morphism and $\text{Im}(\ ,f)(C) = H$. Since $J = \text{Im}(\ ,h)(C) \subset H$ we know there is a commutative exact diagram

$$\begin{array}{ccccccccc} 0 & \to & X & \to & Y & \xrightarrow{f} & C & \to & 0 \\ & & \downarrow & & \downarrow & & \| & & \\ 0 & \to & DTrC & \to & B & \xrightarrow{f} & C & \to & 0 \end{array}$$

with $0 \to DTrC \to B \xrightarrow{f} C \to 0$ not splittable.

b) By a) we know there is a commutative exact diagram

$$\begin{array}{ccccccccc} 0 & \to & X & \to & Y & \xrightarrow{f} & C & \to & 0 \\ & & \downarrow t & & \downarrow u & & \| & & \\ 0 & \to & DTrC & \to & B & \to & C & \to & 0 \end{array}$$

where $0 \longrightarrow DTrC \longrightarrow B \longrightarrow C \longrightarrow 0$ does not split. Hence $0 \longrightarrow Imt \longrightarrow Imu \longrightarrow C \longrightarrow 0$ is a nonsplit exact sequence. Since $DTrC$ is in art Λ, we know that the submodule Imt of $DTrC$ is also in art Λ. Hence Imt is isomorphic to a finite sum $\coprod A_i$ where each A_i is an indecomposable module in art Λ. Therefore the fact that $0 \longrightarrow Imt \longrightarrow Imn \longrightarrow C \longrightarrow 0$ does not split means there is a projection $Imt \longrightarrow A_i$ for some i such that the exact push out diagram

$$\begin{array}{ccccccccc}
0 & \longrightarrow & Im\,t & \longrightarrow & Im\,u & \longrightarrow & C & \longrightarrow & 0 \\
 & & \downarrow & & \downarrow & & \| & & \\
0 & \longrightarrow & A_i & \longrightarrow & B & \longrightarrow & C & \longrightarrow & 0 \\
 & & \downarrow & & \downarrow & & & & \\
 & & 0 & & 0 & & & &
\end{array}$$

has the property $0 \longrightarrow A_i \longrightarrow B \longrightarrow C \longrightarrow 0$ does not split. Thus we have the exact commutative diagram

$$\begin{array}{ccccccccc}
0 & \longrightarrow & X & \longrightarrow & Y & \longrightarrow & C & \longrightarrow & 0 \\
 & & \downarrow & & \downarrow & & \| & & \\
0 & \longrightarrow & A_i & \longrightarrow & B & \longrightarrow & C & \longrightarrow & 0 \\
 & & \downarrow & & \downarrow & & & &
\end{array}$$

which finishes the proof of b).

As a consequence of this result we have the following.

Proposition 10.4: Let $0 \longrightarrow X \xrightarrow{g} Y \xrightarrow{f} Z \longrightarrow 0$ be a nonsplit exact sequence on Mod Λ. Then the following statements are equivalent.

a) There is a finitely generated submodule Z' of Z such that the exact sequence $0 \longrightarrow X \longrightarrow f^{-1}(Z') \longrightarrow Z' \longrightarrow 0$ does not split.

b) There is a submodule X' of X such that X/X' is in art Λ and the exact sequence $0 \longrightarrow X/X' \longrightarrow Y/g(X') \longrightarrow Z \longrightarrow 0$ does not split.

<u>Proof</u>:

a) implies b) Let Z' be a finitely generated submodule of Z such that the element $x\colon 0 \longrightarrow X \longrightarrow f^{-1}(Z') \longrightarrow Z' \longrightarrow 0$ in $\mathrm{Ext}^1(Z', X)$ does not split. Then by Proposition 10.3 we know there is an epimorphism $t\colon X \longrightarrow A$ with A in art Λ such that $\mathrm{Ext}^1_\Lambda(Z', t)(x)$ in $\mathrm{Ext}^1_\Lambda(Z', A)$ is not zero. Letting $0 \longrightarrow X \longrightarrow Y \longrightarrow Z \longrightarrow 0$ be the element y in $\mathrm{Ext}^1_\Lambda(Z, X)$ we have the commutative diagram

$$\begin{array}{ccc} \mathrm{Ext}^1_\Lambda(Z, X) & \xrightarrow{\mathrm{Ext}^1_\Lambda(Z,\, t)} & \mathrm{Ext}^1_\Lambda(Z, A) \\ \downarrow{\scriptstyle \mathrm{Ext}^1_\Lambda(h,\, X)} & & \downarrow{\scriptstyle \mathrm{Ext}^1_\Lambda(h,\, A)} \\ \mathrm{Ext}^1_\Lambda(Z', X) & \xrightarrow[\mathrm{Ext}^1_\Lambda(Z',\, t)]{} & \mathrm{Ext}^1_\Lambda(Z', A) \end{array}$$

where $h\colon Z' \longrightarrow Z$ is the inclusion. Since $\mathrm{Ext}^1_\Lambda(Z', t)(\mathrm{Ext}^1_\Lambda(h, X))(y) = \mathrm{Ext}^1_\Lambda(Z', t)(x) \neq 0$, it follows that $\mathrm{Ext}^1_\Lambda(Z, t)(y)$ in $\mathrm{Ext}^1_\Lambda(Z, A)$ is not zero. This shows that a) implies b).

b) implies a) Suppose $0 \longrightarrow X/X' \xrightarrow{h} Y/g(X') \longrightarrow Z \longrightarrow 0$ does not split and X/X' is in art Λ. Then $X/X' \approx D\,D(X/X')$ and so by Proposition 10.1 X/X' is pure injective. Again by Proposition 10.1, there is a finitely generated submodule Z' of Z such that $0 \longrightarrow X/X' \longrightarrow j^{-1}(Z') \longrightarrow Z' \longrightarrow 0$ does not split. By an argument similar to that used in part a), it is not difficult to show that this implies that

$0 \longrightarrow X \longrightarrow f^{-1}(Z') \longrightarrow Z' \longrightarrow 0$ does not split. This completes the proof of the proposition.

This result enables us to establish the following.

Proposition 10.5: Let $0 \longrightarrow A \xrightarrow{g} B \xrightarrow{f} C \longrightarrow 0$ be an exact sequence of Λ-modules with C in noeth Λ. Suppose X is a Λ-module. The morphism g is left X-determined if given any morphism $h: A \longrightarrow Y$ with Y in art Λ such that $\text{Im}(h, \ )(X) \subset \text{Im}(g, \ )(X)$, there is a morphism $s: B \longrightarrow Y$ such that $sg = h$.

Proof: Let $t: A \longrightarrow L$ in Mod Λ have the property $(\text{Im}(t, \ ))(X) \subset (\text{Im}(g, \ ))(X)$. To prove the proposition we must show there is a morphism $\upsilon: B \longrightarrow L$ such that $\upsilon g = t$.

Suppose no such $\upsilon: B \longrightarrow L$ exists. This implies that if x in $\text{Ext}^1_\Lambda(C, A)$ is the extension $0 \longrightarrow A \longrightarrow B \longrightarrow C \longrightarrow 0$, then $y = \text{Ext}^1_\Lambda(C, t)(x)$ in $\text{Ext}^1_\Lambda(C, L)$ is not zero. Since C is in noeth Λ, we have by Proposition 10.3 that there is an epimorphism $u: L \longrightarrow Y$ with Y in art Λ such that $\text{Ext}^1_\Lambda(C, u)(y)$ in $\text{Ext}^1_\Lambda(C, Y)$ is not zero. Hence the morphism $ut: A \longrightarrow Y$ has the property that $\text{Ext}^1_\Lambda(C, ut)(x)$ in $\text{Ext}^1_\Lambda(C, Y)$ is not zero. Because Y is in art Λ we have by the hypothesis of the proposition that $\text{Im}(ut, \ )(X)$ is not contained in $\text{Im}(g, \ )(X)$. But $\text{Im}(t, \ )(X) \supset \text{Im}(ut, \ )(X)$. This contradiction shows that if $\text{Im}(t, \ )(X) \subset \text{Im}(g, \ )(X)$, then there is a $\upsilon: B \longrightarrow L$ such that $t = \upsilon g$. This finishes the proof of the proposition.

Combining these results we obtain the result

announced in the beginning of this section.

Theorem 10.6: Let $0 \longrightarrow A \xrightarrow{g} B \xrightarrow{f} C \longrightarrow 0$ be an exact sequence in Mod Λ with A in $\text{art}_I\Lambda$ and C in $\text{noeth}_P\Lambda$. Then

a) f is right TrDA-determined in Mod Λ.

b) g is left DTrC-determined in Mod Λ.

Proof:

a) See Section 3.

b) We have already seen in Section 6 that g is left TrDC-determined in arno Λ. In particular, we know that if $j: A \longrightarrow Y$ is a morphism in art Λ such that $\text{Im}(j, \ )(\text{TrD}(C))$ is contained in $\text{Im}(g, \ )(\text{TrDC})$, then there is a morphism $s: B \longrightarrow Y$ such that $sg = j$. Since C is in noeth Λ, it follows from Proposition 10.5, that g is left TrDC-determined in Mod Λ.

Finally we point out, that as a consequence of this result, Theorem 6.6 remains valid if the phrases "determined in arno Λ" are changed to "determined in Mod Λ."

§11. Appendix. Our purpose in this appendix is to give a proof of Proposition 3.8 which was used in establishing one of our basic existence theorems, Theorem 3.9.

Suppose X is a finitely presented Λ-module and $\Gamma = \text{End } X$. Then X is a Γ-module in the obvious way. Since the operations of Λ and Γ commute, X is a $\Lambda - \Gamma$ bimodule. Let I be an injective module. Then $\text{Hom}_\Gamma(X, I)$ is a Λ^{op}-module by means of the operation of Λ on X. Now we have the morphism of functors

$$\psi: \mathrm{Hom}_\Gamma(X, I) \otimes_\Lambda \longrightarrow \mathrm{Hom}_\Gamma(\mathrm{Hom}_\Lambda(\ , X), I)$$

which for each Y in $\mathrm{Mod}\,\Lambda$ is given by $\psi_Y(f \otimes y)(g) = f(g(y))$ for all f in $\mathrm{Hom}_\Gamma(X, I)$, y in Y, and g in $\mathrm{Hom}_\Lambda(Y, X)$. It is well known (see [14, VI, Proposition 5.2]) that since I is Γ-injective, ψ_Y is an isomorphism for all finitely presented Λ-modules Y. In particular, if $Y = X$, we have the isomorphism

$$\psi_X: \mathrm{Hom}_\Gamma(X, I) \otimes_\Lambda X \longrightarrow \mathrm{Hom}_\Gamma(\mathrm{Hom}_\Lambda(X, X), I)$$

which is an isomorphism of Γ-modules where $\mathrm{Hom}_\Gamma(X, I) \otimes_\Lambda X$ is viewed as a Γ-module by means of the operation of Γ on X and the operation on $\mathrm{Hom}_\Gamma(\mathrm{Hom}_\Lambda(X, X), I)$ is given by the operation of Γ^{op} on $\mathrm{Hom}_\Lambda(X, X)$ defined by $\gamma f(x) = f(\gamma x)$ for all γ in Γ^{op}, x in X and f in $\mathrm{Hom}_\Lambda(X, X)$. Since the morphism $\mathrm{Hom}_\Gamma(\mathrm{Hom}_\Lambda(X, X), I) \longrightarrow I$ given by $f \longmapsto f(1_X)$ is an isomorphism of Γ-modules, we obtain the Γ-module isomorphism

$$\alpha: \mathrm{Hom}_\Gamma(X, I) \otimes_\Lambda X \longrightarrow I$$

given explicitely by $\alpha(f \otimes x) = f(x)$, which we will usually consider an identification.

The following result will be used constantly in the rest of this section.

<u>Proposition 11.1</u>: Let A be an arbitrary Λ^{op}-module. Then considering $A \otimes_\Lambda X$ and $\mathrm{Hom}_\Gamma(X, I) \otimes_\Lambda X$ as Γ-modules by means of the operation of Γ

on X we have that the natural morphism

$$\beta: \operatorname{Hom}_{\Lambda^{op}}(A, \operatorname{Hom}_\Gamma(X, I)) \longrightarrow \operatorname{Hom}_\Gamma(A \underset{\Lambda}{\otimes} X, \operatorname{Hom}_\Gamma(X, I) \underset{\Lambda}{\otimes} X)$$

given by $\beta(f) = f \underset{\Lambda}{\otimes} X$ is an isomorphism which is functorial in A.

Proof: We have the standard isomorphism

$\varphi: \operatorname{Hom}_{\Lambda^{op}}(A, \operatorname{Hom}_\Gamma(X, I)) \longrightarrow \operatorname{Hom}_\Gamma(A \underset{\Lambda}{\otimes} X, I)$ which is functorial in A. We also have our identification $\alpha: \operatorname{Hom}_\Gamma(X, I) \underset{\Lambda}{\otimes} X \longrightarrow I$ of Γ-modules. Hence we obtain a diagram

$$\begin{array}{ccc}
\operatorname{Hom}_{\Lambda^{op}}(A, \operatorname{Hom}_\Gamma(X, I)) & \xrightarrow{\beta} & \operatorname{Hom}_\Gamma(A \underset{\Lambda}{\otimes} X, \operatorname{Hom}_\Gamma(X, I) \underset{\Lambda}{\otimes} X) \\
& \searrow^{\varphi} & \downarrow \operatorname{Hom}_\Gamma(A \underset{\Lambda}{\otimes} X, \alpha) \\
& & \operatorname{Hom}_\Gamma(A \underset{\Lambda}{\otimes} X, I)
\end{array}$$

Since φ and $\operatorname{Hom}_\Gamma(A \underset{\Lambda}{\otimes} X, \alpha)$ are isomorphisms, we will have our desired result that β is an isomorphism if we show that the diagram commutes. But this is a routine calculation we leave to the reader.

As an application of this result we obtain the following proof of Proposition 3.8.

Proposition 11.2: The restriction of the functor $\operatorname{Hom}_\Gamma(X, \ ): \operatorname{Mod} \Gamma \longrightarrow \operatorname{Mod} \Lambda^{op}$ to $\underline{\underline{I}}$, the category of injective Γ-modules, is fully faithful. So letting $\underline{\underline{C}}$ be the full subcategory of $\operatorname{Mod} \Lambda^{op}$ consisting of all Λ^{op}-modules isomorphic to modules of the form $\operatorname{Hom}_\Gamma(X, I)$ for some I in $\underline{\underline{I}}$, the functor $\underline{\underline{I}} \longrightarrow \underline{\underline{C}}$ given by $I \longmapsto \operatorname{Hom}_\Gamma(X, I)$ is an equivalence of categories.

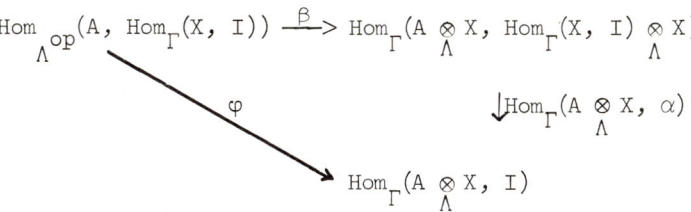

Proof: Let I_1 and I_2 be two objects in $\underline{\underline{I}}$. Then we have the following composition of morphisms

$$\mathrm{Hom}_\Gamma(I_1, I_2) \longrightarrow \mathrm{Hom}_{\Lambda^{op}}(\mathrm{Hom}_\Gamma(X, I_1), \mathrm{Hom}_\Gamma(X, I_2)) \xrightarrow{\sim}$$
$$\mathrm{Hom}_\Gamma(\mathrm{Hom}_\Gamma(X, I_1) \underset{\Lambda}{\otimes} X, \mathrm{Hom}_\Gamma(X, I_2) \underset{\Lambda}{\otimes} X) \xrightarrow{\sim} \mathrm{Hom}_\Gamma(I_1, I_2)$$

which it is easy to check is the identity. Hence

$\mathrm{Hom}_\Gamma(I_1, I_2) \longrightarrow \mathrm{Hom}_{\Lambda^{op}}(\mathrm{Hom}_\Gamma(X, I_1), \mathrm{Hom}_\Gamma(X, I_2))$ is an isomorphism.

As an immediate consequence of this result we have the following.

Corollary 11.3:

a) If I_1 and I_2 are two injective Γ-modules, then $I_1 \approx I_2$ if and only if the Λ^{op}-modules $\mathrm{Hom}_\Gamma(X, I_1)$ and $\mathrm{Hom}_\Gamma(X, I_2)$ are isomorphic Λ^{op}-modules.

b) If I is an injective Γ-module, then the natural morphism $\mathrm{End}\, I \longrightarrow \mathrm{End}\, \mathrm{Hom}_\Gamma(X, I)$ is an isomorphism of rings.

c) Let S be a simple Γ-module and I an injective envelope of S. Then

i) $f: I \longrightarrow I$ is not an isomorphism if and only if $f(S) = 0$, so

ii) $\mathrm{End}\, I \approx \mathrm{End}\, \mathrm{Hom}_\Gamma(X, I)$ are local rings.

Chapter II

SIMPLE FUNCTORS

This chapter is devoted to describing the simple functors in $(\underline{\underline{C}}^{op}, Ab)$ and $(\underline{\underline{C}}, Ab)$ for various additive categories $\underline{\underline{C}}$. Particular attention is given to when these simple functors are finitely presented in the case $\underline{\underline{C}}$ is one of the categories of modules considered in Chapter I. Most of the results given in this section amount to little more than interpretations of the results on morphisms determined by modules developed in the previous chapter.

§1. Simple Functors. Let $\underline{\underline{C}}$ be a preadditve category. We recall that a functor F in $(\underline{\underline{C}}^{op}, Ab)$ is said to be a simple functor if a) $F \neq 0$ and b) (0) and F are the only subfunctors of F. The simple functors in $(\underline{\underline{C}}^{op}, Ab)$ constitute the main concern of this chapter. This section is devoted to applying some of the results of Sections 1 and 2 of Chapter I to a preliminary study of the simple functors in $(\underline{\underline{C}}^{op}, Ab)$. A somewhat different approach to this subject can be found in [7].

As our first description of the simple functors in $(\underline{\underline{C}}^{op}, Ab)$ we have the following.

Proposition 1.1: A nonzero functor S in $(\underline{\underline{C}}^{op}, Ab)$ is simple if and only if for each X in $\underline{\underline{C}}$ such that $S(X) \neq 0$ we have

a) $S(X)$ is a simple $(\text{End } X)^{op}$-module.

b) The subfunctor (0) of S is determined by X.

Proof: Suppose S is a simple functor and suppose $S(X) \neq 0$. Let x be a nonzero element of $S(X)$. Then there is a morphism $\alpha:(\ ,X) \longrightarrow S$ such that $\alpha_X(1_X) = x$ (Yoneda's lemma). Since α is not zero, the fact that S is simple implies that α is an epimorphism. In particular, the $(\text{End } X)^{op}$-morphism α_X: $(X, X) \longrightarrow S(X)$ is surjective and so x in $S(X)$ generates the $(\text{End } X)^{op}$-module $S(X)$. Since this is true for each nonzero element x in $S(X)$, we have that $S(X)$ is a simple $(\text{End } X)^{op}$-module.

Next let T be a subfunctor of S with the property that $T(X) = (0)$. Because $S(X) \neq (0)$, $T \neq S$ and so $T = (0)$ since S is simple. Thus if S is simple and $S(X) \neq 0$, then the zero subfunctor of S is determined by X.

Now suppose that S is a nonzero functor in $(\underline{\underline{C}}^{op}, \text{Ab})$ such that if $S(X) \neq 0$, then a) and b) of the proposition hold. We want to show that S is simple. Suppose T is a proper subfunctor of S. Then there is an X in $\underline{\underline{C}}$ such that $T(X) \neq S(X)$. Hence $T(X) = 0$, since $T(X)$ is a proper submodule of the simple $(\text{End } X)^{op}$-module $S(X)$. Therefore $T = 0$, since by b) the zero subfunctor of S is determined by X. Thus S is simple which finishes the proof of the proposition.

Recalling that a subfunctor F of G is said to be a <u>maximal subfunctor</u> of G if G/F is simple, we have the following easily verified consequence of Proposition 2.1 (Hint: Use I, Proposition 1.3).

Corollary 1.2: Let F be a proper subfunctor of G in $(\underline{\underline{C}}^{op}, \text{Ab})$. Then the following are equivalent where $S = G/F$.

a) F is a maximal subfunctor of G.

b) If $S(X) \neq 0$, then $F(X)$ is a maximal $(\text{End } X)^{op}$-submodule of $G(X)$ and F is determined by X, or equivalently, $F = G_{F(X)}$.

We now apply these results to obtain a description of the maximal subfunctors of the representable functors $(\,,C)$ for each C in $\underline{\underline{C}}$. To simplify terminology we observe that the $(\text{End } C)^{op}$-submodules of $(\,,C)(C)$ are the right ideals of End C.

Proposition 1.3: Let C be in $\underline{\underline{C}}$ and F a subfunctor of $(\,,C)$ in $(\underline{\underline{C}}^{op}, \text{Ab})$.

a) $F = (\,,C)$ if and only if $1 \in F(C)$.

b) F is a maximal subfunctor of $(\,,C)$ if and only if

i) $F(C)$ is a maximal right ideal of End C and

ii) F is determined by C, or equivalently, $F = (\,,C)_{F(C)}$.

Proof:

a) Obviously $1 \in F(C)$ if $F = (\,,C)$. Suppose now that 1 is in $F(C)$. Let $(\,,C) \longrightarrow F$ be the morphism such that 1 in (C, C) is carried to 1 in $F(C)$. Then the composition $(\,,C) \longrightarrow F \longrightarrow (\,,C)$ is the identity where $F \longrightarrow (\,,C)$ is the inclusion. Thus $F = (\,,C)$.

b) That F being a maximal subfunctor of $(\,,C)$ implies i) and ii), follows from a) and Corollary 1.2. Suppose F is a subfunctor of $G = (\,,C)$ satisfying i) and ii). Since $F(C)$ is a proper ideal of (C, C) by i) we know by part a) that F is a proper subfunctor of G. Suppose $F \subset F' \subset G$ and $F' \neq G$. Then $F(C) \subset F'(C)$ and so $F(C) = F'(C)$ since by a) the fact that $F' \neq G$ implies $F'(C) \neq G(C)$ and $F(C)$ is a maximal right ideal of End C. Thus the assumption that F is a subfunctor of $(\,,C)$ determined by C implies $F' \subset F$, or equivalently $F' = F$. This shows that F is a maximal subfunctor of G which finishes the proof of b) and the proof of the proposition.

As immediate consequences of this result we have the following.

Corollary 1.4: Let C be in $\underline{\underline{C}}$. The map which associates with each right ideal H of End C the subfunctor $(\ ,C)_H$ of $(\ ,C)$ induces a bijection between the maximal right ideals of End C and the maximal subfunctors of $(\ ,C)$.

Corollary 1.5: Let F be a proper subfunctor of $(\ ,C)$ for some C in $\underline{\underline{C}}$. Then F is contained in a maximal subfunctor of $(\ ,C)$.

Proof: We know by Proposition 1.3 that $F(C)$ is a proper right ideal of End C since $F \neq (\ ,C)$. Let H be a maximal right ideal of End C containing $F(C)$. Then $(\ ,C)_H$ is a maximal subfunctor of G such that $(\ ,C)_H(C) = H$ (see Proposition 1.3). Since $(\ ,C)_H$ is a subfunctor of $(\ ,C)$ determined by C and $H \supset F(C)$, it follows that $(\ ,C)_H \supset F$, giving us our desired result.

We now apply these results to obtain more information about the simple functors in $(\underline{\underline{C}}^{op}, Ab)$.

Proposition 1.6: Let C be in $\underline{\underline{C}}$.

a) For each maximal right ideal H of End C, the functor $S = (\ ,C)/(\ ,C)_H$ is a simple functor with the property $S(C) \neq (0)$.

b) If S is a simple functor such that $S(C) \neq 0$, then

i) $S(C)$ is a simple $(\text{End } C)^{op}$-module.

ii) If x is a nonzero element of $S(C)$ and $\operatorname{ann}(x)$ is the annihilator of x in $(\operatorname{End} C)^{op}$, then $\operatorname{ann}(x)$ is a maximal right ideal of $\operatorname{End} C$ and $S \approx (\ ,C)/(\ ,C)_{\operatorname{ann}(x)}$.

c) If S_1 and S_2 are simple functors not vanishing on C, then $S_1 \approx S_2$ if and only if $S_1(C)$ and $S_2(C)$ are isomorphic simple $(\operatorname{End} C)^{op}$-modules.

d) The map which associates with each simple functor S not vanishing on C the simple $(\operatorname{End} C)^{op}$-module $S(C)$, induces a bijection between the isomorphism classes of simple functors not vanishing on C and the isomorphism classes of simple $(\operatorname{End} C)^{op}$-modules.

Proof:

a) Follows from Proposition 1.3.

b) Let x be a nonzero element of $S(C)$ and $\alpha\colon (\ ,C) \longrightarrow S$ the morphism with the property $\alpha_C(1) = x$ in $S(C)$. Since S is a simple functor and α is not zero, we know that $\alpha\colon(\ ,C) \longrightarrow S$ is an epimorphism. Moreover by Proposition 1.1, $S(C)$ is a simple $(\operatorname{End} C)^{op}$-module and so $\operatorname{ann}(x)$ is a maximal right ideal in $\operatorname{End} C$. Since $\operatorname{ann}(x)$ is the kernel of $\alpha_C\colon (C, C) \longrightarrow S(C)$ it follows that $(\operatorname{Ker} \alpha)(C) = \operatorname{ann}(x)$. Hence the maximal subfunctor $\operatorname{Ker} \alpha$ of $(\ ,C)$ is $(\ ,C)_{\operatorname{ann}(x)}$ by Proposition 1.3. Thus $S \approx (\ ,C)/(\ ,C)_{\operatorname{ann}(x)}$.

c) Suppose S_1 and S_2 are simple functors not vanishing on C and that $f\colon S_1(C) \longrightarrow S_2(C)$ is an isomorphism of the simple $(\operatorname{End} C)^{op}$-modules. Let x be a nonzero element in $S_1(C)$. Then $f(x)$ is a nonzero element of $S_2(C)$. Since $\operatorname{ann}(x) = \operatorname{ann} f(x)$, it follows from b) that $S_1 \approx (\ ,C)/(\ ,C)_{\operatorname{ann}(x)} \approx S_2$, and so $S_1 \approx S_2$.

d) Follows trivially from a) through c).

As an immediate consequence of Proposition 1.5 we have the following which completes our description of the simple contravariant functors from $\underline{\underline{C}}$ to Ab.

Corollary 1.7: Let S be a simple functor in $(\underline{\underline{C}}^{op}, \text{Ab})$. Then $S(C) \neq (0)$ for some C in $\underline{\underline{C}}$ and so $S \approx (\ ,C)/(\ ,C)_H$ where H is a maximal right ideal of End C.

We end this discussion of simple functors in $(\underline{\underline{C}}^{op}, \text{Ab})$ by recalling the definition of the radical of a functor [12]. In analogy with the situation for rings, the <u>radical of a functor</u> G in $(\underline{\underline{C}}^{op}, \text{Ab})$, which we denote by $\underline{r}G$, is defined to be the intersection of the maximal subfunctors of G. Obviously $\underline{r}G = G$ if and only if G has no maximal subfunctors. The following general observations are useful in describing the radical of a functor.

Let $\{G_i\}_{i\varepsilon I}$ be a family of subfunctors of G in $(\underline{\underline{C}}^{op}, \text{Ab})$. For each C in $\underline{\underline{C}}$ define A_C to be the subgroup $\bigcap_{i\varepsilon I} G_i(C)$ of $G(C)$. It is easily checked that for each morphism $f\colon X \longrightarrow Y$ in $\underline{\underline{C}}$ we have $G(f)(A_Y) \subset A_X$. Thus by I, Lemma 1.1, there is a unique subfunctor F of G such that $F(C) = A_C$ for all C in $\underline{\underline{C}}$. This subfunctor F of G is called the intersection of the G_i and is denoted by $\bigcap_{i\varepsilon I} G_i$.

Proposition 1.9: Suppose G is in $(\underline{\underline{C}}^{op}, \text{Ab})$ and X is in $\underline{\underline{C}}$. Let $\{H_i\}_{i\varepsilon I}$ be a family of $(\text{End } X)^{op}$-submodules of $G(X)$. Then

$$\bigcap_{i\varepsilon I} G_{H_i} = G_{\bigcap_{i\varepsilon I} H_i}.$$

Proof: Since $(\bigcap_{i\varepsilon I} G_{H_i})(X) = \bigcap_{i\varepsilon I} H_i = G_{\cap H_i}(X)$, the fact that $G_{\cap H_i}$ is determined by X shows that $\bigcap_{i\varepsilon I} G_{H_i} \subset G_{\cap H_i}$. On the other hand, since for each i in I we have $G_{H_i}(X) = H_i \supset G_{\bigcap_{i\varepsilon I} H_i}(X)$, the fact that each G_{H_i} is determined by X implies that

$G_{H_i} \supset G_{\bigcap_{i\varepsilon I} H_i}$ for each i in I. Thus $\bigcap_{i\varepsilon I} G_{H_i} \supset G_{\bigcap_{i\varepsilon I} H_i}$, which finishes the proof that $\bigcap_{i\varepsilon I} G_{H_i} = G_{\bigcap_{i\varepsilon I} H_i}$.

Returning to our discussion of the radical of a functor $(\ ,C)$ we have as an easy consequence of Proposition 1.8 and Proposition 1.3 the following description of the radical of a representable functor $(\ ,C)$ in $(\underline{\underline{C}}^{op}, Ab)$.

Proposition 1.8: Let C be in $\underline{\underline{C}}$. Then

a) $\underline{r}(\ ,C) = (\ ,C)_{\text{rad End } C}$.

b) The following are equivalent.

i) End C is local.

ii) $(\ ,C)$ has a unique maximal subfunctor.

iii) $\underline{r}(\ ,C)$ is a maximal subfunctor.

Proof:

a) Let $\{\underline{m}_i\}_{i\varepsilon I}$ be the family of maximal right ideals of End C. Then, by Proposition 1.3, $\{(\ ,C)_{\underline{m}_i}\}_{i\varepsilon I}$ is the family of maximal subfunctors of $(\ ,C)$. Hence $\underline{r}(\ ,C) = \bigcap_{i\varepsilon I} (\ ,C)_{\underline{m}_i}$. By Proposition 1.8 we know that $\bigcap_{i\varepsilon I} (\ ,C)_{\underline{m}_i} = (\ ,C)_{\cap \underline{m}_i}$. Since rad End $C = \cap \underline{m}_i$ we have that $\underline{r}(\ ,C) = (\ ,C)_{\text{rad End } C}$.

b) Follows easily from a).

We end this preliminary discussion of simple functors by pointing out when they have projective covers.

Proposition 1.10: Let S be a simple functor in $(\underline{\underline{C}}^{op}, Ab)$ and $\alpha: (\ , C) \longrightarrow S$ a nonzero morphism. Then the following are equivalent.

a) $\alpha: (\ ,C) \longrightarrow S$ is a projective cover.

b) End C is local.

c) Ker $\alpha = \underline{r}(\ ,C)$.

Proof:

a) implies b). To show that End C is local, it suffices to show that each f in End C which is not in the maximal right ideal $(\text{Ker}\ \alpha)(C)$ has a right inverse (see [14, VIII, Proposition 2.1]. Suppose f is not in $(\text{Ker}\ \alpha)(C)$. Then the composition $\alpha(\ ,f): (\ ,C) \longrightarrow S$ is not zero since $(\alpha(\ ,f))_C: (C, C) \longrightarrow S(C)$ is not zero. From this it follows that there is a morphism $h: C \longrightarrow C$ such that the diagram

$$\begin{array}{ccc} (\ ,C) & \xrightarrow{\alpha} & S \\ {\scriptstyle (\ ,h)}\downarrow & & \| \\ (\ ,C) & \xrightarrow{\alpha(\ ,f)} & S \end{array}$$

commutes. Therefore $\alpha = \alpha(\ ,f)(\ ,h) = \alpha(\ ,fh)$ and so $fh: C \longrightarrow C$ is an isomorphism since $\alpha: (\ ,C) \longrightarrow S$ is a projective cover. Hence h is a right inverse for f. This completes the proof that a) implies b).

b) implies a). Suppose End C is a local ring. Let $f: C \longrightarrow C$ be an endomorphism such that $\alpha(\ ,f) = \alpha$. In particular $\alpha(\ ,f)_C$: End $C \longrightarrow S(C)$ is not zero so that f is not in the maximal ideal of the local ring End C. Thus f is an isomorphism. This shows by I, Proposition 2.8 that $\alpha: (\ ,C) \longrightarrow S$ is a projective cover.

b) equivalent to c). Special case of Proposition 1.8.

As an immediate consequence of this result and the uniqueness properties of projective covers we have the following.

Corollary 1.11: The map which associates with each C in $\underline{\underline{C}}$ having the property End C is local the simple functor (,C)/$\underline{r}$(,C) induces a bijection between the isomorphism classes of objects in $\underline{\underline{C}}$ with local endomorphism rings and the isomorphism classes of simple functors in ($\underline{\underline{C}}^{op}$, Ab) having projective covers.

As usual, definitions and results analogous to those given for simple functors in ($\underline{\underline{C}}^{op}$, Ab) can be given for simple functors in ($\underline{\underline{C}}$, Ab). Free use of these analogies will be made throughout the rest of the paper.

§2. Presentations of Simple Functors. Throughout this section we assume that $\underline{\underline{C}}$ is an additive category. We have already shown in the previous section that all simple functors in ($\underline{\underline{C}}^{op}$, Ab) are finitely generated, i.e. if S is a simple functor, then there is an epimorphism (,C) ⟶ S ⟶ 0 for some C in $\underline{\underline{C}}$. This section is devoted to studying what it means for a simple functor S in ($\underline{\underline{C}}^{op}$, Ab) to be finitely presented, i.e. there is an exact sequence (,B) ⟶ (,C) ⟶ S ⟶ 0 with B and C in $\underline{\underline{C}}$.

Since a morphism f: B ⟶ C has the property Coker((,B) $\xrightarrow{(,f)}$ (,C)) is simple if and only if Im(,f) is a maximal subfunctor of (,C), it is important to have criteria for when a morphism f: B ⟶ C has the property Im(,f) is a maximal subfunctor of $(\ ,C)$. The following is a particularly useful result along these lines.

Proposition 2.1: Let $f: B \longrightarrow C$ be a morphism in $\underline{\underline{C}}$. Then the following statements are equivalent.

a) $\text{Im}(\ ,f)$ is a maximal subfunctor of $(\ ,C)$.

b) $f: B \longrightarrow C$ is right C-determined with $(\text{Im}(\ ,f))(C)$ a maximal right ideal of End C.

c) One of the following holds for a morphism $g: L \longrightarrow C$.

i) There is an $h: L \longrightarrow B$ such that $fh = g$.

ii) The induced morphism $B \coprod L \longrightarrow C$ is a splittable epimorphism.

Proof:

a) implies b). We know by Proposition 1.3 that if $\text{Im}(\ ,f)$ is a maximal subfunctor of $(\ ,C)$, then $\text{Im}(\ ,f)$ is determined by C and $(\text{Im}(\ ,f))(C)$ is a maximal right ideal of End C. Thus a) implies b).

b) implies c). Let $g: L \longrightarrow C$ be a morphism. If $(\text{Im}(\ ,g))(C) \subset (\text{Im}(\ ,f))(C)$, then there is a morphism $h: L \longrightarrow B$ such that $fh = g$ since $f: B \longrightarrow C$ is determined by C. If $(\text{Im}(\ ,g))(C)$ is not contained in $(\text{Im}(\ ,f))(C)$, then the right ideals $\text{Im}(\ ,g)(C)$ and $(\text{Im}(\ ,f))(C)$ generate all of End C since $(\text{Im}(\ ,f))(C)$ is a maximal right ideal of End C. From this it follows that there are morphisms $u: C \longrightarrow B$ and $v: C \longrightarrow L$ such that $fu + gv = 1$ in End C, or equivalently, the induced morphism $B \coprod L \longrightarrow C$ is a splittable epimorphism.

c) implies a). It is easily verified that $\text{Im}(\ ,f)$ is a maximal subfunctor of $(\ ,C)$ if given a morphism $g: L \longrightarrow C$, then $\mathrm{Im}(\ ,g)$ and $\mathrm{Im}(\ ,f)$ together generate $(\ ,C)$ whenever $\mathrm{Im}(\ ,g)$ is not contained in $\mathrm{Im}(\ ,f)$. The fact that c) implies a) is a trivial consequence of this observation.

For the convenience of the reader we state the covariant version of the above proposition.

Proposition 2.2: Let $g: A \longrightarrow B$ be a morphism in $\underline{\underline{C}}$. Then the following are equivalent.

a) $\mathrm{Im}(g, \ )$ is a maximal subfunctor of $(A, \ )$.

b) g is left A-determined and $\mathrm{Im}(g, \ )(A)$ is a maximal left ideal of End A.

c) One of the following holds for a morphism $h: A \longrightarrow L$:

i) There is an $f: B \longrightarrow L$ such that $fg = h$.

ii) The induced morphism $A \longrightarrow B \coprod L$ is a splittable monomorphism.

We now point out that right and left almost split morphisms are special cases of the types of morphisms described in the above propositions. For the convenience of the reader we begin by recalling the definitions of right and left almost split morphisms.

A morphism $f: B \longrightarrow C$ is said to be right almost split if a) f is not a splittable epimorphism and b) given any morphism $h: L \longrightarrow C$ which is not a splittable epimorphism then there is a morphism $g: L \longrightarrow B$ such that $fg = h$. A morphism $g: A \longrightarrow B$ is said to be left almost split if a) g is not a splittable monomorphism and b) given any morphism $h: A \longrightarrow L$ which is not a splittable monomorphism, then there is an $f: B \longrightarrow L$ such that $fg = h$.

Proposition 2.3: The following are equivalent for a morphism $f: B \longrightarrow C$ in $\underline{\underline{C}}$.

a) Im(,f) is a maximal subfunctor of (,C) and End C is local.

b) f is right almost split.

Proof:

a) implies b). Since there is a one-to-one correspondence between the maximal subfunctors of (,C) and the maximal right ideals of End C (see Corollary 1.4), the fact that End C is local (or equivalently, End C has a unique maximal right ideal) implies that Im(,f) is the unique maximal subfunctor of (,C). In particular, f is not a splittable epimorphism.

Suppose now that $g: L \longrightarrow C$ is not a splittable epimorphism, then Im(,g) $\neq$ (C). Hence Im(,g) $\subset$ Im(,f), the unique maximal subfunctor of (,C), since every proper subfunctor of a representable functor is contained in some maximal subfunctor (see Corollary 1.5). The fact that (,L) is a projective functor implies there exists some morphism $h: L \longrightarrow B$ such that (,f)(,h) = (,g) or equivalently, $fh = g$. This shows that $f: B \longrightarrow C$ is right almost split and finishes the proof of a) implies b).

b) implies a). We know by [7, Corollary 2.6] and [11, Lemma 2.3] that if $f: B \longrightarrow C$ is right almost split, then End C is local and Coker(,f) is simple, which shows that b) implies a).

For the convenience of the reader we state the analogue of Proposition 2.3 for covariant functors.

Proposition 2.4: The following are equivalent for a morphism $g: A \longrightarrow B$ in $\underline{\underline{C}}$.

a) $\mathrm{Im}(g, \ )$ is a maximal subfunctor of $(A, \ )$ and End A is local.

b) g is left almost split.

We now turn our attention to when simple functors are finitely presented. Recall that a functor F in $(\underline{\underline{C}}^{op}, \mathrm{Ab})$ is said to be finitely presented if there is an exact sequence $(\ ,B) \longrightarrow (\ ,C) \longrightarrow F \longrightarrow 0$ with B and C in $\underline{\underline{C}}$. Just as in the case of modules over a ring, it is not difficult to show the following.

Lemma 2.5: Let $(\ ,C) \longrightarrow F \longrightarrow 0$ be exact in $(\underline{\underline{C}}^{op}, \mathrm{Ab})$. Then F is finitely presented if and only if there is a morphism $B \longrightarrow C$ such that the induced sequence $(\ ,B) \longrightarrow (\ ,C) \longrightarrow F \longrightarrow 0$ is exact.

Applying our remarks to simple functors in $(\underline{\underline{C}}^{op}, \mathrm{Ab})$ we obtain

Proposition 2.6: Let $\alpha: (\ ,C) \longrightarrow S \longrightarrow 0$ be exact with S a simple functor in $(\underline{\underline{C}}^{op}, \mathrm{Ab})$. Then S is finitely presented if and only if there is a right C-determined morphism $f: B \longrightarrow C$ such that $\mathrm{Im}(\ ,f)(C)$ is the maximal right ideal $(\mathrm{Ker}\ \alpha)(C)$ of End C. If such an $f: B \longrightarrow C$ exists, then the exact sequence $(\ ,B) \xrightarrow{(\ ,f)} (\ ,C) \xrightarrow{\alpha} S \longrightarrow 0$ is a projective presentation of S.

Thus we see that the problem of when simple functors are finitely presented is intimately related to the existence of morphisms determined by objects in $\underline{\underline{C}}$. We further emphasize this point in the proposition which follows the following definitions.

A morphism $f: B \longrightarrow C$ in $\underline{\underline{C}}$ is said to be <u>minimal right almost split</u> if it is right minimal and right almost split. The morphism $f: B \longrightarrow C$ is said to be <u>minimal left almost split</u> if it is left minimal and left almost split.

<u>Proposition 2.7</u>: Let S be a simple functor in $(\underline{\underline{C}}^{op}, Ab)$ and suppose $f: B \longrightarrow C$ is a morphism in $\underline{\underline{C}}$ such that $(\ ,B) \xrightarrow{(\ ,f)} (\ ,C) \longrightarrow S \longrightarrow 0$ is exact. Then the following are equivalent.

a) $(\ ,B) \xrightarrow{(\ ,f)} (\ ,C) \longrightarrow S \longrightarrow 0$ is a minimal projective presentation of S.

b) f is a minimal right almost split morphism.

As in previous sections analogues exist in $(\underline{\underline{C}}, Ab)$ for all results and definitions given here in $(\underline{\underline{C}}^{op}, Ab)$, even when not given explicitly in the text. These analogues will be used freely in the rest of the paper.

§<u>3</u>. <u>A Duality Theorem</u>. Throughout this section we assume that we are given a fixed abelian category $\underline{\underline{A}}$. Our first concern is to develop a duality theory between certain covariant and contravariant functors from $\underline{\underline{A}}$ to abelian groups. This duality theory will then be used in the next section to obtain more information about morphisms which are determined by objects. These results in turn will be used in later sections in connection with establishing certain existence theorems. We begin with a review of some of the results in [2], where a more detailed discussion of this material is available.

It is well known that unless $\underline{\underline{A}}$ is skeletally small (i.e. the isomorphism classes of objects in $\underline{\underline{A}}$ form a set) the collections of functors $(\underline{\underline{A}}^{op}, Ab)$ and $(\underline{\underline{A}}, Ab)$ are not necessarily a category since the collections of morphisms between functors need not be a set. However f. p. (A^{op}, Ab), the collection of finitely presented functors, do form a category since the morphisms from a finitely generated functor F to an arbitrary functor G do form a set. For if $(\ ,C) \longrightarrow F \longrightarrow 0$ is an epimorphism, then (F, G) is isomorphic to a subgroup of $((\ ,C), G) = G(C)$. Thus it makes sense to talk about the categories f. p. $(\underline{\underline{A}}^{op}, Ab)$ and f. p. $(\underline{\underline{A}}, Ab)$.

Suppose F is in f. p. $(\underline{\underline{A}}^{op}, Ab)$. Then there is an exact sequence $(\ ,B) \xrightarrow{(\ ,f)} (\ ,C) \longrightarrow F \longrightarrow 0$. If we let $A \xrightarrow{g} B$ be a kernel for f, then $0 \longrightarrow (\ ,A) \longrightarrow (\ ,B) \longrightarrow (\ ,C) \longrightarrow F \longrightarrow 0$ is exact. From this fact one can deduce the following (see [2, Proposition 2.1]).

<u>Proposition 3.1</u>:

a) Let $0 \longrightarrow F_1 \longrightarrow F_2 \longrightarrow F_3 \longrightarrow F_4 \longrightarrow 0$ be an exact sequence in $(\underline{\underline{A}}^{op}, Ab)$. If F_2 and F_3 are in f. p. $(\underline{\underline{A}}^{op}, Ab)$, then F_1 and F_4 are in f. p. $(\underline{\underline{A}}^{op}, Ab)$.

b) If $0 \longrightarrow F_1 \longrightarrow F_2 \longrightarrow F_3 \longrightarrow 0$ is an exact sequence in $(\underline{\underline{A}}^{op}, Ab)$ and F_1 and F_3 are in f. p. $(\underline{\underline{A}}^{op}, Ab)$, then F_2 is in f. p. $(\underline{\underline{A}}^{op}, Ab)$.

c) Hence f. p. $(\underline{\underline{A}}^{op}, Ab)$ is an abelian category with enough projectives of global dimension at most 2.

d) By duality, f. p. $(\underline{\underline{A}}, Ab)$ has the same properties a), b), c).

Let f. p. $(\underline{\underline{A}}^{op}, Ab)_o$ be the full subcategory consisting of those F in f. p. $(\underline{\underline{A}}^{op}, Ab)$ having the property that there is an exact sequence $0 \longrightarrow (\ ,A) \xrightarrow{(\ ,g)} (\ ,B) \xrightarrow{(\ ,f)} (\ ,C) \longrightarrow F \longrightarrow 0$ with $0 \longrightarrow A \xrightarrow{g} B \xrightarrow{f} C \longrightarrow 0$ exact. Similarly, let f. p. $(\underline{\underline{A}}, Ab)_o$ be the full subcategory of f. p. $(\underline{\underline{A}}, Ab)$ consisting of those G in f. p. $(\underline{\underline{A}}^{op}, Ab)$ having the property that there is an exact sequence $0 \longrightarrow (C,\ ) \xrightarrow{(f,\ )} (B,\ ) \xrightarrow{(g,\ )} (A,\ ) \longrightarrow G \longrightarrow 0$ with $0 \longrightarrow A \xrightarrow{g} B \xrightarrow{f} C \longrightarrow 0$ exact. The following description of the objects in f. p. $(\underline{\underline{A}}^{op}, Ab)_o$ is quite useful (see [2, Proposition 3.2] for a proof).

Proposition 3.2: The following statements are equivalent for an F in f. p. $(\underline{\underline{A}}^{op}, Ab)$.

a) F is in f. p. $(\underline{\underline{A}}^{op}, Ab)_o$.

b) If $0 \longrightarrow A \longrightarrow B \longrightarrow C$ is exact in $\underline{\underline{A}}$ such that $0 \longrightarrow (\ ,A) \longrightarrow (\ ,B) \longrightarrow (\ ,C) \longrightarrow F \longrightarrow 0$ is exact, then $0 \longrightarrow A \longrightarrow B \longrightarrow C \longrightarrow 0$ is exact.

c) Let G in $(\underline{\underline{A}}^{op}, Ab)$ be left exact, i.e. if $X \longrightarrow Y \longrightarrow Z \longrightarrow 0$ is exact in $\underline{\underline{A}}$, then $0 \longrightarrow G(Z) \longrightarrow G(Y) \longrightarrow G(X)$ is exact. Then $\text{Ext}^i(F, G) = 0$ for $i = 0, 1$.

d) $(F, G) = 0$ for any G in $(\underline{\underline{A}}^{op}, Ab)$ which is left exact.

e) $(F, (\ ,A)) = 0$ for all A in $\underline{\underline{A}}$.

An analogous description of the objects in f. p. $(\underline{\underline{A}}, Ab)_o$ can be obtained by duality. As an easy consequence of Proposition 3.2 we have the following.

Corollary 3.3: Let $0 \longrightarrow F_1 \longrightarrow F_2 \longrightarrow F_3 \longrightarrow 0$ be an exact sequence in f. p. $(\underline{\underline{A}}^{op}, Ab)$.

a) F_2 is in f. p. $(\underline{\underline{A}}^{op}, Ab)_o$ if and only if F_1 and F_3 are in f. p. $(\underline{\underline{A}}^{op}, Ab)_o$.

b) f. p. $(\underline{\underline{A}}^{op}, Ab)_o$ is an abelian category with the property that the inclusion functor f. p. $(\underline{\underline{A}}^{op}, Ab)_o \longrightarrow$ f. p. $(\underline{\underline{A}}^{op}, Ab)$ is exact.

Again a result analogous to this corollary holds for f. p. $(\underline{\underline{A}}, Ab)_o$. Our aim now is to describe a duality between f. p. $(\underline{\underline{A}}^{op}, Ab)_o$ and f. p. $(\underline{\underline{A}}, Ab)_o$. To this end we introduce the contravariant functor V: f. p. $(\underline{\underline{A}}^{op}, Ab) \longrightarrow$ f. p. $(\underline{\underline{A}}, Ab)$ and W: f. p. $(\underline{\underline{A}}, Ab) \longrightarrow$ f. p. $(\underline{\underline{A}}^{op}, Ab)$ as follows.

For each F in f. p. $(\underline{\underline{A}}^{op}, Ab)$ define $V(F)$ to be the covariant functor $\underline{\underline{A}} \longrightarrow Ab$ given by $V(F)(X) = \text{Ext}^2(F, (\ ,X))$ for all X in $\underline{\underline{A}}$. The functor $V(F)$ can be described as follows. Let $0 \longrightarrow A \xrightarrow{g} B \xrightarrow{f} C$ be an exact sequence in $\underline{\underline{A}}$ such that $0 \longrightarrow (\ ,A) \xrightarrow{(\ ,g)} (\ ,B) \xrightarrow{(\ ,f)} (\ ,C) \longrightarrow F \longrightarrow 0$ is exact. Since this exact sequence is a projective resolution of F, we have that $\text{Ext}^2(F, (\ ,X)) = \text{Coker}((\ ,B), (\ ,X)) \longrightarrow ((\ ,A), (\ ,X)) = \text{Coker}((B, X) \longrightarrow (A, X))$ for all X in $\underline{\underline{A}}$. Thus $(B,\ ) \longrightarrow (A,\ ) \longrightarrow V(F) \longrightarrow 0$ is exact. Hence $V(F)$ is in f. p. $(\underline{\underline{A}}, Ab)$. Moreover $V(F)$ is in f. p. $(\underline{\underline{A}}, Ab)_o$ if F is in f. p. $(\underline{\underline{A}}^{op}, Ab)_o$. Therefore we obtain the contravariant functor V: f. p. $(\underline{\underline{A}}^{op}, Ab) \longrightarrow$ f. p. (A, Ab) given by $F \longmapsto V(F)$ for all F in f. p. $(\underline{\underline{A}}^{op}, Ab)$. Since $V(F)$ is in f. p. $(A, Ab)_o$ whenever F is in f. p. $(\underline{\underline{A}}^{op}, Ab)_o$, V induces a contravariant functor V_o: f. p. $(\underline{\underline{A}}^{op}, Ab)_o \longrightarrow$ f. p. $(\underline{\underline{A}}, Ab)_o$.

We define the contravariant functors

$W: \text{f. p.}(\underline{\underline{A}}, Ab) \longrightarrow \text{f. p.}(\underline{\underline{A}}^{op}, Ab)$ and

$W_o: \text{f. p.}(\underline{\underline{A}}, Ab)_o \longrightarrow \text{f. p.}(\underline{\underline{A}}^{op}, Ab)_o$ analogously. Namely, for G in $\text{f. p.}(\underline{\underline{A}}, Ab)$ define the contravariant functor $W(G): \underline{\underline{A}} \longrightarrow Ab$ by $W(G)(X) = Ext^2(G, (X, \;))$ for all X in $\underline{\underline{A}}$. The functor $W(G)$ can be described as follows. Let $A \xrightarrow{g} B \xrightarrow{f} C \longrightarrow 0$ be an exact sequence in $\underline{\underline{A}}$ such that $0 \longrightarrow (C, \;) \xrightarrow{(f, \;)} (B, \;) \xrightarrow{(g, \;)} (A, \;) \longrightarrow G \longrightarrow 0$ is exact. Since this exact sequence is a projective resolution of G, we have that $Ext^2(G, (X, \;)) = \text{Coker}((B, \;), (X, \;)) \longrightarrow ((C, \;), (X, \;)) = \text{Coker}\,(X, B) \longrightarrow (X, C)$ for all X in $\underline{\underline{A}}$. Thus $(\;, B) \longrightarrow (\;, C) \longrightarrow W(G) \longrightarrow 0$ is exact. Hence $W(G)$ is in $\text{f. p.}(\underline{\underline{A}}^{op}, Ab)$. Moreover, $W(G)$ is in $\text{f. p.}(\underline{\underline{A}}^{op}, Ab)_o$ if G is in $\text{f. p.}(\underline{\underline{A}}, Ab)_o$. Therefore we obtain the contravariant functor $W: \text{f. p.}(\underline{\underline{A}}, Ab) \longrightarrow \text{f. p.}(\underline{\underline{A}}^{op}, Ab)$ given by $G \longmapsto W(G)$ for all G in $\text{f. p.}(\underline{\underline{A}}, Ab)$. Since $W(G)$ is in $\text{f. p.}(\underline{\underline{A}}^{op}, Ab)_o$ whenever G is in $\text{f. p.}(\underline{\underline{A}}, Ab)_o$, W induces a contravariant functor $W_o: \text{f. p.}(\underline{\underline{A}}, Ab) \longrightarrow \text{f. p.}(\underline{\underline{A}}^{op}, Ab)_o$.

We can now state and prove our desired duality theorem.

Theorem 3.4: The contravariant functors

$V_o: \text{f. p.}(\underline{\underline{A}}^{op}, Ab)_o \longrightarrow \text{f. p.}(\underline{\underline{A}}, Ab)_o$ and

$W_o: \text{f. p.}(\underline{\underline{A}}, Ab)_o \longrightarrow \text{f. p.}(\underline{\underline{A}}^{op}, Ab)_o$ have the properties

$W_oV_o: \text{f. p.}(\underline{\underline{A}}^{op}, Ab)_o \longrightarrow \text{f. p.}(\underline{\underline{A}}^{op}, Ab)_o$ and

$V_oW_o: \text{f. p.}(\underline{\underline{A}}, Ab)_o \longrightarrow \text{f. p.}(\underline{\underline{A}}, Ab)_o$ are isomorphic to identity functors and so V_o and W_o are dualities which are inverses of each other.

Proof: We will just outline a proof that $V_o\colon \text{f. p. }(\underline{\underline{A}}^{op}, Ab)_o \longrightarrow \text{f. p. }(\underline{\underline{A}}, Ab)_o$ is dense and fully faithful, in other words, that V_o is a duality. The rest of the theorem is left to the reader.

We first show that V_o is dense, i.e. given G in $\text{f. p. }(\underline{\underline{A}}, Ab)_o$ there is an F in $\text{f. p. }(\underline{\underline{A}}^{op}, Ab)_o$ such that $V_o(F) \approx G$. Since G is in $\text{f. p. }(\underline{\underline{A}}, Ab)_o$ there is an exact sequence $0 \longrightarrow A \longrightarrow B \longrightarrow C \longrightarrow 0$ in $\underline{\underline{A}}$ such that $0 \longrightarrow (C, \ ) \longrightarrow (B, \ ) \longrightarrow (A, \ ) \longrightarrow G \longrightarrow 0$ is exact. Let F in $\text{f. p. }(\underline{\underline{A}}^{op}, Ab)_o$ be such that $0 \longrightarrow (\ ,A) \longrightarrow (\ ,B) \longrightarrow (\ ,C) \longrightarrow F \longrightarrow 0$ is exact. Then from the description of $V(F)$ in terms of resolutions it is clear that $V(F) \approx G$.

We now want to show that V_o is fully faithful. To do this it is useful to describe the action of V_o on morphisms in terms of resolutions. Suppose F, F' are in $\text{f. p. }(\underline{\underline{A}}^{op}, Ab)_o$ and $0 \longrightarrow (\ ,A) \xrightarrow{(\ ,g)} (\ ,B) \xrightarrow{(\ ,f)} (\ ,C) \longrightarrow F \longrightarrow 0$ and $0 \longrightarrow (\ ,A') \xrightarrow{(\ ,g')} (\ ,B') \xrightarrow{(\ ,f')} (\ ,C') \longrightarrow F' \longrightarrow 0$ are exact and hence projective resolutions. Let $\alpha\colon F \longrightarrow F'$ be a morphism. Then there are morphisms $h_1\colon A \longrightarrow A'$, $h_2\colon B \longrightarrow B'$, and $h_3\colon C \longrightarrow C'$ such that the exact diagram

$$\begin{array}{ccccccccccc}
0 & \longrightarrow & (\ ,A) & \xrightarrow{(\ ,g)} & (\ ,B) & \xrightarrow{(\ ,f)} & (\ ,C) & \longrightarrow & F & \longrightarrow & 0 \\
 & & \downarrow{\scriptstyle (\ ,h_1)} & & \downarrow{\scriptstyle (\ ,h_2)} & & \downarrow{\scriptstyle (\ ,h_3)} & & \downarrow{\scriptstyle \alpha} & & \\
0 & \longrightarrow & (\ ,A') & \xrightarrow{(\ ,g')} & (\ ,B') & \xrightarrow{(\ ,f')} & (\ ,C') & \longrightarrow & F' & \longrightarrow & 0
\end{array}$$

commutes. Hence

$$\begin{array}{ccccccccc}
0 & \longrightarrow & A & \xrightarrow{g} & B & \xrightarrow{f} & C & \longrightarrow & 0 \\
 & & \downarrow{\scriptstyle h_1} & & \downarrow{\scriptstyle h_2} & & \downarrow{\scriptstyle h_3} & & \\
0 & \longrightarrow & A' & \xrightarrow{g'} & B' & \xrightarrow{f'} & C' & \longrightarrow & 0
\end{array}$$

commutes, which gives rise to the exact commutative diagram

$$\begin{array}{ccccccccc} 0 \longrightarrow & (C', \ ) & \longrightarrow & (B', \ ) & \longrightarrow & (C', \ ) & \longrightarrow & V_o(F') & \longrightarrow 0 \\ & \downarrow (h_3, \ ) & & \downarrow (h_2, \ ) & & \downarrow (h_3, \ ) & & & \\ 0 \longrightarrow & (C, \ ) & \longrightarrow & (B, \ ) & \longrightarrow & (C, \ ) & \longrightarrow & V_o(F) & \longrightarrow 0 \end{array} .$$

Hence there is a unique morphism $V_o(F') \longrightarrow V_o(F)$ which makes this diagram commute. That this morphism is $V_o(\alpha)$ is not difficult to check. Using this description of the functor V_o, it follows from standard properties of projective resolutions that $V_o: (F, F') \longrightarrow (V(F), V(F'))$ is an isomorphism for all F and F' in f. p. $(\underline{\underline{A}}^{op}, Ab)_o$. This completes the outline of the proof that V_o is a duality.

We now apply this duality to some special situations of particular interest to us.

Throughout the rest of this section we assume that $\underline{\underline{C}}$ is a full subcategory of $\underline{\underline{A}}$ closed under extensions (i.e. if $0 \longrightarrow A \longrightarrow B \longrightarrow C \longrightarrow 0$ is exact with A and C in $\underline{\underline{C}}$, then B is in $\underline{\underline{C}}$) and summands. Let f. p. $(\underline{\underline{C}}^{op}, Ab)$ be the category of finitely presented contravariant functors from $\underline{\underline{C}}$ to Ab. We observe that every contravariant functor F in f. p. $(\underline{\underline{C}}^{op}, Ab)$ has a canonical extension to $\underline{\underline{A}}$ as follows. Let $(\ ,B) \longrightarrow (\ ,C) \longrightarrow F \longrightarrow 0$ be an exact sequence in f. p. $(\underline{\underline{C}}^{op}, Ab)$. Then define $\tilde{F}: \underline{\underline{A}} \longrightarrow Ab$ by $\tilde{F}(X) = \text{Coker}((X, B) \longrightarrow (X, C))$ for all X in $\underline{\underline{A}}$. Then $\tilde{F}$ is in f. p. $(\underline{\underline{A}}^{op}, Ab)$ and is independent (up to canonical isomorphism) of the projective presentation $(\ ,B) \longrightarrow (\ ,C) \longrightarrow F \longrightarrow 0$ used. It is easily checked that we obtain a fully faithful functor f. p. $(\underline{\underline{C}}^{op}, Ab) \longrightarrow$ f. p. $(\underline{\underline{A}}^{op}, Ab)$ by sending F to $\tilde{F}$ for all

F in f. p. $(\underline{\underline{C}}^{op}, Ab)$. Moreover the fully faithful functor f. p. $(\underline{\underline{C}}^{op}, Ab) \longrightarrow$ f. p. $(\underline{\underline{A}}^{op}, Ab)$ is right exact in the sense that if $0 \longrightarrow F_1 \longrightarrow F_2 \longrightarrow F_3 \longrightarrow 0$ is exact in f. p. $(\underline{\underline{C}}^{op}, Ab)$ (i.e. $0 \longrightarrow F_1(X) \longrightarrow F_2(X) \longrightarrow F_3(X) \longrightarrow 0$ is exact for all X in $\underline{\underline{C}}$), then $\tilde{F}_1 \longrightarrow \tilde{F}_2 \longrightarrow \tilde{F}_3 \longrightarrow 0$ is exact in f. p. $(\underline{\underline{A}}^{op}, Ab)$ (see [6, Proposition 3.1]). By duality there is an analogous fully faithful right exact functor f. p. $(\underline{\underline{C}}, Ab) \longrightarrow$ f. p. $(\underline{\underline{A}}, Ab)$. Thus we can consider f. p. $(\underline{\underline{C}}^{op}, Ab)$ and f. p. $(\underline{\underline{C}}, Ab)$ full subcategories of f. p. $(\underline{\underline{A}}^{op}, Ab)$ and f. p. $(\underline{\underline{A}}, Ab)$ respectively.

Now define f. p. $(\underline{\underline{C}}^{op}, Ab)_o$ to be the full subcategory of f. p. $(\underline{\underline{A}}, Ab)_o$ consisting of all F in f. p. $(\underline{\underline{A}}^{op}, Ab)_o$ which have a projective resolution $0 \longrightarrow (\ ,A) \longrightarrow (\ ,B) \longrightarrow (\ ,C) \longrightarrow F \longrightarrow 0$ with A, B, C in $\underline{\underline{C}}$. Define the subcategory f. p. $(\underline{\underline{C}}, Ab)_o$ of f. p. $(\underline{\underline{A}}, Ab)_o$ similarly. It is then obvious from the definition of the dualities V_o: f. p. $(\underline{\underline{A}}^{op}, Ab)_o \longrightarrow$ f. p. $(\underline{\underline{A}}, Ab)_o$ and W_o: f. p. $(\underline{\underline{A}}, Ab)_o \longrightarrow$ f. p. $(\underline{\underline{A}}^{op}, Ab)_o$, that $V_o(F)$ is in f. p. $(\underline{\underline{C}}, Ab)_o$ wnenever F is in f. p. $(\underline{\underline{C}}^{op}, Ab)_o$ and $W_o(G)$ is in f. p. $(\underline{\underline{C}}^{op}, Ab)_o$ whenever G is in f. p. $(\underline{\underline{C}}, Ab)_o$. As an easily verified consequence of these observations we have the following.

<u>Proposition 3.5</u>: The inverse dualities
V_o: f. p. $(\underline{\underline{A}}^{op}, Ab)_o \longrightarrow$ f. p. $(\underline{\underline{A}}, Ab)_o$ and
W_o: f. p. $(\underline{\underline{A}}, Ab)_o \longrightarrow$ f. p. $(\underline{\underline{A}}^{op}, Ab)_o$ induce inverse dualities
V_o: f. p. $(\underline{\underline{C}}^{op}, Ab)_o \longrightarrow$ f. p. $(\underline{\underline{C}}, Ab)_o$ and
W_o: f. p. $(\underline{\underline{C}}, Ab)_o \longrightarrow$ f. p. $(\underline{\underline{C}}^{op}, Ab)_o$. These dualities have the properties:

a) If $0 \longrightarrow (\ ,A) \xrightarrow{(\ ,g)} (\ ,B) \xrightarrow{(\ ,f)} (\ ,C) \longrightarrow F \longrightarrow 0$ is exact in f. p. $(\underline{\underline{C}}^{op}, Ab)_o$, then

$0 \longrightarrow (C, \ ) \xrightarrow{(f, \ )} (B, \ ) \xrightarrow{(g, \ )} (A, \ ) \longrightarrow V_o(F) \longrightarrow 0$ is exact in f. p. $(\underline{\underline{C}}, Ab)_o$.

b) If $0 \longrightarrow (C, \ ) \xrightarrow{(f, \ )} (B, \ ) \xrightarrow{(g, \ )} (A, \ ) \longrightarrow G \longrightarrow 0$ is exact in f. p. $(\underline{\underline{C}}, Ab)_o$, then

$0 \longrightarrow (\ ,A) \xrightarrow{(\ ,g)} (\ ,B) \xrightarrow{(\ ,f)} (\ ,C) \longrightarrow W_o(G) \longrightarrow 0$ is exact in f. p. $(\underline{\underline{C}}^{op}, Ab)_o$.

Our applications of these dualities to morphisms determined by objects are based on the following.

Proposition 3.6:

a) A functor F in f. p. $(\underline{\underline{C}}^{op}, Ab)_o$ is simple in f. p. $(\underline{\underline{C}}^{op}, Ab)_o$ if and only if F is simple in $(\underline{\underline{C}}^{op}, Ab)$.

b) A functor G in f. p. $(\underline{\underline{C}}, Ab)_o$ is simple in f. p. $(\underline{\underline{C}}, Ab)_o$ if and only if G is simple in $(\underline{\underline{C}}, Ab)$.

Proof:

a) Suppose F is simple in f. p. $(\underline{\underline{C}}^{op}, Ab)_o$. To show that it is simple in $(\underline{\underline{C}}^{op}, Ab)$ it suffices to show that if X is in $\underline{\underline{C}}$ and $\alpha: (\ ,X) \longrightarrow F$ is not zero, then it is an epimorphism. Since F is in f. p. $(\underline{\underline{C}}^{op}, Ab)_o$, there is an exact sequence $0 \longrightarrow A \xrightarrow{g} B \xrightarrow{f} C \longrightarrow 0$ in $\underline{\underline{C}}$ (i.e. $0 \longrightarrow A \xrightarrow{g} B \xrightarrow{f} C \longrightarrow 0$ is exact in $\underline{\underline{A}}$ with A, B, C in $\underline{\underline{C}}$) such that

$0 \longrightarrow (\ ,A) \xrightarrow{(\ ,g)} (\ ,B) \xrightarrow{(\ ,f)} (\ ,C) \xrightarrow{\beta} F \longrightarrow 0$ is exact in f. p. $(\underline{\underline{C}}^{op}, Ab)_o$ and, consequently, in $(\underline{\underline{C}}^{op}, Ab)$.

Let X be in $\underline{\underline{C}}$ and $\alpha: (\ ,X) \longrightarrow F$ a nonzero morphism. Then there is a morphism $h: X \longrightarrow C$ such that $\alpha = \beta(\ ,h)$.

Consider the pull back diagram in $\underline{\underline{A}}$

$$\begin{array}{ccccccccc} 0 & \longrightarrow & A & \longrightarrow & B \times_C X & \longrightarrow & X & \longrightarrow & 0 \\ & & \| & & \downarrow j & & \downarrow h & & \\ 0 & \longrightarrow & A & \longrightarrow & B & \longrightarrow & C & \longrightarrow & 0 \end{array}$$

Since X and A are in $\underline{\underline{C}}$, we have that $B \times_C X$ is in $\underline{\underline{C}}$. Hence we obtain the exact commutative diagram in f. p. $(\underline{\underline{C}}^{op}, Ab)_0$

$$\begin{array}{ccccccccccc} 0 & \longrightarrow & (\ ,A) & \longrightarrow & (\ ,B \times_C X) & \longrightarrow & (\ ,X) & \longrightarrow & G & \longrightarrow & 0 \\ & & \| & & \downarrow (\ ,j) & & \downarrow (\ ,h) & & & & \\ 0 & \longrightarrow & (\ ,A) & \longrightarrow & (\ ,B) & \longrightarrow & (\ ,C) & \xrightarrow{\beta} & F & \longrightarrow & 0 \end{array} .$$

Let $\gamma: G \longrightarrow F$ be the unique morphism which makes the diagram commute. Consequently $\operatorname{Im} \gamma = \operatorname{Im} \beta(\ ,h) = \operatorname{Im} \alpha$. Thus the fact that $\alpha \neq 0$ implies $\gamma \neq 0$. Since F is simple in f. p. $(\underline{\underline{C}}^{op}, Ab)_0$ and $\gamma: G \longrightarrow F$ is not zero with G in f. p. $(\underline{\underline{C}}^{op}, Ab)_0$, we have that $\operatorname{Im} \gamma = F$. Hence $\operatorname{Im} \alpha = \operatorname{Im} \gamma = F$. Since this holds for all nonzero morphisms $\alpha: (\ ,X) \longrightarrow F$ with X in $\underline{\underline{C}}$, it follows that F is simple in $(\underline{\underline{C}}^{op}, Ab)$.

That F in f. p. $(\underline{\underline{C}}^{op}, Ab)_0$ being simple in $(\underline{\underline{C}}^{op}, Ab)$ implies F is simple in f. p. $(\underline{\underline{C}}^{op}, Ab)_0$ is obvious.

b) Same as a).

§4. Almost Split Sequences. As in Section 3 we assume throughout this section that $\underline{\underline{C}}$ is a full subcategory of an abelian category A and that $\underline{\underline{C}}$ is closed under extensions and summands. We now apply the duality theory given in Section 3 to studying certain types of morphisms determined by objects in $\underline{\underline{C}}$ including almost split morphisms.

We say that a sequence $C_1 \xrightarrow{f_1} C_2 \xrightarrow{f_2} C_3$ of morphisms in $\underline{\underline{C}}$ is exact in $\underline{\underline{C}}$ if it is exact in $\underline{\underline{A}}$.

Proposition 4.1: Let $0 \longrightarrow A \xrightarrow{g} B \xrightarrow{f} C \longrightarrow 0$ be an exact sequence in $\underline{\underline{C}}$. Then Coker (,f) is simple in $(\underline{\underline{C}}^{op}, Ab)$ if and only if Coker (g,) is simple in $(\underline{\underline{C}}, Ab)$.

Proof: Since $0 \longrightarrow A \xrightarrow{g} B \xrightarrow{f} C \longrightarrow 0$ is an exact sequence in $\underline{\underline{C}}$, it follows that Coker (,f) is in f. p. $(\underline{\underline{C}}^{op}, Ab)_o$ and Coker (g,) is in f. p. $(\underline{\underline{C}}, Ab)_o$. Moreover V_o(Coker (,f)) = Coker (g,) where V_o is the duality V_o: f. p. $(\underline{\underline{C}}^{op}, Ab)_o \longrightarrow$ f. p. $(\underline{\underline{C}}, Ab)_o$ described in Section 3. Therefore Coker (,f) is simple in f. p. $(\underline{\underline{C}}^{op}, Ab)_o$ if and only if Coker (g,) is simple in f. p. $(\underline{\underline{C}}, Ab)_o$. Since by Proposition 3.6 we know that Coker (,f) is simple in f. p. $(\underline{\underline{C}}^{op}, Ab)_o$ if and only if it is simple in f. p. $(\underline{\underline{C}}^{op}, Ab)$ and Coker (g,) is simple in f. p. $(\underline{\underline{C}}, Ab)_o$ if and only if it is simple in f. p. $(\underline{\underline{C}}, Ab)$, it follows that Coker (,f) is simple in $(\underline{\underline{C}}^{op}, Ab)$ if and only if Coker (g,) is simple in $(\underline{\underline{C}}, Ab)$.

The following application of Proposition 4.1 is of particular interest to us.

Proposition 4.2: Let $0 \longrightarrow A \xrightarrow{g} B \xrightarrow{f} C \longrightarrow 0$ be exact in $\underline{\underline{C}}$.

a) Suppose Coker (,f) is simple in $(\underline{\underline{C}}^{op}, Ab)$. Then the following are equivalent.

i) f is right minimal.

ii) g is left almost split in $\underline{\underline{C}}$.

iii) End A is local.

b) Suppose Coker $(g, \ )$ is simple in $(\underline{\underline{C}}, Ab)$. Then the following are equivalent.

i) g is left minimal.

ii) f is right almost split in $\underline{\underline{C}}$.

iii) End C is local.

<u>Proof</u>:

a) Suppose f is right minimal. To show that $g: A \longrightarrow B$ is left almost split in $\underline{\underline{C}}$, we have to show that a morphism $h: A \longrightarrow X$ is a splittable monomorphism provided there does not exist a morphism $j: B \longrightarrow X$ such that $jg = h$.

Let the exact commutative diagram

$$\begin{array}{ccccccccc} 0 & \longrightarrow & A & \longrightarrow & B & \longrightarrow & C & \longrightarrow & 0 \\ & & \downarrow & & \downarrow & & \| & & \\ 0 & \longrightarrow & X & \longrightarrow & B \overset{A}{\times} X & \longrightarrow & C & \longrightarrow & 0 \end{array}$$

be a push out diagram in $\underline{\underline{A}}$. Since X and C are in $\underline{\underline{C}}$, the bottom row is an exact sequence in $\underline{\underline{C}}$. The hypothesis that there is no $j: B \longrightarrow X$ such that $jf = h$ is equivalent to the bottom row not splitting. From this exact commutative diagram we deduce the exact commutative diagram of functors in $(\underline{\underline{C}}^{op}, Ab)$

$$\begin{array}{ccccccccccc} 0 & \longrightarrow & (\ ,A) & \longrightarrow & (\ ,B) & \longrightarrow & (\ ,C) & \longrightarrow & F & \longrightarrow & 0 \\ & & \downarrow(\ ,h) & & \downarrow(\ ,\nu) & & \| & & \downarrow\beta & & \\ 0 & \longrightarrow & (\ ,X) & \longrightarrow & (\ ,B \overset{A}{\times} X) & \longrightarrow & (\ ,C) & \longrightarrow & G & \longrightarrow & 0 \end{array} .$$

Since $0 \longrightarrow X \longrightarrow B \overset{A}{\times} X \longrightarrow C \longrightarrow 0$ does not split, $G \neq 0$.

Since β is an epimorphism and F is simple by hypothesis, $\beta: F \longrightarrow G$ is an isomorphism. Therefore the subfunctor $H = \text{Im}((\ ,B) \longrightarrow (\ ,C))$ of $(\ ,C)$ is the same as $\text{Im}((\ ,B \overset{A}{\times} X) \longrightarrow (\ ,C))$. Since the exact sequences $0 \longrightarrow (\ ,A) \longrightarrow (\ ,B) \longrightarrow H \longrightarrow 0$ and $0 \longrightarrow (\ ,X) \longrightarrow (\ ,B \overset{A}{\times} X) \longrightarrow H \longrightarrow 0$ are projective resolutions, we have an exact commutative diagram

$$\begin{array}{ccccccccc}
0 & \longrightarrow & (\ ,A) & \xrightarrow{(\ ,g)} & (\ ,B) & \longrightarrow & H & \longrightarrow & 0 \\
 & & \downarrow(\ ,h) & & \downarrow(\ ,v) & & \| & & \\
0 & \longrightarrow & (\ ,X) & \longrightarrow & (\ ,B \overset{A}{\times} X) & \longrightarrow & H & \longrightarrow & 0 \\
 & & \downarrow(\ ,w) & & \downarrow(\ ,u) & & \| & & \\
0 & \longrightarrow & (\ ,A) & \longrightarrow & (\ ,B) & \longrightarrow & H & \longrightarrow & 0
\end{array}$$

Since $(\ ,B)$ is a projective cover of H, we know that $(\ ,u)(\ ,v) = (\ ,uv)$ is an isomorphism and hence that $(\ ,w)(\ ,h) = (\ ,wh)$ is an isomorphism. Therefore the composition $A \xrightarrow{h} X \xrightarrow{w} A$ is an isomorphism, which gives us our desired result that h is a splittable monomorphism. This finishes the proof that i) implies ii).

ii) implies iii). The fact that $g: A \longrightarrow B$ being left almost split implies that End A is local was shown in Proposition 2.4.

iii) implies i). This is a trivial consequence of the following more general result.

<u>Lemma 4.3</u>: Let $0 \longrightarrow A \xrightarrow{g} B \xrightarrow{f} C \longrightarrow 0$ be a nonsplit exact sequence in $\underline{\underline{C}}$.

a) If End A is local, then $f: B \longrightarrow C$ is right minimal in $\underline{\underline{C}}$.

b) If End C is local, then $g: A \longrightarrow B$ is left minimal.

Proof:

a) Suppose we are given $h: B \longrightarrow B$ such that $fh = f$. Then we have an exact commutative diagram in $\underline{\underline{C}}$

$$\begin{array}{ccccccccc} 0 & \longrightarrow & A & \xrightarrow{g} & B & \xrightarrow{f} & C & \longrightarrow & 0 \\ & & \downarrow j & & \downarrow h & & \| & & \\ 0 & \longrightarrow & A & \longrightarrow & B & \xrightarrow{f} & C & \longrightarrow & 0 \end{array} .$$

Since $0 \longrightarrow A \longrightarrow B \longrightarrow C \longrightarrow 0$ does not split, we know that it represents a nonzero element x of the abelian group $\operatorname{Ext}^1(C, A)$ viewing $\operatorname{Ext}^1(C, A)$ as a module over End A by means of the operation End A on A. The fact that the above diagram commutes means that the endomorphism j has the property that $jx = x$. Hence j is not in the rad End A. Thus j is an isomorphism since End A is local. From the fact that j is an isomorphism, it follows that $h: B \longrightarrow B$ is an isomorphism, which shows that $f: B \longrightarrow C$ is right minimal.

b) Dual to a).

Note: We have tacitly assumed in the above proof that the objects in $\underline{\underline{C}}$ have the property that $\operatorname{Ext}^1(C, A)$ is a set for all A and C in $\underline{\underline{C}}$. This will always be true in the situations of interest to us (for instance when there are enough projectives in $\underline{\underline{A}}$).

Returning to the proof of Proposition 4.2, it only remains to remark that b) is simply the dual of a) to finish the proof.

We now recall the definition of an almost split sequence (see [11, §2]).

We say that a morphism $f: B \longrightarrow C$ is *right almost split in* $\underline{\underline{C}}$ if for each morphism $h: X \longrightarrow C$ in $\underline{\underline{C}}$ which is not a splittable epimorphism there is a morphism $s: X \longrightarrow B$ such that $fg = h$. A morphism $g: A \longrightarrow B$ in $\underline{\underline{C}}$ is said to be *left almost split in* $\underline{\underline{C}}$ if for each morphism $h: A \longrightarrow Y$ in $\underline{\underline{C}}$ which is not a splittable monomorphism there is a morphism $t: B \longrightarrow Y$ such that $tg = h$. Finally, we say that an exact sequence $0 \longrightarrow A \xrightarrow{g} B \xrightarrow{f} C \longrightarrow 0$ in $\underline{\underline{C}}$ is an almost split sequence in $\underline{\underline{C}}$ if f is right almost split in $\underline{\underline{C}}$ and g is left almost split in $\underline{\underline{C}}$.

We now describe some of the basic properties of almost split sequences.

<u>Proposition 4.4</u>: Let $0 \longrightarrow A \xrightarrow{g} B \xrightarrow{f} C \longrightarrow 0$ be an exact sequence in $\underline{\underline{C}}$. Then the following are equivalent.

a) $0 \longrightarrow A \xrightarrow{g} B \xrightarrow{f} C \longrightarrow 0$ is an almost split sequence in $\underline{\underline{C}}$.

b) f is minimal right almost split in $\underline{\underline{C}}$.

c) f is right almost split in $\underline{\underline{C}}$ and End A is local.

d) g is minimal left almost split in $\underline{\underline{C}}$.

e) g is left almost split in $\underline{\underline{C}}$ and End C is local.

f) i) Coker $(\ ,f)$ is simple in $(\underline{\underline{C}}^{op}, \text{Ab})$ and the exact sequence

$$0 \longrightarrow (\ ,A) \longrightarrow (\ ,B) \longrightarrow (\ ,C) \longrightarrow \text{Coker} (\ ,f) \longrightarrow 0$$

is a minimal projective resolution of Coker $(\ ,f)$ in $(\underline{\underline{C}}^{op}, \text{Ab})$ and

ii) Coker$(g, \ )$ is simple in $(\underline{\underline{C}}, \text{Ab})$ and

$$0 \longrightarrow (C, \ ) \longrightarrow (B, \ ) \longrightarrow (A, \ ) \longrightarrow \text{Coker}(g, \ ) \longrightarrow 0$$

is a minimal projective resolution of Coker $(g, \ )$ in $(\underline{\underline{C}}, \text{Ab})$.

Proof: The equivalence of a) through e) is an immediate consequence of Proposition 4.2 and Lemma 4.3. The equivalence of a) and f) follows easily from Proposition 2.7 and 3.8 and the equivalence of a) through e).

Corollary 4.5:

a) A simple functor S in f. p. $(\underline{\underline{C}}^{op}, Ab)_0$ has a minimal projective resolution in $(\underline{\underline{C}}^{op}, Ab)$ if and only if there is an exact sequence $0 \longrightarrow (\ ,A) \xrightarrow{(\ ,g)} (\ ,B) \xrightarrow{(\ ,f)} (\ ,C) \longrightarrow S \longrightarrow 0$ with $0 \longrightarrow A \xrightarrow{g} B \xrightarrow{f} C \longrightarrow 0$ an almost split sequence in $\underline{\underline{C}}$.

b) A simple functor T in f. p. $(\underline{\underline{C}}, Ab)_0$ has a minimal projective resolution in f. p. $(\underline{\underline{C}}, Ab)$ if and only if there is an exact sequence $0 \longrightarrow (C,\ ) \xrightarrow{(f,\ )} (B,\ ) \xrightarrow{(g,\ )} (A,\ ) \longrightarrow T \longrightarrow 0$ with $0 \longrightarrow A \xrightarrow{g} B \xrightarrow{f} C \longrightarrow 0$ an almost split sequence.

Proof:

a) If there is an almost split sequence $0 \longrightarrow A \xrightarrow{g} B \xrightarrow{f} C \longrightarrow 0$ in $\underline{\underline{C}}$ such that $0 \longrightarrow (\ ,A) \longrightarrow (\ ,B) \longrightarrow (\ ,C) \longrightarrow S \longrightarrow 0$ is exact, then we know by Proposition 4.4 that this exact sequence of functors is a minimal projective resolution.

On the other hand, suppose S is in f. p. $(\underline{\underline{C}}^{op}, Ab)_0$ and has a minimal projective resolution in $(\underline{\underline{C}}^{op}, Ab)$, $0 \longrightarrow (\ ,A) \xrightarrow{(\ ,g)} (\ ,B) \xrightarrow{(\ ,f)} (\ ,C) \longrightarrow S \longrightarrow 0$. Then $0 \longrightarrow A \xrightarrow{g} B \longrightarrow C \longrightarrow 0$ is exact in $\underline{\underline{C}}$ by the definition of f. p. $(\underline{\underline{C}}^{op}, Ab)_0$ and is not split since $S \neq 0$. Since $(\ ,B) \xrightarrow{(\ ,f)} (\ ,C) \longrightarrow S \longrightarrow 0$ is a minimal projective presentation in $(\underline{\underline{C}}^{op}, Ab)$ we know that $f: B \longrightarrow C$ is minimal right almost split in $\underline{\underline{C}}$ (see Proposition 2.7). By Proposition 4.2 we know that this implies that End A is local. Thus $0 \longrightarrow A \xrightarrow{g} B \xrightarrow{f} C \longrightarrow 0$ is almost split in $\underline{\underline{C}}$ by Proposition 4.4.

As our final comment on almost split sequences in this section, we point out the following uniqueness theorem.

Theorem 4.6: Let $0 \longrightarrow A \xrightarrow{g} B \xrightarrow{f} C \longrightarrow 0$ and $0 \longrightarrow A' \xrightarrow{g'} B' \xrightarrow{f'} C' \longrightarrow 0$ be two almost split sequences in $\underline{\underline{C}}$. Then the following are equivalent:

a) The sequences are isomorphic.

b) Coker (,f) $\approx$ Coker (,f').

c) $C \approx C'$.

d) Coker(g,) $\approx$ Coker (g',).

e) $A \approx A'$.

Proof:

a) equivalent to b). Since by Proposition 4.4 the exact sequences

$$0 \longrightarrow (\ ,A) \xrightarrow{(\ ,g)} (\ ,B) \xrightarrow{(\ ,f)} (\ ,C) \longrightarrow \text{Coker}(\ ,f) \longrightarrow 0$$

and

$$0 \longrightarrow (\ ,A') \xrightarrow{(\ ,g')} (\ ,B') \xrightarrow{(\ ,f')} (\ ,C') \longrightarrow \text{Coker}(\ ,f') \rightarrow 0$$

are minimal projective resolutions, the equivalence of a) and b) follows from the uniqueness of minimal projective resolutions.

a) equivalent to c). Since a) obviously implies c) we only have to show that c) implies a). Suppose C and C' are isomorphic. Since Coker(,f) and Coker (,f') are simple and End C and End C' are local, we know that Coker (,f) $\cong$ (,C)/$\underline{r}$(,C) and

Coker $(\ ,f') \cong (\ ,C')/\underline{r}(\ ,C')$. The fact that $C \approx C'$ now implies Coker $(\ ,f) \approx$ Coker $(\ ,f')$ which implies a). The equivalence of a), c), and d) is the dual of the above argument.

As our final result of this section we give the following generalization of the uniqueness of almost split sequences.

Proposition 4.7: Let $0 \longrightarrow A \xrightarrow{g} B \xrightarrow{f} C \longrightarrow 0$ and $0 \longrightarrow A' \xrightarrow{g'} B' \xrightarrow{f'} C' \longrightarrow 0$ be two exact sequences in $\underline{\underline{C}}$.

a) Suppose f and f' are right minimal with the property that Coker $(\ ,f)$ and Coker $(\ ,f')$ are simple. Then the following are equivalent:

i) Coker $(\ ,f) \approx$ Coker $(\ ,f')$;

ii) Coker $(g, \ ) \approx$ Coker $(g', \ )$;

iii) $A \approx A'$.

b) Suppose g and g' are left minimal with the property that Coker $(g, \ )$ and Coker $(g', \ )$ are simple. Then the following are equivalent:

i) Coker $(g, \ ) \approx$ Coker$(g', \ )$;

ii) Coker $(f, \ ) \approx$ Coker $(f', \ )$;

iii) $C \approx C'$.

Proof:

a) i) equivalent to ii). Trivial consequence of the duality V_o: f. p. $(\underline{\underline{C}}^{op}, Ab)_o \longrightarrow$ f. p. $(\underline{\underline{C}}, Ab)_o$ since V_o(Coker $(\ ,f)$) = Coker $(g, \ )$ and V_o(Coker $(\ ,f')$) = Coker $(g', \ )$
ii) equivalent to iii). Since f and f' are right minimal and Coker $(\ ,f)$ and Coker $(\ ,f')$ are simple, we know by Proposition

4.2 that g and g' are left almost split. Hence by Proposition 2.7 we know that (A,) ⟶ Coker (g,) and (A',) ⟶ Coker (g',) are projective covers with kernels $\underline{r}$(A,) and $\underline{r}$(A',) respectively. The equivalence ii) and iii) now follows trivially.

b) Dual to a).

§5. Simple Functors Not Vanishing on mod Λ. Let Λ be an arbitrary ring. This section is devoted to applying the results of the previous sections to studying the simple functors in $((\text{Mod } \Lambda)^{op}, \text{Ab})$ which do not vanish on some finitely presented Λ-module. Analogous results for noetherian algebras over complete local rings and orders will be given later on. In the following proposition we summarize some of our previous results in a form particularly well suited to our present purposes.

Proposition 5.1: Let C be a nonprojective Λ-module in mod Λ and H a maximal right ideal in End C containing P(C, C). Let Γ = End TrC and I an injective Γ-module which is an injective envelope for the simple Γ-module End C/H (remember that we have a ring isomorphism $(\underline{\text{End}}\ \text{TrC})^{op} \cong \underline{\text{End}}\ C$).

Then there is a unique (up to isomorphism) exact sequence $0 \longrightarrow \text{Hom}_\Gamma(\text{TrC}, I) \xrightarrow{g} B \xrightarrow{f} C \longrightarrow 0$ of Λ-modules having the property f is a minimal right C-determined morphism such that (Im(,f))(C) = H. Moreover this uniquely determined exact sequence has the following properties.

a) Coker (,f) is a simple functor in $((\text{Mod } \Lambda)^{op}, \text{Ab})$ which vanishes on all projective Λ-modules.

b) g is left almost split and so

i) $\operatorname{Coker}(g, \ )$ is a simple functor in $((\operatorname{Mod} \Lambda)^{op}, \operatorname{Ab})$ which vanishes on all injective modules, and

ii) $\operatorname{End}_\Lambda(\operatorname{Hom}_\Gamma(\operatorname{Tr}C, I)) \approx \operatorname{End}_\Gamma(I)$ is local.

iii) $(\operatorname{Hom}_\Gamma(\operatorname{Tr}C, I), \ ) \longrightarrow \operatorname{Coker}(g, \ ) \longrightarrow 0$ is a projective cover.

c) $0 \longrightarrow \operatorname{Hom}_\Gamma(\operatorname{Tr}C, I) \xrightarrow{g} B \xrightarrow{f} C \longrightarrow 0$ is an almost split sequence if and only if $\operatorname{End} C$ is local.

Proof: The existence and uniqueness of the exact sequence $0 \longrightarrow \operatorname{Hom}_\Gamma(\operatorname{Tr}C, I) \xrightarrow{g} B \xrightarrow{f} C \longrightarrow 0$ follows from I, Theorem 3.9 and I, Proposition 3.12.

a) Since H is a maximal right ideal in $\operatorname{End} C$ and $f: B \longrightarrow C$ is right C-determined with $\operatorname{Im}(\ ,f)(C) = H$, it follows that $\operatorname{Coker}(\ ,f)$ is a simple functor in $((\operatorname{Mod} \Lambda)^{op}, \operatorname{Ab})$. The fact that f is an epimorphism clearly implies that $\operatorname{Coker}(\ ,f)(P) = 0$ for all projective modules P.

b) Since I is the Γ-injective envelope of the simple Γ-module $\operatorname{End} C/H$ we know that f is a minimal right C-determined morphism (see I, Theorem 3.9). This implies that g is left almost split as well as $\operatorname{End}_\Lambda(\operatorname{Hom}_\Gamma(\operatorname{Tr}C, I))$ is local (see Proposition 4.2). The fact that g is left almost split implies that $\operatorname{Coker}(g, \ )$ is a simple functor in $(\operatorname{Mod} \Lambda, \operatorname{Ab})$ (see Proposition 2.4). Also $\operatorname{Coker}(g, \ )$ vanishes on injectives since g is a monomorphism. Finally, because $\operatorname{End}_\Lambda(\operatorname{Hom}_\Gamma(\operatorname{Tr}C, I))$ is local $(\operatorname{Hom}_\Gamma(\operatorname{Tr}C, I), \ ) \longrightarrow \operatorname{Coker}(g, \ ) \longrightarrow 0$ is a projective cover.

c) Because g is left almost split, it follows from Proposition 4.4 that the exact sequence $0 \longrightarrow \operatorname{Hom}_\Gamma(\operatorname{Tr}C, I) \xrightarrow{g} B \xrightarrow{f} C \longrightarrow 0$ is almost split if and only if $\operatorname{End} C$ is local.

As a consequence of this proposition we have the following results.

Proposition 5.2: Let S be a simple functor in $((\text{Mod }\Lambda)^{op}, \text{Ab})$ vanishing on Λ.

a) Suppose C is in $\text{mod }\Lambda$ such that $S(C) \neq 0$. Then

i) $S(C)$ is a simple $(\text{End }C)^{op}$-module isomorphic to $\text{End }C/H$ for some maximal right ideal H containing $P(C, C)$.

ii) Let $\Gamma = \text{End TrC}$ and I a Γ-injective envelope for the simple Γ-module $S(C)$ and $0 \longrightarrow \text{Hom}_\Gamma(\text{TrC}, I) \xrightarrow{g} B \xrightarrow{f} C \longrightarrow 0$ an exact Λ-sequence such that f is minimal right C-determined with $\text{Im}(\ ,f)(C) = H$. Then

$$0 \longrightarrow (\ ,\text{Hom}_\Gamma(\text{TrC}, I)) \xrightarrow{(\ ,g)} (\ ,B) \xrightarrow{(\ ,f)} (\ ,C) \to S \to 0$$

is exact.

iii) The exact sequence

$$0 \longrightarrow (\ ,\text{Hom}_\Gamma(\text{TrC}, I)) \longrightarrow (\ ,B) \longrightarrow (\ ,C) \longrightarrow S \longrightarrow 0$$ is

a minimal projective resolution of S is and only if $\text{End }C$ is local or equivalently

$0 \longrightarrow \text{Hom}_\Gamma(\text{TrC}, I) \xrightarrow{g} B \xrightarrow{f} C \longrightarrow 0$ is almost split.

b) If $0 \longrightarrow A' \xrightarrow{g'} B' \xrightarrow{f'} C' \longrightarrow 0$ is an exact sequence with f' a minimal right C'-determined morphism such that

$$0 \longrightarrow (\ ,A') \xrightarrow{(\ ,g')} (\ ,B') \xrightarrow{(\ ,f')} (\ ,C') \longrightarrow S$$ is exact

then $A' \approx \text{Hom}_\Gamma(\text{TrC}, I)$.

Proof:

a) i) Since S is a simple functor, the fact that $S(C) \neq 0$ implies that $S(C)$ is a simple $(\text{End }C)^{op}$-module. Since $S(\Lambda) = 0$, we know that $S(P) = 0$ for all finitely generated projective Λ-modules P. Because C is finitely generated, it follows from this that $P(C, C)S(C) = 0$. Thus if H is a maximal right ideal in End C such that End $C/H \approx S(C)$ it follows that $H \supset P(C, C)$.

ii) and iii) follow from Proposition 5.1.

b) Follows from Proposition 4.7.

As an immediate consequence of Proposition 5.2 we have:

Corollary 5.3: Let S be a simple functor in $((\text{Mod } \Lambda)^{op}, \text{Ab})$. If $S(C) \neq 0$ for some C in mod Λ and $S(\Lambda) = 0$, then S is finitely presented. Further S has a minimal projective presentation in $((\text{Mod } \Lambda)^{op}, \text{Ab})$ if and only if there is a C in mod Λ with End C local such that $S(C) \neq 0$.

In view of these results, in order to finish our description of the simple functors S in $((\text{Mod } \Lambda)^{op}, \text{Ab})$, not vanishing on some C in mod Λ, it suffices to consider those S such that $S(P) \neq 0$ for some projective module P in mod Λ. We begin with the following somewhat more general result.

Proposition 5.4: Let M be a submodule of the projective Λ-module P (P not necessarily finitely generated). Then the following are equivalent.

a) M is a maximal submodule of P.

b) The subfunctor $(\ ,M)$ of $(\ ,P)$ is a maximal subfunctor.

c) The inclusion $f: M \longrightarrow P$ is a right P-determined morphism with $\text{Im}(\ ,f)(P)$ a maximal right ideal of End P.

Proof:

a) equivalent to b). By Proposition 2.1 we know that $(\ ,M)$ is a maximal subfunctor of $(\ ,P)$ if and only if the inclusion $f\colon M \longrightarrow P$ has the property that given a morphism $g\colon L \longrightarrow P$ either there exists $h\colon L \longrightarrow M$ such that $g = fh$ or the induced morphism $M \coprod L \longrightarrow P$ is a splittable epimorphism. It is easily checked that since P is projective this latter property of $f\colon M \longrightarrow P$ is equivalent to M being a maximal submodule of P.

b) equivalent to c). Special case of Proposition 2.1.

We now apply this result to obtain the following description of the maximal subfunctors of $(\ ,P)$ with P a finitely generated projective Λ-module.

Proposition 5.5: Let P be a finitely generated projective Λ-module and F a subfunctor of $(\ ,P)$. Then F is a maximal subfunctor of $(\ ,P)$ if and only if there is a maximal submodule M of P such that $F = (\ ,M)$. Thus there is a bijection between the maximal submodules of P and the maximal subfunctors of $(\ ,P)$ given by $M \longmapsto (\ ,M)$ for each maximal submodule M of P.

Proof: We have already seen in Proposition 5.3 that if M is a maximal submodule of P, then $(\ ,M)$ is a maximal subfunctor of $(\ ,P)$.

Suppose F is a maximal subfunctor of $(\ ,P)$. Then we know that $F(P)$ is a maximal right ideal H of $\operatorname{End} P$. Let N be the submodule of P generated by the submodules $f(P)$ as f runs through H. We claim that N is a proper submodule of P. For suppose $N = P$. Then because P is finitely generated there would be a finite number of morphisms $f_1, \ldots, f_n$ in H such that the induced morphism $\coprod_{i=1}^{n} P \longrightarrow P$ is an epimorphism. The fact that P is projective implies that there is a morphism $P \longrightarrow \coprod_{i=1}^{u} P$ such that the composition $P \longrightarrow \coprod_{i=1}^{u} P \longrightarrow P$ is the identity on P. But this is impossible since it implies that 1 is in $\operatorname{Im}((P, \coprod_{i=1}^{u} P) \longrightarrow (P, P)) \subset H$, which contradicts the fact that H is a right proper ideal of End P. Thus we have that N is a proper submodule of P.

Next it is obvious from the definition of N and the fact that N is a proper submodule of P that $H \subset \operatorname{Hom}_\Lambda(P, N) \neq \operatorname{End} P$. Since H is a maximal right ideal of End P, it follows that $H = (P, N)$. Because P is finitely generated, we know there is a maximal submodule M of P containing N. Therefore $(P, M) = (P, N)$ since (P, M) is a proper right ideal of End P containing the maximal right ideal $(P, N) = H$. Therefore by Proposition 5.1 we know that $(\ ,M)$ is a P-determined subfunctor of $(\ ,P)$ with $(P, M) = H = F(P)$. Therefore $F \subset (\ ,M)$ which implies that $F = (\ ,M)$ since F is a maximal subfunctor of $(\ ,P)$. This concludes the proof of the first part of Proposition 5.5.

The rest of the proof of the proposition now follows easily.

As a consequence of this description of the maximal subfunctors of $(\ ,P)$ with P a finitely generated projective Λ-module, we have the following.

Proposition 5.6: Let S be a simple functor in $((\operatorname{Mod} \Lambda)^{op}, \operatorname{Ab})$.

a) The following are equivalent:

i) $S(\Lambda) \neq 0$.

ii) $S(P) \neq 0$ for some finitely generated projective Λ-module P.

iii) There exists a maximal submodule M of P in mod Λ such that there is an exact sequence

$0 \longrightarrow (\ ,M) \longrightarrow (\ ,P) \longrightarrow S \longrightarrow 0$.

b) Suppose $S(P) \neq 0$ for some projective Λ-module P in mod Λ. Then

i) The projective dimension of S in $((\text{Mod } \Lambda)^{op}, \text{Ab})$ is at most 1.

ii) S is projective if and only if there is some simple projective Λ-module Q such that $S \approx (\ ,Q)$.

iii) S has a projective cover if and only if there is a projective Λ-module Q in mod Λ with End Q local such that $S \approx (\ ,Q)/(\ ,M)$ where M is the unique maximal submodule of Q.

Proof: Only parts ii) and iii) of b) do not follow immediately from the results cited before the statement of the proposition.

b) ii) Suppose Q is a simple projective module, then $(\ ,Q)$ is simple since (0) is the maximal submodule of Q (see Proposition 5.4). Suppose S is a projective simple functor and $S(P) \neq 0$ for some P in mod Λ. Then there is an epimorphism $(\ ,P) \longrightarrow S \longrightarrow 0$ which implies that S is isomorphic to a summand of $(\ ,P)$. Thus $S \approx (\ ,Q)$ for some summand Q of P. From the fact that S is simple it follows trivially that Q is simple.

b) iii) Suppose $(\ ,C) \longrightarrow S \longrightarrow 0$ is a projective cover for S. Since $(\ ,P) \longrightarrow S \longrightarrow 0$ is exact, it follows that $(\ ,C)$ is isomorphic to a summand of P and so C is projective. By Proposition 1.10 we know that End C is local since $(\, ,C) \longrightarrow S \longrightarrow 0$ is a projective cover of the simple functor S. It is not difficult to see that this combined with the fact that C is a projective module implies that C has a unique maximal submodule M. Thus by Proposition 5.5 we have that $S \approx (\, ,C)/(\, ,M)$. The rest of the proof is a consequence of Proposition 1.10.

An easily verified consequence of Proposition 5.6 is the following.

<u>Corollary 5.7</u>: The map $S \longrightarrow S(\Lambda)$ establishes a bijection between the isomorphism classes of simple functors S in $((\mathrm{Mod}\ \Lambda)^{op}, \mathrm{Ab})$ not vanishing on Λ and the isomorphism classes of simple Λ-modules. This correspondence has the following properties.

a) S has a projective cover in $((\mathrm{Mod}\ \Lambda)^{op}, \mathrm{Ab})$ if and only if the simple Λ-module $S(\Lambda)$ has a projective cover P. Furthermore, P is a projective cover for $S(\Lambda)$ if and only if $(\, ,P)$ is a projective cover for S.

b) S is projective if and only if $S(\Lambda)$ is a simple projective Λ-module. If $S(\Lambda)$ is simple projective, then $S \approx (\, ,S(\Lambda))$.

We now apply these results concerning simple functors S in $((\mathrm{Mod}\ \Lambda)^{op}, \mathrm{Ab})$ which do not vanish on Λ to finish our description of the simple factors of $(\, ,C)$ with C in $\mathrm{mod}\ \Lambda$. It has been shown that the simple factors of $(\, ,C)$ are of the form $(\, ,C)/(\, ,C)_H$ for some maximal right ideal H of $\mathrm{End}\ C$. A projective resolution for $(\, ,C)/(\, ,C)_H$ in case $H \supset P(C, C)$ was described in Proposition 5.1. Our aim now is to give a projective resolution for $(\, ,C)/(\, ,C)_H$ in case H is a maximal right ideal in $\mathrm{End}\ C$ not containing $P(C, C)$.

Proposition 5.8: Let C be in $\operatorname{mod} \Lambda$ and H a maximal right ideal in $\operatorname{End} C$, and S the simple functor $(\ ,C)/(\ ,C)_H$ in $((\operatorname{Mod} \Lambda)^{op}, Ab)$.

a) $S(\Lambda) \neq 0$ if and only if H does not contain $P(C, C)$.

b) $S(\Lambda) \neq 0$ if and only if there is a submodule M of C such that $0 \longrightarrow (\ ,M) \longrightarrow (\ ,C) \longrightarrow S \longrightarrow 0$ is exact, i.e., $(\ ,M) = (\ ,C)_H$.

c) Let M be a submodule of C. Then the following statements are equivalent.

i) $(\ ,C)/(\ ,M)$ is simple.

ii) If $h: L \longrightarrow C$ is a morphism with $\operatorname{Im} h \not\subset M$, then the induced morphism $M \amalg L \longrightarrow C$ is a splittable epimorphism.

Proof:

a) We have already shown in Proposition 5.1 that if $S(\Lambda) = 0$, then $H \supset P(C, C)$. Hence if H does not contain $P(C, C)$, then $S(\Lambda) \neq 0$.

Suppose now that $S(\Lambda) \neq 0$. Then there is an epimorphism $(\ ,\Lambda) \longrightarrow S \longrightarrow 0$. Since we also have the epimorphism $(\ ,C) \longrightarrow S \longrightarrow 0$, it follows that there are morphisms $g: C \longrightarrow \Lambda$ and $h: \Lambda \longrightarrow C$ such that the diagram

$$\begin{array}{ccccc} (\ ,C) & \longrightarrow & S & \longrightarrow & 0 \\ \downarrow (\ ,g) & & \| & & \\ (\ ,\Lambda) & \longrightarrow & S & \longrightarrow & 0 \\ \downarrow (\ ,h) & & \| & & \\ (\ ,C) & \longrightarrow & S & \longrightarrow & 0 \end{array}$$

is exact and commutes. From this it follows that the endomorphism hg: C ——> C in P(C, C) has the property Im(,hg) is not contained in $(,C)_H$. Hence $(\mathrm{Im}(,hg))(C) \not\subset H$ since $(,C)_H$ is determined by C. In particular $hg \notin H$. Since hg is in P(C, C), it follows that H does not contain P(C, C). This finishes the proof of a).

b) Suppose $S(\Lambda) \neq 0$. Then there is an exact sequence $0 \longrightarrow (,\underline{m}) \longrightarrow (,\Lambda) \longrightarrow S \longrightarrow 0$ with $\underline{m}$ a maximal left ideal in Λ (see Proposition 5.6). Then there is a morphism $f: C \longrightarrow \Lambda$ which makes the exact diagram

$$\begin{array}{ccccc} (,C) & \longrightarrow & S & \longrightarrow & 0 \\ \downarrow & & \| & & \\ (,\Lambda) & \longrightarrow & S & \longrightarrow & 0 \end{array}$$

commute. From the exact commutative diagram

$$\begin{array}{ccccc} & & 0 & & 0 \\ & & \downarrow & & \downarrow \\ & & \mathrm{Ker}\ f & = & \mathrm{Ker}\ f \\ & & \downarrow & & \downarrow \\ 0 & \longrightarrow & f^{-1}(\underline{m}) & \longrightarrow & C \\ & & \downarrow h & & \downarrow f \\ 0 & \longrightarrow & \underline{m} & \xrightarrow{inc} & \Lambda \end{array}$$

we deduce the exact commutative diagram

$$
(*)\qquad
\begin{array}{ccccccccc}
 & & 0 & & 0 & & & & \\
 & & \downarrow & & \downarrow & & & & \\
0 & \longrightarrow & (,\mathrm{Ker}\ f) & = & (,\mathrm{Ker}\ f) & & & & \\
 & & \downarrow & & \downarrow & & & & \\
0 & \longrightarrow & (,f^{-1}(\underline{m})) & \longrightarrow & (,C) & \longrightarrow & F & \longrightarrow & 0 \\
 & & \downarrow{\scriptstyle (,h)} & & \downarrow{\scriptstyle (,f)} & & \downarrow & & \\
0 & \longrightarrow & (,\underline{m}) & \longrightarrow & (,\Lambda) & \longrightarrow & S & \longrightarrow & 0 \\
 & & \downarrow{\scriptstyle \alpha} & & \downarrow{\scriptstyle \beta} & & & & \\
 & & G & \xrightarrow[\gamma]{} & H & & & & \\
 & & \downarrow & & \downarrow & & & & \\
 & & 0 & & 0 & & & &
\end{array}
$$

Suppose x is in $\mathrm{Ker}(G(L) \xrightarrow{\gamma_L} H(L))$ for some L in $\mathrm{Mod}\ \Lambda$. Let $u: L \longrightarrow \underline{m}$ be such that $\alpha_L(u) = x$. The fact that $\gamma_L(x) = 0$ implies there is a $v: L \longrightarrow C$ such that $fv = \mathrm{inc}\ u$. Then $\mathrm{Im}\ v \subset f^{-1}(\underline{m})$ and so there is $w: L \longrightarrow f^{-1}(\underline{m})$ such that $v = \mathrm{inc}\ w$. Thus $hw = u$ and so $0 = \alpha_L(u) = x$. Thus $\gamma: G \longrightarrow H$ is a monomorphism. It is well known, because of the commutativity and exactness of the diagram (*) that this implies that $F \longrightarrow S$ is a monomorphism. Since the composition $(,C) \xrightarrow{(,f)} (,\Lambda) \longrightarrow S$ is an epimorphism, it follows that $F \longrightarrow S$ is an epimorphism. Thus $F \longrightarrow S$ is an isomorphism. Therefore $0 \longrightarrow (,f^{-1}(\underline{m})) \longrightarrow (,C) \longrightarrow S \longrightarrow 0$ is exact and so $(,f^{-1}(\underline{m})) = (,C)_H$, which completes the proof of b), since the other implication is trivial.

c) This is a special case of Proposition 2.1.

As an immediate consequence of this proposition we have the following.

Corollary 5.9: Let C be in mod Λ. The the two sided ideal $P(C, C)$ of End C is not contained in the radical of End C if and only if there is a submodule M of C having the property that if $h: L \longrightarrow C$ is a morphism with Im h not contained in M, then the induced morphism $M \coprod L \longrightarrow C$ is a splittable epimorphism.

§6. Simple Functors in the Complete Case.

Throughout this section we assume that R is a complete noetherian local ring and Λ is a noetherian R-algebra. Our purpose is to give some more information about the simple functors in $((\text{Mod } \Lambda)^{op}, \text{Ab})$ which do not vanish on some module in noeth Λ.

Suppose S is a simple functor in $((\text{Mod } \Lambda)^{op}, \text{Ab})$ whose restriction to noeth Λ is not zero. Since S is additive and every module is a finite sum of indecomposable modules, S does not vanish on some indecomposable modules. Suppose C is an indecomposable module in noeth Λ such that $S(C) \neq 0$. Then there is a nonzero morphism $\alpha: (\ ,C) \longrightarrow S$ which must be an epimorphism since S is simple. The fact that Λ is a noetherian R-algebra with R a complete local ring implies that every indecomposable C in noeth Λ has a local endomorphism ring. Since $\alpha: (\ ,C) \longrightarrow S$ is an epimorphism with End C local, we know that $\alpha: (\ ,C) \longrightarrow S$ is a projective cover of S (see Proposition 1.10). Moreover $0 \longrightarrow \underline{r}(\ ,C) \longrightarrow (\ ,C) \longrightarrow S \longrightarrow 0$ is exact, i.e. $(\ ,C)/\underline{r}(\ ,C) \approx S$.

On the other hand, suppose C is indecomposable in noeth Λ. Since End C is local, $r(\ ,C)$ is the unique maximal subfunctor of $(\ ,C)$. Therefore $(\ ,C)/\underline{r}(\ ,C)$ is a simple functor and the natural epimorphism $(\ ,C) \longrightarrow (\ ,C)/\underline{r}(\ ,C)$ is a projective cover. Finally suppose C and C' are indecomposable modules in noeth Λ such that the simple functors

$(,C)/\underline{r}(,C)$ and $(,C')/\underline{r}(,C')$ are isomorphic. Then the functors $(,C)$ and $(,C')$ are isomorphic since they are projective covers of the isomorphic functors $(,C)/\underline{r}(,C)$ and $(,C')/\underline{r}(,C')$ respectively. Hence two indecomposable modules C and C' in noeth Λ are isomorphic if and only if the simple functors $(,C)/\underline{r}(,C)$ and $(,C')/\underline{r}(,C')$ are isomorphic.

We summarize this discussion as follows.

Proposition 6.1: The map noeth $\Lambda \longrightarrow ((\text{Mod }\Lambda)^{op}, \text{Ab})$ given by $C \longmapsto (,C)/\underline{r}(,C)$ for all C in noeth Λ induces a bijection between the isomorphism classes of indecomposable modules in noeth Λ and the isomorphism classes of simple functors in $((\text{Mod }\Lambda)^{op}, \text{Ab})$ which do not vanish on noeth Λ. Moreover a simple functor S is isomorphic to $(,C)/\underline{r}(,C)$ for some indecomposable C in noeth Λ if and only if $S(C) \neq 0$.

Having determined the projective covers for simple functors in $((\text{Mod }\Lambda)^{op}, \text{Ab})$ which do not vanish on noeth Λ we now show that these functors have minimal projective resolutions. We first consider those simple functors which do not vanish on some projective in noeth Λ.

Proposition 6.2: Let P be an indecomposable projective Λ-module and $\underline{r}$ the radical of Λ.

a) $\underline{r}P$ is the unique maximal submodule of P.

b) The inclusion $\underline{r}P \longrightarrow P$ is a minimal right almost split morphism.

c) $0 \longrightarrow (,\underline{r}P) \longrightarrow (,P) \longrightarrow (,P)/\underline{r}(,P) \longrightarrow 0$ is a minimal projective resolution of the simple functor $(,P)/\underline{r}(,P)$.

Suppose S is a simple functor in $((\mathrm{Mod}\ \Lambda)^{op}, \mathrm{Ab})$ such that $S(P) \neq 0$.

d) Then $S \cong (\ ,P)/\underline{r}(\ ,P)$ and there is an exact sequence $0 \longrightarrow (\ ,\underline{r}P) \longrightarrow (\ ,P) \longrightarrow S \longrightarrow 0$ which is a minimal projective resolution of S.

e) $\mathrm{pd}\ S \leq 1$ and $\mathrm{pd}\ S = 0$ if and only if $\underline{r}P = 0$, i.e. P is a simple projective. If $\mathrm{pd}\ S = 0$, then $S \cong (\ ,P)$ with P a simple projective.

Proof: Follows easily by either direct calculations or by results given in Section 5.

Having described minimal projective resolutions for simple functors in $((\mathrm{Mod}\ \Lambda)^{op}, \mathrm{Ab})$ not vanishing on an indecomposable projective in noeth Λ, we now turn our attention to describing minimal projective resolutions for the simple functors which vanish on the projectives in noeth Λ but not on all of noeth Λ.

Proposition 6.3: Let C be an indecomposable module in $\mathrm{noeth}_P\Lambda$.

a) There exists an almost split sequence $0 \longrightarrow \mathrm{DTr}C \xrightarrow{g} B \xrightarrow{f} C \longrightarrow 0$. This sequence has the following properties.

i)

$$0 \longrightarrow (\ ,\mathrm{DTr}C) \xrightarrow{(\ ,g)} (\ ,B) \xrightarrow{(\ ,f)} (\ ,C) \longrightarrow \mathrm{Coker}\ (\ ,f) \longrightarrow 0$$

is a minimal projective resolution of the simple functor $\mathrm{Coker}\ (\ ,f)$ which does not vanish on C.

ii)

$$0 \longrightarrow (\ ,C) \xrightarrow{(\ ,f)} (\ ,B) \xrightarrow{(\ ,g)} (\mathrm{DTr}C,\ ) \longrightarrow \mathrm{Coker}\ (g,\ ) \longrightarrow 0$$

is a minimal projective resolution of the simple functor Coker (g,) which does not vanish on DTrC.

b) Suppose S is a simple functor in $((\text{Mod } \Lambda)^{op}, \text{Ab})$ such that $S(C) \neq 0$. Then $S \approx \text{Coker}(\ ,f)$, so there is an exact sequence $0 \longrightarrow (\ ,\text{DTrC}) \xrightarrow{(\ ,g)} (\ ,B) \xrightarrow{(\ ,f)} (\ ,C) \longrightarrow S \longrightarrow 0$ which is a minimal projective resolution of S.

c) Suppose T is a simple functor in (Mod Λ, Ab) such that $T(\text{DTrC}) \neq 0$. Then $T \approx \text{Coker}(g,\ )$, so there is an exact sequence $0 \longrightarrow (C,\ ) \longrightarrow (B,\ ) \longrightarrow (\text{DTrC},\ ) \longrightarrow T \longrightarrow 0$ which is a minimal projective resolution of T.

Proof:

a) Since End C is local, we have by Proposition 5.1 that there is an almost split sequence $0 \longrightarrow \text{Hom}_\Gamma(\text{TrC}, I) \longrightarrow B \longrightarrow C \longrightarrow 0$ where $\Gamma = \text{End TrC}$ and I is an injective envelope of the unique simple Γ-module (remember Γ is a local ring). But $I = D(\Gamma)$ where $D(\Gamma) = \text{Hom}_R(\Gamma, I(R/m))$. Thus $\text{Hom}_\Gamma(\text{TrC}, I) \cong \text{Hom}_R(\text{TrC}, I(R/m)) = \text{DTrC}$. Thus the existence of the almost split sequence is established.

The rest of the proposition is just a recapitulation of earlier results.

Having described the minimal projective resolutions of simple functors in $((\text{Mod } \Lambda)^{op}, \text{Ab})$ which do not vanish on noeth Λ, we now turn our attention to describing the minimal projective resolutions for the simple functors in (Mod Λ, Ab) which do not vanish on art Λ. For the most part, proofs will be omitted since these results are essentially restatements of earlier results as were the results given in this section so far. Those readers wishing to supply proofs should consult the obvious analogues already established in this section.

We begin with the following analogue of Proposition 6.1.

<u>Proposition 6.4</u>:

a) For each indecomposable A in $\operatorname{art} \Lambda$, we have that $(A, \ )/\underline{r}(A, \ )$ is a simple functor not vanishing on A and that the natural morphism $(A, \ ) \longrightarrow (A, \ )/\underline{r}(A, \ )$ is a projective cover.

b) The map $\operatorname{art} \Lambda \longrightarrow (\operatorname{Mod} \Lambda, \operatorname{Ab})$ given by $A \longrightarrow (A, \ )/\underline{r}(A, \ )$ for all A in $\operatorname{art} \Lambda$ induces a bijection between the isomorphism classes of indecomposable modules in $\operatorname{art} \Lambda$ and the isomorphism classes of simple functors in $(\operatorname{Mod} \Lambda, \operatorname{Ab})$ which do not vanish on $\operatorname{art} \Lambda$. Moreover, a simple functor T is isomorphic to $(A, \ )/r(A, \ )$ for some indecomposable A in $\operatorname{art} \Lambda$ if and only if $T(A) \neq 0$.

Having determined the projective covers for the simple functors in $(\operatorname{Mod} \Lambda, \operatorname{Ab})$ not vanishing on $\operatorname{art} \Lambda$, we now show that these functors have minimal projective resolutions. We first consider those simple functors which do not vanish on some injective module in $\operatorname{art} \Lambda$.

<u>Proposition 6.5</u>: Let I be an indecomposable injective module in $\operatorname{art} \Lambda$ and $\operatorname{soc} I$, the socle of I, which is a simple Λ-module.

a) The epimorphism $I \longrightarrow I/\operatorname{soc} I$ is a minimal left almost split morphism.

b) $0 \longrightarrow (I/\text{soc } I, \ ) \longrightarrow (I, \ ) \longrightarrow (I, \ )/\underline{r}(I, \ ) \longrightarrow 0$ is a minimal projective resolution of the simple functor $(I, \ )/\underline{r}(I, \ )$.

Suppose T is a simple functor in $(\text{Mod } \Lambda, \text{Ab})$ such that $T(I) \neq 0$.

c) Then $T \approx (I, \ )/(I/\text{soc } I, \ )$ and so there is an exact sequence $0 \longrightarrow (I/\text{soc } I, \ ) \longrightarrow (I, \ ) \longrightarrow T \longrightarrow 0$ which is a minimal projective resolution of T.

d) $\text{pd } T \leq 1$ and $\text{pd } T = 0$ if and only if I is simple. If $\text{pd } T = 0$, then $T \approx (I, \ )$ with I a simple injective.

Proof: Essentially the dual of Proposition 6.2.

Having described minimal projective resolutions for simple functors in $(\text{Mod } \Lambda, \text{Ab})$ which do not vanish on some indecomposable injective in $\text{art } \Lambda$, we now describe the minimal projective resolutions of simple functors in $(\text{Mod } \Lambda, \text{Ab})$ which do not vanish on $\text{art}_I \Lambda$.

Proposition 6.6: Let A be an indecomposable module in $\text{art}_I \Lambda$.

a) There exists an almost split sequence $0 \longrightarrow A \xrightarrow{g} B \xrightarrow{f} \text{TrDA} \longrightarrow 0$. This sequence has the following properties:

i)

$$0 \longrightarrow (\text{TrDA}, \ ) \xrightarrow{(f, \ )} (B, \ ) \xrightarrow{(g, \ )} (A, \ ) \longrightarrow \text{Coker } (g, \ ) \longrightarrow 0$$

is a minimal projective resolution of the simple functor $\text{Coker } (g, \ )$ which does not vanish on A.

ii)

$$0 \longrightarrow (\ ,A) \xrightarrow{(\ ,g)} (\ ,B) \xrightarrow{(\ ,f)} (\ ,\text{TrDA}) \longrightarrow \text{Coker } (\ ,f) \longrightarrow 0$$

is a minimal projective resolution for the simple functor $\text{Coker } (\ ,f)$ which does not vanish on TrDA.

b) Suppose T is a simple functor in $(\text{Mod}\ \Lambda, \text{Ab})$ such that $T(A) \neq 0$. Then $T \approx \text{Coker}\ (g, \ )$, so there is an exact sequence

$$0 \longrightarrow (\text{TrDA}, \ ) \xrightarrow{(f, \ )} (B, \ ) \xrightarrow{(g, \ )} (A, \ ) \longrightarrow T \longrightarrow 0$$

which is a minimal projective resolution of T.

c) Suppose S is a simple functor in $((\text{Mod}\ \Lambda)^{op}, \text{Ab})$ such that $S(\text{TrDA}) \neq 0$. Then $S \approx \text{Coker}\ (\ ,f)$, so there is an exact sequence

$$0 \longrightarrow (\ ,A) \xrightarrow{(\ ,g)} (\ ,B) \xrightarrow{(\ ,f)} (\ ,\text{TrDA}) \longrightarrow S \longrightarrow 0$$

which is a minimal projective resolution of S.

Proof: Simply a restatement of Proposition 6.3 using the fact that $A \approx \text{DTr}(\text{TrDA})$.

§7. Simple Functors for Orders. We assume throughout this section that Λ is an R-order with $d = \dim R$. Our aim is to study the simple functors in $(L(\Lambda)^{op}, \text{Ab})$ and $(L(\Lambda), \text{Ab})$. We begin with the simple functors in $(L(\Lambda)^{op}, \text{Ab})$ and $(L(\Lambda), \text{Ab})$ which vanish on the projective and the injective lattices respectively.

The following is simply a restatement of earlier results which we recall for the convenience of the reader.

Lemma 7.1:

a) Let S be a simple functor in $(L(\Lambda)^{op}, \text{Ab})$. Then the following are equivalent.

i) $S(P) = 0$ for all projectives P in $L(\Lambda)$.

ii) $S(C) \neq 0$ for some nonprojective C in $L(\Lambda)$.

iii) $S(C) \neq 0$ if and only if there is a maximal right ideal H

of End C containing P(C, C) such that $S \approx (\ ,C)/(\ ,C)_H$.

b) Let T be a simple functor in $(L(\Lambda), Ab)$. Then the following are equivalent.

i) $T(I) = 0$ for all injective lattices I in $L(\Lambda)$.

ii) $T(A) \neq (0)$ for some noninjective lattice C in $L(\Lambda)$.

iii) $T(A) \neq 0$ if and only if there is a maximal left ideal H of End A containing I(A, A) such that $T \approx (A,\ )/(A,\ )_H$

We now describe projective resolutions for the simple functors satisfying the above equivalent conditions.

Proposition 7.2: Let C be a nonprojective lattice and H a maximal right ideal of End C containing P(C, C). There is an exact sequence $0 \longrightarrow \mathrm{Hom}_R(\mathrm{Tr}_L C, R) \xrightarrow{g} B \xrightarrow{f} C \longrightarrow 0$ in $L(\Lambda)$ having the following properties.

a) f is right C-determined in $L(\Lambda)$ with $\mathrm{Im}(\ ,f)(C) = H$.

b) The exact sequence

$$0 \longrightarrow (\ ,\mathrm{Hom}_R(\mathrm{Tr}_L C, R)) \xrightarrow{(\ ,g)} (\ ,B) \xrightarrow{(\ ,f)} (\ ,C) \longrightarrow \mathrm{Coker}\ (f,\ ) \longrightarrow 0$$

is a projective resolution of the simple functor Coker (,f).

c)

$$0 \longrightarrow (C,\ ) \xrightarrow{(f,\ )} (B,\ ) \xrightarrow{(g,\ )} (\mathrm{Hom}_R(\mathrm{Tr}_L C, R),\ ) \longrightarrow \mathrm{Coker}\ (g,\ ) \longrightarrow 0$$

is a projective resolution of the simple functor Coker (g,).

d) g is a left determined $\mathrm{Hom}_R(\mathrm{Tr}_L C, R)$-morphism in $L(\Lambda)$ with $\mathrm{Im}(g,\ )(\mathrm{Hom}_R(\mathrm{Tr}_L C, R))$ a maximal left ideal in $\mathrm{End}(\mathrm{Hom}_R(\mathrm{Tr}_L C, R))$.

e) If S is a simple functor in $(L(\Lambda)^{op}, Ab)$ such that $S \approx (\ ,C)/(\ ,C)_H$, then there is an exact sequence

$$0 \longrightarrow (\ ,\mathrm{Hom}_R(\mathrm{Tr}_L C, R)) \xrightarrow{(\ ,g)} (\ ,B) \xrightarrow{(\ ,f)} (\ ,C) \longrightarrow S \longrightarrow 0$$

which is a projective resolution of S. In particular, S is finitely presented in $(L(\Lambda)^{op}, Ab)$.

Proof:

a) Since $(C, C)/H$ is a simple End(C)-module, it follows from the basic existence theorems that there is an exact sequence $0 \longrightarrow Hom_R(Tr_L C, R) \xrightarrow{g} B \xrightarrow{f} C \longrightarrow 0$ in $L(\Lambda)$ with the desired properties.

b) Since $f: B \longrightarrow C$ is a right C-determined morphism with $Im(\ ,f)(C)$ a maximal right ideal of End C, we know that Coker $(\ ,f)$ is simple (see Proposition 2.1). The rest of b) follows trivially from this remark.

c) Since $0 \longrightarrow Hom_R(Tr_L C, R) \xrightarrow{g} B \xrightarrow{f} C \longrightarrow 0$ is an exact sequence in $L(\Lambda)$, we know by Proposition 4.1 that Coker $(\ ,f)$ is simple in $(L(\Lambda)^{op}, Ab)$ if and only if Coker $(g, \ )$ is simple in $(L(\Lambda), Ab)$. Thus the fact that Coker $(\ ,f)$ is simple implies Coker $(g, \ )$ is simple.

d) Trivial consequence of c).

e) Trivial.

Combining these results, we have the following.

Proposition 7.3: Each simple functor S in $(L(\Lambda)^{op}, Ab)$ vanishing on projectives has a projective resolution of the form

$$0 \longrightarrow (\ ,Hom_R(Tr_L C, R)) \xrightarrow{(\ ,g)} (\ ,B) \xrightarrow{(\ ,f)} (\ ,C) \longrightarrow S \longrightarrow 0$$

where $0 \longrightarrow Hom_R(Tr_L C, R) \xrightarrow{g} B \xrightarrow{f} C \longrightarrow 0$ is exact in $L(\Lambda)$.

As usual we have the dual results.

Proposition 7.4: Let A be a noninjective lattice and H a maximal left ideal of $\operatorname{End} A$ containing $I(A, A)$. There is an exact sequence $0 \longrightarrow A \xrightarrow{g} B \xrightarrow{f} \operatorname{Tr}_L\operatorname{Hom}_R(A, R) \longrightarrow 0$ in $L(\Lambda)$ having the following properties.

a) g is left A-determined with $\operatorname{Im}(g, \ )(A) = H$.

b) The exact sequence

$$0 \longrightarrow (\operatorname{Tr}_L\operatorname{Hom}_R(A, R)) \xrightarrow{(f,\ )} (B, \ ) \xrightarrow{(g,\ )} (A, \ ) \to \operatorname{Coker}(g, \ ) \to 0$$

is a projective resolution of the simple functor $\operatorname{Coker}(g, \ )$.

c)

$$0 \longrightarrow (\ ,A) \xrightarrow{(\ ,g)} (\ ,B) \xrightarrow{(\ ,f)} (\ ,\operatorname{Tr}_L\operatorname{Hom}_R(A, R)) \longrightarrow \operatorname{Coker}(\ ,f) \longrightarrow 0$$

is a projective resolution of the simple functor $\operatorname{Coker}(\ ,f)$.

d) f is right $\operatorname{Tr}_L(\operatorname{Hom}_R(A, R))$-determined with $\operatorname{Im}(f, \ )(\operatorname{Tr}_L(\operatorname{Hom}_R(A, R))$ a maximal right ideal of $\operatorname{End}\operatorname{Tr}_L(\operatorname{Hom}_R(A, R))$.

e) If T is a simple functor in $(L(\Lambda), \mathrm{Ab})$ such that $T \approx (A, \ )/(A, \ )_H$, then there is an exact sequence $0 \longrightarrow (\operatorname{Tr}_L\operatorname{Hom}_R(A, R), \ ) \xrightarrow{(f,\ )} (B, \ ) \xrightarrow{(g,\ )} (A, \ ) \longrightarrow T \longrightarrow 0$ which is a projective resolution of T.

Corollary 7.5: Let T be a simple functor in $(L(\Lambda), \mathrm{Ab})$ which vanishes on injective lattices. Then there is a projective resolution $0 \longrightarrow (\operatorname{Tr}_L\operatorname{Hom}_R(A, R), \ ) \xrightarrow{(f,\ )} (B, \ ) \xrightarrow{(g,\ )} (A, \ ) \longrightarrow T \longrightarrow 0$ with $0 \longrightarrow A \xrightarrow{g} B \xrightarrow{f} \operatorname{Tr}_L\operatorname{Hom}_R(A, R) \longrightarrow 0$ exact in $L(\Lambda)$. In particular T is finitely presented.

We now consider simple functors in $(L(\Lambda)^{op}, \mathrm{Ab})$ which do not vanish on projectives. If S is a simple functor such that $S(P) \neq 0$, then we have an epimorphism $(\ ,P) \longrightarrow S \longrightarrow 0$. So it would be helpful to be able to determine the maximal subfunctors in $(L(\Lambda)^{op}, Ab)$ of $(,P)$. Suppose we consider $(,P)$ a functor from all of Mod Λ to Ab. Then in Proposition 5.5 we saw that the maximal subfunctors of $(,P)$ are the inclusion morphisms $(,M) \longrightarrow (,P)$ as M ranges over all maximal submodules of P. This result suggests the following considerations.

Suppose $\dim R = d = 0$. Then each submodule of P is a lattice since $L(\Lambda) = \text{noeth}\ \Lambda$ in this case. In particular, each maximal submodule M of P is a lattice and it is obvious that each of the inclusions $0 \longrightarrow (,M) \longrightarrow (,P)$ is a maximal subfunctor of $(,P)$ in $(L(\Lambda)^{op}, Ab)$. The proof of Proposition 5.5 can be used to show these are precisely all the maximal subfunctors of $(,P)$ in $(L(\Lambda)^{op}, Ab)$ in this case.

Suppose now that $\dim R = d = 1$. Let C' be a nonzero submodule of the lattice C. Then C' has no nonzero submodules of finite length since this is true for C. Thus C' is Cohen-Macauley since $\dim R = 1$. Hence we know that C' is a lattice if $(C')_{\underline{p}}$ is $\Lambda_{\underline{p}}$-projective for all nonmaximal prime ideals $\underline{p}$ of R. Suppose C/C" is a module of finite length. Then $(C/C")_{\underline{p}} = 0$ for all nonmaximal prime ideals $\underline{p}$ in R. Hence $C"_{\underline{p}} = C_{\underline{p}}$ for all nonmaximal primes $\underline{p}$ or R and so $C"_{\underline{p}}$ is $\Lambda_{\underline{p}}$-projective for all nonmaximal prime ideals $\underline{p}$ of Λ. Hence if C' is a submodule of the lattice C such that C/C' has finite length then C' is a lattice. In particular, every maximal submodule of a lattice is a lattice. This observation enables us to show, as in Proposition 5.5, that if P is a projective lattice, then the maximal subfunctors of $(,P)$ in $(L(\Lambda)^{op}, Ab)$ are precisely the subfunctors of $(,P)$ of the form $\text{Im}((,M) \longrightarrow (,P))$ given by the inclusions of the maximal sublattices M of P.

Summarizing this discussion we have the following.

Proposition 7.6: Let Λ be an R-order with $\dim R \leq 1$.

a) If C' is a submodule of the lattice C such that C/C' has finite length, then C' is a lattice.

b) If P is a projective lattice, then the maximal subfunctors of $(\ ,P)$ in $(L(\Lambda)^{op}, Ab)$ are precisely the subfunctors $\mathrm{Im}((\ ,M) \longrightarrow (\ ,P))$ as $M \longrightarrow P$ ranges over all inclusion morphisms of maximal sublattices M of P.

c) If S is a simple functor in $(L(\Lambda)^{op}, Ab)$ such that $S(P) \neq 0$ for some projective lattice P, then there is a maximal sublattice M of P such that the sequence $0 \longrightarrow (\ ,M) \longrightarrow (\ ,P) \longrightarrow S \longrightarrow 0$ is exact and hence a projective resolution of S in $(L(\Lambda)^{op}, Ab)$. In particular, S is finitely presented in $(L(\Lambda)^{op}, Ab)$ and $\mathrm{pd}\, S \leq 1$.

Continuing with our assumption that $\dim R = d \leq 1$, it is possible, using the duality $(\ ,R): L(\Lambda) \longrightarrow L(\Lambda^{op})$, to deduce from Proposition 7.7 analogous results concerning the simple functors in $(L(\Lambda), Ab)$ not vanishing on injective lattices.

Suppose A in $L(\Lambda)$ is an injective lattice. Then $\mathrm{Hom}_R(A, R)$ in $L(\Lambda^{op})$ is projective. Suppose $0 \longrightarrow M \xrightarrow{g} \mathrm{Hom}_R(A, R)$ is a maximal sublattice of $\mathrm{Hom}_R(A, R)$. This induces a morphism of lattices $A \xrightarrow{\mathrm{Hom}_R(g, R)} \mathrm{Hom}_R(M, R)$ which is an epimorphism with simple kernel if $\dim R = 0$ and is a monomorphism with a simple Λ-module as cokernel if $\dim R = 1$. But in either case $\mathrm{Im}((\mathrm{Hom}_R(M, R), \ ) \longrightarrow (A, \ ))$ is a maximal subfunctor of $(A, \ )$ in $(L(\Lambda), Ab)$, as we now show.

Since $\mathrm{Hom}_R(\ ,R): (L(\Lambda) \longrightarrow L(\Lambda^{op})$ is a duality, we know by I, Lemma 2.3 and Proposition 2.4 that $A \xrightarrow{\mathrm{Hom}_R(g, R)} \mathrm{Hom}_R(M, R)$ is left A-determined with $\operatorname{Im}(\operatorname{Hom}_R(g, R),)(A)$ a maximal left ideal of End A because $M \xrightarrow{g} \operatorname{Hom}_R(A, R)$ is right $\operatorname{Hom}_R(A, R)$-determined with $\operatorname{Im}(,g)(\operatorname{Hom}_R(A, R))$ a maximal right ideal of End $\operatorname{Hom}_R(A, R)$. Therefore $\operatorname{Im}(\operatorname{Hom}_R(g, R),)$ is a maximal subfunctor of $(A,)$ by Proposition 2.1. We now show that we obtain all the maximal subfunctors of $(A,)$ in this way.

Suppose $F \subset (A,)$ is a maximal subfunctor of $(A,)$. Then we know by Corollary 1.2 that $F(A)$ is a maximal left ideal of End A and F is determined by A. Under the ring isomorphism End $A \longrightarrow (\text{End } \operatorname{Hom}_R(A,R))^{op}$ we know that $\operatorname{Hom}_R(F(A), R)$ is a maximal right ideal of $(\operatorname{End}(\operatorname{Hom}_R(A, R))^{op}$. Thus $(,\operatorname{Hom}_R(A, R))_{\operatorname{Hom}_R(F(A), R)}$ is a maximal subfunctor of $(,\operatorname{Hom}_R(A, R))$. Hence by Proposition 7.6, there is a maximal sublattice $0 \longrightarrow M \xrightarrow{g} \operatorname{Hom}_R(A, R)$ of $\operatorname{Hom}_R(A, R)$ which is right $\operatorname{Hom}_R(A, R)$-determined with $\operatorname{Im}(g,)(\operatorname{Hom}_R(A, R)) = DF(A)$. Hence by I, Lemma 2.3 and Proposition 2.4, $A \xrightarrow{\operatorname{Hom}_R(g, R)} \operatorname{Hom}_R(M, R)$ is left A-determined with $\operatorname{Im}(\operatorname{Hom}_R(g, R),)(A) = F(A)$. Therefore $\operatorname{Im}(\operatorname{Hom}_R(g, R),) = F$.

Summarizing this discussion we have the following.

Proposition 7.7: Let Λ be an R-order with $\dim R \leq 1$.

a) Let A be an injective lattice. A subfunctor F of $(A,)$ is maximal if and only if there is a morphism of lattices $A \xrightarrow{g} B$ satisfying both of the following:

i) $0 \longrightarrow \operatorname{Hom}_R(B, R) \xrightarrow{\operatorname{Hom}_R(g, R)} \operatorname{Hom}_R(A, R)$ is exact with $\operatorname{Im}(\operatorname{Hom}_R(g, R))$ a maximal sublattice of $\operatorname{Hom}_R(A, R)$.

b) $0 \longrightarrow (B,) \xrightarrow{(g,)} (A,)$ is exact and $\operatorname{Im}(g,) = F$.

c) If $\dim R = 0$, then g is an epimorphism with Ker g simple.

If $\dim R = 1$, then g is a monomorphism with Coker g simple.

If T is a simple functor in $(L(\Lambda), Ab)$ such that $T(A) \neq 0$, then there is a morphism $A \xrightarrow{g} B$ in $L(\Lambda)$ such that $0 \to (B, \ ) \to (A, \ ) \to T \to 0$ is a projective resolution of T. Thus T is a finitely presented functor in $(L(\Lambda), Ab)$ and pd $T \leq 1$.

Proof: The only parts of this proposition that need any explanation are parts b) and c).

By a) we have that (*) $0 \to \text{Hom}_R(B, R) \to \text{Hom}_R(A, R) \to V \to 0$ is exact with V a simple Λ-module. If $\dim R = 0$, then R is self injective. Applying $\text{Hom}_R(\ ,R)$ to (*) we have the exact sequence $0 \to \text{Hom}_R(V, R) \to A \xrightarrow{g} B \to 0$ in noeth Λ. Since $\text{Hom}_R(\ ,R)$: noeth $\Lambda \to$ noeth Λ^{op} we know that $\text{Hom}_R(V, R)$ is a simple Λ-module since V is a simple Λ-module. Also $0 \to (A, \ ) \to (B, \ )$ is obviously exact in $(L(\Lambda), Ab)$.

Suppose $\dim R = 1$. Then applying $\text{Hom}_R(\ ,R)$ to (*) we obtain the exact sequence $0 \to A \xrightarrow{g} B \to \text{Ext}^1_R(V, R) \to 0$ since $\text{Hom}_R(V, R) = 0$ because V is of finite length and depth $R = 1$ and $\text{Ext}^1_R(X, R) = 0$ for all lattices X. Because R is a Gorenstein ring of dimension 1, the functor $\text{Ext}^1_R(\ ,R)$: noeth $\Lambda^{op} \to$ noeth Λ is a duality on modules of finite length. Hence the fact that V is a simple Λ^{op}-module implies $\text{Ext}^1_R(V, R)$ is a simple Λ-module. Hence $0 \to (B, X) \to (A, X)$ is exact for all X in $L(\Lambda)$, since $\text{Hom}_\Lambda(\text{Ext}^1_\Lambda(V, R), X) = 0$ for all lattices X because $\text{Ext}^1_\Lambda(V, R)$ has finite length and no lattice has a nonzero submodule of finite length because $\dim R = 1$. Thus $0 \to (B, \ ) \xrightarrow{(g,\ )} (A, \ )$ is exact in $(L(\Lambda), Ab)$.

Summarizing this discussion, we have the following.

<u>Corollary 7.8</u>: Let Λ be an R-order with $\dim R \leq 1$. Then the simple functors in $(L(\Lambda)^{op}, Ab)$ not vanishing on projectives and the simple functors in $(L(\Lambda), Ab)$ not vanishing on injectives are finitely presented and have projective dimension at most 1.

We next point out the following analogue of Proposition 5.8, which we state without proof since it can either be easily deduced from Proposition 5.8 or proven using essentially the same proof as that given for Proposition 5.8.

<u>Proposition 7.9</u>: Let Λ be an R-order with $\dim R \leq 1$. Suppose C is in $L(\Lambda)$ and H a maximal right ideal in End C and S the simple functor $(\ ,C)/(\ ,C)_H$ in $(L(\Lambda)^{op}, Ab)$.

a) $S(\Lambda) \neq 0$ if and only if H does not contain $P(C, C)$.

b) $S(\Lambda) \neq 0$ if and only if there is a sublattice M of C such that $0 \longrightarrow (\ ,M) \longrightarrow (\ ,C) \longrightarrow S \longrightarrow 0$ is exact in $(L(\Lambda)^{op}, Ab)$, i.e. $(\ ,M) = (\ ,C)_H$.

c) Let M be a sublattice of C. Then the following statements are equivalent.

   i) $(\ ,C)/(\ ,M)$ is simple.

   ii) If $h: L \longrightarrow C$ is a morphism in $L(\Lambda)$ with $\operatorname{Im} h \not\subset M$, then the induced morphism $M \coprod L \longrightarrow C$ is a splittable epimorphism.

The following is an immediate consequence of this proposition.

<u>Corollary 7.10</u>: Suppose Λ is an R-order with $\dim R \leq 1$. Let C be in

$L(\Lambda)$. The two sided ideal $P(C, C)$ of End C is not contained in the radical of End C if and only if there is a sublattice M of C having the property that if $h: L \longrightarrow C$ is a morphism in $L(\Lambda)$ with $\operatorname{Im} h \not\subset M$, then the induced morphism $M \coprod L \longrightarrow C$ is a splittable epimorphism.

There are obvious analogues of Proposition 7.9 and Corollary 7.10 dealing with the two sided ideal $I(A, A)$ of End A for A in $L(\Lambda)$ where Λ is an R-order such that $\dim R \leq 1$. The statements and proofs of these analogues are left to the reader to give.

§8. Simple Functors for Complete Orders. Throughout this section we assume that Λ is an R-order with R a complete local ring. Our main purpose is to point out the influence the special properties of lattices have when R is a complete local ring on the properties of simple functors in $(L(\Lambda)^{op}, Ab)$ and $(L(\Lambda), Ab)$. Most of these special features of the theory in this case are obvious analogues of results given in Section 6 for the case Λ is a noetherian R-algebra with R a complete local ring. We omit those proofs which are essentially the same as those given in Section 6 if one keeps in mind the results of Section 7.

Let S be a simple functor in $(L(\Lambda)^{op}, Ab)$. Since S is additive and every lattice is a finite sum of indecomposable lattices, S does not vanish on some indecomposable lattice C. Hence there is a nonzero morphism $\alpha: (\ ,C) \longrightarrow S$ which must be an epimorphism since S is simple. Because R is complete, we know that End C is local. Therefore the epimorphism $\alpha: (\ ,C) \longrightarrow S$ is a projective cover of S (see Proposition 1.10). Moreover $0 \longrightarrow \underline{r}(\ ,C) \longrightarrow (\ ,C) \longrightarrow S \longrightarrow 0$ is exact, i.e. $(\ ,C)/\underline{r}(\ ,C) \approx S$.

On the other hand, suppose C is indecomposable in $L(\Lambda)$. Since $\text{End } C$ is local, $\underline{r}(\ ,C)$ is the unique maximal subfunctor of $(\ ,C)$ in $(L(\Lambda)^{op}, Ab)$. Therefore $(\ ,C)/\underline{r}(\ ,C)$ is a simple functor and the natural epimorphism $(\ ,C) \longrightarrow (\ ,C)/\underline{r}(\ ,C)$ is a projective cover. Finally, suppose C and C' are indecomposable lattices such that the simple functors $(\ ,C)/\underline{r}(\ ,C)$ and $(\ ,C')/\underline{r}(\ ,C')$ in $(L(\Lambda)^{op}, Ab)$ are isomorphic. Then the functors $(\ ,C)$ and $(\ ,C')$ are isomorphic since they are projective covers of the isomorphic functors $(\ ,C)/\underline{r}(\ ,C)$ and $(\ ,C')/\underline{r}(\ ,C')$ respectively. Hence two indecomposable lattices C and C' are isomorphic if and only if the simple functors $(\ ,C)/\underline{r}(\ ,C)$ and $(\ ,C')/\underline{r}(\ ,C')$ are isomorphic.

We summarize this discussion as follows.

Proposition 8.1: The map $L(\Lambda) \longrightarrow (L(\Lambda)^{op}, Ab)$ given by $C \longrightarrow (\ ,C)/\underline{r}(\ ,C)$ for all C in $L(\Lambda)$ induces a bijection between the isomorphism classes of indecomposable lattices and the isomorphism classes of simple functors in $(L(\Lambda)^{op}, Ab)$. Moreover a simple functor S in $(L(\Lambda)^{op}, Ab)$ is isomorphic to $(\ ,C)/\underline{r}(\ ,C)$ for some indecomposable C in $L(\Lambda)$ if and only if $S(C) \neq 0$.

Having determined the projective covers for simple functors in $(L(\Lambda)^{op}, Ab)$, we now show that these functors have minimal projective resolutions, at least in certain cases. We first consider those simple functors which vanish on projectives.

Proposition 8.2: Let C be indecomposable in $L_P(\Lambda)$.

a) There exists an almost split sequence

$$0 \longrightarrow DTr_L C \xrightarrow{g} B \xrightarrow{f} C \longrightarrow 0 \quad (\text{where } D(X) = \text{Hom}_R(X, R))$$

in $L(\Lambda)$. This sequence has the following properties:

i)

$$0 \longrightarrow (\ ,DTr_L C) \xrightarrow{(\ ,g)} (\ ,B) \xrightarrow{(\ ,f)} (\ ,C) \longrightarrow \text{Coker}\ (\ ,f) \longrightarrow 0$$

is a minimal projective resolution in $(L(\Lambda)^{op}, Ab)$ of the simple functor Coker $(\ ,f)$ which does not vanish on C.

ii)

$$0 \longrightarrow (C,\ ) \xrightarrow{(f,\ )} (B,\ ) \xrightarrow{(g,\ )} (DTr_L C,\ ) \longrightarrow \text{Coker}\ (g,\ ) \longrightarrow 0$$

is a minimal projective resolution in $(L(\Lambda), Ab)$ of the simple functor Coker $(g,\ )$ which does not vanish on $DTr_L C$.

b) Suppose S is a simple functor in $(L(\Lambda)^{op}, Ab)$ such that $S(C) \neq 0$. Then $S \approx \text{Coker}\ (\ ,f)$, so there is an exact sequence $0 \longrightarrow (\ ,DTr_L C) \xrightarrow{(\ ,g)} (\ ,B) \xrightarrow{(\ ,f)} S \longrightarrow 0$ which is a minimal projective resolution of S in $(L(\Lambda)^{op}, Ab)$.

c) Suppose T is a simple functor in $(L(\Lambda), Ab)$ such that $T(DTr_L C) \neq 0$. Then $T \approx \text{Coker}\ (g,\ )$, so there is an exact sequence $0 \longrightarrow (C,\ ) \longrightarrow (B,\ ) \longrightarrow (DTr_L C) \longrightarrow T \longrightarrow 0$ which is a minimal projective resolution of T in $(L(\Lambda), Ab)$.

Proof: Analogous to the proof of Proposition 6.3.

We now turn our attention to the simple functors in $(L(\Lambda)^{op}, Ab)$ which do not vanish on projectives. We restrict our discussion to the classical situations of $\dim R \leq 1$.

Proposition 8.3: Suppose $\dim R \leq 1$ and $\underline{r}$ is the radical of Λ. Let P be an indecomposable projective Λ-lattice.

a) $\underline{r}P$ is the unique maximal sublattice of P.

b) The inclusion $\underline{r}P \longrightarrow P$ is a minimal right almost split morphism in $L(\Lambda)$.

c) $0 \longrightarrow (\ ,\underline{r}P) \longrightarrow (\ ,P) \longrightarrow (\ ,P)/\underline{r}(\ ,P) \longrightarrow 0$ is a minimal projective resolution in $(L(\Lambda)^{op}, Ab)$ of the simple functor $(\ ,P)/\underline{r}(\ ,P)$.

Suppose S is a simple functor in $(L(\Lambda)^{op}, Ab)$ such that $S(P) \neq 0$.

d) Then $S \approx (\ ,P)/\underline{r}(\ ,P)$ and there is an exact sequence $0 \longrightarrow (\ ,\underline{r}P) \longrightarrow (\ ,P) \longrightarrow S \longrightarrow 0$ which is a minimal projective resolution of S in $(L(\Lambda)^{op}, Ab)$.

e) pd $S \leq 1$.

Proof: The proof is the same as that for Proposition 6.2, if one keeps in mind the fact established in Proposition 7.6 that a submodule C' of a lattice C is a sublattice provided C/C' is a module of finite length.

Having described the minimal projective resolutions of simple functors in $(L(\Lambda)^{op}, Ab)$ (at least in case $\dim R \leq 1$), we now turn our attention to describing the minimal projective resolutions for the simple functors in $(L(\Lambda), Ab)$. We begin with the following analogue of Proposition 8.1.

Proposition 8.4:

a) For each indecomposable A in $L(\Lambda)$, we have that $(A,\ )/\underline{r}(A,\ )$ is a simple functor not vanishing on A and that the natural morphism $(A,\ ) \longrightarrow (A,\ )/\underline{r}(A,\ )$ is a projective cover.

b) The map $L(\Lambda) \longrightarrow (L(\Lambda), Ab)$ given by $A \longrightarrow (A,\ )/\underline{r}(A,\ )$ for all A in $L(\Lambda)$ induces a bijection between the isomorphism classes of indecomposable lattices and the isomorphism classes of simple functors in $(L(\Lambda), Ab)$. Moreover a simple functor T is isomorphic to $(A,\ )/\underline{r}(A,\ )$ for an indecomposable A if and only if $T(A) \neq 0$.

Having determined the projective covers for simple functors in $(L(\Lambda), Ab)$ we now show that those which vanish on injective lattices have minimal projective resolutions in $(L(\Lambda), Ab)$. We do this in the following analogue of Proposition 6.6.

Proposition 8.5: Let A be an indecomposable lattice in $L_I(\Lambda)$.

a) There exists an almost split sequence

$0 \longrightarrow A \xrightarrow{g} B \xrightarrow{f} Tr_L DA \longrightarrow 0$ in $L(\Lambda)$. This sequence has the following properties.

i)

$$0 \longrightarrow (Tr_L DA, \ ) \xrightarrow{(\ ,f)} (B, \ ) \xrightarrow{(g, \ )} (A, \ ) \longrightarrow \operatorname{Coker}(g, \ ) \longrightarrow 0$$

is a minimal projective resolution of the simple functor $\operatorname{Coker}(g, \ )$ which does not vanish on A.

ii)

$$0 \longrightarrow (\ , A) \xrightarrow{(\ ,g)} (\ ,B) \xrightarrow{(\ ,f)} (\ ,Tr_L DA) \longrightarrow \operatorname{Coker}(\ ,f) \longrightarrow 0$$

is a minimal projective resolution for the simple functor $\operatorname{Coker}(\ ,f)$ which does not vanish on $Tr_L DA$.

b) Suppose T is a simple functor in $(L(\Lambda), Ab)$ such that $T(A) \neq 0$. Then $T \approx \operatorname{Coker}(g, \ )$, so there is an exact sequence

$$0 \longrightarrow (Tr_L DA, \ ) \xrightarrow{(f, \ )} (B, \ ) \xrightarrow{(g, \ )} (A, \ ) \longrightarrow T \longrightarrow 0$$

which is a projective resolution of T.

c) Suppose S is a simple functor in $(L(\Lambda)^{op}, Ab)$ such that $S(Tr_L DA) \neq 0$. Then $S \approx \operatorname{Coker}(\ ,f)$, so there is an exact sequence

$$0 \longrightarrow (\ ,A) \xrightarrow{(\ ,g)} (\ ,B) \xrightarrow{(\ ,f)} (\ ,Tr_L DA) \longrightarrow S \longrightarrow 0$$

which is a minimal projective resolution of S.

Having described minimal projective resolutions for simple functors T in $(L(\Lambda), Ab)$ vanishing on injective lattices, we now describe minimal projective resolutions for those that do not vanish on injective lattices in the case $\dim R \leq 1$. This is accomplished in the following analogue of Proposition 6.5.

<u>Proposition 8.6</u>: Suppose $\dim R \leq 1$. Let I be an indecomposable injective lattice. Then $D(I)$ is an indecomposable projective Λ^{op}-lattice with unique maximal sublattice $\underline{r}D(I)$. Let $I_o = D(\underline{r}D(I))$ and let $I \longrightarrow I_o$ be the dual of the inclusion $\underline{r}D(I) \longrightarrow D(I)$. Then the morphism $I \longrightarrow I_o$ has the following properties.

a) $I \longrightarrow I_o$ is a minimal left almost split morphism.

b) $0 \longrightarrow (I_o, \ ) \longrightarrow (I, \ ) \longrightarrow (I, \ )/\underline{r}(I, \ ) \longrightarrow 0$ is a minimal projective resolution of the simple functor $(I, \ )/\underline{r}(I, \ )$ in $(L(\Lambda), Ab)$.

Suppose T is a simple functor in $(L(\Lambda), Ab)$ such that $T(I) \neq 0$.

c) Then $T \approx (I, \ )/(I_o, \ )$ and so there is an exact sequence $0 \longrightarrow (I_o, \ ) \longrightarrow (I, \ ) \longrightarrow T \longrightarrow 0$ which is a minimal projective resolution of T.

<u>Proof</u>: Easily deduced from Proposition 7.7.

Chapter III

SOME SPECIAL ORDERS

This chapter is devoted mainly to pointing out some special features of our general theory of lattices for particular types of orders including Gorenstein orders and R-orders where R is a complete local ring of dimension 1 or 2. In the last section we return to some general considerations which put in a more general setting are our basic existence theorems.

§1. Gorenstein Orders. Let R be an equidimensional Gorenstein ring. A noetherian R-algebra Λ is said to be a Gorenstein R-order if a) Λ is a Cohen-Macauley R-module and b) $\Lambda \approx \mathrm{Hom}_R(\Lambda, R)$ as two-sided Λ-modules. It is obvious that a Gorenstein R-order is an R-order. The following are some important examples of Gorenstein R-orders.

a) R itself is a Gorenstein R-order.

b) If G is a finite group, then $R[G]$ is a Gorenstein R-order.

c) If Λ is a symmetric finite dimensional algebra over a field k, then Λ is a Gorenstein k-order.

This section is devoted primarily to pointing out some of the special features the category $L(\Lambda)$ has when Λ is a Gorenstein R-order. These properties will be used in this and the next section to prove some special properties of lattices over Gorenstein R-orders. We assume throughout this section that Λ is a Gorenstein R-order and that $\alpha\colon \Lambda \longrightarrow \mathrm{Hom}_R(\Lambda, R)$ is a fixed two-sided isomorphism. We begin our discussion with the following preliminaries.

Proposition 1.1:

a) The R-morphism $\alpha^{op}: \Lambda^{op} \longrightarrow \mathrm{Hom}_R(\Lambda^{op}, R)$ given by $\alpha^{op}(\lambda) = \alpha(\lambda)$ for all $\lambda \in \Lambda$, is a two-sided Λ^{op}-isomorphism which we also denote by α. Thus Λ^{op} is a Gorenstein R-order.

b) The functors $\mathrm{Hom}_\Lambda(\ ,\Lambda)$, $\mathrm{Hom}_R(\ ,R)$: noeth $\Lambda \longrightarrow$ noeth Λ^{op} are isomorphic.

c) If M is in $L(\Lambda)$, then $\mathrm{Hom}_\Lambda(M, \Lambda) \approx \mathrm{Hom}_R(M, R)$ is in $L(\Lambda^{op})$.

d) If M is in $L(\Lambda)$, then the natural morphism $M \longrightarrow \mathrm{Hom}_{\Lambda^{op}}(\mathrm{Hom}_\Lambda(M, \Lambda), \Lambda^{op})$ is an isomorphism of Λ-lattices.

e) The functors $\mathrm{Hom}_\Lambda(\ ,\Lambda): L(\Lambda) \longrightarrow L(\Lambda^{op})$ and $\mathrm{Hom}_{\Lambda^{op}}(\ ,\Lambda^{op}): L(\Lambda^{op}) \longrightarrow L(\Lambda)$ are inverse dualities.

f) If M is in $L(\Lambda)$, then $\mathrm{Ext}^i_\Lambda(M, \Lambda) = 0 = \mathrm{Ext}^i_{\Lambda^{op}}(\mathrm{Hom}_\Lambda(M, \Lambda), \Lambda^{op})$ for all $i > 0$.

g) For each $i > 0$, the R-algebras $\underline{\mathrm{End}}\,(M)$ and $\underline{\mathrm{End}}\,(\Omega^i M)$ are isomorphic R-algebras.

h) Let $0 \longrightarrow \Omega^2(M) \longrightarrow P_1 \longrightarrow P_0 \longrightarrow M \longrightarrow 0$ be exact with the P_i finitely generated projective Λ-modules and M a lattice.

   i) $\Omega^2(M)$ is in $L(\Lambda)$.

   ii) $\mathrm{Tr}(M) \approx \mathrm{Hom}_\Lambda(\Omega^2(M), \Lambda)$ and is therefore in $L(\Lambda)$.

   iii) $\mathrm{Hom}_R(\mathrm{Tr}M, R) \approx \mathrm{Hom}_\Lambda(\mathrm{Tr}(M), \Lambda) \approx \Omega^2(M)$.

   iv) Λ is an injective as well as projective lattice. Thus $P(\Lambda, B) = I(\Lambda, B)$ for all lattices and so $\underline{L}(\Lambda) = \overline{L}(\Lambda)$.

Proof:

a) Easily checked.

b) Let $\alpha: \Lambda \longrightarrow \mathrm{Hom}_R(\Lambda, R)$ be our given two-sided isomorphism. Then for each M in noeth Λ, the isomorphism α induces an isomorphism of Λ^{op}-modules $\mathrm{Hom}_\Lambda(M, \Lambda) \longrightarrow \mathrm{Hom}_\Lambda(M, \mathrm{Hom}_R(\Lambda, R))$ which is functorial in M. Standard associative laws give isomorphisms $\mathrm{Hom}_\Lambda(M, \mathrm{Hom}_R(\Lambda, R)) \cong \mathrm{Hom}_R(M, R)$ functorial in M. Hence we obtain isomorphisms $\mathrm{Hom}_\Lambda(M, \Lambda) \cong \mathrm{Hom}_R(M, R)$ functorial in M.

c) Let M be in $L(\Lambda)$. Then $\mathrm{Hom}_R(M, R)$ is a Λ^{op}-lattice. Since by b) we have that $\mathrm{Hom}_R(M, R) \approx \mathrm{Hom}_\Lambda(M, \Lambda)$, it follows that $\mathrm{Hom}_\Lambda(M, \Lambda)$ is a lattice.

d) If M is in $L(\Lambda)$, then we know that the natural morphism $M \longrightarrow \mathrm{Hom}_R(\mathrm{Hom}_R(M, R), R)$ is an isomorphism. From this it follows, using our fixed two-sided isomorphism $\Lambda \longrightarrow \mathrm{Hom}_R(\Lambda, R)$ and $\Lambda^{op} \longrightarrow \mathrm{Hom}_R(\Lambda^{op}, R)$, that $M \longrightarrow \mathrm{Hom}_{\Lambda^{op}}(\mathrm{Hom}_\Lambda(M, \Lambda), \Lambda^{op})$ is an isomorphism.

e) Easy consequence of d).

f) Let $0 \longrightarrow R \longrightarrow I_0 \longrightarrow I_1 \longrightarrow \cdots \longrightarrow I_d \longrightarrow 0$ be minimal injective resolution of R. Since $\mathrm{Hom}_R(\Lambda, R) \cong \Lambda$ is an R-lattice, we know that $\mathrm{Ext}^i_R(\mathrm{Hom}_R(\Lambda, R), R) = 0$ for all $i > 0$. Thus

(*) $$0 \longrightarrow \mathrm{Hom}_R(\mathrm{Hom}_R(\Lambda, R), R) \longrightarrow \mathrm{Hom}_R(\mathrm{Hom}_R(\Lambda, R), I_0) \longrightarrow \cdots \longrightarrow \mathrm{Hom}_R(\mathrm{Hom}_R(\Lambda, R), I_d) \longrightarrow 0$$

is exact. Since $\mathrm{Hom}_R(\Lambda, R) \cong \Lambda$ is a two-sided Λ-module, it follows that $\mathrm{Hom}_R(\mathrm{Hom}_R(\Lambda, R), I_2)$ is an injective Λ-module for all $i > 0$. Thus (*) is an injective Λ-resolution of Λ. Since $\mathrm{Hom}_\Lambda(M, \mathrm{Hom}_R(\mathrm{Hom}_R(\Lambda, R), I_i)) \cong \mathrm{Hom}_R(\mathrm{Hom}_R(\Lambda, R) \otimes_\Lambda M, I_i)$ for all i and $\mathrm{Hom}_R(\Lambda, R) \cong \Lambda$ as a Λ^{op}-module, we have that $\mathrm{Hom}_\Lambda(M, \mathrm{Hom}_R(\mathrm{Hom}_R(\Lambda, R), I_i)) \cong \mathrm{Hom}_R(M, I_i)$ for all i. Therefore $\mathrm{Ext}^i_\Lambda(M, \Lambda) \cong \mathrm{Ext}^i_R(M, R)$ for all $i > 0$. The fact that M is an R-lattice, implies that $\mathrm{Ext}^i_R(M, R) = 0$ for $i > 0$. Hence $\mathrm{Ext}^i_\Lambda(M, \Lambda) = 0$ for all $i > 0$.

The rest of f) follows trivially from what has been proven

and the fact that $\text{Hom}_\Lambda(M, \Lambda)$ is in $L(\Lambda^{op})$ if M is in $L(\Lambda)$.

g) Follows from the fact that $\text{Ext}^i_\Lambda(M, \Lambda) = 0$ for all $i > 0$ (see I, Proposition 7.4).

h) i) Since $0 \longrightarrow \Omega^2(M) \longrightarrow P_1 \longrightarrow P_0 \longrightarrow M \longrightarrow 0$ is exact with the P_i and M in $L(\Lambda)$, it follows that $\Omega^2(M)$ is in $L(\Lambda)$.

ii) Since $\text{Ext}^i_\Lambda(M, \Lambda) = 0$ for $i > 0$, it follows that

$$0 \to \text{Hom}_\Lambda(M, \Lambda) \to \text{Hom}_\Lambda(P_0, \Lambda) \to \text{Hom}_\Lambda(P_1, \Lambda) \to \text{Hom}_\Lambda(\Omega^2(M), \Lambda) \longrightarrow 0$$

is an exact sequence of Λ^{op}-modules. Thus $\text{Hom}_\Lambda(\Omega^2(M), \Lambda) \cong \text{TrM}$.

iii) Since $\text{Hom}_R(\text{TrM}, R) \cong \text{Hom}_\Lambda(\text{TrM}, \Lambda)$, it follows that $\text{Hom}_R(\text{TrM}, R) \cong \text{Hom}_{\Lambda^{op}}(\text{Hom}_\Lambda(\Omega^2(M, \Lambda), \Lambda^{op}) \cong \Omega^2(M)$.

iv) Obvious from the fact that $\Lambda \cong \text{Hom}_R(\Lambda, R)$ as a two-sided Λ-module.

In order to state the main point of this dicussion of lattices over Gorenstein R-orders, it is convenient to make the following definitions.

Suppose we have chosen for each M in noeth Λ an epimorphism $P_M \longrightarrow M \longrightarrow 0$ with P_M a finitely generated projective Λ-module. Suppose we have done a similar thing for Λ^{op}-modules in noeth Λ^{op}. This enables us to construct particular projective resolutions for noetherian Λ and Λ^{op}-modules. For each M in noeth Λ and noeth Λ^{op}, we define $\Omega^i(M)$ using these projective resolutions. In particular, if M is in $L(\Lambda)$, then we know that $\Omega^i(M)$ is in $L(\Lambda)$ for all $i \geq 0$. Thus given M in $L(\Lambda)$, we have the Λ^{op}-lattices $\Omega^i(\text{Hom}_\Lambda(M, \Lambda))$ associated with M for all $i > 0$. We define $\Omega^{-j}(M)$ to be the Λ-lattice $\text{Hom}_{\Lambda^{op}}(\Omega^j(\text{Hom}_\Lambda(M, \Lambda), \Lambda^{op})$ for all $j \geq 0$. The following alternative description of $\Omega^{-j}(M)$ will perhaps make a little clearer what is going on.

Let $\cdots \xrightarrow{g_{i+1}} Q_i \longrightarrow \cdots \xrightarrow{g_1} Q_0 \longrightarrow \mathrm{Hom}_\Lambda(M, \Lambda) \longrightarrow 0$ be our chosen projective resolution of the Λ^{op}-lattice $\mathrm{Hom}_\Lambda(M, \Lambda)$. Since $\mathrm{Ext}^i_{\Lambda^{op}}(\mathrm{Hom}_\Lambda(M, \Lambda), \Lambda^{op}) = 0$ for all $i > 0$, the sequence

$$0 \longrightarrow \mathrm{Hom}_\Lambda(\mathrm{Hom}_\Lambda(M, \Lambda), \Lambda^{op}) \longrightarrow \mathrm{Hom}_{\Lambda^{op}}(Q_0, \Lambda^{op}) \longrightarrow \cdots$$
$$\xrightarrow{\mathrm{Hom}_{\Lambda^{op}}(g_i, \Lambda^{op})} \mathrm{Hom}_{\Lambda^{op}}(Q_i, \Lambda^{op}) \longrightarrow \cdots$$

is an exact sequence with the $\mathrm{Hom}_{\Lambda^{op}}(Q_i, \Lambda^{op})$ projective Λ-modules in $L(\Lambda)$. Then $\mathrm{Ker}(g_i, \Lambda^{op}) = \Omega^{1-i}(M)$ for all $i > 0$. In this way we get associated with the lattice M an exact sequence of finitely generated projective Λ-modules

$$(*) \qquad \cdots \longrightarrow P_1 \longrightarrow \cdots \xrightarrow{f_2} P_1 \xrightarrow{f_1} P_0$$
$$\longrightarrow \mathrm{Hom}_{\Lambda^{op}}(Q_0, \Lambda^{op}) \xrightarrow{\mathrm{Hom}(g_1, \Lambda^{op})} \mathrm{Hom}_{\Lambda^{op}}(Q_1, \Lambda^{op}) \longrightarrow \cdots$$

with the following properties:

a) $\mathrm{Im} f_i = \Omega^i(M)$ for all $i \geq 0$.

b) $\mathrm{Im}(g_i, \Lambda^{op}) = \Omega^{-i}(M)$ for all $i > 0$.

c) The sequence (*) remains exact if we apply the functor $\mathrm{Hom}_\Lambda(\ ,\Lambda)$ to it.

Using this description of the $\Omega^i(M)$ for all i in Z, the group of all integers, it is not difficult to show the following.

<u>Lemma 1.2</u>: For each M in $L(\Lambda)$ we have

a) $\Omega^j(\Omega^k(M)) \cong \Omega^{j+k}(M)$ in $\underline{L}(\Lambda)$ for all j and k in Z, the group of integers;

b) $\Omega^j(\mathrm{Hom}_\Lambda(A, \Lambda)) \cong \mathrm{Hom}_\Lambda(\Omega^{-j}A, \Lambda)$ in $\underline{L}(\Lambda)$ for all integers j.

We now come to the main point of this section.

Proposition 1.3: Suppose $\dim R = d$ and A is in $L(\Lambda)$.

a) $\mathrm{Hom}_R(\mathrm{Tr}_L A, R) \cong \Omega^{2-d}A$ in $\underline{L}(\Lambda)$.

b) $\mathrm{Tr}_L(\mathrm{Hom}_R(A, R)) \cong \Omega^{d-2}A$ in $\underline{L}(\Lambda)$.

Proof:

a) We recall (see I, Section 7) that $\mathrm{Tr}_L\colon \underline{L}(\Lambda) \longrightarrow \underline{L}(\Lambda^{op})$ is defined to be the composition $\underline{L}(\Lambda) \xrightarrow{\mathrm{Tr}} \underline{J}(\Lambda^{op}) \xrightarrow{\Omega^d} \underline{L}(\Lambda^{op})$. Let $0 \longrightarrow \Omega^2 A \longrightarrow P_1 \longrightarrow P_0 \longrightarrow A \longrightarrow 0$ be exact with the P_i finitely generated projective Λ-modules. Then

$$0 \to \mathrm{Hom}_\Lambda(A, \Lambda) \to \mathrm{Hom}_\Lambda(P_0, \Lambda) \to \mathrm{Hom}_\Lambda(P_1, \Lambda) \to \mathrm{Hom}_\Lambda(\Omega^2 A, \Lambda) \longrightarrow 0$$

is exact since $\mathrm{Ext}^i_\Lambda(A, \Lambda) = 0$ for all $i > 0$. Thus
$\mathrm{Tr}(A) = \Omega^{-2}\mathrm{Hom}_\Lambda(A, \Lambda)$. Hence
$\mathrm{Tr}_L(A) = \Omega^d(\Omega^{-2}\mathrm{Hom}_\Lambda(A, \Lambda)) = \Omega^{d-2}(\mathrm{Hom}_\Lambda(A, \Lambda))$. Thus
$\mathrm{Hom}_R(\mathrm{Tr}_L(A), R) \cong \mathrm{Hom}_\Lambda(\Omega^{d-2}(\mathrm{Hom}_\Lambda(A, \Lambda), \Lambda)$ (see Proposition 1.1).
But by Lemma 1.2, we know that
$\mathrm{Hom}_\Lambda(\Omega^{d-2}(\mathrm{Hom}_\Lambda(A, \Lambda)), \Lambda) \cong \Omega^{2-d}(\mathrm{Hom}_{\Lambda^{op}}\mathrm{Hom}_\Lambda(A, \Lambda), \Lambda^{op})$. Since A is reflexive it follows that $\mathrm{Hom}_R(\mathrm{Tr}_L(A), R) \cong \Omega^{2-d}(A)$ in $\underline{L}(\Lambda)$.

b) By a) $\mathrm{Hom}_R(\mathrm{Tr}_L(\mathrm{Hom}_R(A, R)), R) \cong \Omega^{2-d}(\mathrm{Hom}_R(A, R))$. But $\mathrm{Hom}_R(\Omega^{d-2}(A), R) \cong \Omega^{2-d}(\mathrm{Hom}_R(A, R))$ by Lemma 1.2. Thus $\mathrm{Tr}_L(\mathrm{Hom}_R(A, R)) \cong \Omega^{d-2}(A)$ in $\underline{L}(\Lambda)$.

We now point out a few consequences of Proposition 1.3.

Proposition 1.4: Suppose $\dim R = d \geq 2$ and A is a nonprojective Λ-lattice. Then $\mathrm{Ext}_{\Lambda}^{d-1}(A, A) \cong \mathrm{Hom}_R(\underline{\mathrm{End}}(A), I_d) \neq 0$ as an $\underline{\mathrm{End}}(A)$-$\underline{\mathrm{End}}(A)^{op}$-bimodule.

Proof: First we observe that $\mathrm{Ext}_{\Lambda}^{d-1}(A, A) = \mathrm{Ext}_{\Lambda}^{1}(\Omega^{d-2}A, A)$. By I, Proposition 8.7 we have that

$\mathrm{Ext}_{\Lambda}^{1}(\Omega^{d-2}A, A) \cong \mathrm{Hom}_R(\underline{\mathrm{Hom}}_{\Lambda}(\mathrm{Tr}_L\mathrm{Hom}_R(A, R), \Omega^{d-2}A), I_d)$. By Proposition 1.3 we have $\mathrm{Tr}_L\mathrm{Hom}_R(A, R) \cong \Omega^{d-2}A$. Thus

$\mathrm{Ext}_{\Lambda}^{1}(\Omega^{d-2}A, A) \cong \mathrm{Hom}_R(\underline{\mathrm{Hom}}_{\Lambda}(\Omega^{d-2}A, \Omega^{d-2}A), I_d)$. But $\underline{\mathrm{End}}(\Omega^{d-2}A) \cong \underline{\mathrm{End}}(A)$ by Proposition 1.1. Since all these isomorphisms are $\mathrm{End}(A)$-$(\mathrm{End}A)^{op}$-bimodule isomorphisms, we are done.

As a consequence of this proposition we have the following.

Theorem 1.5: Suppose $\dim R = d \geq 2$. A Cohen-Macauley Λ-module A is Λ-projective if and only if it satisfies both of the following conditions.

a) $A_{\underline{p}}$ is $\Lambda_{\underline{p}}$-projective for all prime ideals $\underline{p}$ of R such that $\dim R_{\underline{p}} \geq 1$,

b) $\mathrm{Ext}_{\Lambda}^{i}(A, A) = 0$ for $i = 1,\ldots,d - 1$.

Proof: It is easy to see that if Λ is a Gorenstein R-order, then $\Lambda_{\underline{p}}$ is a Gorenstein $R_{\underline{p}}$-order for all prime ideals $\underline{p}$ of R. Furthermore a Λ-module A is Cohen-Macauley and satisfies a) and b) if and only if $A_{\underline{p}}$ is a Cohen-Macauley $\Lambda_{\underline{p}}$-module satisfying a) and b) for all prime ideals $\underline{p}$ of R. Finally, these observations together with the fact that a module A in noeth Λ is projective if and only if $A_{\underline{p}}$ is projective for all prime ideals $\underline{p}$ of R, show that we may assume without loss of generality that R is a local ring with $\dim R \geq 2$. We now proceed by induction on $d = \dim R$.

Assume $d = 2$. Since A is Cohen-Macauley and $A_{\underline{p}}$ is $\Lambda_{\underline{p}}$-free for all nonmaximal ideals $\underline{p}$ of R (since $\dim R_{\underline{p}} \leq 1$ if $\underline{p}$ is not maximal) it follows that A is a Λ-lattice. The fact that $\mathrm{Ext}^1_{\Lambda}(A, A) = 0$ then implies by Proposition 1.4 that A is Λ-projective. The rest of the inductive proof proceeds in the same way.

Before giving our final application of Proposition 1.3, it is convenient to make a few observations. Suppose $\dim R = 1$ and M is an R-module of finite length. Then $\mathrm{ann}(M)$, the annihilator of M, contains some ideal $\underline{a}$ such that $R/\underline{a}$ is a Gorenstein artin ring. For instance, since $R/\mathrm{ann}(M)$ is an artin ring (remember M is a module of finite length) $\mathrm{ann}(M)$ must contain some R-regular element x and R/xR is a Gorenstein artin ring. In particular if C is a Λ-lattice, then $\underline{\mathrm{End}}\ C$ is an R-module of finite length and so there is an ideal $\underline{a}$ of R such that $\underline{a}\ \underline{\mathrm{End}}\ C = 0$ and $R/\underline{a}$ is a Gorenstein artin ring.

Theorem 1.6: Suppose $\dim R = 1$, Λ is a Gorenstein R-order, and C is nonprojective lattice over Λ. Let $\underline{a}$ be an ideal in R such that $\underline{a}\ \underline{\mathrm{End}}\ C = 0$ and $R/\underline{a}$ is an artin Gorenstein ring. Then $\underline{\mathrm{End}}\ C$ is a Gorenstein R/a-algebra.

Proof: By our remarks preceeding the statement of the proposition we know that such an ideal $\underline{a}$ exists. Since $R/\underline{a}$ is selfinjective it follows that $\mathrm{Hom}_R(R/\underline{a}, I_1) \cong R/\underline{a}$ as R-modules. Hence $\mathrm{Hom}_R(\underline{\mathrm{End}}\ C, I_1) \cong \mathrm{Hom}_{R/\underline{a}}(\underline{\mathrm{End}}\ C, R/\underline{a})$ as a two-sided $\underline{\mathrm{End}}$ C-module. Therefore to prove the proposition it suffices to show that $\underline{\mathrm{End}}\ C$ is isomorphic to $\mathrm{Hom}_R(\underline{\mathrm{End}}\ C, I_1)$ as a two-sided $\underline{\mathrm{End}}$ C-module.

Since $\mathrm{Hom}_R(\ ,I_1)$ is a duality on the category of R-modules of finite length, we know that $\mathrm{Hom}_R(\underline{\mathrm{End}}\ C, I_1)$ and $\underline{\mathrm{End}}\ C$ have the same length over R. Now because Λ is a Gorenstein R-order we know by Proposition 1.3 that $\mathrm{Hom}_R(\mathrm{Tr}_L C, R) = \Omega C$ and so by I, Proposition 8.8 we have an isomorphism $\mathrm{Ext}^1_\Lambda(C, \Omega(C)) \cong \mathrm{Hom}_\Lambda(\underline{\mathrm{Hom}}_\Lambda(C, C), I_1)$ functorial in both variables. Thus this isomorphism is a two-sided $\underline{\mathrm{End}}$ C-isomorphism under the following operations.

The operation of $\underline{\mathrm{End}}\ C^{op}$ is that induced by End C on C. The operation of $\underline{\mathrm{End}}$ C on $\mathrm{Ext}^1_\Lambda(C, \Omega C)$ is given by the isomorphicm Ω: $\underline{\mathrm{End}}\ C \longrightarrow \underline{\mathrm{End}}\ \Omega C$ and the operation of $\underline{\mathrm{End}}\ \Omega C$ induced by the natural operation of End ΩC on ΩC (remember that since injective and projective lattices over a Gorenstein order are the same, $P(C, C) \cdot (\mathrm{Ext}^1_\Lambda(C, \Omega C)) = 0$). Thus to show that $\underline{\mathrm{End}}$ C and $\mathrm{Hom}_R(\underline{\mathrm{End}}\ (C), I_1)$ are isomorphic bimodules over $\underline{\mathrm{End}}\ C^{op}$-$\underline{\mathrm{End}}$ C, it suffices to show that $\underline{\mathrm{End}}$ C and $\mathrm{Ext}^1_\Lambda(C, \Omega(C))$ are isomorphic $(\underline{\mathrm{End}}\ C)^{op}$-$\underline{\mathrm{End}}$ C-bimodules.

Let $x: 0 \longrightarrow \Omega C \longrightarrow P \longrightarrow C \longrightarrow 0$ be the extension in $\mathrm{Ext}^1_\Lambda(C, \Omega C)$ with P projective. Since P is projective, given any extension $0 \longrightarrow \Omega C \longrightarrow E \longrightarrow C \longrightarrow 0$, there exists a commutative diagram

$$\begin{array}{ccccccccc} 0 & \longrightarrow & \Omega E & \longrightarrow & P & \longrightarrow & C & \longrightarrow & 0 \\ & & f\downarrow & & \downarrow & & \| & & \\ 0 & \longrightarrow & \Omega E & \longrightarrow & E & \longrightarrow & C & \longrightarrow & 0 \end{array}$$

sc that x generates $\mathrm{Ext}^1_\Lambda(C, \Omega C)$ over $\underline{\mathrm{End}}\ \Omega C$ and hence over $\underline{\mathrm{End}}$ C. Since P is also an injective lattice, a similar argument shows that x generates $\mathrm{Ext}^1_\Lambda(C, \Omega C)$ over $(\underline{\mathrm{End}}\ C)^{op}$. Hence the End C morphism g: $\underline{\mathrm{End}}\ C \longrightarrow \mathrm{Ext}^1_\Lambda(C, \Omega C)$ given by $g(\underline{f})(x) = \mathrm{Ext}^1_\Lambda(C, \Omega \underline{f})(x)$ is an epimorphism and thus an isomorphism since $\underline{\text{End}}\ C$ and $\text{Ext}^1_\Lambda(C, \Omega C)$ have the same length over R. A similar argument shows that the $(\underline{\text{End}}\ C)^{op}$-morphism $g\colon (\text{End}\ C)^{op} \longrightarrow \text{Ext}^1(C, \Omega(C))$ given by $g(f) = \text{Ext}^1_\Lambda(f, \Omega C)(x)$ is also an isomorphism. Therefore if we show for each f in $\underline{\text{End}}\ C$ that $\text{Ext}^1_\Lambda(f, \Omega C)(x) = \text{Ext}^1_\Lambda(C, \Omega\underline{f})(x)$, then the maps g and h would be the same and so $h\colon \underline{\text{End}}\ C \longrightarrow \text{Ext}^1_\Lambda(C, \Omega C)$ would be a $(\underline{\text{End}}\ C)^{op}$-$\underline{\text{End}}\ C$-bimodule isomorphism, which is our desired result.

We first show that if s and t in End C are such that $\text{Ext}^1_\Lambda(s, \Omega(C))(x)$ and $\text{Ext}^1_\Lambda(C, \Omega(\underline{t}))(x)$ have the same value y, then $\underline{s} = \underline{t}$. Let $0 \longrightarrow \Omega C \longrightarrow E \longrightarrow C \longrightarrow 0$ represent y. Then our hypothesis means that there is a commutative exact diagram

$$\begin{array}{ccccccccc} 0 & \longrightarrow & \Omega C & \longrightarrow & P & \longrightarrow & C & \longrightarrow & 0 \\ & & \downarrow{\scriptstyle \Omega(t)} & & \downarrow & & \| & & \\ 0 & \longrightarrow & \Omega C & \longrightarrow & E & \longrightarrow & C & \longrightarrow & 0 \\ & & \| & & \downarrow & & \downarrow{\scriptstyle s} & & \\ 0 & \longrightarrow & \Omega C & \longrightarrow & P & \longrightarrow & C & \longrightarrow & 0 \end{array}$$

from which we deduce the commutative exact diagram

$$\begin{array}{ccccccccc} 0 & \longrightarrow & \Omega C & \longrightarrow & P & \longrightarrow & C & \longrightarrow & 0 \\ & & \downarrow{\scriptstyle \Omega(t)} & & \downarrow & & \downarrow{\scriptstyle s} & & \\ 0 & \longrightarrow & \Omega C & \longrightarrow & P & \longrightarrow & C & \longrightarrow & 0 \end{array}$$

which shows that $\Omega(\underline{s}) = \Omega(\underline{t})$. Since $\Omega\colon \text{End}\ C \longrightarrow \underline{\text{End}}\ (\Omega C)$ is an isomorphism we know that $\underline{s} = \underline{t}$. Hence if s and t are in $\underline{\text{End}}\ C$ such that $\text{Ext}^1_\Lambda(s, \Omega(C))(x) = \text{Ext}^1_\Lambda(C, \Omega(t))(x)$, then $\underline{s} = \underline{t}$.

To see that $\text{Ext}^1_\Lambda(s, \Omega(C))(x) = \text{Ext}^1_\Lambda(C, \Omega(s))(x)$ for all s in $\underline{\text{End}}\ C$ is now trivial. For we know that

$\mathrm{Ext}^1_\Lambda(s, \Omega(C))(x) = \mathrm{Ext}^1_\Lambda(C, \Omega(\underline{t}))(x)$ for some t. Then by our previous result we know that $\underline{t} = \underline{s}$. So $\mathrm{Ext}^1_\Lambda(s, \Omega(C))(x) = \mathrm{Ext}^1(C, \Omega(\underline{s})(x)$ for all s in $\underline{\mathrm{End}}$ C. This finishes the proof that $\underline{\mathrm{End}}$ (C), $\mathrm{Ext}^1_\Lambda(C, \Omega(C))$, and $\mathrm{Hom}_{R/\underline{a}}(\underline{\mathrm{End}}\ (C), R/\underline{a})$ are all isomorphic $(\underline{\mathrm{End}}\ C)^{op}$-$\underline{\mathrm{End}}$ C-bimodules and thus the fact that $\underline{\mathrm{End}}$ C is a Gorenstein $R/\underline{a}$-algebra is complete.

We end this section with a few remarks about lattices over Gorenstein R-orders in case R is a complete local ring.

Suppose Λ is a Gorenstein R-order with R a complete local ring of dimension d. As usual, we assume that we use only minimal projective resolutions in making our definitions and computations. With these conventions in mind we have the following result.

Proposition 1.7: Suppose Λ is a Gorenstein R-order with R a complete local ring of dimension d. Let M be in $L_P(\Lambda)$ and $\cdots \longrightarrow P_i \xrightarrow{f_i} P_{i-1} \longrightarrow \cdots \longrightarrow P_0 \xrightarrow{f_0} M \longrightarrow 0$ a minimal projective resolution of M.

a) $\Omega^i(M) = \mathrm{Im} f_i$ is in $L_P(\Lambda)$ for all $i \geq 0$.

b) The following statements are equivalent:

i) M is indecomposable.

ii) $\Omega^j(M)$ is indecomposable for some $j \geq 0$.

iii) $\Omega^j(M)$ is indecomposable for all $j \geq 0$.

c) $\mathrm{Hom}_\Lambda(M, \Lambda)$ is in $L_P(\Lambda^{op})$.

d) Let $\cdots \xrightarrow{g_i} Q_i \xrightarrow{g_{i-1}} \cdots \longrightarrow Q_0 \xrightarrow{g_0} \mathrm{Hom}_\Lambda(M, \Lambda) \longrightarrow 0$ be a minimal projective Λ^{op}-resolution of $\mathrm{Hom}_\Lambda(M, \Lambda)$. Then the exact sequence of Λ-modules

$0 \longrightarrow \mathrm{Hom}_\Lambda(\mathrm{Hom}_\Lambda(M, \Lambda), \Lambda) \xrightarrow{\mathrm{Hom}_\Lambda(g_0, \Lambda)} \mathrm{Hom}_\Lambda(Q_0, \Lambda) \longrightarrow \cdots$

$\xrightarrow{\mathrm{Hom}_\Lambda(g_i, \Lambda)} \mathrm{Hom}_\Lambda(Q_i, \Lambda) \longrightarrow \cdots$ has the following properties:

i) The epimorphisms $\mathrm{Hom}_\Lambda(Q_i, \Lambda) \longrightarrow \mathrm{Im}\,\mathrm{Hom}_\Lambda(g_{i+1}, \Lambda)$ are projective covers for all $i \geq 0$.

ii) $\mathrm{Im}\,\mathrm{Hom}_\Lambda(g_i, \Lambda)$ is in $L_P(\Lambda)$ for all $i \geq 0$.

Proof: All these statements follow from the fact that $\mathrm{Ext}^i_\Lambda(M, \Lambda) = 0$ for all $i > 0$ using arguments similar to those given in I, Section 9. See in particular I, Lemma 9.1.

As a readily verified consequence of this result we have the following strengthening of Lemma 1.2 and Proposition 1.3.

Proposition 1.8: Let Λ be an R-order with R a complete local ring of dimension d. For each Λ-lattice M in $L_P(\Lambda)$ we have

a) $\Omega^j(\Omega^k(M)) \approx \Omega^{j+k}(M)$ in $L(\Lambda)$ for all integers j and k.

b) $\Omega^j(\mathrm{Hom}_\Lambda(M, \Lambda)) \approx \mathrm{Hom}_\Lambda(\Omega^{-j}(M), \Lambda)$ in $L(\Lambda)$ for all integers j.

c) $\mathrm{Hom}_R(\mathrm{Tr}_L M, R) \cong \Omega^{2-d}M$ in $L(\Lambda)$.

d) $\mathrm{Tr}_L(\mathrm{Hom}_R(M, R)) \cong \Omega^{d-2}M$ in $L(\Lambda)$.

§2. Orders Over Complete One Dimensional Local Rings. Throughout this section we assume that R is a complete one dimensional Gorenstein local ring and Λ is an R-order. Our purpose in this section is to develop some results concerning $L(\Lambda)$ which are direct analogues of results already extablished for finitely generated modules over artin algebras (see [7] and [11]). For other analogies between these theories the reader should consult some of the recent work of K. Roggenkamp and J. Schmidt. Our first aim is to give a criterion for when $L(\Lambda)$ is of finite representation type, (i.e. when $L(\Lambda)$ has only a finite number of nonisomorphic indecomposable objects) in terms of the category of functors in $(L(\Lambda)^{op}, \mathrm{Ab})$ vanishing on projectives. We begin with the following preliminary result.

Proposition 2.1: Let Λ be an R-order.

a) $\underline{r}$, the radical of Λ, is a lattice.

b) Let A be in $L(\Lambda)$. Then the restriction of the functor $\text{Ext}^1_\Lambda(\ ,A)$: Mod $\Lambda \longrightarrow$ Ab to $L(\Lambda)$ is a finitely presented functor in $(L(\Lambda)^{op}, \text{Ab})$ which vanishes on the projective Λ-lattices.

c) A lattice C is projective if and only if $\text{Ext}^1_\Lambda(C, \underline{r}) = 0$.

d) If F is a finitely generated functor in $(L(\Lambda)^{op}, \text{Ab})$ which vanishes on projectives, then F(A) is an $\underline{\text{End}}(A)^{op}$-module of finite length for all A in $L(\Lambda)$.

Proof:

a) It was shown in II, Proposition 7.6 that if $\dim R \leq 1$ and C' is a submodule of a lattice C such that C/C' has finite length, then C' is a lattice. Since Λ is a lattice and $\Lambda/\underline{r}$ has finite length, it follows that $\underline{r}$ is a lattice.

b) This is an easy consequence of I, Proposition 8.2.

c) It is well known and easily shown that a module C in noeth Λ is projective if $\text{Ext}^1_\Lambda(C, \underline{r}) = 0$. In particular a lattice C is projective if $\text{Ext}^1_\Lambda(C, \underline{r}) = 0$.

d) Since F in $(L(\Lambda)^{op}, \text{Ab})$ is finitely generated there is an epimorphism φ: $(\ ,C) \longrightarrow F \longrightarrow 0$. The fact that $F(P) = 0$ for all projective P in $L(\Lambda)$ implies that if f: $P \longrightarrow C \longrightarrow 0$ is exact with P projective, then the composition $(\ ,P) \longrightarrow (\ ,C) \longrightarrow F \longrightarrow 0$ is zero. Hence the epimorphism $(\ ,C) \longrightarrow F \longrightarrow 0$ induces an epimorphism $\underline{\text{Hom}}_\Lambda(\ ,C) \longrightarrow F \longrightarrow 0$. Let X be in $L(\Lambda)$. We know that $\underline{\text{Hom}}_\Lambda(X, C)$ is an R-module of finite length and consequently an

(End $X)^{op}$-module of finite length. Therefore $F(X)$ is an (End $X)^{op}$-module of finite length for all X in $L(\Lambda)$.

We now give our first application in this section. We know that an artin algebra Λ is of finite representation type if and only if every finitely presented functor in $((\text{mod } \Lambda)^{op}, \text{Ab})$ has finite length (see [7]). Based on this result, it is not difficult to show that Λ is of finite representation type if and only if every finitely presented functor in $((\text{mod } \Lambda)^{op}, \text{Ab})$ vanishing on projectives has finite length. We now prove an analogue to this latter result for when $L(\Lambda)$ is of finite representation type. To this end, it is convenient to have the following result.

Lemma 2.2: A finitely generated functor F in $(L(\Lambda)^{op}, \text{Ab})$ which vanishes on projectives is of finite length if and only if there are only a finite number of nonisomorphic indecomposable A in $L(\Lambda)$ such that $F(A) \neq 0$.

Proof: By [7, Theorem 2.12] we know that F is of finite length if and only if $F(A)$ is of finite length over (End $A)^{op}$ for each indecomposable A in $L(\Lambda)$ and there are only a finite number of nonisomorphic indecomposable A in $L(\Lambda)$ such that $F(A) \neq 0$. Since $F(A)$ is of finite length over (End $A)^{op}$ for all A in $L(\Lambda)$ by Proposition 1.1, the lemma is proven.

Theorem 2.3: The following statements are equivalent for an R-order Λ.

a) $L(\Lambda)$ is of finite representation type.

b) Hom (,X) is a functor in $(L(\Lambda)^{op}, \text{Ab})$ of finite length for each X in $L(\Lambda)$.

c) Each finitely generated functor F in $(L(\Lambda)^{op}, Ab)$ which vanishes on projectives is of finite length.

d) $\mathrm{Ext}^1_\Lambda(\ ,\underline{r})$ is of finite length.

Proof:

a) implies b). Since for each X in $L(\Lambda)$, $\underline{\mathrm{Hom}}(\ ,X)$ is a finitely generated functor vanishing on projectives, the fact that $L(\Lambda)$ is of finite representation implies that there are only a finite number of nonisomorphic indecomposable A in $L(\Lambda)$ such that $\underline{\mathrm{Hom}}(A, X) \neq 0$. Hence by Lemma 1.2, $\underline{\mathrm{Hom}}(\ ,X)$ is of finite length.

b) implies c). Since F is finitely generated, there is an epimorphism $(\ ,X) \longrightarrow F \longrightarrow 0$. Since F vanishes on projectives, $(\ ,X) \longrightarrow F \longrightarrow 0$ induces an epimorphism $\underline{\mathrm{Hom}}_\Lambda(\ ,X) \longrightarrow F \longrightarrow 0$. The fact that $\underline{\mathrm{Hom}}_\Lambda(\ ,X)$ is of finite length obviously implies that F is of finite length.

c) implies d). By Proposition 1.1 we know that $\mathrm{Ext}^1_\Lambda(\ ,\underline{r})$ is a finitely generated functor in $(L(\Lambda)^{op}, Ab)$ vanishing on projectives. Hence $\mathrm{Ext}^1_\Lambda(\ ,\underline{r})$ is of finite length.

d) implies a). Assume that $\mathrm{Ext}^1_\Lambda(\ ,\underline{r})$ has finite length. Thus there are only a finite number of nonisomorphic nonprojective indecomposable A in $L(\Lambda)$ such that $\mathrm{Ext}^1_\Lambda(A, \underline{r}) \neq 0$. Since A in $L(\Lambda)$ has $\mathrm{Ext}^1_\Lambda(A, \underline{r}) = 0$ only if A is projective (see Proposition 1.1), there are only a finite number of nonisomorphic nonprojective indecomposable A in $L(\Lambda)$. But there are only a finite number of nonisomorphic indecomposable projective modules in $L(\Lambda)$. Therefore there are only a finite number of nonisomorphic indecomposable A in $L(\Lambda)$ if $\mathrm{Ext}^1_\Lambda(\ ,\underline{r})$ in $(L(\Lambda)^{op}, Ab)$

is a functor of finite length.

Before giving our final result of this section we recall the notion of an irreducible morphism. A morphism $f: A \longrightarrow B$ in $L(\Lambda)$ is said to be <u>irreducible in $L(\Lambda)$</u> if a) f is not a splittable monomorphism or epimorphism and b) if f is the composition $A \xrightarrow{g} X \xrightarrow{h} B$ in $L(\Lambda)$, then either h is a splittable epimorphism or g is a splittable monomorphism. The reader is referred to [11] for a more complete discussion of irreducible morphisms. The only facts concerning irreducible morphisms we will need we give in the following result.

<u>Proposition 2.4</u>: Let Λ be an R-order.

a) Let $g: A \longrightarrow B$ be a minimal left almost split morphism in $L(\Lambda)$. A morphism $h: A \longrightarrow X$ with $X \neq 0$ is irreducible if and only if h can be written as a composition $A \xrightarrow{g} B \xrightarrow{t} X$ with t a splittable epimorphism.

b) Let $f: B \longrightarrow C$ be a minimal right almost split morphism in $L(\Lambda)$. A morphism $h: X \longrightarrow C$ with $X \neq 0$ is irreducible if and only if h can be written as a composition $X \xrightarrow{t} B \xrightarrow{f} C$ with t a splittable monomorphism.

c) Suppose $0 \longrightarrow A \xrightarrow{g} B \xrightarrow{f} C \longrightarrow 0$ is an almost split sequence in $L(\Lambda)$. Then B has an indecomposable projective summand if and only if A is isomorphic to a noninjective summand of $\underline{r}P$ for some indecomposable projective P.

<u>Proof</u>:

a) and b). See [11, Theorem 2.4].

c) We include a proof of c) for the convenience of the reader even though it is the same as the proof of [11, Theorem 4.2].

Suppose $0 \longrightarrow A \xrightarrow{g} B \xrightarrow{f} C \longrightarrow 0$ is an almost split sequence in $L(\Lambda)$. Since the sequence does not split, A is not an injective lattice. Let P be an indecomposable projective summand of B and suppose $t: B \longrightarrow P$ is an epimorphism. Then by a), the composition $tg: A \longrightarrow P$ is irreducible. Since by II, Proposition 8.3 the inclusion $\underline{r}P \longrightarrow P$ is a minimal right almost split morphism, there is by b) a splittable monomorphism $u: A \longrightarrow \underline{r}P$ such that tg is inc u. Thus A is isomorphic to a noninjective summand of $\underline{r}P$.

On the other hand, suppose A is isomorphic to a noninjective summand of $\underline{r}P$ for some indecomposable projective P. Then there is an irreducible morphism $h: A \longrightarrow P$. Hence by a) the morphism h can be written as a composition $A \xrightarrow{g} B \xrightarrow{t} P$ such that t is a splittable epimorphism. Thus B has the indecomposable projective module P as a summand.

With these preliminary results in mind, we now prove the following.

Theorem 2.5: Let Λ be an R-order. Suppose C is an indecomposable nonprojective lattice. Then the following are equivalent.

a) If $0 \longrightarrow A \xrightarrow{g} B \xrightarrow{f} C \longrightarrow 0$ is an almost split sequence in $L(\Lambda)$, then B has an indecomposable projective summand.

b) If $X \longrightarrow C$ is an epimorphism with X in $L(\Lambda)$ which is not a splittable epimorphism, then X has an indecomposable projective summand.

c) A is isomorphic to a noninjective summand of $\underline{r}P$ for some indecomposable projective P.

Proof:

a) implies b). Suppose P is an indecomposable projective summand of B. Then $h = f|P$ is an irreducible morphism $P \longrightarrow C$. Suppose $u: X \longrightarrow C$ is an epimorphism with X in $L(\Lambda)$ which is not a splittable epimorphism. Because u is an epimorphism and P is projective, there is a morphism $s: P \longrightarrow X$ such that $us = h$. Since u is not a splittable epimorphism and h is irreducible, $s: P \longrightarrow X$ is a splittable monomorphism. Hence X has an indecomposable projective summand.

b) implies a). Trivial.

a) equivalent to c). Contained in Proposition 3.11.

Theorem 2.6: Let Λ be an R-order.

a) There is a one-to-one correspondence between the isomorphism classes of indecomposable nonprojective C in $L(\Lambda)$ having the property if X is in $L(\Lambda)$ and $X \longrightarrow C$ is an epimorphism which is not splittable, then X has an indecomposable projective summand and the isomorphism classes of noninjective indecomposable summands of $\underline{r}$ given by $C \longmapsto (Tr_L C)^* = Hom_R(Tr_L C, R)$.

b) Λ is not an hereditary ring if and only if there is an indecomposable nonprojective C in $L(\Lambda)$ such that an X in $L(\Lambda)$ must have an indecomposable projective summand if there is an epimorphism $X \longrightarrow C$ which is not splittable.

Proof:

a) Since C is indecomposable and not projective, we know that there is an almost split sequence in $L(\Lambda)$ $0 \longrightarrow (Tr_L C)^* \longrightarrow B \longrightarrow C \longrightarrow 0$. By Theorem 2.5 we know C

satisfies the condition of a) if and only if $(Tr_L C)^*$ is isomorphic to a noninjective summand of $\underline{r}$. Since, up to isomorphism, C is uniquely determined by $(Tr_L(C))^*$, the map $C \longmapsto (Tr_L(C))^*$ gives an injection between the isomorphism classes of C satisfying the condition of a) and noninjective indecomposable summands of $\underline{r}$. On the other hand, if A is a noninjective indecomposable summand of $\underline{r}$, then the almost split sequence $0 \longrightarrow A \longrightarrow B \longrightarrow C \longrightarrow 0$ has the property that B has an indecomposable projective summand (see Theorem 2.5). Since $(Tr_L(C))^* \approx A$, the map $C \longmapsto (Tr_L(C))^*$ gives a bijection between the isomorphism classes of C in $L(\Lambda)$ satisfying the condition of a) and the isomorphism classes of noninjective indecomposable summands of $\underline{r}$.

b) Since Λ is a semi-perfect noetherian ring, it is well known that gl. dim. Λ = p.d. $\Lambda/\underline{r}$. Therefore Λ is hereditary if and only if $\underline{r}$ is a projective Λ-lattice. Because all Λ-lattices are submodules of projective Λ-modules, it follows that Λ is hereditary if and only if every Λ-lattice is projective. But we know that a Λ-lattice C is projective if and only if $\mathrm{Ext}^1_\Lambda(C, \underline{r}) = 0$ (see Proposition 1.1). Therefore Λ is hereditary if and only if $\mathrm{Ext}^1_\Lambda(C, \underline{r}) = 0$ for all Λ-lattices C. But $\mathrm{Ext}^1_\Lambda(C, \underline{r}) = 0$ for all Λ-lattices C if and only if $\underline{r}$ is an injective lattice. Thus by a), Λ is not hereditary if and only if there is an indecomposable nonprojective Λ-lattice C such that an X in $L(\Lambda)$ has an indecomposable projective summand whenever there is an epimorphism $X \longrightarrow C$ which is not splittable.

As an immediate consequence of this result we have the following result about complete commutative one-dimensional local integral domains.

Corollary 2.7: Let S be a complete commutative one-dimensional local domain which is not a discrete valuation ring. Then there is a unique, up to isomorphism, indecomposable nonprojective torsion-free module C such that if $X \longrightarrow C$ is an epimorphism which is not a splittable epimorphism with X a torsion-free module, then X has S as a summand.

Proof: It is well known that there is a complete discrete valuation ring R which is a subring of S such that S is a finitely generated free R-module (see[17, Corollary 31.6]). Hence S is an R-order. S is not hereditary since it is not a discrete valuation ring. The result is now a trivial consequence of Theorem 2.6.

Added in Proof:

Suppose Λ is an R-order with R an arbitrary 1-dimensional Gorenstein ring (not necessarily local or complete). Let C be a Λ-lattice. In general, while a Λ-submodule B of C is Cohen-Macauley since $\dim R = 1$, it need not be a Λ-lattice since $B_{\underline{p}}$ need not be $\Lambda_{\underline{p}}$-projective for all nonmaximal prime ideals $\underline{p}$ of R. However if $\Lambda_{\underline{p}}$ is semi-simple for all nonmaximal prime ideals $\underline{p}$ of R, then this difficulty disappears and we have that every Λ-submodule B of a Λ-lattice C is again a Λ-lattice. For the rest of this discussion we assume that the R-order Λ has the property $\Lambda_{\underline{p}}$ is semi-simple for all nonmaximal prime ideals $\underline{p}$ or R. Clearly the R-order Λ^{op} also has the property $(\Lambda^{op})_{\underline{p}} = (\Lambda_{\underline{p}})^{op}$ is semi-simple for all nonmaximal prime ideals $\underline{p}$ of R. Hence $L(\Lambda)$ and $L(\Lambda^{op})$ are closed under taking submodules. This fact enables us to prove the following analogue of I, Theorem 3.19.

Theorem 2.8: Let X and C be in $L(\Lambda)$. Suppose H is an $(\text{End } X)^{op}$-submodule of (X, C). Then there is an exact sequence $0 \longrightarrow A \xrightarrow{g} B \xrightarrow{f} C$ of Λ-lattices with f a right X-determined morphism in $L(\Lambda)$ such that $\text{Im}(\,,f)(X) = H$. Moreover, if we let $C' = \text{Im} f$ and let $0 \longrightarrow A \xrightarrow{g} B \xrightarrow{f'} C' \longrightarrow 0$ be the induced exact sequence, then

a) H is an $(\text{End } X)^{op}$-submodule of (X, C') containing $P(X, C')$.

b) f' is right X-determined in $L(\Lambda)$ with $\text{Im}(\,,f')(X) = H$.

Proof: The proof is entirely analogous to that given for I, Theorem 3.19 and is left to the reader.

Using the usual duality between $L(\Lambda)$ and $L(\Lambda^{op})$ we obtain the following dual of Theorem 2.8.

Theorem 2.9: Let Y and A be in $L(\Lambda)$ and H an End Y-submodule (A, Y). Then there is an exact sequence $A \xrightarrow{g} B \xrightarrow{f} C \longrightarrow 0$ in $L(\Lambda)$ with g left Y-determined in $L(\Lambda)$ such that $\text{Im}(g, \,)(Y) = H$. Moreover, if we let $\text{Im } g = A'$, and let $0 \longrightarrow A' \xrightarrow{g'} B \longrightarrow C \longrightarrow 0$ be the induced exact sequence in $L(\Lambda)$, then

a) H is an End Y-submodule of (A', Y) containing $I(A', Y)$.

b) g' is left Y-determined in $L(\Lambda)$ with $\text{Im}(g', \,)(Y) = H$.

It should be noted that if R is a complete local ring, then the morphism $f: B \longrightarrow C$ in Theorem 2.8 can be chosen to be right minimal while the morphism $g: A \longrightarrow B$ in Theorem 2.9 can be chosen to be left minimal.

§3. Orders Over Two Dimensional Rings. Suppose Λ is an R-order with dim R = d. In II, Section 7, we showed that if S is a simple functor in $(L(\Lambda)^{op}, Ab)$ vanishing on projectives, then there is an exact sequence $0 \longrightarrow A \longrightarrow B \longrightarrow C \longrightarrow 0$ in $L(\Lambda)$ such that $0 \longrightarrow (\ ,A) \longrightarrow (\ ,B) \longrightarrow (\ ,C) \longrightarrow S \longrightarrow 0$ is exact. Similarly, if T is a simple functor in $(L(\Lambda), Ab)$ vanishing on injective lattices, we showed that there is an exact sequence $0 \longrightarrow A \longrightarrow B \longrightarrow C \longrightarrow 0$ in $L(\Lambda)$ such that $0 \longrightarrow (C, \ ) \longrightarrow (B, \ ) \longrightarrow (A, \ ) \longrightarrow T \longrightarrow 0$ is exact. This leaves open the question of describing projective resolutions for simple functors in $(L(\Lambda)^{op}, Ab)$ and $(L(\Lambda), Ab)$ which do not vanish on projectives or injective lattices respectively. This question was solved in II, Section 7 in the casè dim R $\leq$ 1. In this section we give a solution of the problem for dim R = 2. As far as we know, the question is open for dim R $\geq$ 3.

Let S be a simple functor in $(L(\Lambda)^{op}, Ab)$ not vanishing on a projective lattice P. Then there is an epimorphism α: $(\ ,P) \longrightarrow S$. This induces an epimorphism of Λ-modules $\alpha_\Lambda: (\Lambda, P) \rightarrow S(\Lambda) \rightarrow 0$. Since S is a simple functor and $S(\Lambda) \neq 0$, we know that $S(\Lambda)$ is a simple Λ-module. Hence Ker α_Λ is a maximal submodule M of $(\Lambda, P) = P$. We have seen in II, Section 2, that if dim R $\leq$ 1, then M is a lattice. However, M fails to be a lattice for dim R $\geq$ 2 because, although it has the property $M_{\underline{p}}$ is $\Lambda_{\underline{p}}$-projective for all nonmaximal prime ideals $\underline{p}$ of R, it is not a Cohen-Macauley module. Nonetheless, the restriction of the representable functor $(\ ,M)$ in $((\mathrm{Mod}\ \Lambda)^{op}, Ab)$ to $L(\Lambda)$, which we also denote by $(\ ,M)$, plays an important role in describing a projective presentation for S since, as we now show, the sequence $0 \longrightarrow (\ ,M) \xrightarrow{\beta} (\ ,P) \xrightarrow{\alpha} S \longrightarrow 0$ is exact in $(L(\Lambda)^{op}, Ab)$.

Obviously, β is a monomorphism. Suppose $\alpha\beta$: $(\ ,M) \longrightarrow S$ is not zero. Then it is an epimorphism since S is simple. In particular $\alpha\beta_\Lambda: (\Lambda, M) \longrightarrow S(\Lambda)$ is an epimorphism, i.e. the composition $(\Lambda, M) \xrightarrow{\beta_\Lambda} (\Lambda, P) \xrightarrow{\alpha_\Lambda} S(\Lambda)$ is an epimorphism. But this contradicts the fact that $0 \longrightarrow (\Lambda, M) \longrightarrow (\Lambda, P) \xrightarrow{\alpha_\Lambda} S(\Lambda) \longrightarrow 0$ is exact since $M = \operatorname{Ker} \alpha_\Lambda$. Therefore $\alpha\beta = 0$ or equivalently $\operatorname{Im} \beta \subset \operatorname{Ker} \alpha$. Thus we are done if we show that $\operatorname{Im} \beta$ is a maximal subfunctor of $(\ ,P)$. To see this, we only have to show that if $X \longrightarrow P$ is a morphism in $L(\Lambda)$ such that $\operatorname{Im}((\ ,X) \longrightarrow (\ ,P))$ is not contained in $\operatorname{Im} \beta$, then the induced morphism $(\ ,X) \coprod (\ ,M) \longrightarrow (\ ,P)$ is an epimorphism. But such an $f: X \longrightarrow P$ has to have the property that $\operatorname{Im} f \not\subset M$. Since M is a maximal submodule of P, this implies that the induced morphism of modules $X \coprod M \longrightarrow P$ is an epimorphism and thus a splittable epimorphism because P is projective. Therefore $(\ ,X) \coprod (\ ,M) \longrightarrow (\ ,P)$ is an epimorphism, which is our desired result. As a consequence of this discussion we have the following.

Proposition 3.1:

a) Suppose M is a maximal submodule of the finitely generated projective module P and $(\ ,M)$ denotes the restriction of the representable functor $(\ ,M)$ in $((\operatorname{Mod} \Lambda)^{op}, Ab)$ to $L(\Lambda)$. Then $\operatorname{Coker}((\ ,M) \longrightarrow (\ ,P))$ is a simple functor in $(L(\Lambda)^{op}, Ab)$.

b) Suppose S is a simple functor in $(L(\Lambda)^{op}, Ab)$ and P a projective lattice such that $S(P) \neq 0$. Then there is a maximal submodule M of P such that there is an exact sequence $0 \longrightarrow (\ ,M) \longrightarrow (\ ,P) \longrightarrow S \longrightarrow 0$ in $(L(\Lambda)^{op}, Ab)$, where $(\ ,M) \longrightarrow (\ ,P)$ is given by the inclusion $M \longrightarrow P$.

As an easily verified consequence of this result we have the following.

Corollary 3.2: Every simple functor S in $(L(\Lambda)^{op}, Ab)$ not vanishing on projectives is finitely presented if and only if given any maximal submodule M of a projective lattice P, there is a lattice E and an epimorphism $E \longrightarrow M$ of modules such that $\mathrm{Hom}_\Lambda(X, E) \to \mathrm{Hom}_\Lambda(X, M) \to 0$ is exact for all lattices X. Moreover any morphism $E \longrightarrow M$ such that $\mathrm{Hom}_\Lambda(X, E) \longrightarrow \mathrm{Hom}_\Lambda(X, M) \longrightarrow 0$ is exact for all lattices X is an epimorphism of modules.

Our aim now is to show that if $\dim R = 2$ and M is a maximal submodule of a projective lattice P, then there is an epimorphism $E \longrightarrow M \longrightarrow 0$ with E a lattice such that $(X, E) \longrightarrow (X, M) \longrightarrow 0$ is exact for all lattices X.

Suppose such an epimorphism $E \xrightarrow{f} M \longrightarrow 0$ exists. This gives rise to an exact sequence $0 \longrightarrow A \xrightarrow{g} E \xrightarrow{f} P \longrightarrow P/M \longrightarrow 0$. Since $\dim R = 2$ and P and E are Cohen-Macauley modules, it follows that A is Cohen-Macauley. Also since P/M is a simple Λ-module and P and E are lattices, it follows that $A_{\underline{p}}$ is $\Lambda_{\underline{p}}$-projective for all nonmaximal prime ideals $\underline{p}$ of R. Thus A is a lattice with the property that $0 \longrightarrow (X, A) \longrightarrow (X, E) \longrightarrow (X, M) \longrightarrow 0$ is exact for all lattices X, or equivalently, the monomorphism $0 \longrightarrow A \xrightarrow{g} E$ of lattices has the property that $0 \longrightarrow \mathrm{Ext}^1_\Lambda(X, A) \longrightarrow \mathrm{Ext}^1_\Lambda(X, E)$ is exact for all lattices X. Since this latter condition is satisfied if $\mathrm{Ext}^1_\Lambda(X, A) = 0$ for all lattices X, i.e. if A is an injective lattice, this analysis suggests trying to find an exact sequence
$0 \longrightarrow A \xrightarrow{g} E \xrightarrow{f} P \longrightarrow P/M \longrightarrow 0$ of Λ-modules with A and E lattices with A an injective lattice. On the basis of our preceding

discussion, it is not difficult to see that such an exact sequence has the property that the induced epimorphism $E \longrightarrow M$ satisfies the condition that $(X, E) \longrightarrow (X, M) \longrightarrow 0$ is exact for all lattices X. We now proceed to show that such an exact sequence exists.

First we show that if $\dim R = 2$ and M is a maximal submodule of a projective lattice P, then there is an exact sequence $0 \longrightarrow A \xrightarrow{g} E \longrightarrow P \longrightarrow P/M \longrightarrow 0$ of Λ-modules with A an injective lattice such that g is not a splittable monomorphism. We will then show that such an exact sequence has our desired properties by showing that E is a lattice. The existence of such an exact sequence will follow from the following more general result.

Proposition 3.3: Suppose $\dim R = d \geq 1$. If X is a lattice and A is a Λ-module of finite length, then $\mathrm{Ext}^i_\Lambda(A, X) = 0$ for $i = 0,\ldots,d - 1$ and $\mathrm{Ext}^d_\Lambda(A, X) \cong \mathrm{Hom}_R(\mathrm{Hom}_R(X, R) \otimes_\Lambda A, I_d)$. Thus $\mathrm{Ext}^d_\Lambda(A, X) = 0$ if and only if $\mathrm{Hom}_R(X, R) \otimes_\Lambda A = 0$.

Proof: Let $0 \longrightarrow R \longrightarrow I_0 \longrightarrow \ldots \longrightarrow I_d \longrightarrow 0$ be a minimal injective resolution of R over R. Let $Y = \mathrm{Hom}_R(X, R)$. Then $0 \longrightarrow \mathrm{Hom}_R(Y, R) \longrightarrow \mathrm{Hom}_R(Y, I_0) \longrightarrow \ldots \longrightarrow \mathrm{Hom}_R(Y, I_d) \longrightarrow 0$ is exact with the $\mathrm{Hom}_R(Y, I_j)$ injective Λ-modules for $j = 0,\ldots,d - 1$ since Y is a Λ^{op}-lattice (see I, Lemma 8.4). Now $\mathrm{Hom}_\Lambda(A, \mathrm{Hom}_R(Y, I_j)) \cong \mathrm{Hom}_R(Y \otimes_\Lambda A, I_j)$ for all $j = 0,\ldots,d$. Since A is of finite length, $Y \otimes_\Lambda A$ is an R-module of finite length and so $\mathrm{Hom}_R(Y \otimes_\Lambda A, I_j) = 0$ for $j = 0,\ldots,d - 1$ (see I, Section 7). From this it follows $\mathrm{Ext}^i_\Lambda(A, \mathrm{Hom}_R(Y, R)) = 0$ for $i = 0,\ldots,d - 1$ and $\mathrm{Ext}^d_\Lambda(A, \mathrm{Hom}_R(Y, R)) \cong \mathrm{Hom}_R(Y \otimes_\Lambda A, I_d)$. The first part of the proposition now follows since $\mathrm{Hom}_R(Y, R) \cong X$ and $Y = \mathrm{Hom}_R(X, R)$.

Since I_d is an injective cogenerator for R, we know that $\mathrm{Hom}_R(X, R) \otimes_\Lambda A = 0$ if and only if $\mathrm{Hom}_R(\mathrm{Hom}_R(X, R) \otimes_\Lambda A, I_d) = 0$. This finishes the proof of the proposition.

As an easy consequence of this proposition, we have the following result.

<u>Corollary 3.4</u>: Suppose $\dim R = 2$, S is a simple Λ-module, and P is a projective lattice such that there is an epimorphism $\varepsilon: P \longrightarrow S \longrightarrow 0$. Then there is an exact sequence $0 \longrightarrow A \xrightarrow{g} E \xrightarrow{f} P \xrightarrow{\varepsilon} S \longrightarrow 0$ such that g is not a splittable monomorphism and A is the injective lattice $\mathrm{Hom}_R(\mathrm{Hom}_\Lambda(P, \Lambda), R)$.

<u>Proof</u>: Since P is a projective Λ-module, the elements of $\mathrm{Ext}^2_\Lambda(S, A)$ correspond to exact sequences $0 \longrightarrow A \xrightarrow{g} E \xrightarrow{f} P \xrightarrow{\varepsilon} S \longrightarrow 0$ with $0 \longrightarrow A \xrightarrow{g} E \xrightarrow{f} P \xrightarrow{\varepsilon} S \longrightarrow 0$ corresponding to the zero element of $\mathrm{Ext}^2_\Lambda(S, A)$ if and only if g is a splittable monomorphism. Now $\mathrm{Hom}_R(A, R) \approx \mathrm{Hom}_\Lambda(P, \Lambda)$, while by Proposition 3.3 we know that $\mathrm{Ext}^2_\Lambda(S, A) \neq 0$ if $\mathrm{Hom}_R(A, R) \otimes_\Lambda S \neq 0$. But $\mathrm{Hom}_\Lambda(P, \Lambda) \otimes_\Lambda S = \mathrm{Hom}_\Lambda(P, S)$ since P is a finitely generated projective Λ-module. Therefore $\mathrm{Ext}^2(S, A) \neq 0$ since the epimorphism $\varepsilon: P \longrightarrow S$ is in $\mathrm{Hom}_\Lambda(P, S)$. Consequently there is an exact sequence $0 \longrightarrow A \xrightarrow{g} E \xrightarrow{f} P \xrightarrow{\varepsilon} S \longrightarrow 0$ with g not a splittable monomorphism.

We now proceed to the next step of the proof of our main result.

<u>Proposition 3.5</u>: Suppose $\dim R = 2$ and $0 \longrightarrow A \xrightarrow{g} B \xrightarrow{f} P \xrightarrow{\varepsilon} S \longrightarrow 0$ is an exact sequence in noeth Λ

with g not a splittable monomorphism. Then B is a lattice if A and P are lattices with P projective and S a simple Λ-module.

Proof: Since S is a simple Λ-module, then S is an R-module of finite length so $S_{\underline{p}} = 0$ for all nonmaximal prime ideals $\underline{p}$ of R. Thus $0 \longrightarrow A_{\underline{p}} \longrightarrow B_{\underline{p}} \longrightarrow P_{\underline{p}} \longrightarrow 0$ is an exact sequence of $\Lambda_{\underline{p}}$-modules for all nonmaximal prime ideals $\underline{p}$ of R. Since $P_{\underline{p}}$ and $A_{\underline{p}}$ are $\Lambda_{\underline{p}}$-projective for all nonmaximal prime ideals of R (remember A and P are lattices), $B_{\underline{p}}$ is $\Lambda_{\underline{p}}$-projective for all nonmaximal prime ideals of R. Therefore to show that B is a lattice, we must show that B is a Cohen-Macauley module over R. Because $\dim R = 2$, this amounts to showing that B is a reflexive R-module.

We first split the exact sequence of Λ-modules $0 \longrightarrow A \xrightarrow{g} B \xrightarrow{f} P \xrightarrow{\varepsilon} S \longrightarrow 0$ into the two exact sequences $0 \longrightarrow A \xrightarrow{g} B \longrightarrow M \longrightarrow 0$ and $0 \longrightarrow M \longrightarrow P \longrightarrow S \longrightarrow 0$ where $M = \operatorname{Im} f = \operatorname{Ker} \varepsilon$, which is a maximal submodule of P since S is simple. The fact that S is a simple Λ-module implies that S is an R-module of finite length. Therefore $\operatorname{Ext}^i_R(S, R) = 0$ for $i = 0, 1$ since R is a Gorenstein equidimensional ring of dimension 2. Consequently the inclusion $0 \longrightarrow M \longrightarrow P$ induces an isomorphism $\operatorname{Hom}_R(P, R) \xrightarrow{\sim} \operatorname{Hom}_R(M, R)$. Also because S is an R-module of finite length we have that $M_{\underline{p}} \cong P_{\underline{p}}$ and is therefore $\Lambda_{\underline{p}}$-projective for all nonmaximal prime ideals $\underline{p}$ of R.

On the other hand the exact sequence $0 \longrightarrow A \longrightarrow B \longrightarrow M \longrightarrow 0$ induces the exact sequence

$0 \longrightarrow \operatorname{Hom}_R(M, R) \longrightarrow \operatorname{Hom}_R(B, R) \longrightarrow \operatorname{Hom}_R(A, R) \longrightarrow \operatorname{Ext}^1_R(M, R)$. Since $M_{\underline{p}}$ is $\Lambda_{\underline{p}}$-projective and therefore a Cohen-Macauley $R_{\underline{p}}$-module for all nonmaximal prime ideals $\underline{p}$ of R, we have that

$\operatorname{Ext}^1_R(M, R)_{\underline{p}} \cong \operatorname{Ext}^1_{R_{\underline{p}}}(M_{\underline{p}}, R_{\underline{p}}) = 0$ for all nonmaximal prime ideals $\underline{p}$ of

R. Therefore $\mathrm{Ext}_R^1(M, R)$ is an R-module of finite length. As in the argument above, this implies that

$$0 \longrightarrow \mathrm{Hom}_R(\mathrm{Hom}_R(A, R), R)) \longrightarrow \mathrm{Hom}_R(\mathrm{Hom}_R(B, R), R) \longrightarrow \mathrm{Hom}_R(\mathrm{Hom}_R(M, R), R)$$

is exact. Using the fact that the monomorphism $M \longrightarrow P$ induces an isomorphism $\mathrm{Hom}_R(P, R) \longrightarrow \mathrm{Hom}_R(M, R)$ and therefore an isomorphism $\mathrm{Hom}_R(\mathrm{Hom}_R(M, R), R) \longrightarrow \mathrm{Hom}_R(\mathrm{Hom}_R(P, R), R)$ and the fact that A and P are R-reflexive we have the following exact commutative diagram of Λ-modules

$$\begin{array}{ccccccccc}
 & & & & 0 & & 0 & & \\
 & & & & \downarrow & & \downarrow & & \\
0 & \longrightarrow & A & \xrightarrow{g} & B & \xrightarrow{u} & M & \longrightarrow & 0 \\
 & & \| & & \downarrow & & \downarrow & & \\
0 & \longrightarrow & A & \longrightarrow & \mathrm{Hom}_R(\mathrm{Hom}_R(B, R), R) & \xrightarrow{t} & P & &
\end{array}$$

Now if t were onto, then the bottom row would split since P is Λ-projective. But this would imply that g is a splittable monomorphism, which contradicts the hypothesis of the proposition. Thus T is not onto. Since $M \subset \mathrm{Im}\, t \subset P$ and M is a maximal submodule of P, this means that $M = \mathrm{Im}\, t$. It then follows from the commutative diagram that $B \longrightarrow \mathrm{Hom}_R(\mathrm{Hom}_R(B, R), R)$ is an isomorphism or in other words B is R-reflexive. This establishes the proposition.

Combining our previous remarks we obtain our desired result.

Theorem 3.6: Suppose $\dim R = 2$.

a) Let M be a maximal submodule of a projective lattice P. Then there is an exact sequence of Λ-modules

$0 \longrightarrow A \xrightarrow{g} E \xrightarrow{f} P \longrightarrow P/M \longrightarrow 0$ having the following properties.

i) A, E, P are in $L(\Lambda)$.

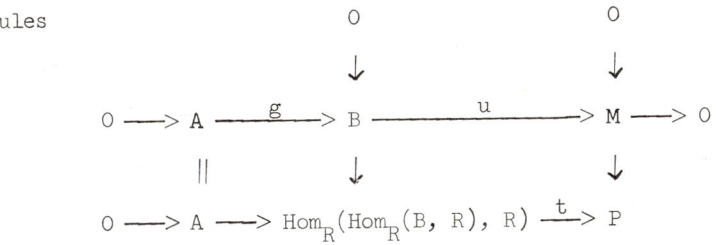

ii) A is isomorphic to the injective lattice

$\mathrm{Hom}_R(\mathrm{Hom}_\Lambda(P, \Lambda), R)$.

iii) g is not a splittable monomorphism.

iv) The exact sequence

$$0 \longrightarrow (\ ,A) \xrightarrow{(\ ,g)} (\ ,E) \xrightarrow{(\ ,f)} (\ ,P) \longrightarrow \mathrm{Coker}\ (\ ,f) \longrightarrow 0$$

in $(L(\Lambda)^{op}, Ab)$ has the property that Coker $(\ ,f)$ is a simple functor not vanishing on P.

b) Let S be a simple functor in $(L(\Lambda)^{op}, Ab)$ such that $S(P) \neq 0$ for some projective P in $L(\Lambda)$. Then there is an exact sequence $0 \longrightarrow A \xrightarrow{g} E \xrightarrow{f} P$ in $L(\Lambda)$ with g not a splittable monomorphism with $A \approx \mathrm{Hom}_R(\mathrm{Hom}_\Lambda(P, \Lambda), R)$ such that there is an exact sequence

$0 \longrightarrow (\ ,A) \xrightarrow{(\ ,g)} (\ ,E) \xrightarrow{(\ ,f)} (\ ,P) \longrightarrow S \longrightarrow 0$ in $(L(\Lambda)^{op}, Ab)$.

We now study the projective resolutions of simple functors in $(L(\Lambda)^{op}, Ab)$ which do not vanish on some injectives.

Proposition 3.7: Suppose dim R = 2. Let $0 \longrightarrow A \xrightarrow{g} E \xrightarrow{f} P \xrightarrow{\varepsilon} S \longrightarrow 0$ be an exact sequence in noeth Λ having the following properties:

i) P is projective and S is simple.

ii) $A \cong \mathrm{Hom}_R(\mathrm{Hom}_\Lambda(P, \Lambda), R)$.

Then the sequence

$0 \longrightarrow (P,\ ) \xrightarrow{(f,\ )} (E,\ ) \xrightarrow{(g,\ )} (A,\ ) \longrightarrow \mathrm{Coker}\ (g,\ ) \longrightarrow 0$ of functors in $(L(\Lambda), Ab)$ is exact with Coker $(g,\ )$ a simple functor which does not vanish on the injective lattice A.

Proof: We first show that the sequence

$0 \longrightarrow (P,\ ) \xrightarrow{(f,\ )} (E,\ ) \xrightarrow{(g,\ )} (A,\ )$ is exact. Let X be a Λ-lattice. Since S is a simple Λ-module, we have by Proposition 3.3 that $\mathrm{Ext}^i_\Lambda(S, X) = 0$ for $i = 0, 1$. From this it follows that the inclusion $\mathrm{Im}\, f \longrightarrow P$ induces an isomorphism $(P, X) \longrightarrow (\mathrm{Im}\, f, X)$. Since the exact sequence $0 \longrightarrow A \xrightarrow{g} E \longrightarrow \mathrm{Im}\, f \longrightarrow 0$ induces the exact sequence $0 \longrightarrow (\mathrm{Im}\, f, X) \longrightarrow (E, X) \longrightarrow (A, X)$, we have that $0 \longrightarrow (P, X) \xrightarrow{(f, X)} (E, X) \longrightarrow (A, X)$ is exact. The fact that this is true for all X in $L(\Lambda)$ shows that $0 \longrightarrow (P,\ ) \xrightarrow{(f,\ )} (E,\ ) \xrightarrow{(g,\ )} (A,\ )$ is exact in $(L(\Lambda), \mathrm{Ab})$. We now show that $\mathrm{Coker}\,(g,\ )$ is a simple functor.

We know by II, Proposition 2.2, that a morphism $g\colon A \longrightarrow E$ which is not a splittable monomorphism has the property $\mathrm{Coker}\,(g,\ )$ is simple if given any morphism $h\colon A \longrightarrow X$ in $L(\Lambda)$ such that there is no morphism $t\colon E \longrightarrow X$ such that $tg = h$, then the induced morphism $A \longrightarrow E \amalg X$ is a splittable monomorphism. Suppose $h\colon A \longrightarrow X$ is a morphism in $L(\Lambda)$ such that there is no morphism $t\colon E \longrightarrow X$ such that $tg = h$. Consider the exact push out diagram

$$\begin{array}{ccccccccc}
0 & \longrightarrow & A & \xrightarrow{g} & E & \longrightarrow & \mathrm{Im}\, f & \longrightarrow & 0 \\
 & & \downarrow{\scriptstyle h} & & \downarrow{\scriptstyle v} & & \| & & \\
0 & \longrightarrow & X & \xrightarrow{w} & E \overset{A}{\times} X & \longrightarrow & \mathrm{Im}\, f & \longrightarrow & 0
\end{array}$$

By the hypothesis on h, the bottom row does not split. Since $0 \longrightarrow X \xrightarrow{w} E \overset{A}{\times} X \longrightarrow P \xrightarrow{\varepsilon} S \longrightarrow 0$ is exact with w a nonsplittable monomorphism, X a lattice, P a projective lattice, and S simple, it follows from Proposition 3.5 that $E \overset{A}{\times} X$ is a lattice. Therefore there exists an exact commutative diagram

$$\begin{array}{ccccccccc} 0 & \longrightarrow & X & \xrightarrow{w} & E \overset{A}{\times} X & \longrightarrow & \text{Im } f & \longrightarrow & 0 \\ & & \downarrow j & & \downarrow k & & \| & & \\ 0 & \longrightarrow & A & \xrightarrow{g} & E & \longrightarrow & \text{Im } f & \longrightarrow & 0 \end{array}$$

which gives rise to the commutative diagram

$$\begin{array}{ccccccccc} 0 & \longrightarrow & A & \xrightarrow{g} & E & \longrightarrow & \text{Im } f & \longrightarrow & 0 \\ & & jh\downarrow & & kv\downarrow & & \| & & \\ 0 & \longrightarrow & A & \longrightarrow & E & \longrightarrow & \text{Im } f & \longrightarrow & 0 \end{array}$$

From this it follows that there is a morphism $s: E \longrightarrow A$ such that $1_A - jh = sg$. The fact that $1_A = jh + sg$ shows that the composition

$$A \xrightarrow{(g,\ h)} E \coprod X \xrightarrow{(s,\ j)} A$$

is the identity, or in other words, $A \longrightarrow E \coprod X$ is a splittable monomorphism. This shows that Coker $(g,\ )$ is simple which finishes the proof of the proposition.

Proposition 3.7 describes projective resolutions for certain simple functors in $(L(\Lambda), Ab)$ which do not vanish on injective lattices. We now show that in fact one actually gets projective resolutions for all simple functors in $(L(\Lambda), Ab)$ not vanishing on injective lattices in this way.

Proposition 3.8: Let T be a simple functor in $(L(\Lambda), Ab)$ which does not vanish on an injective lattice A. Then there is an exact sequence $0 \longrightarrow A \xrightarrow{g} E \xrightarrow{f} P \xrightarrow{\varepsilon} S \longrightarrow 0$ with E a lattice, P the

projective Λ-module $\text{Hom}_{\Lambda^{op}}(\text{Hom}_R(A, R), \Lambda^{op})$ and S a simple Λ-module such that there is an exact sequence
$0 \longrightarrow (P, \ ) \longrightarrow (E, \ ) \longrightarrow (A, \ ) \longrightarrow T \longrightarrow 0$ in $(L(\Lambda), Ab)$.

Proof: Since $T(A) \neq 0$, we know that there is a maximal left ideal H of End A such that $T \cong (A, \ )/(A, \ )_H$. Let $D: L(\Lambda) \longrightarrow L(\Lambda^{op})$ be the duality given by $\text{Hom}_R(\ ,R)$. Then $D(A)$ is a projective Λ^{op}-module and $D(H)$ is a maximal right ideal in End D A. Let S be the simple functor $(\ ,D(A))/(\ ,D(A))_H$ in $(L(\Lambda)^{op}, Ab)$. Therefore by Proposition 3.6 there is an exact sequence

$$(*) \qquad 0 \longrightarrow U \xrightarrow{g} V \xrightarrow{f} D(A) \xrightarrow{\varepsilon} W \longrightarrow 0$$

with $U \cong \text{Hom}_R(\text{Hom}_{\Lambda^{op}}(D(A), \Lambda^{op}), R)$, V a lattice and W a simple Λ^{op}-module such that $0 \longrightarrow (\ ,U) \longrightarrow (\ ,V) \longrightarrow (\ ,D(A)) \longrightarrow S \longrightarrow 0$ is exact. Then by the usual duality arguments it follows that $A \xrightarrow{D(f)} D(V)$ has the property that $(D(V), \ ) \xrightarrow{(D f, \ )} (A, \ ) \longrightarrow T \longrightarrow 0$ is exact. Since W is an R-module of finite length and $D(A)$ is a Λ^{op}-lattice, we have that $\text{Ext}_R^i(W, R) = 0$ for $i = 0, 1$ and $\text{Ext}_R^i(D(A), R) = 0$ for $i > 0$. Therefore the exact sequence $0 \longrightarrow \text{Im } f \longrightarrow D(A) \longrightarrow W \longrightarrow 0$ has the property $D^2(A) \longrightarrow D(\text{Im } f)$ is an isomorphism and $\text{Ext}_R^1(\text{Im } f, R) \cong \text{Ext}_R^2(W, R)$. Therefore applying D to the exact sequence $(*)$ we obtain the exact sequence

$$(**) \qquad 0 \longrightarrow A \longrightarrow D(V) \longrightarrow D(U) \longrightarrow \text{Ext}_R^2(W, R) \longrightarrow 0 \ .$$

Since $U = \text{Hom}_R(\text{Hom}_{\Lambda^{op}}(D(A), \Lambda^{op}), R)$, we have that

$DU = \operatorname{Hom}_{\Lambda^{op}}(D(A), \Lambda^{op}) = P$. Finally, since the functor $\operatorname{Ext}^2_R(\ ,R)$: noeth $\Lambda^{op} \longrightarrow$ noeth Λ induces a duality between the Λ^{op}- and Λ-modules of finite length because R is a Gorenstein ring of dim 2, we have that W being a simple Λ^{op}-module implies that $\operatorname{Ext}^2_R(W, R)$ is a simple Λ-module. This finishes the proof of the proposition since we have already seen that the properties of (**) imply that $0 \longrightarrow (D(U),\ ) \longrightarrow (D(V),\ ) \longrightarrow (A,\ ) \longrightarrow T \longrightarrow 0$ is exact in $(L(\Lambda), Ab)$ (see Proposition 3.7).

Throughout the rest of this section we assume that Λ is an R-order with $\dim R = 2$. Our aim is to show that in the case R is a complete local ring then $L(\Lambda)$ is of finite representation type if and only if the finitely presented functors in $(L(\Lambda)^{op}, Ab)$ which vanish on Λ have finite length. This result is based on the following observation.

Let $\{\underline{m}_i\}_{i\varepsilon I}$ be a family of maximal left ideals such that $\{\Lambda/\underline{m}_i\}_{i\varepsilon I}$ is a complete set of nonisomorphic simple Λ-modules. Then by Theorem 3.6, we know that for each i in I there is an epimorphism $t_i\colon E_i \longrightarrow \underline{m}_i \longrightarrow 0$ of Λ-modules with E_i a lattice such that $(X, E_i) \longrightarrow (X, \underline{m}_i) \longrightarrow 0$ is exact for all lattices X. We now point out some other properties of these lattices $\{E_i\}_{i\varepsilon I}$ and the epimorphisms $t_i\colon E_i \longrightarrow \underline{m}_i$.

<u>Proposition 3.9</u>: Let $\{E_i\}_{i\varepsilon I}$ be the family of lattices described above. Let X be a nonzero lattice. Then

a) $(X, E_i) \neq 0$ for each i in I.

b) X is projective if and only if $\underline{\operatorname{Hom}}_\Lambda(X, E_i) = 0$ for all i in I.

Proof:

a) Since R is a Gorenstein ring, we know that the prime ideals associated to (0) are all minimal prime ideals. Combining this with the fact that R is equidimensional of dimension 2, we have that each maximal ideal contains a regular element of R. Because each $\underline{m}_i$ is a maximal left ideal of Λ, we know that $\underline{m}_i \cap R$ is a maximal ideal of R and therefore contains an R-regular element r_i. Since Λ is a lattice and therefore a reflexive R-module, we know that each r_i is a regular in Λ, i.e. $0 \longrightarrow \Lambda \xrightarrow{r_i} \Lambda$ is exact.

Suppose now that X is a nonzero lattice. Since $\dim R = 2$, we know that X is a reflexive Λ-module. From this it follows that $\mathrm{Hom}_\Lambda(X, \Lambda) \neq 0$. Let $f: X \longrightarrow \Lambda$ be a nonzero morphism. Then for each i in I, the Λ-morphism $r_i f: X \longrightarrow \Lambda$ is not zero with $\mathrm{Im}(r_i f) \subset \Lambda r_i \subset m_i$. Thus we have shown that for each i in I, there is a nonzero morphism $f_i: X \longrightarrow \underline{m}_i$. Since $(X, E_i) \xrightarrow{(X, t_i)} (X, \underline{m}_i) \longrightarrow 0$ is exact for each i in I, we have that $(X, E_i) \neq 0$ for each i in I.

b) Suppose X is a lattice such that $\underline{\mathrm{Hom}}_\Lambda(X, E_i) = 0$. Since the epimorphism $t_i: E_i \longrightarrow \underline{m}_i$ has the property that $(X, E_i) \xrightarrow{(X, t_i)} (X, \underline{m}_i) \longrightarrow 0$ is exact, the fact that $\underline{\mathrm{Hom}}_\Lambda(X, E_i) = 0$ implies that $\underline{\mathrm{Hom}}_\Lambda(X, \underline{m}_i) = 0$. But $\underline{\mathrm{Hom}}_\Lambda(X, \underline{m}_i) \simeq \mathrm{Tor}_1^\Lambda(\mathrm{Tr}X, \underline{m}_i)$. Since $\mathrm{Tor}_1^\Lambda(\mathrm{Tr}X, \underline{m}_i) \cong \mathrm{Tor}_2^\Lambda(\mathrm{Tr}X, \Lambda/\underline{m}_i)$, we have that $\mathrm{Tor}_2^\Lambda(\mathrm{Tr}X, \Lambda/\underline{m}_i) = 0$.

Let $P_1 \longrightarrow P_0 \longrightarrow X \longrightarrow 0$ be a projective presentation of X used to construct $\mathrm{Tr}X$. Then we have the exact sequence

$$0 \longrightarrow (X, \Lambda) \longrightarrow (P_0, \Lambda) \longrightarrow (P_1, \Lambda) \longrightarrow \mathrm{Tr}X \longrightarrow 0$$

Hence if we assume that $\underline{\text{Hom}}_\Lambda(X, E_i) = 0$ for all i, then $\text{Tor}_1^\Lambda(V, \Lambda/\underline{m}_i) = 0$ for all i in I where $V = \text{Ker}((P_1, \Lambda) \longrightarrow \text{Tr}X)$. Thus V is a projective Λ^{op}-module since $\text{Tor}_1^\Lambda(V, S)$ being zero for all simple Λ-modules S implies that V is projective. Since $0 \longrightarrow (X, \Lambda) \longrightarrow (P_0, \Lambda) \longrightarrow V \longrightarrow 0$ is exact, it follows that (X, Λ) is a projective Λ^{op}-module. This implies that $X \cong ((X, \Lambda), \Lambda)$ is projective. Hence if $\underline{\text{Hom}}_\Lambda(X, E_i) = 0$ for all i in I, then X is projective. Because it is obvious that $\underline{\text{Hom}}_\Lambda(X, E_i) = 0$ for X projective, the proof of b) is complete.

We now apply this result to prove our previously cited result.

Theorem 3.10: For an R-order Λ with R a complete local ring of dimension 2, the following statements are equivalent.

a) Λ is of finite lattice type.

b) For each X in $L(\Lambda)$, the functor $\underline{\text{Hom}}_\Lambda(\ ,X)$ in $(L(\Lambda)^{op}, \text{Ab})$ is of finite length.

c) Each finitely generated F in $(L(\Lambda)^{op}, \text{Ab})$ such that $F(\Lambda) = 0$ is of finite length.

d) $\underline{\text{Hom}}_\Lambda(\ ,E_i)$ is of finite length, where $\{E_i\}_{i\varepsilon I}$ is the family of lattices described prior to the statement of Proposition 3.9.

Proof:

a) implies b). Since for Y in $L(\Lambda)$, we have that $\underline{\text{Hom}}_\Lambda(Y, X)$ is an $(\text{End } Y)^{op}$-submodule of finite length, it follows that $\underline{\text{Hom}}_\Lambda(\ ,X)$ is of finite length if and only if there are only a finite number of nonisomorphic indecomposable Y in L(Λ) such that $\underline{\mathrm{Hom}}_\Lambda(Y, X) \neq 0$ ([7, Proposition 2.11]). Thus if L(Λ) is of finite type, then $\underline{\mathrm{Hom}}_\Lambda(\ ,X)$ is of finite length for all X in L(Λ).

b) implies c). Let F be a finitely generated functor such that $F(\Lambda) = 0$. Since F is finitely generated there is an epimorphism $(\ ,X) \longrightarrow F \longrightarrow 0$ for some X in L(Λ). Since $F(\Lambda) = 0$, it is not hard to see that $(\ ,X) \longrightarrow F \longrightarrow 0$ factors through $\underline{\mathrm{Hom}}_\Lambda(\ ,X)$ and hence there is an epimorphism $\underline{\mathrm{Hom}}_\Lambda(\ ,X) \longrightarrow F \longrightarrow 0$. Therefore F is of finite length since $\underline{\mathrm{Hom}}_\Lambda(\ ,X)$ is of finite length.

c) implies d). Obvious.

d) implies a). Since R is complete, there are only a finite number of simple Λ-modules. Thus the set I is finite. Since each $\underline{\mathrm{Hom}}_\Lambda(\ ,E_i)$ is of finite length, there are only a finite number of nonisomorphic indecomposable lattices X such that $\underline{\mathrm{Hom}}_\Lambda(X, E_i) \neq 0$ for each i in I. Therefore there are only a finite number of nonisomorphic indecomposable lattices X such that $\underline{\mathrm{Hom}}_\Lambda(X, E_i) \neq 0$ for some i in I. This means there are only a finite number of nonisomorphic nonprojective lattices since by Proposition 3.9 we know that a lattice X is projective if and only if $\underline{\mathrm{Hom}}_\Lambda(X, E_i) = 0$ for all i in I. Combining this with the fact that R being a complete local ring implies there are only a finite number of nonisomorphic indecomposable projective Λ-modules, we have that Λ is of finite lattice type, which finishes the proof that d) implies a).

§4. A Generalization. Let Λ be an arbitrary ring, X a module in mod Λ, the category of finitely presented Λ-modules. In addition, suppose the finitely presented Λ^{op}-module TrX is a Λ^{op}-Σ bimodule and I is an injective Σ-module. Considering $\mathrm{Hom}_{\Sigma}(\mathrm{TrX}, I)$ a Λ-module in the obvious way, we saw in I, Section 3, that for each Λ-module C the duality Tr: $\underline{\mathrm{mod}}\ \Lambda \longrightarrow \underline{\mathrm{mod}}\ \Lambda^{op}$ gave rise to a Σ-module structure on $\underline{\mathrm{Hom}}_{\Lambda}(X, C)$ and that we have isomorphisms

$$\mathrm{Ext}^1_{\Lambda}(C, \mathrm{Hom}_{\Sigma}(\mathrm{TrX}, I)) \longrightarrow \mathrm{Hom}_{\Sigma}(\underline{\mathrm{Hom}}_{\Lambda}(X, C), I)$$

which are functorial in C and I (see I, Proposition 3.4). These functorial isomorphisms played a fundamental role in establishing our basic existence theorems for morphisms determined by modules. It is our aim in this section to give some applications in a very different direction.

In II, Section 3, we established a duality D: f. p. $((\mathrm{Mod}\ \Lambda)^{op}, \mathrm{Ab})_o \longrightarrow$ f. p. $(\mathrm{Mod}\ \Lambda, \mathrm{Ab})_o$. Recall that F is in f. p. $((\mathrm{Mod}\ \Lambda)^{op}, \mathrm{Ab})_o$ if and only if F is finitely presented and vanishes on all projective Λ-modules. Suppose, as above, X is in mod Λ and TrX is a Λ^{op}-Σ bimodule. Since $F(P) = 0$ for all projective Λ-modules P, we have that the $(\mathrm{End}_{\Lambda}X)^{op}$-module F(X) is annihilated by P(X, X) and so is an $(\underline{\mathrm{End}}_{\Lambda}X)^{op}$-module. Also the duality Tr: $\underline{\mathrm{mod}}\ \Lambda \longrightarrow \underline{\mathrm{mod}}\ \Lambda^{op}$ gives rise to a ring isomorphism Tr: $(\underline{\mathrm{End}}_{\Lambda}X)^{op} \longrightarrow \underline{\mathrm{End}}_{\Lambda}\mathrm{TrX}$. Therefore the ring morphism $\Sigma \longrightarrow \mathrm{End}_{\Lambda}\mathrm{TrX}$ given by the bimodule structure Λ^{op}-Σ of TrX induces a ring morphism $\Sigma \longrightarrow (\underline{\mathrm{End}}_{\Lambda}X)^{op}$. Hence F(X) is a Σ-module in a natural way. As already noted $\mathrm{Hom}_{\Sigma}(\mathrm{TrX}, I)$ is a Λ-module in a natural way. With these definitions and notations in mind, we establish the following relationship between F and DF for all F in f. p. $((\mathrm{Mod}\ \Lambda)^{op}, \mathrm{Ab})$.

Theorem 4.1: Let X be in mod Λ and suppose that the finitely presented

Λ^{op}-module TrX is a Λ^{op}-Σ-bimodule. Then for each F in f. p. $((\mathrm{Mod}\ \Lambda)^{op}, \mathrm{Ab})$ and each injective Σ-module I we have isomorphisms

$$\mathrm{Hom}_\Sigma(F(X), I) \approx DF(\mathrm{Hom}_\Sigma(\mathrm{Tr}X, I))$$

which are functorial in F and I.

Proof: Since F is in f. p. $((\mathrm{Mod}\ \Lambda)^{op}, \mathrm{Ab})$, there is an exact sequence $0 \longrightarrow A \longrightarrow B \longrightarrow C \longrightarrow 0$ of Λ-modules such that $0 \longrightarrow (\ ,A) \longrightarrow (\ ,B) \longrightarrow (\ ,C) \longrightarrow F \longrightarrow 0$ is exact. Now there is an exact commutative diagram

$$\begin{array}{ccccccccc}
0 & \longrightarrow & P_0 & \longrightarrow & P_1 & \longrightarrow & P_2 & \longrightarrow & 0 \\
 & & \downarrow & & \downarrow & & \downarrow & & \\
0 & \longrightarrow & A & \longrightarrow & B & \longrightarrow & C & \longrightarrow & 0 \\
 & & \downarrow & & \downarrow & & \downarrow & & \\
 & & 0 & & 0 & & 0 & &
\end{array}$$

with the P_i projective Λ-modules. This gives rise to the following exact commutative diagram of functors

$$\begin{array}{ccccccccc}
0 & \longrightarrow & (\ ,P_0) & \longrightarrow & (\ ,P_1) & \longrightarrow & (\ ,P_2) & \longrightarrow & 0 \\
 & & \downarrow & & \downarrow & & \downarrow & & \\
0 & \longrightarrow & (\ ,A) & \longrightarrow & (\ ,B) & \longrightarrow & (\ ,C) & \longrightarrow & 0 \\
 & & \downarrow & & \downarrow & & \downarrow & & \\
 & & \underline{\mathrm{Hom}}_\Lambda(\ ,A) & \longrightarrow & \underline{\mathrm{Hom}}_\Lambda(\ ,B) & \longrightarrow & \underline{\mathrm{Hom}}_\Lambda(\ ,C) & & \\
 & & \downarrow & & \downarrow & & \downarrow & & \\
 & & 0 & & 0 & & 0 & &
\end{array}$$

which gives rise to the exact sequence

$$\underline{\mathrm{Hom}}_\Lambda(\ ,B) \longrightarrow \underline{\mathrm{Hom}}_\Lambda(\ ,C) \longrightarrow F \longrightarrow 0 \ .$$

Thus $\underline{\text{Hom}}_\Lambda(X, B) \longrightarrow \underline{\text{Hom}}_\Lambda(X, C) \longrightarrow F(X) \longrightarrow 0$ is an exact sequence of Σ-modules by means of the ring morphism $\Sigma \longrightarrow (\underline{\text{End}}\ X)^{op}$ described earlier. Thus we obtain the exact sequence

$$0 \longrightarrow \text{Hom}_\Sigma(F(X), I) \longrightarrow \text{Hom}_\Sigma(\underline{\text{Hom}}_\Lambda(X, C), I) \longrightarrow \text{Hom}_\Sigma(\underline{\text{Hom}}_\Lambda(X, B), I).$$

Using the isomorphism of functors $\text{Ext}^1_\Lambda(\ , \text{Hom}_\Sigma(\text{Tr}X, I)) \longrightarrow \text{Hom}_\Sigma(\underline{\text{Hom}}_\Lambda(X,\ ), I)$ we obtain the exact sequence

(*)
$$0 \to \text{Hom}_\Sigma(F(X), I) \to \text{Ext}^1_\Lambda(C, \text{Hom}_\Sigma(\text{Tr}X, I)) \to \text{Ext}^1_\Lambda(B, \text{Hom}_\Sigma(\text{Tr}X, I)).$$

Since $0 \longrightarrow A \longrightarrow B \longrightarrow C \longrightarrow 0$ is exact, we have that

$$(B,\ ) \longrightarrow (A,\ ) \longrightarrow \text{Ext}^1_\Lambda(C,\ ) \longrightarrow \text{Ext}^1_\Lambda(B,\ )$$

is exact, which implies that

$$0 \longrightarrow DF \longrightarrow \text{Ext}^1_\Lambda(C,\ ) \longrightarrow \text{Ext}^1_\Lambda(B,\ )$$

is exact, since $(B,\ ) \longrightarrow (A,\ ) \longrightarrow DF \longrightarrow 0$ is exact (see II, Section 3). Therefore

(**)
$$0 \longrightarrow DF(\text{Hom}_\Sigma(\text{Tr}X, I)) \to \text{Ext}^1_\Lambda(C, \text{Hom}_\Sigma(\text{Tr}X, I)) \to \text{Ext}^1_\Lambda(B, \text{Hom}_\Sigma(\text{Tr}X, I))$$

is exact. Combining (*) and (**) we obtain an isomorphism $\text{Hom}_\Sigma(F(X), I) \approx DF(\text{Hom}_\Sigma(\text{Tr}X, I))$. The fact that this isomorphism is functorial in F and I can be verified in a straightforward way by checking through the proof and seeing that it is functorial in F and I at each stage.

As an application of this result we have the following.

<u>Corollary 4.2</u>: Let $0 \longrightarrow A \xrightarrow{g} B \xrightarrow{f} C \longrightarrow 0$ be an exact sequence of Λ-modules. Suppose X is in mod Λ and $\text{Tr}X$ in mod Λ^{op} is a

Λ^{op}-Σ bimodule. Further suppose I is an injective cogenerator for Σ. Then the following statements are equivalent.

a) For each $h: X \longrightarrow C$ there is a $s: X \longrightarrow B$ such that $fs = h$

b) For each $j: A \longrightarrow \mathrm{Hom}_\Sigma(\mathrm{Tr}X, I)$ there is a $t: B \longrightarrow \mathrm{Hom}_\Sigma(\mathrm{Tr}X, I)$ such that $j = tg$.

Proof: From the exact sequences $(\ ,B) \longrightarrow (\ ,C) \longrightarrow F \longrightarrow 0$ and $(B, \ ) \longrightarrow (A, \ ) \longrightarrow DF \longrightarrow 0$ we deduce that a) is equivalent to $F(X) = 0$ and b) is equivalent to $DF(\mathrm{Hom}_\Sigma(\mathrm{Tr}X, I)) = 0$. Since $\mathrm{Hom}_\Sigma(F(X), I) \approx DF(\mathrm{Hom}_\Sigma(\mathrm{Tr}X, I))$ by Theorem 4.1 and I is injective cogenerator, it follows that $F(X) = 0$ if and only if $DF(\mathrm{Hom}_\Sigma(\mathrm{Tr}X, I)) = 0$. This finishes the proof of the corollary.

As another application of Theorem 4.1, we have the following.

Corollary 4.3: Let X be in mod Λ and suppose TrX in mod Λ^{op} is a Λ^{op}-Σ bimodule. Then for A in Mod Λ and injective Σ-module I we have isomorphisms

$$\mathrm{Hom}_\Sigma(\mathrm{Ext}^1_\Lambda(X, A), I) \approx \overline{\mathrm{Hom}}_\Lambda(A, \mathrm{Hom}_\Sigma(\mathrm{Tr}X, I))$$

which are functorial in A and I.

Proof: Let $0 \longrightarrow A \longrightarrow B \longrightarrow C \longrightarrow 0$ be an exact sequence of Λ-modules with B an injective Λ-module. Then $(\ ,B) \longrightarrow (\ ,C) \longrightarrow \mathrm{Ext}^1_\Lambda(\ ,A) \longrightarrow 0$ and $(B, \ ) \longrightarrow (A, \ ) \longrightarrow \overline{\mathrm{Hom}}_\Lambda(A, \ ) \longrightarrow 0$ are exact since B is injective. Therefore $D(\mathrm{Ext}^1_\Lambda(\ ,A)) = \overline{\mathrm{Hom}}_\Lambda(A, \ )$ and so applying Theorem 4.1 we have the isomorphism

$$\mathrm{Hom}_\Sigma(\mathrm{Ext}^1_\Lambda(X, A), I) \cong \overline{\mathrm{Hom}}_\Lambda(A, \mathrm{Hom}_\Sigma(\mathrm{Tr}X, I))$$

which is functorial in A and I.

The rest of this section is devoted to pointing out the analogues of these results for lattices over an R-order.

Suppose Λ is an R-order. Let f. p. $(L(\Lambda)^{op}, Ab)$ denote the category of finitely presented functors from $L(\Lambda)^{op}$ to Ab. Further let f. p. $(L(\Lambda)^{op}, Ab)_o$ denote the full subcategory of f. p. $(L(\Lambda)^{op}, Ab)$ consisting of all F in f. p. $(L(\Lambda)^{op}, Ab)$ such that $F(P) = 0$ for projective lattices P. It is easily checked that a functor F is in f. p. $(L(\Lambda)^{op}, Ab)$ if and only if there is an exact sequence of lattices $0 \longrightarrow A \longrightarrow B \longrightarrow C \longrightarrow 0$ such that $0 \longrightarrow (\ ,A) \longrightarrow (\ ,B) \longrightarrow (\ ,C) \longrightarrow F \longrightarrow 0$ is exact.

Similarly we denote by f. p. $(L(\Lambda), Ab)$ the category of finitely presented functors from $L(\Lambda)$ to Ab and denote by f. p. $(L(\Lambda), Ab)_o$ the full subcategory of f. p. $(L(\Lambda), Ab)$ consisting of all G such that $G(I) = 0$ for all injective lattices I. It is easily checked that a functor G is in f. p. $(L(\Lambda), Ab)_o$ if and only if there is an exact sequence of lattices $0 \longrightarrow A \longrightarrow B \longrightarrow C \longrightarrow 0$ such that $0 \longrightarrow (C,\ ) \longrightarrow (B,\ ) \longrightarrow (A,\ ) \longrightarrow G \longrightarrow 0$ is exact.

Now we define a duality D: f. p. $(L(\Lambda)^{op}, Ab)_o \longrightarrow$ f. p. $(L(\Lambda), Ab)_o$ as follows. For each F in f..p. $(L(\Lambda)^{op}, Ab)_o$ choose an exact sequence $0 \longrightarrow A \longrightarrow B \longrightarrow C \longrightarrow 0$ in $L(\Lambda)$ such that $0 \longrightarrow (\ ,A) \longrightarrow (\ ,B) \longrightarrow (\ ,C) \longrightarrow F \longrightarrow 0$ is exact. Then define DF by the exact sequence $0 \longrightarrow (C,\ ) \longrightarrow (B,\ ) \longrightarrow (A,\ ) \longrightarrow DF \longrightarrow 0$. It is not difficult to see that D is a duality which, up to a unique isomorphism, is independent of the particular exact sequences $0 \longrightarrow A \longrightarrow B \longrightarrow C \longrightarrow 0$ of lattices chosen for the F in f. p. $(L(\Lambda)^{op}, Ab)$. With these preliminaries in mind, we have the following analogue of Theorem 4.1.

Theorem 4.4: Let Λ be an R-order. For each F in f. p. $(L(\Lambda)^{op}, Ab)$ and X in $L(\Lambda)$ we have isomorphisms

$$\mathrm{Hom}_R(F(X), I_d^n) \cong DF(\mathrm{Hom}_R(\mathrm{Tr}_L X, R^n))$$

which are functorial in F and X.

Proof: Recall that for lattices C and X we have isomorphisms

$$\mathrm{Ext}^1_\Lambda(C, \mathrm{Hom}_R(\mathrm{Tr}_L X, R^n)) \longrightarrow \mathrm{Hom}_R(\underline{\mathrm{Hom}}_\Lambda(X, C), I_d^n)$$

which are functorial in C and X (see I, Proposition 8.9). Using these functorial isomorphisms, the proof of the theorem proceeds almost verbatim the same way as the proof of Theorem 4.1.

As an analogue of Corollary 4.2 we have the following.

Corollary 4.5: Let Λ be an R-order and let $0 \longrightarrow A \xrightarrow{g} B \xrightarrow{f} C \longrightarrow 0$ be an exact sequence of Λ-lattices. Then for a Λ-lattice X, the following statements are equivalent.

a) For each $h: X \longrightarrow C$ there is an $s: X \longrightarrow B$ such that $h = fs$.

b) For each $j: A \longrightarrow \mathrm{Hom}_R(\mathrm{Tr}_L X, R)$, there is a $t: B \longrightarrow \mathrm{Hom}_R(\mathrm{Tr}_L X, R)$ such that $j = tg$.

Proof: Analogous to proof of Corollary 4.2 once one observes that for all

F in f. p. $(L(\Lambda)^{op}, Ab)$ we have that $F(X)$ is an R-module of finite length and so $F(X) = 0$ if and only if $\mathrm{Hom}_R(F(X), I_d) = 0$.

As our final result, we have the following analogue of Corollary 4.3.

Corollary 4.6: Let Λ be an R-order. For all lattices X and A we have isomorphisms

$$\mathrm{Hom}_R(\mathrm{Ext}^1_\Lambda(X, A), I_d) \cong \overline{\mathrm{Hom}}_\Lambda(A, \mathrm{Hom}_R(\mathrm{Tr}_L X, R))$$

functorial in X and A.

REFERENCES

1. Auslander, M. "Applications of almost split sequences," these Proceedings.
2. ___________. "Coherent functors," Proceedings of the Conference on Categorical Algebra , Springer-Verlag, Berlin, Heidelberg, New York, 1966.
3. ___________. "Existence theorems for almost split sequences," Ring Theory II: Proceedings of the Second Oklahoma Conference, Marcel Dekker, New York and Basel, 1977.
4. ___________. "Large modules over Artin algebras," Algebra, Topology and Categories, Academic Press, New York, 1976.
5. ___________. "Representation dimension of Artin algebras," Queen Mary College Mathematics Notes, London.
6. ___________. "Representation theory of Artin algebras, I," Comm. in Algebra (1974), 177-268.
7. ___________. "Representation theory of Artin algebras, II," Comm. in Algebra (1974), 269-310.
8. Auslander, M. and Bridger, M. "Stable module theory," Memoirs of the American Mathematical Society, No. 94, 1969.
9. Auslander, M. and Platzeck, M. I. "Representation theory of hereditary Artin algebras," these Proceedings.
10. Auslander, M. and Reiten, I. "Representation theory of Artin algebras, III," Comm. in Algebra (1975), 239-294.

11. ________________________. "Representation theory of Artin Algebras, IV," Comm. in Algebra, (1977) 443-518.

12. ________________________. "Stable equivalence of dualizing R-varieties," Advances in Math. (1974), 300-366.

13. Bass, H. "On the ubiquity of Gorenstein rings," Math. Zeitschr. (1963), 8-28.

14. Eilenberg, S. and Cartan, E. Homological Algebra, Princeton University Press, Princeton, New Jersey, 1956.

15. Eckman, B. and Hilton, P. J. "Homotopy groups of maps and exact sequences," Comment. Math. Helv. (1960) 271-304.

16. Herzog, J. and Kung, E. Der Kanonische Module eines Cohen-Macauley Rings, Springer-Verlag, New York, 1971.

17. Nagata, M. Local Rings, Interscience Publishers, New York, 1962.

APPLICATIONS OF MORPHISMS DETERMINED BY MODULES

Maurice Auslander
Brandeis University
Waltham, Massachusetts

In [2] a general theory of morphisms determined by modules was developed. This paper begins with a summary of this general theory specialized to artin algebras. Basic definitions and the important results are given, but no effort is made to duplicate proofs already given in [2].

After these preliminaries, various applications are given of morphisms determined by modules to the study of finitely generated modules. In particular, much of the paper is devoted to showing the connections between almost split morphisms, a notion introduced in [4], and morphisms determined by modules. The paper ends with a discussion of some variations of the Brauer-Thrall Conjecture II.

§1. The Transpose.

We assume throughout this section that R is a commutative artin ring. By an *artin R-algebra* we mean an R-algebra Λ which is a finitely generated R-module. Obviously an artin R-algebra is a two-sided artin ring. Our main purpose in this section is to recall some basic facts concerning the category mod Λ, the category of finitely generated left Λ-modules. Of particular concern is the equivalence of categories

DTr: $\underline{\text{mod}}\,\Lambda \longrightarrow \overline{\text{mod}}\,\Lambda$ where $\underline{\text{mod}}\,\Lambda$ denotes the category mod Λ modulo projectives and $\overline{\text{mod}}\,\Lambda$ denotes the category mod Λ modulo injectives. In addition to recalling the basic properties of DTr we give various elementary applications. The basic facts concerning DTr are given without proof. The reader is referred to [5] and [6] for more details.

Since R is a commutative artin ring, $I = I(R/\text{rad } R)$, the injective envelope of $R/\text{rad } R$, is a finitely generated R-module which is an injective cogenerator for Mod R, the category of all R-modules.

Suppose Λ is an artin R-algebra. Then Λ^{op}, the opposite ring of Λ, is also an artin R-algebra in an obvious way. Then the functor $D: \text{mod } \Lambda \longrightarrow \text{mod } \Lambda^{op}$ given by $D(M) = \text{Hom}_R(M, I)$ is well known to be a duality with dual inverse $D: \text{mod } \Lambda^{op} \longrightarrow \text{mod } \Lambda$. Since the objects in mod Λ^{op} have projective covers, it follows from the duality $D: \text{mod } \Lambda^{op} \longrightarrow \text{mod } \Lambda$ that every object in mod Λ has an injective envelope in mod Λ.

Associated with the category mod Λ are the categories $\underline{\text{mod}}\,\Lambda$ and $\overline{\text{mod}}\,\Lambda$, called respectively the <u>category mod Λ modulo projectives</u> and <u>mod Λ modulo injectives</u>. Since these categories $\underline{\text{mod}}\,\Lambda$ and $\overline{\text{mod}}\,\Lambda$ play such an important role in the entire paper, we briefly recall their definitions (see [5] and [6]).

For each pair of objects A and B in mod Λ, denote by $P(A, B)$ the R-submodule of $(A, B) = \text{Hom}_\Lambda(A, B)$ consisting of all morphisms $f: A \longrightarrow B$ which can be factored through a projective in mod Λ, i.e. $f: A \longrightarrow B$ is in $P(A, B)$ if and only if f is a composition $A \xrightarrow{g} P \xrightarrow{h} B$ with P a projective module in mod Λ. It is easily checked that given morphisms $f: A \longrightarrow B$ and $g: B \longrightarrow C$ in mod Λ, then gf is in $P(A, C)$ if either f is in $P(A, B)$ or g is in $P(B, C)$. For each pair of objects A and B in mod Λ we denote the R-module $(A, B)/P(A, B)$ by $\underline{\text{Hom}}_\Lambda(A, B)$ and for each f in (A, B) we denote its image in $\underline{\mathrm{Hom}}_\Lambda(A, B)$ under the canonical epimorphism $(A, B) \longrightarrow (A, B)/P(A, B)$ by $\underline{f}$. Then by our previous remarks it is clear that given A, B, C in $\mathrm{mod}\,\Lambda$, the R-bilinear map $(A, B) \times (B, C) \longrightarrow (A, C)$ given by $(f, g) \longmapsto gf$ induces an R-bilinear map $\underline{\mathrm{Hom}}_\Lambda(B, C) \longrightarrow \underline{\mathrm{Hom}}_\Lambda(A, C)$ given by $(\underline{f}, \underline{g}) \longmapsto \underline{fg}$. Then $\underline{\mathrm{mod}}\,\Lambda$ is defined to be the category given by the following data. The objects of $\underline{\mathrm{mod}}\,\Lambda$ are the same as those of $\mathrm{mod}\,\Lambda$. For each pair of objects A, B, in $\underline{\mathrm{mod}}\,\Lambda$, the morphisms from A to B in $\underline{\mathrm{mod}}\,\Lambda$ are $\underline{\mathrm{Hom}}_\Lambda(A, B)$. Finally, the composition in $\underline{\mathrm{mod}}\,\Lambda$ is given by the R-bilinear maps $\underline{\mathrm{Hom}}_\Lambda(A, B) \times \underline{\mathrm{Hom}}_\Lambda(B, C) \longrightarrow \underline{\mathrm{Hom}}_\Lambda(A, C)$ defined above by $(\underline{f}, \underline{g}) \longmapsto \underline{gf}$. It is not difficult to check that $\underline{\mathrm{mod}}\,\Lambda$ has the following properties.

Proposition 1.1:

a) $\underline{\mathrm{mod}}\,\Lambda$ is an additive category.

b) The functor $F\colon \mathrm{mod}\,\Lambda \longrightarrow \underline{\mathrm{mod}}\,\Lambda$ given by $F(A) = A$ for all A in $\mathrm{mod}\,\Lambda$ and $F\colon (A, B) \longrightarrow \underline{\mathrm{Hom}}_\Lambda(F(A), F(B))$ is the canonical morphism $(A, B) \longrightarrow (A, B)/P(A, B)$ for all A, B in $\mathrm{mod}\,\Lambda$ has the following properties:

i) F is a full additive functor.

ii) $F(A) \approx 0$ in $\underline{\mathrm{mod}}\,\Lambda$ if and only if A is projective.

iii) $F(A) \approx F(B)$ if and only if there are projective modules P and Q in $\mathrm{mod}\,\Lambda$ such that $A \amalg P \approx B \amalg Q$.

iv) If $\underline{\underline{C}}$ is an additive category and $G\colon \mathrm{mod}\,\Lambda \longrightarrow \underline{\underline{C}}$ is an additive functor such that $G(P) = 0$ for all projective modules P in $\mathrm{mod}\,\Lambda$, then there is a unique (up to isomorphism) additive functor $H\colon \underline{\mathrm{mod}}\,\Lambda \longrightarrow \underline{\underline{C}}$ such that $G = HF$.

Next we recall that mod Λ is a Krull-Schmidt category, i.e. each indecomposable object in mod Λ has a local endomorphism ring and each object A in mod Λ can be written as a finite sum $A_1 \amalg \cdots \amalg A_n$ of indecomposable modules. From this it follows that if $A_1 \amalg \cdots \amalg A_m \approx B_1 \amalg \cdots \amalg B_n$ with the A_i and B_j indecomposable, then $m = n$ and after suitable relabeling $A_i \approx B_i$ for all $i = 1,\ldots,n$. Thus if we let $\text{mod}_P\Lambda$ be the full subcategory of mod Λ consisting of all A in mod Λ with no projective summands, then $\text{mod}_P\Lambda$ is closed under summands and finite sums and every A in mod Λ can be written uniquely (up to isomorphism) as $A_1 \amalg P$ with A_1 in $\text{mod}_P\Lambda$ and P a projective in mod Λ. From this it follows that if we denote by $\underline{\text{mod}}_P\Lambda$ the full subcategory of $\underline{\text{mod}}\ \Lambda$ whose objects are the objects in $\text{mod}_P\Lambda$, then the inclusion $\underline{\text{mod}}_P\Lambda \longrightarrow \underline{\text{mod}}\ \Lambda$ is an equivalence of categories which we will often consider an identification.

Clearly the functor F: mod $\Lambda \longrightarrow \underline{\text{mod}}\ \Lambda$ induces a full functor F: $\text{mod}_P\Lambda \longrightarrow \underline{\text{mod}}_P\Lambda$. It is not difficult to show that for each A in $\text{mod}_P\Lambda$, the kernel of the surjective R-algebra morphism F: End A $\longrightarrow$ $\underline{\text{End}}$ A is contained in the radical of End A (see [1, Proposition 2.5]). Thus f in End A is an isomorphism if and only if $\underline{f}$ in $\underline{\text{End}}$ A is an isomorphism. Further, combining this observation with the fact that End A is an artin ring since it is an artin R-algebra in a natural way, it is not difficult to show the following.

<u>Proposition 1.2</u>: Let F: $\text{mod}_P\Lambda \longrightarrow \underline{\text{mod}}_P\Lambda$ be the canonical functor.

a) If A and B are in $\text{mod}_P\Lambda$, then $A \approx B$ if and only if $F(A) \approx F(B)$.

b) If e in $\underline{\text{End}}$ F(A) is an idempotent, then there is an idempotent f in End A such that $\underline{f} = e$.

c) For A in $\text{mod}_P\Lambda$, the following are equivalent:

i) A is indecomposable.

ii) End A is local.

iii) $\underline{\text{End}}\ A$ is local.

iv) $F(A)$ is indecomposable in $\underline{\text{mod}}_P\Lambda$.

d) For each A in $\text{mod}_P\Lambda$, we have that $A \approx A_1 \amalg \dots \amalg A_n$ $\text{mod}_P\Lambda$ if and only if $F(A) \approx F(A_1) \amalg \dots \amalg F(A_n)$ in $\underline{\text{mod}}_P\Lambda$ and the A_i are indecomposable in mod Λ if and only if the $F(A_i)$ are indecomposable in $\underline{\text{mod}}\ \Lambda$.

We next recall that there is a duality $\text{Tr}: \underline{\text{mod}}\ \Lambda \longrightarrow \underline{\text{mod}}\ \Lambda^{op}$ called the transpose which has the following property. Let M be in $\text{mod}\ \Lambda$ and suppose that $P_1 \xrightarrow{f} P_0 \longrightarrow M \longrightarrow 0$ is a minimal projective presentation of M. Then denoting by X^* the Λ^{op}-module $\text{Hom}_\Lambda(X, \Lambda)$ for all X in mod Λ, we have that $\text{TrM} \approx \text{Coker}(P_0^* \xrightarrow{f^*} P_1^*)$ in mod Λ^{op}. Combining this property of $\text{Tr}: \underline{\text{mod}}\ \Lambda \longrightarrow \underline{\text{mod}}\ \Lambda^{op}$ with the properties of the functor $F: \text{mod}\ \Lambda \longrightarrow \underline{\text{mod}}\ \Lambda$ cited above, it is not difficult to show that $\text{Tr}: \underline{\text{mod}}\ \Lambda \longrightarrow \underline{\text{mod}}\ \Lambda^{op}$ has the following properties.

Proposition 1.3:

a) If A is in mod Λ, then $\text{TrA} = 0$ in mod Λ if and only if A is projective.

b) Given a finite family $\{A_i\}_{i\varepsilon I}$ of modules in mod Λ, then $\text{Tr}(\coprod_{i\varepsilon I} A_i) \approx \coprod_{i\varepsilon I} \text{TrA}_i$.

c) TrA is in $\text{mod}_P\Lambda^{op}$ for each A in mod Λ.

d) For each A in mod Λ, we have that $A \approx \text{TrTrA} \amalg P$ where TrTrA is in $\text{mod}_P\Lambda$ and P is projective.

e) The induced duality $\text{Tr}: \underline{\text{mod}}_P\Lambda \longrightarrow \underline{\text{mod}}_P\Lambda^{op}$ has the following properties:

i) $\operatorname{TrTr}A \approx A$ in $\operatorname{mod}_P\Lambda$ for all A in $\operatorname{mod}_P\Lambda$.

ii) $A \approx B$ in $\operatorname{mod}_P\Lambda$ if and only if $\operatorname{Tr}A \approx \operatorname{Tr}B$ in $\operatorname{mod}_P\Lambda^{op}$.

iii) If $\{A_i\}_{i\varepsilon I}$ is a finite family of objects in $\operatorname{mod}_P\Lambda$, then $\operatorname{Tr}(\coprod_{i\varepsilon I} A_i) \approx \coprod_{i\varepsilon I} \operatorname{Tr}A_i$ in $\operatorname{mod}_P\Lambda^{op}$.

iv) A in $\operatorname{mod}_P\Lambda$ is indecomposable if and only if $\operatorname{Tr}A$ is indecomposable in $\operatorname{mod}_P\Lambda^{op}$.

With this preliminary discussion of the category $\underline{\operatorname{mod}}\ \Lambda$ in mind, we turn our attention to the category $\overline{\operatorname{mod}}\ \Lambda$, the category of $\operatorname{mod}\Lambda$ modulo injectives. The objects of $\overline{\operatorname{mod}}\ \Lambda$ are the same as those of $\operatorname{mod}\Lambda$. For each pair of objects A, B in $\operatorname{mod}\Lambda$, the R-module of morphisms from A to B in $\overline{\operatorname{mod}}\ \Lambda$ which we denote by $\overline{\operatorname{Hom}}_\Lambda(A, B)$ is defined to be $\operatorname{Hom}_\Lambda(A, B)/I(A, B)$, where $I(A, B)$ is the R-submodule of $\operatorname{Hom}_\Lambda(A, B)$ consisting of those morphisms $f: A \longrightarrow B$ which can be factored through an injective Λ-module. The composition of morphisms in $\overline{\operatorname{mod}}\ \Lambda$ is given by the formula $\overline{g}\,\overline{f} = \overline{gf}$ where by $\overline{f}$ we mean the image of f in (A, B) in $\overline{\operatorname{Hom}}_\Lambda(A, B)$ under the canonical epimorphism $(A, B) \longrightarrow (A, B)/I(A, B)$.

In analogy with the category $\operatorname{mod}_P\Lambda$, we define $\operatorname{mod}_I\Lambda$ to be the full subcategory of $\operatorname{mod}\Lambda$ consisting of all A in $\operatorname{mod}\Lambda$ having no nonzero injective summands. Then denoting by $\overline{\operatorname{mod}}_I\Lambda$ the full subcategory of $\overline{\operatorname{mod}}\ \Lambda$ whose objects are those of $\operatorname{mod}_I\Lambda$, we have that the inclusion $\overline{\operatorname{mod}}_I\Lambda \longrightarrow \overline{\operatorname{mod}}\ \Lambda$ is an equivalence of categories which we sometines consider an identification.

Next we observe that the duality $D: \operatorname{mod}\Lambda \longrightarrow \operatorname{mod}\Lambda^{op}$ induces dualities $D: \operatorname{mod}_P\Lambda \longrightarrow \operatorname{mod}_I\Lambda$ and $D: \underline{\operatorname{mod}}\ \Lambda \longrightarrow \overline{\operatorname{mod}}\ \Lambda^{op}$. These dualities enable us to establish properties of $\mathrm{mod}_I\Lambda$ and $\overline{\mathrm{mod}}_I\Lambda$ analogous to the results cited above for $\mathrm{mod}_P\Lambda$ and $\underline{\mathrm{mod}}_P\Lambda$. For this reason we do not state these analogues explicitly even though they will be used freely throughout the paper.

Since the functors $\mathrm{Tr}\colon \underline{\mathrm{mod}}_P\Lambda \longrightarrow \underline{\mathrm{mod}}_P\Lambda^{op}$ and $D\colon \underline{\mathrm{mod}}_P\Lambda^{op} \longrightarrow \overline{\mathrm{mod}}_I\Lambda$ are dualities, it is obvious that their composition $\mathrm{DTr}\colon \underline{\mathrm{mod}}_P\Lambda \longrightarrow \overline{\mathrm{mod}}_I\Lambda$ is an equivalence of categories. Similarly the functor $\mathrm{TrD}\colon \overline{\mathrm{mod}}_I\Lambda \longrightarrow \underline{\mathrm{mod}}_P\Lambda$ which is the composition $\overline{\mathrm{mod}}_I\Lambda \xrightarrow{D} \underline{\mathrm{mod}}_P\Lambda^{op} \xrightarrow{\mathrm{Tr}} \underline{\mathrm{mod}}_P\Lambda$ is an equivalence of catagories Furthermore, $\mathrm{TrD}(\mathrm{DTr})\colon \underline{\mathrm{mod}}_P\Lambda \longrightarrow \underline{\mathrm{mod}}_P\Lambda$ is isomorphic to the identity on $\underline{\mathrm{mod}}_P\Lambda$ while $\mathrm{DTr}(\mathrm{TrD})\colon \overline{\mathrm{mod}}_I\Lambda \longrightarrow \overline{\mathrm{mod}}_I\Lambda$ is isomorphic to the identity on $\overline{\mathrm{mod}}_I\Lambda$. These various dualities and equivalences play a critical role throughout the paper. In fact, the whole paper is essentially devoted to the study of these functors and their connections with the module theory of artin algebras. The rest of this section is devoted to pointing out some of their more elementary properties.

<u>Proposition 1.4</u>: Let Λ be an artin R-algebra with $\underline{r}$ the radical of Λ.

a) Let M be in $\mathrm{mod}_P\Lambda$ and let $0 \longrightarrow \Omega M \longrightarrow P_0 \longrightarrow M \longrightarrow 0$ be exact with $P \longrightarrow M \longrightarrow 0$ a projective cover.

  i) $D(\Omega M/\underline{r}\Omega M) \approx \mathrm{Tr}M/\underline{r}\mathrm{Tr}M$ and M is an indecomposable Λ-module if and only if $\mathrm{Tr}M$ is an indecomposable Λ^{op}-module.

  ii) $\mathrm{Soc}(\mathrm{DTr}M) \approx \Omega M/\underline{r}\Omega M$ and M is an indecomposable Λ-module if and only if $\mathrm{DTr}M$ is an indecomposable Λ-module.

b) Let M be in $\text{mod}_I \Lambda$ and let

$0 \longrightarrow M \longrightarrow I_0 \longrightarrow \Omega^{-1}M \longrightarrow 0$ be exact with $0 \longrightarrow M \longrightarrow I$ an injective envelope of M. Then

$\text{Soc}(\Omega^{-1}M) \approx \text{TrDM}/\underline{r}\text{TrDM}$. Furthermore, M is indecomposable if and only if TrDM is indecomposable.

Proof:

a) i) Let $P_1 \longrightarrow \Omega M \longrightarrow 0$ be a projective cover for ΩM. Then $P_1/\underline{r}P_1 \approx \Omega M/\underline{r}\Omega M$ and $P_1 \longrightarrow P_0 \longrightarrow M \longrightarrow 0$ is a minimal projective presentation for M. From the fact that $P_1 \longrightarrow P_0 \longrightarrow M \longrightarrow 0$ is a minimal projective presentation for M, it follows easily that $P_0^* \longrightarrow P_1^* \longrightarrow \text{TrM} \longrightarrow 0$ is a minimal projective presentation for TrM. Hence $\text{TrM}/\underline{r}\text{TrM} \approx P_1^*/\underline{r}P_1^*$. But it is well known that $P^*/\underline{r}P^* \approx D(P/rP)$ for all projective Λ-modules P. Hence $\text{TrM}/\underline{r}\text{TrM} \approx D(P_1/rP_1) \simeq D(\Omega M/\underline{r}\Omega M)$. The fact that M is an indecomposable Λ-module if and only if TrM is an indecomposable Λ^{op}-module follows from Proposition 1.3.

a) ii) Follows from a)i) using the duality $D: \text{mod } \Lambda^{op} \longrightarrow \text{mod } \Lambda$.

b) Follows from a) using the duality $D: \text{mod } \Lambda \longrightarrow \text{mod } \Lambda^{op}$.

It is worthwhile observing that Proposition 1.4 can be used to show that certain types of indecomposable modules exist. We illustrate this point in the following easily verified consequence of Proposition 1.4.

Corollary 1.5:

a) Let $\underline{a}$ be a proper nonzero submodule of an indecomposable projective module P. Then there is an indecomposable Λ-module M such that $\text{SocM} \approx \underline{a}/\underline{r}\,\underline{a}$, for instance $M = \text{DTr}(P/\underline{a})$.

b) Let M be a proper nonzero submodule of an indecomposable injective Λ-module I. Then there is an indecomposable Λ-module N such that $N/\underline{r}N \approx \mathrm{Soc}(I/M)$, for instance $N \approx \mathrm{TrD}(M)$.

We now give some estimates concerning the relationship between $L(M)$, the length of M, and $L(\mathrm{Tr}M)$ which has some interesting applications.

Proposition 1.6: Let Λ be an artin R-algebra and suppose m is the maximum of $L(\Lambda)$ and $L(\Lambda^{op})$.

a) If M is a Λ- module and $n = L(M/\underline{r}M)$, then $L(\mathrm{Tr}M) \leq nm^2$.

b) Suppose $\{M_i\}_{i\varepsilon I}$ is a family of Λ-modules in $\mathrm{mod}_P\Lambda$. Then the following statements are equivalent:

i) $\{L(M_i)\}_{i\varepsilon I}$ is bounded.

ii) $\{(\mathrm{Tr}M_i)\}_{i\varepsilon I}$ is bounded.

iii) $\{L(\mathrm{DTr}M_i)\}_{i\varepsilon I}$ is bounded.

iv) $\{L(\mathrm{TrD}M_i)\}_{i\varepsilon I}$ is bounded.

Proof:

a) Let $0 \longrightarrow \Omega M \longrightarrow P \longrightarrow M \longrightarrow 0$ be exact with $P \longrightarrow M \longrightarrow 0$ a projective cover. Then $L(P) \leq L(M/\underline{r}M)L(\Lambda) \leq nm$. Since $L(\Omega M/\underline{r}\Omega M) \leq L(\Omega M) \leq L(P) \leq nm$, the fact that $D(\Omega M/\underline{r}\Omega M) \approx \mathrm{Tr}M/\underline{r}\mathrm{Tr}M$ implies that $L(\mathrm{Tr}M/\underline{r}\mathrm{Tr}M) \leq nm$. Hence $L(\mathrm{Tr}M) \leq nm^2$, our desired result.

b) i) implies ii), follows from a).

ii) implies i). Since each M_i is in $\mathrm{mod}_P\Lambda$, we know that $\mathrm{Tr}(\mathrm{Tr}M_i) \approx M_i$ for all i in I. Hence ii) implies i) also follows from a).

ii) equivalent to iii). Easy consequence of previous results.

i) equivalent to iv). Easy consequence of previous results.

As a straighforward application of this result we have the following.

Corollary 1.7:

a) Let $\{M_i\}_{i\varepsilon I}$ be a family of Λ-modules in $\text{mod}_P\Lambda$ and let $0 \longrightarrow \Omega M_i \longrightarrow P_i \longrightarrow M_i \longrightarrow 0$ be exact sequences with the $P_i \longrightarrow M_i \longrightarrow 0$ projective covers. Then $\{L(M_i)\}_{i\varepsilon I}$ is bounded if and only if $\{L(\Omega M_i)\}_{i\varepsilon I}$ is bounded.

b) Let $\{M_i\}_{i\varepsilon I}$ be a family of Λ-modules in $\text{mod}_I\Lambda$ and let $0 \longrightarrow M_i \longrightarrow I_i \longrightarrow \Omega^{-1}M_i \longrightarrow o$ be exact with the $0 \longrightarrow M_i \longrightarrow I_i$ injective envelopes. Then $\{L(M_i)\}_{i\varepsilon I}$ is bounded if and only if $\{L(\Omega^{-1}M_i)\}_{i\varepsilon I}$ is bounded.

Proof:

a) It is obvious that if $\{L(M_i)\}_{i\varepsilon I}$ is bounded, then $\{L(\Omega M_i)\}_{i\varepsilon I}$ is bounded. Assume now that $\{L(\Omega M_i)\}_{i\varepsilon I}$ is bounded. Then $\{L(\Omega M_i/\underline{r}\Omega M_i)\}_{i\varepsilon I}$ is bounded and so $\{L(\text{Soc}(DTrM_i)\}_{i\varepsilon I}$ is bounded since $\Omega M_i/\underline{r}\Omega M_i \approx \text{Soc}(DTrM_i)$ for all i. Hence $\{L(DTrM_i)\}_{i\varepsilon I}$ is bounded which implies by Proposition 1.6 that $\{L(M_i)\}_{i\varepsilon I}$ is bounded since each of the M_i are in $\text{mod}_P\Lambda$. Thus if $\{L(\Omega M_i)\}_{i\varepsilon I}$ is bounded, then $\{L(M_i)\}_{i\varepsilon I}$ is bounded which finished the proof of a).

b) Just the dual of a).

We conclude this section by pointing out that in the case of symmetric artin algebras the functors DTr and TrD are familiar functors. In order to do this it is convenient to recall the definitions of the functors $\Omega\colon \underline{\text{mod}}\ \Lambda \longrightarrow \underline{\text{mod}}\ \Lambda$ and $\Omega^{-1}\colon \overline{\text{mod}}\ \Lambda \longrightarrow \overline{\text{mod}}\ \Lambda$.

The functor $\Omega\colon \underline{\text{mod}}\ \Lambda \longrightarrow \underline{\text{mod}}\ \Lambda$ is defined on objects as follows. For each M in mod Λ choose an exact sequence

$0 \longrightarrow K_M \longrightarrow P_M \longrightarrow M \longrightarrow 0$ with $P \longrightarrow M \longrightarrow 0$ a projective cover. Then $\Omega M = K$ by definition. We now define Ω on morphisms. Given a morphism $f: M \longrightarrow N$ in mod Λ there is a commutative diagram

$$\begin{array}{ccccccccc} 0 & \longrightarrow & \Omega M & \longrightarrow & P_M & \longrightarrow & M & \longrightarrow & 0 \\ & & \downarrow f_1 & & \downarrow f_0 & & \downarrow f & & \\ 0 & \longrightarrow & \Omega N & \longrightarrow & P_N & \longrightarrow & N & \longrightarrow & 0 . \end{array}$$

While the morphism $f_1: \Omega M \longrightarrow \Omega N$ is uniquely determined by $f_0: P_M \longrightarrow P_N$, the morphism $f_0: P_M \longrightarrow P_N$ is not uniquely determined by $f: M \longrightarrow N$. However the different morphisms $\Omega M \longrightarrow \Omega N$ which result from different choices of $P_M \longrightarrow P_N$, which give a commutative diagram, differ from f_1 by morphisms in $P(\Omega M, \Omega N)$. Thus if given $f: M \longrightarrow N$ we define $\Omega(f)$ in $\underline{\mathrm{Hom}}_\Lambda(\Omega M, \Omega N)$ to be $\underline{f}_1$, then $\Omega(f)$ is well defined and we obtain a morphism of groups $\Omega: (M, N) \longrightarrow \underline{\mathrm{Hom}}_\Lambda(\Omega M, \Omega N)$ given by $f \longrightarrow \Omega(f)$ for all f in (M, N). Since it is easily seen that $\mathrm{Ker}\ \Omega \supset P(M, N)$, the morphism $\Omega: (M, N) \longrightarrow \underline{\mathrm{Hom}}_\Lambda(\Omega M, \Omega N)$ induces a morphism $\Omega: \underline{\mathrm{Hom}}_\Lambda(M, N) \longrightarrow \underline{\mathrm{Hom}}_\Lambda(\Omega M, \Omega N)$. The fact that the morphisms $\Omega: \underline{\mathrm{Hom}}_\Lambda(M, N) \longrightarrow \underline{\mathrm{Hom}}_\Lambda(\Omega M, \Omega N)$ satisfy the requirements for Ω to be a functor is readily verified. Having defined the functor $\Omega: \underline{\mathrm{mod}}\ \Lambda \longrightarrow \underline{\mathrm{mod}}\ \Lambda$, we define $\Omega^i: \underline{\mathrm{mod}}\ \Lambda \longrightarrow \underline{\mathrm{mod}}\ \Lambda$ for all nonnegative integers i by induction as follows: $\Omega^0: \underline{\mathrm{mod}}\ \Lambda \longrightarrow \underline{\mathrm{mod}}\ \Lambda$ is the identity and Ω^{i+1} is defined to be the composition $\underline{\mathrm{mod}}\ \Lambda \xrightarrow{\Omega^i} \underline{\mathrm{mod}}\ \Lambda \xrightarrow{\Omega} \underline{\mathrm{mod}}\ \Lambda$.

By analogy we define $\Omega^{-1}: \overline{\mathrm{mod}}\ \Lambda \longrightarrow \overline{\mathrm{mod}}\ \Lambda$ as follows. For each M we choose an exact sequence $0 \longrightarrow M \longrightarrow I_M \longrightarrow L_M \longrightarrow 0$ with $0 \longrightarrow M \longrightarrow I_M$ an injective envelope. We define $\Omega^{-1}M$ to be L_M for each M in mod Λ so Ω^{-1} is defined on objects. The morphisms

$\Omega^{-1}: \overline{\mathrm{Hom}}_\Lambda(M, N) \longrightarrow \overline{\mathrm{Hom}}_\Lambda(\Omega^{-1}M, \Omega^{-1}N)$ are defined in a manner dual to the definition of Ω. Similarly we define $\Omega^{-i}: \overline{\mathrm{mod}}\ \Lambda \longrightarrow \overline{\mathrm{mod}}\ \Lambda$ for all nonnegative integers i by induction on i as follows: $\Omega^{-0}: \overline{\mathrm{mod}}\ \Lambda \longrightarrow \overline{\mathrm{mod}}\ \Lambda$ is the identity and define $\Omega^{-(i+1)}$ to be the composition $\Omega^{-1}\Omega^{-i}$.

Suppose now that Λ is selfinjective. Then the projective and injective Λ-modules are the same. Hence $P(M, N) = I(M, N)$ for all M and N in mod Λ, so $\underline{\mathrm{mod}}\ \Lambda = \overline{\mathrm{mod}}\ \Lambda$. Also it is well known that under these circumstances $\Omega: \underline{\mathrm{mod}}\ \Lambda \longrightarrow \underline{\mathrm{mod}}\ \Lambda$ and $\Omega^{-1}: \underline{\mathrm{mod}}\ \Lambda \longrightarrow \underline{\mathrm{mod}}\ \Lambda$ are inverse equivalences of categories. Further it is well known that since Λ is selfinjective, the functor $(\ ,\Lambda)$: mod $\Lambda \longrightarrow$ mod Λ^{op} is a duality with inverse duality $(\ ,\Lambda^{op})$: mod $\Lambda^{op} \longrightarrow$ mod Λ. In general the duality $(\ ,\Lambda)$: mod $\Lambda \longrightarrow$ mod Λ^{op} need not be isomorphic to the duality D: mod $\Lambda \longrightarrow$ mod Λ^{op}. In fact, it is not difficult to see that the dualities $(\ ,\Lambda)$ and D: mod $\Lambda \longrightarrow$ mod Λ^{op} are isomorphic functors if and only if Λ is a symmetric R-algebra, i.e. there is a two-sided Λ isomorphism $\Lambda \longrightarrow \mathrm{Hom}_R(\Lambda, I)$. Straightforward calculations now suffice to show the following.

<u>Proposition 1.8</u>: Let Λ be a symmetric R-algebra. Then the functors DTr, Ω^2: $\underline{\mathrm{mod}}\ \Lambda \longrightarrow \underline{\mathrm{mod}}\ \Lambda$ are isomorphic and the functors TrD, Ω^{-2}: $\underline{\mathrm{mod}}\ \Lambda \longrightarrow \underline{\mathrm{mod}}\ \Lambda$ are isomorphic.

§2. Morphisms Determined by Modules.

We assume as usual that R is a commutative artin ring and Λ is an artin R-algebra. Also, unless stated to the contrary, we assume that all Λ-modules are finitely generated. Our aim in this section is to study the question: Given morphisms $f: B \longrightarrow C$ and $h: L \longrightarrow C$, when does there exist a morphism $g: L \longrightarrow B$ such that $fg = h$? The answer to this question that we give is in terms of the notion of a morphism being determined by a module, a notion we introduced in [2]. In fact, this section consists of a summary of some of the results in [2] specialized to artin algebras, with a somewhat different organization. While few proofs are given in detail, some outlines of proofs are discussed. The material in this section will be used freely in the rest of the paper.

Suppose we are given a pair of morphisms $f: B \longrightarrow C$ and $h: L \longrightarrow C$. A morphism $g: L \longrightarrow B$ is called a <u>lifting of h to B</u> if $fg = h$. Suppose $g: L \longrightarrow B$ is a lifting of h to B. Then given a morphism $t: X \longrightarrow L$ the composition $ht: X \longrightarrow C$ can be lifted to B since $ht = fgt$. On the other hand, if $h: L \longrightarrow C$ has the property that for each $t: X \longrightarrow L$, the composition ht can be lifted to B, then $h: L \longrightarrow B$ can be lifted to B. Simply let $X = L$ and $t: L \longrightarrow L$ be the identity. Then the morphism $u: L \longrightarrow B$ such that $ht = fu$ is our desired lifting of h to B.

These observations suggest the following question. For a morphism $f: B \longrightarrow C$ does there exist a fixed module X such that a morphism $h: L \longrightarrow C$ can be lifted to B if for each $t: X \longrightarrow L$ the composition ht can be lifted to B? In connection with this question, it is convenient to make the following definitions.

A morphism $f: B \longrightarrow C$ in mod Λ is said to be <u>right determined by a module X</u> in mod Λ if a morphism $h: L \longrightarrow C$ in mod Λ can be lifted to B whenever for each $t: X \longrightarrow L$ the composition $ht: X \longrightarrow B$ can be lifted to B.

We restate this definition in more compact form using the map $(X, d): (X, E) \longrightarrow (X, F)$ induced for each X in mod Λ by a morphism

$d: E \longrightarrow F$ in mod Λ. A morphism $f: B \longrightarrow C$ is right determined by a module X if and only if a morphism $h: L \longrightarrow C$ can be lifted to B whenever $\mathrm{Im}(X, f) \supset \mathrm{Im}(X, h)$. We say that a morphism $f: B \longrightarrow C$ is right X-determined to mean that f is right determined by the module X. Thus our original question becomes in this language: Is every morphism $f: B \longrightarrow C$ in mod Λ right X-determined for some X in mod Λ?

In order to discuss the notion dual to a morphism being right determined by a module, it is convenient to introduce the following definitions.

Suppose $f: A \longrightarrow B$ and $h: A \longrightarrow L$ are morphisms in mod Λ. A morphism $g: B \longrightarrow L$ is called an extension of h to B if $gf = h$. A morphism $f: A \longrightarrow B$ in mod Λ is said to be left determined by a module Y in mod Λ if a morphism $h: A \longrightarrow L$ can be extended to B whenever given $t: L \longrightarrow Y$ the composition th can be extended to B.

We restate this definition in a more compact form using the map $(d, Y): (F, Y) \longrightarrow (E, Y)$ induced for each Y in mod Λ by a morphism $d: E \longrightarrow F$ in mod Λ. A morphism $f: A \longrightarrow B$ is left determined by a module Y in mod Λ if and only if a morphism $h: A \longrightarrow L$ can be extended to B whenever $\mathrm{Im}(f, Y) \supset \mathrm{Im}(h, Y)$.

We say that a morphism $f: A \longrightarrow B$ is left Y-determined to mean that $f: A \longrightarrow B$ is left determined by the module Y in mod Λ. In view of our previous discussion it is natural to ask: Is every $f: A \longrightarrow B$ left Y-determined for some Y in mod Λ?

Surprisingly, every morphism $f: B \longrightarrow C$ in mod Λ is right X-determined for some X in mod Λ as well as left Y-determined for some Y in mod Λ. That this is the case is shown in Chapter I of [2]. We briefly outline the main elements of the proof for the convenience of the reader.

We first point out the following easily verified results.

Lemma 2.1: Let $f: B \longrightarrow C$ be a morphism in mod Λ. Then

a) f is right X-determined if and only if the Λ^{op}-morphism $D(f): D(C) \longrightarrow D(B)$ is left $D(X)$-determined.

b) If f is a monomorphism, then f is right P-determined where P is a projective cover for Coker f.

c) If f is an epimorphism, then f is left I-determined where I is an injective envelope of Ker f.

d) If the induced epimorphism $B \longrightarrow \mathrm{Im} f$ is right X-determined, then $f: B \longrightarrow C$ is right $X \coprod P$-determined where P is projective cover for Coker f.

e) If the induced monomorphism $\mathrm{Im} f \longrightarrow C$ is left Y-determined, then $f: B \longrightarrow C$ is left $Y \coprod I$-determined where I is an injective envelope for Ker f.

As a consequence of parts d) and e) of this lemma, we have that in order to show that every morphism in mod Λ is both right and left determined by modules in mod Λ, it suffices to show that if $0 \longrightarrow A \xrightarrow{g} B \xrightarrow{f} C \longrightarrow 0$ is exact in mod Λ, then g is left Y-determined and f is right X-determined for some Λ-modules Y and X. This follows from the following somewhat more explicit result.

Proposition 2.2: Let $0 \longrightarrow A \xrightarrow{g} B \xrightarrow{f} C \longrightarrow 0$ be an exact sequence in mod Λ. Then

a) g is left determined by DTrC.

b) f is right determined by TrDA.

The following are some useful consequences of this result.

Proposition 2.3: Let $0 \longrightarrow A \xrightarrow{g} B \xrightarrow{f} C \longrightarrow 0$ be an exact sequence in mod Λ.

a) The following statements are equivalent:

i) $0 \longrightarrow A \xrightarrow{g} B \xrightarrow{f} C \longrightarrow 0$ splits.

ii) $(g, DTrC): (B, DTrC) \longrightarrow (A, DTrC)$ is an epimorphism.

iii) $(TrDA, f): (TrDA, B) \longrightarrow (TrDA, C)$ is an epimorphism.

b) Let X be in mod Λ. The following are equivalent:

i) $(X, f): (X, B) \longrightarrow (X, C)$ is an epimorphism.

ii) $(g, DTrX): (B, DTrX) \longrightarrow ((A, DTrX))$ is an epimorphism.

Proof:

a) That i) implies ii) and iii) is obvious.

ii) implies i). By Proposition 2.2, we know that $A \xrightarrow{g} B$ is left DTrC-determined. Therefore, the identity morphism $id_A: A \longrightarrow A$ has the property that there is an $s: B \longrightarrow A$ such that $sg = id$ if $Im(g, DTrC) \supset (A, DTrC)$, i.e. if the morphism $(g, DTrC): (B, DTrC) \longrightarrow (A, DTrC)$ is an epimorphism. Since we are assuming by ii) that $(g, DTrC)$ is an epimorphism, we have that there is an $s: B \longrightarrow A$ such that $sg = id_A$, or equivalently, the exact sequence $0 \longrightarrow A \xrightarrow{g} B \xrightarrow{f} C \longrightarrow 0$ splits. Thus we have shown that ii) implies i).

iii) implies i). The proof proceeds in an obviously analogous way to the proof that ii) implies i).

b) i) implies ii). Suppose $(X, f): (X, B) \longrightarrow (X, C) \longrightarrow 0$ is exact and let h be in $(A, DTrX)$. Then we have the following exact pushout diagram

$$\begin{array}{ccccccccc} 0 & \longrightarrow & A & \xrightarrow{g} & B & \xrightarrow{f} & C & \longrightarrow & 0 \\ & & \downarrow h & & \downarrow t & & \| & & \\ 0 & \longrightarrow & DTrX & \xrightarrow{g'} & B' & \xrightarrow{f'} & C & \longrightarrow & 0 \end{array}$$

From this it follows that (X, f') is an epimorphism since (X, f) is an epimorphism. Because $TrD(DTrX) \coprod P \approx X$ for some projective P (possibly zero), the fact that (X, f') is an epimorphism implies that $(TrD(DTrX), f')$ is an epimorphism. Hence by part a), we have that

$0 \longrightarrow DTrX \xrightarrow{g'} B' \longrightarrow C \longrightarrow 0$ splits. Thus there is an $s: B \longrightarrow DTrX$ such that $sg = h$. Since h was an arbitrary element of $(A, DTrX)$, this argument shows that (X, f) being an epimorphism implies that $(g, DTrX)$ is an epimorphism. Thus we have shown that i) implies ii).

ii) implies i). The proof is an obvious analogue of the proof given for i) implies ii).

In connection with this result, we point out the following generalization of Proposition 2.3, which was established in Chapter III, Section 4 of [2].

Proposition 2.4: Let $0 \longrightarrow A \xrightarrow{g} B \xrightarrow{f} C \longrightarrow 0$ be an exact sequence in mod Λ. For each X in mod Λ, there is an isomorphism

$$\text{Hom}_R(\text{Coker}(X, f), I) \approx \text{Coker}(g, DTrX)$$

which is functorial in X, where $I = I(R/\text{rad } R)$.

Before proceeding further, it is convenient to introduce the following notation. Suppose M is in $\text{mod}\ \Lambda$ and $M \approx \coprod_{i=1}^{k} M_i^{n_i}$ with the M_i indecomposable and the n_i positive integers (where M^n denotes the sum of n copies of M). Then we denote the sum $\coprod_{i=1}^{k} M_i$ by $[M]$. The following easily verified results make clear the reason for introducing this notation.

<u>Lemma 2.5</u>: Let $f\colon B \longrightarrow C$ be a morphism in $\text{mod}\ \Lambda$.

a) If f is right X-determined, then f is right $X \coprod V$-determined for any V in $\text{mod}\ \Lambda$.

b) f is right X-determined if and only if X is right $[X]$-determined.

c) If f is left Y-determined, then f is left $Y \coprod V$-determined for any V in $\text{mod}\ \Lambda$.

d) f is left Y-determined if and only if f is left $[Y]$-determined.

Summarizing and extending slightly some of our results so far, we have the following.

<u>Theorem 2.6</u>:

a) Let $0 \longrightarrow A \xrightarrow{g} B \xrightarrow{f} C$ be exact in $\text{mod}\ \Lambda$. Then f is right $[\text{TrD}A \coprod P]$-determined, where P is a projective cover for $\text{Coker}\ f$.

b) Let $A \xrightarrow{g} B \xrightarrow{f} C \longrightarrow 0$ be exact in $\text{mod}\ \Lambda$. Then g is left $[\text{DTr}C \coprod I]$-determined, where I is an injective envelope for $\text{Ker}\ g$.

While most of the results given so far follow fairly easily from the basic definitions involved, the proof of Proposition 2.2 is not so straightforward. For this reason we now outline its proof.

Let X be in mod Λ. For each C in mod Λ we view (X, C) as an $(\text{End } X)^{op}$-module by means of the action of End X on X. Since it is easily verified that $P(X, C)$ is an $(\text{End } X)^{op}$-submodule of (X, C) (remember that $P(X, C)$ denotes those morphisms from X to C which factor through projectives), we have that $\underline{\text{Hom}}_\Lambda(X, C) = (X, C)/P(X, C)$ is an $(\text{End } X)^{op}$-module. But the two sided ideal $P(X, X)$ of $(\text{End } X)^{op}$ is contained in the annihilator of $\underline{\text{Hom}}_\Lambda(X, C)$, as is easily verified. Hence $\underline{\text{Hom}}_\Lambda(X, C)$ is an $(\underline{\text{End}}\ X)^{op}$-module. Since the duality $\text{Tr}: \underline{\text{mod}}\ \Lambda \longrightarrow \underline{\text{mod}}\ \Lambda^{op}$ gives a ring isomorphism $(\underline{\text{End}}\ X)^{op} \longrightarrow \underline{\text{End}}\ \text{Tr}X$ we have that $\underline{\text{Hom}}_\Lambda(X, C)$ is an $\underline{\text{End}}$ TrX-module. Therefore the ring surjection End TrX $\longrightarrow \underline{\text{End}}$ TrX makes $\underline{\text{Hom}}_\Lambda(X, C)$ an End TrX-module for each C in mod Λ. With these observations in mind we have the following fundamental result (see [2, Chapter I, Proposition 3.4]).

<u>Proposition 2.7</u>: Let C and X be in mod Λ and let E be an injective Γ = End TrX-module. Then there are isomorphisms

$$v\colon \text{Ext}^1_\Lambda(C, \text{Hom}_\Gamma(\text{Tr}X, E)) \longrightarrow \text{Hom}_\Gamma(\underline{\text{Hom}}_\Lambda(X, C), E)$$

which are functorial in C and E, where the operation of Λ on $\text{Hom}_\Gamma(\text{Tr}X, E)$ is the one induced by the Λ^{op}-module structure on TrX. Moreover if x in $\text{Ext}^1_\Lambda(C, \text{Hom}_\Gamma(\text{Tr}X, E))$ is represented by the exact sequence $0 \longrightarrow \text{Hom}_\Gamma(\text{Tr}X, E) \xrightarrow{g} B \xrightarrow{f} C \longrightarrow 0$ then

a) f is right X-determined.

b) $\text{Im}(X, f) \supset P(X, C)$ and $\text{Im}(X, f)/P(X, C) = \text{Ker } v(x)$.

As one application of this result, we obtain a proof of Proposition 2.2. For suppose $0 \longrightarrow A \xrightarrow{g} B \xrightarrow{f} C \longrightarrow 0$ is an exact sequence in mod Λ. Then $A = A_0 \coprod I$, where A_0 is in $\text{mod}_I \Lambda$ and I is an injective Λ-module. Hence $0 \longrightarrow A \xrightarrow{g} B \xrightarrow{f} C \longrightarrow 0$ is isomorphic to

$$\begin{array}{ccccccccc} & & I & = & I & & & & \\ & & \amalg & & \amalg & & & & \\ 0 & \longrightarrow & A_o & \xrightarrow{g_o} & B_o & \xrightarrow{f_o} & C & \longrightarrow & 0. \end{array}$$

From this it follows that $\mathrm{Im}(X, f) = \mathrm{Im}(X, f_o)$ for all X in $\mathrm{mod}\,\Lambda$ and that X in $\mathrm{mod}\,\Lambda$ has the property that f is right X-determined if and only if f_o is right X-determined. Since $\mathrm{TrD}A = \mathrm{TrD}A_o$, if we show that f_o is right $\mathrm{TrD}A_o$-determined, we will also have shown that f is right TrDA-determined. This will establish part b) or Proposition 2.2.

Hence we can assume without loss of generality that A is in $\mathrm{mod}_I\Lambda$. Then we know $A = \mathrm{DTr}(\mathrm{TrD}A)$. Hence letting $X = \mathrm{TrD}A$, we have that $A = \mathrm{DTr}X$. Thus letting $\Gamma = \mathrm{End}\,\mathrm{Tr}X$ we have that $A = \mathrm{Hom}_R(\Gamma \otimes_\Gamma \mathrm{Tr}X, I) \cong \mathrm{Hom}_\Gamma(\mathrm{Tr}X, \mathrm{Hom}_R(\Gamma, I))$, where $I = I(R/\mathrm{rad}\,R)$. Since $E = \mathrm{Hom}_R(\Gamma, I)$ is an injective Γ-module, $A = \mathrm{Hom}_\Gamma(\mathrm{Tr}X, E)$ with E an injective Γ-module. Thus the exact sequence $0 \longrightarrow A \xrightarrow{g} B \xrightarrow{f} C \longrightarrow 0$ can be written in the form $0 \longrightarrow \mathrm{Hom}_\Gamma(\mathrm{Tr}X, E) \xrightarrow{g} B \xrightarrow{f} C \longrightarrow 0$. It then follows from Proposition 2.7 that f is right X-determined. Substituting $X = \mathrm{TrD}A$, we have that f is right TrDA-determined which extablishes part b) of Proposition 2.2. Since part a) of Proposition 2.2 follows from part b) by duality, the proof of Proposition 2.2 is complete.

Before giving our next application of Proposition 2.7, it is convenient to make the following convention. Suppose Γ is an R-artin algebra. Since $D\colon \mathrm{mod}\,\Gamma \longrightarrow \mathrm{mod}\,\Gamma^{op}$ is a duality, the morphisms $D_{A,B}\colon \mathrm{Hom}_\Gamma(A, B) \longrightarrow \mathrm{Hom}_{\Gamma^{op}}(D(B), D(A))$ given by $D_{A,B}(f) = D(f)$ are isomorphisms functorial in A and B in $\mathrm{mod}\,\Gamma$. We will often consider these isomorphisms identifications. With these remarks in mind, we have the following important consequence of Proposition 2.7.

Proposition 2.8: Suppose X and C are in $\operatorname{mod} \Lambda$ and H is an $(\operatorname{End} X)^{op}$-submodule of (X, C) containing $P(X, C)$. Let E be a finitely generated injective $\Gamma = \operatorname{End} \operatorname{Tr}X$-module such that there is a Γ-morphism $t: \underline{\operatorname{Hom}}_{\Lambda}(X, C) \longrightarrow E$ with $\operatorname{Ker} t = H/P(X, C)$ (since Γ is an artin R-algebra, such an E exists). Then $D(E)$ is a finitely generated projective Γ^{op}-module and the exact sequence

$$x: 0 \longrightarrow \operatorname{Hom}_{\Gamma^{op}}(D(E), D\operatorname{Tr}X) \xrightarrow{g} B \xrightarrow{f} C \longrightarrow 0$$

in $\operatorname{Ext}^1_{\Lambda}(C, \operatorname{Hom}_{\Gamma^{op}}(D(E), D\operatorname{Tr}X))$ corresponding to t under the isomorphism

$$v: \operatorname{Ext}^1_{\Lambda}(C, \operatorname{Hom}_{\Gamma^{op}}(D(E), D\operatorname{Tr}X)) \longrightarrow \operatorname{Hom}_{\Gamma}(\underline{\operatorname{Hom}}_{\Lambda}(X, C), E)$$

described in Proposition 2.5 has the following properties:

a) f is right X-determined.

b) $\operatorname{Im}(X, f) = H$.

c) x splits if and only if $H = (X, C)$.

d) If E is an injective envelope for $\operatorname{Im} t$, then in any commutative diagram

$$\begin{array}{ccccccccc} 0 & \longrightarrow & \operatorname{Hom}_{\Gamma^{op}}(D(E), D\operatorname{Tr}X) & \xrightarrow{g} & B & \xrightarrow{f} & C & \longrightarrow & 0 \\ & & \downarrow & & \downarrow & & \| & & \\ 0 & \longrightarrow & \operatorname{Hom}_{\Gamma^{op}}(D(E), D\operatorname{Tr}X) & \xrightarrow{g} & B & \xrightarrow{f} & C & \longrightarrow & 0 \end{array}$$

the vertical morphisms are isomorphisms.

In connection with this result it should be noted that under the hypothesis of the previous proposition $D(E)$ is a finitely generated projective Γ^{op}-module with the property there is an epimorphism $D(E) \longrightarrow \operatorname{Hom}_R((X, C)/H, I)$ and DE is a projective cover for $\operatorname{Hom}_R((X, C)/H, I)$ if and only if E is an injective envelope for $\operatorname{Im} t$.

So far we have been looking primarily at epimorphisms right determined by a module. We now develop the analogous results for monomorphisms left determined by a module. While much of this information could be obtained by duality, we indicate the exact analogues.

Let A be in module. For each X in mod Λ we view (A, C) as an End X-module by means of the operation of End X on X. Since it is easily checked that I(A, X) is an End X-submodule of (A, X) we have that $\overline{\text{Hom}}_\Lambda(A, X)$ is an End X-module. It is also easily checked that the two sided ideal I(X, X) of End X is contained in the annihilator of $\overline{\text{Hom}}_\Lambda(A, X)$. Thus $\overline{\text{Hom}}_\Lambda(A, X)$ is an $\overline{\text{End}}$ X-module. The equivalence TrD: $\overline{\text{mod}}\ \Lambda \longrightarrow \underline{\text{mod}}\ \Lambda$ gives an isomorphism $\overline{\text{End}}$ X $\longrightarrow$ $\underline{\text{End}}$ TrDX. Hence $\overline{\text{Hom}}_\Lambda(A, X)$ is an ($\underline{\text{End}}$ TrDX)-module and thus an (End TrDX)-module. Letting Σ = End TrDX we have that Σ is an artin R-algebra in a natural way. Hence the duality D: mod $\Sigma \longrightarrow$ mod Σ^{op} induces a duality between the category of finitely generated injective Σ-modules and the category of finitely projective Σ^{op}-modules. With these observations and notations we have the following analogue of Proposition 2.7.

Proposition 2.9: Let A and X be in mod Λ. Let E be a finitely generated injective Σ = (End TrDX)-module. Then there are isomorphisms

$$\omega\colon \text{Ext}^1_\Lambda(D(E) \otimes_\Sigma \text{TrDX}, A) \longrightarrow \text{Hom}_\Sigma(\overline{\text{Hom}}(A, X), E)$$

which are functorial in A and E, where the operation of Λ on $D(E) \otimes_\Sigma \text{TrDX}$ is the one induced by the Λ-module structure on TrDX. Moreover if x in $\text{Ext}^1_\Lambda(D(E) \otimes \text{TrDX}, A)$ is represented by the exact sequence $0 \longrightarrow A \xrightarrow{g} B \longrightarrow D(E) \otimes_\Sigma \text{TrDX} \longrightarrow 0$ then

a) g is left X-determined.

b) $\mathrm{Im}(g, X) \supset I(A, X)$ and $\mathrm{Im}(g, X)/I(A, X) = \mathrm{Ker}\ \omega(X)$.

Proof: We only sketch the proof of the existence of the isomorphisms $\omega\colon \mathrm{Ext}^1_\Lambda(D(E) \otimes_\Sigma \mathrm{TrDX}, A) \longrightarrow \mathrm{Hom}_\Sigma(\overline{\mathrm{Hom}}_\Lambda(A, X), E)$. The rest of the proposition can be proven using arguments similar to those used in Chapter I, Section 3 of [2] to prove Proposition 2.7 of this paper.

Since A is finitely generated we know that $A \approx DD(A)$. Hence by [11, Chapter VI, Proposition 5.1] we have the isomorphism

$$\mathrm{Ext}^1_\Lambda(D(E) \otimes_\Sigma \mathrm{TrDX}, A) \longrightarrow \mathrm{Hom}_R(\mathrm{Tor}^\Lambda_1(DA, D(E) \otimes_\Sigma \mathrm{TrDX}), I).$$

Since $D(E)$ is a projective Σ^{op}-module, we have that $\mathrm{Tor}^\Lambda_1(DA, D(E) \otimes_\Sigma \mathrm{TrDX}) \cong D(E) \otimes_\Sigma \mathrm{Tor}^\Lambda_1(DA, \mathrm{TrDX})$. We also know that $\mathrm{Tor}^\Lambda_1(DA, \mathrm{TrDX}) \cong \underline{\mathrm{Hom}}_\Lambda(\mathrm{TrDA}, \mathrm{TrDX})$ (see [6, Proposition 2.2]). But the equivalence $\mathrm{TrD}\colon \overline{\mathrm{mod}}\ \Lambda \longrightarrow \underline{\mathrm{mod}}\ \Lambda$, gives a Σ-isomorphism, $\underline{\mathrm{Hom}}_\Lambda(\mathrm{TrDA}, \mathrm{TrDX}) \longrightarrow \overline{\mathrm{Hom}}_\Lambda(A, X)$. Hence we have the isomorphism $\mathrm{Hom}_R(\mathrm{Tor}^\Lambda_1(DC, D(E) \otimes_\Sigma \mathrm{TrDX}) \longrightarrow \mathrm{Hom}_R(D(E) \otimes_\Sigma \overline{\mathrm{Hom}}_\Lambda(A, X), I)$. But $\mathrm{Hom}_R(D(E) \otimes_\Sigma \overline{\mathrm{Hom}}_\Lambda(A, X), I) \approx \mathrm{Hom}_\Sigma(\overline{\mathrm{Hom}}_\Lambda(A, X), DDE)$. Since $E \approx DDE$ we have our desired isomorphism

$\omega\colon \mathrm{Ext}^1_\Lambda(D(E) \otimes_\Sigma \mathrm{TrDX}) \longrightarrow \mathrm{Hom}_\Sigma(\overline{\mathrm{Hom}}_\Lambda(A, X), E)$.

As an application of Proposition 2.9 we obtain the following analogue of Proposition 2.8.

Proposition 2.10: Suppose X and A are in mod Λ and H is an End X-submodule of (A, X) containing $I(A, X)$. Let E be a finitely generated injective $\Sigma = \mathrm{End}\ \mathrm{TrDX}$-module such that there is a Σ-morphism $t\colon \overline{\mathrm{Hom}}_\Lambda(A, X) \longrightarrow E$ with $\mathrm{Ker}\ t = H/I(A, X)$ (since Σ is an artin R-algebra such an E exists). Then the exact sequence

$$x: 0 \longrightarrow A \xrightarrow{g} B \xrightarrow{f} D(E) \otimes_{\Sigma} \mathrm{TrDX} \longrightarrow 0$$

in $\mathrm{Ext}^1_{\Lambda}(D(E) \otimes_{\Sigma} \mathrm{TrDX}, A)$ corresponding to t under the isomorphism

$$\omega: \mathrm{Ext}^1_{\Lambda}(D(E) \otimes_{\Sigma} \mathrm{TrDX}, A) \longrightarrow \mathrm{Hom}_{\Sigma}(\overline{\mathrm{Hom}}_{\Lambda}(A, X), I)$$

described in Proposition 2.9 has the following properties:

a) g is left X-determined.

b) $\mathrm{Im}(g, X) = H$.

c) x splits if and only if $H = (A, X)$.

d) If E is an injective envelope for $\mathrm{Im}t$, then in any commutative diagram

$$\begin{array}{ccccccccc} 0 & \longrightarrow & A & \xrightarrow{g} & B & \longrightarrow & D(E) \otimes_{\Sigma} \mathrm{TrDX} & \longrightarrow & 0 \\ & & \| & & \downarrow & & \downarrow & & \\ 0 & \longrightarrow & A & \longrightarrow & B & \longrightarrow & D(E) \otimes_{\Sigma} \mathrm{TrDX} & \longrightarrow & 0 \end{array}$$

the vertical morphisms are isomorphisms.

Proof: Follows by arguments similar to those used in Chapter I, Section 3 of [2] to derive Proposition 2.8 from 2.7.

In connection with this result it should be noted that under the hypothesis of the previous proposition $D(E)$ is a finitely generated projective Σ^{op}-module with the property there is an epimorphism $D(E) \longrightarrow \mathrm{Hom}_R((X, C)/H, I)$ and $D(E)$ is a projective cover for $\mathrm{Hom}_R((X, C)/H, I)$ if and only if E is an injective envelope for $\mathrm{Im}t$.

Suppose we are given a morphism $f: B \longrightarrow C$. We know that f is right X-determined for some X in $\mathrm{mod}\ \Lambda$. In view of our previous discussion, the $(\mathrm{End}\ X)^{op}$-submodule $\mathrm{Im}(X, f)$ of (X, C) is an interesting invariant of f. It is therefore natural to wonder to what extent a morphism $f: B \longrightarrow C$ is determined by knowing that $f: B \longrightarrow C$ is right X-determined and knowing the $(\text{End } X)^{op}$-submodule $\text{Im}(X, f)$ of (X, C). Obviously a morphism $f: B \longrightarrow C$ is not determined by this data, for if B' is any nonzero module and $f': B \coprod B' \longrightarrow C$ is the morphism such that $f'|B = f$ and $f'|B' = 0$, then f' is also right X-determined and $\text{Im}(X, f') = \text{Im}(X, f)$ even though f and f' are certainly not the same. Nonetheless, knowing that a morphism $f: B \longrightarrow C$ is right X-determined together with knowing the $(\text{End } X)^{op}$-submodule $\text{Im}(X, f)$ of (X, C) does tell us quite a bit about the morphism f as we shall soon see. As a preliminary step in this direction, we have the following easily verified result.

Lemma 2.11: Suppose $f: B \longrightarrow C$ and $f': B' \longrightarrow C$ are two morphisms in $\text{mod } \Lambda$ with f right X-determined.

a) $\text{Im}(X, f') \subset \text{Im}(X, f)$ if and only if there is a morphism $g: B' \longrightarrow B$ such that $f' = fg$.

b) If f' is also right X-determined, then $\text{Im}(X, f') = \text{Im}(X, f)$ if and only if there are morphisms $g: B' \longrightarrow B$ and $h: B \longrightarrow B'$ such that $f' = gf$ and $f = f'h$.

Similarly, we have the following analogue for left determined morphisms.

Lemma 2.12: Suppose $g: A \longrightarrow B$ and $g': A \longrightarrow B'$ are two morphisms in $\text{mod } \Lambda$ with g left X-determined.

a) $\text{Im}(g', X) \subset \text{Im}(g, X)$ if and only if there is a morphism $h: B \longrightarrow B'$ such that $hg = g'$.

b) If g' is also left X-determined, then $\operatorname{Im}(g, X) = \operatorname{Im}(g', X)$ if and only if there are morphisms $h: B \longrightarrow B'$ and $h': B' \longrightarrow B$ such that $hg = g'$ and $h'g' = g$.

Suppose now that $0 \longrightarrow A \xrightarrow{g} B \xrightarrow{f} C \longrightarrow 0$ is exact with f right X-determined. Since f is an epimorphism it is easily seen that the $(\operatorname{End} X)^{op}$-submodule $\operatorname{Im}(X, f)$ contains $P(X, C)$. Because (X, C) is a finitely generated $(\operatorname{End} X)^{op}$-module, there is a finitely generated injective $(\operatorname{End} X)^{op}$-module E such that there is a morphism $t: \underline{\operatorname{Hom}}_{\Lambda}(X, C) \longrightarrow E$ with the property that $\operatorname{Ker} t = \operatorname{Im}(X, f)/P(X, C)$ and the induced monomorphism $(X, C)/\operatorname{Im}(X, f) \longrightarrow E$ is an injective envelope. Then by Proposition 2.8 there is an exact sequence in $\operatorname{mod} \Lambda$

$$0 \longrightarrow \operatorname{Hom}_{\Gamma^{op}}(D(E), \operatorname{DTr}X) \xrightarrow{g'} B' \xrightarrow{f'} C \longrightarrow 0$$

with the following properties:

a) f' is right X-determined;

b) $\operatorname{Im}(X, f') = \operatorname{Im}(X, f)$;

c) In any commutative diagram

$$\begin{array}{ccccccccc} 0 & \longrightarrow & \operatorname{Hom}_{\Gamma^{op}}(D(E), \operatorname{DTr}X) & \xrightarrow{g'} & B' & \xrightarrow{f'} & C & \longrightarrow & 0 \\ & & \downarrow & & \downarrow & & \| & & \\ 0 & \longrightarrow & \operatorname{Hom}_{\Gamma^{op}}(D(E), \operatorname{DTr}X) & \xrightarrow{g'} & B' & \xrightarrow{f'} & C & \longrightarrow & 0 \end{array}$$

The vertical morphisms are isomorphisms.

Since f and f' are both right X-determined and $\operatorname{Im}(X, f') = \operatorname{Im}(X, f)$ we have by Lemma 2.11 that there is a commutative exact diagram

$$
\begin{array}{ccccccccc}
0 \longrightarrow & \mathrm{Hom}_{\Gamma^{op}}(D(E),\ DTrX) & \xrightarrow{g'} & B' & \xrightarrow{f'} & C & \longrightarrow 0 \\
& \downarrow s & & \downarrow t & & \| & \\
0 \longrightarrow & A & \xrightarrow{g} & B & \xrightarrow{f} & C & \longrightarrow 0 \\
& \downarrow u & & \downarrow v & & \| & \\
0 \longrightarrow & \mathrm{Hom}_{\Gamma^{op}}(D(E),\ DTrX) & \xrightarrow{g'} & B' & \xrightarrow{f'} & C & \longrightarrow 0
\end{array}
$$

with us and vt isomorphisms. From this it follows that A and B have decompositions $A_1 \coprod A_2$ and $B_1 \coprod B_2$ such that $g(A_1) \subset B_1$ and $g(A_2) = B_2$. This implies that $0 \longrightarrow A_1 \xrightarrow{g|A_1} B_1 \xrightarrow{f|B_1} C \longrightarrow 0$ is exact and $h\colon A_2 \longrightarrow B_2$, given by $h(a_2) = g(a_2)$ for all a_2 in A_2, is an isomorphism. So the exact sequence $0 \longrightarrow A \xrightarrow{g} B \xrightarrow{f} C \longrightarrow 0$ can be written as

$$
\begin{array}{ccccccc}
& A_2 & \xrightarrow{h} & B_2 & & & \\
& \coprod & & \coprod & & & \\
0 \longrightarrow & A_1 & \xrightarrow{g|A_1} & B_1 & \xrightarrow{f|B_1} & C & \longrightarrow 0\ .
\end{array}
$$

Moreover there is a commutative exact sequence

$$
\begin{array}{ccccccc}
0 \longrightarrow & A_1 & \xrightarrow{g|A_1} & B_1 & \xrightarrow{f|B_1} & C & \longrightarrow 0 \\
& \downarrow \omega & & \downarrow z & & \| & \\
0 \longrightarrow & \mathrm{Hom}_{\Gamma}(TrX,\ E) & \xrightarrow{g'} & B' & \xrightarrow{f'} & C &
\end{array}
$$

with the ω and the z isomorphisms. Hence the exact sequence $0 \longrightarrow A_1 \xrightarrow{g|A_1} B_1 \xrightarrow{f|B_1} C \longrightarrow 0$ has the following properties:

a) $f|B_1$ is X-determined.

b) $\mathrm{Im}(X,\ f|B_1) = \mathrm{Im}(X,\ f)$.

c) In any commutative diagram

$$
\begin{array}{ccccccc}
0 \longrightarrow & A_1 & \xrightarrow{g|A_1} & B_1 & \xrightarrow{f|B_1} & C & \longrightarrow 0 \\
& \downarrow & & \downarrow & & \| & \\
0 \longrightarrow & A_i & \xrightarrow{g|A_1} & B_1 & \xrightarrow{f|B_1} & C & \longrightarrow 0
\end{array}
$$

The vertical morphisms are isomorphisms.

It should be observed that c) is equivalent to the following statement. If $h: B_1 \longrightarrow B_1$ has the property $(f|B_1)h = f|B_1$, then h is an isomorphism. This discussion suggests introducing the following definitions (see [7, Section 2]).

A morphism $f: B \longrightarrow C$ is said to be <u>right minimal</u> if each endomorphism $h: B \longrightarrow B$ satisfying $fh = f$ is an isomorphism. A morphism $f: B \longrightarrow C$ which is both right minimal and right X-determined is said to be <u>minimal right X-determined</u>.

A morphism $g: A \longrightarrow B$ is said to be <u>left minimal</u> if each endomorphism $h: B \longrightarrow B$ satisfying $hg = g$ is an isomorphism. A morphism $g: A \longrightarrow B$ which is both left minimal and left X-determined is said to be <u>minimal left X-determined</u>.

Our previous discussion yields the following result in this new terminology.

<u>Proposition 2.13</u>: Let $0 \longrightarrow A \xrightarrow{g} B \xrightarrow{f} C \longrightarrow 0$ be an exact sequence in mod Λ.

a) Suppose f is right X-determined. Then there is an exact sequence $0 \longrightarrow A' \xrightarrow{g'} B' \xrightarrow{f'} C \longrightarrow 0$ in mod Λ with f' a minimal right X-determined morphism such that $\operatorname{Im}(X, f) = \operatorname{Im}(X, f')$. Further there is a commutative diagram

$$\begin{array}{ccccccccc}
0 & \longrightarrow & A' & \xrightarrow{g'} & B' & \xrightarrow{f'} & C & \longrightarrow & 0 \\
 & & \downarrow u & & \downarrow v & & \| & & \\
0 & \longrightarrow & A & \longrightarrow & B & \xrightarrow{f} & C & \longrightarrow & 0 \\
 & & \downarrow \omega & & \downarrow z & & \| & & \\
0 & \longrightarrow & A' & \longrightarrow & B' & \longrightarrow & C & \longrightarrow & 0
\end{array}$$

with ωu and zv isomorphisms.

b) Suppose g is left Y-determined. Then there is an exact sequence $0 \longrightarrow A \xrightarrow{g''} B'' \xrightarrow{f''} C'' \longrightarrow 0$ in mod Λ with g'' a minimal left Y-determined morphism such that $\mathrm{Im}(g'', Y) = \mathrm{Im}(g, Y)$. Further there is a commutative diagram

$$\begin{array}{ccccccccc}
0 & \longrightarrow & A & \xrightarrow{g''} & B'' & \xrightarrow{f''} & C'' & \longrightarrow & 0 \\
 & & \| & & \downarrow s & & \downarrow t & & \\
0 & \longrightarrow & A & \xrightarrow{g} & B & \xrightarrow{f} & C & \longrightarrow & 0 \\
 & & \| & & \downarrow q & & \downarrow r & & \\
0 & \longrightarrow & A & \xrightarrow{g''} & B'' & \xrightarrow{f''} & C'' & \longrightarrow & 0
\end{array}$$

such that qs and rt isomorphisms.

In view of our discussion so far, the following uniqueness result is of interest.

Proposition 2.14:

a) Let $f: B \longrightarrow C$ and $f': B' \longrightarrow C$ be two minimal right X-determined morphisms. Then the following are equivalent:

i) $\mathrm{Im}(X, f) = \mathrm{Im}(X, f)$.

ii) There is an $h: B \longrightarrow B'$ having the property that $f = f'h$, and any such h is an isomorphism.

b) Let $g: A \longrightarrow B$ and $g': A \longrightarrow B'$ be two minimal left Y-determined morphisms. Then the following are equivalent:

i) $\mathrm{Im}(g, Y) = \mathrm{Im}(g', Y)$.

ii) There are $h: B \longrightarrow B'$ such that $hg = g'$, and any such h is an isomorphism.

In view of these results, it is clear that right minimal epimorphisms and left minimal monomorphisms are the basic types of epimorphisms and monomorphisms. For this reason characterizations and properties of such morphisms are of importance. We now give some which have proven to be useful.

<u>Proposition 2.15</u>: Let $0 \longrightarrow A \xrightarrow{g} B \xrightarrow{f} C \longrightarrow 0$ be exact in mod Λ.

a) The following are equivalent:

i) f is right minimal.

ii) If B' is a summand of B such that $f|B' = 0$, then $B' = 0$.

iii) If $A' \neq 0$ and $t: A \longrightarrow A'$ is a splittable epimorphism, i.e. there is an $s: A' \longrightarrow A$ such that $ts = id_{A'}$, then the exact pushout sequence $0 \longrightarrow A' \longrightarrow B \times_A A' \longrightarrow C \longrightarrow 0$ does not split.

b) If f is right minimal, then f is right X-determined if and only if X has a summand isomorphic to [TrDA].

c) The following are equivalent:

i) g is left minimal.

ii) If $B' \subset B$ is a summand of B such that $\operatorname{Im} g \subset B'$, then $B' = B$.

iii) If $C' \subset C$ is a summand of C such that the pullback exact sequence $0 \longrightarrow A \longrightarrow B \times_C C' \longrightarrow C' \longrightarrow 0$ splits, then $C' = 0$.

d) If g is left minimal, then g is left Y-determined if and only if Y has a summand isomorphic to [DTrC].

e) If $0 \longrightarrow A \xrightarrow{g} B \xrightarrow{f} C \longrightarrow 0$ does not split, then

i) f is right minimal if A is indecomposable.

ii) g is left minimal if C is indecomposable.

These results can be readily applied to arbitrary morphisms and not just epimorphisms and monomorphisms once one observes the following.

<u>Lemma 2.16</u>:

a) A morphism $f: B \longrightarrow C$ is right minimal if and only if the induced epimorphism $B \longrightarrow \operatorname{Im} f$ is right minimal.

b) If a morphism $f: B \longrightarrow C$ is right X-determined, then the induced epimorphism $B \longrightarrow \operatorname{Im} f$ is right X-determined.

c) A morphism $g: A \longrightarrow B$ is left minimal if and only if the induced monomorphism $\operatorname{Im} g \longrightarrow B$ is left minimal.

d) If $g: A \longrightarrow B$ is left Y-determined, then the induced monomorphism $\operatorname{Im} g \longrightarrow B$ is left Y-determined.

As a consequence of this lemma and the previous results, we have the following.

<u>Proposition 2.17</u>: Let $0 \longrightarrow A \xrightarrow{g} B \xrightarrow{f} C$ be an exact sequence in mod Λ.

a) There exists an exact sequence $0 \longrightarrow A' \xrightarrow{g'} B' \xrightarrow{f'} C$ with f' right minimal such that there is a commutative diagram

$$\begin{array}{ccccccccc}
0 & \longrightarrow & A' & \xrightarrow{g'} & B' & \xrightarrow{f'} & C & \longrightarrow & 0 \\
 & & \downarrow u & & \downarrow v & & \| & & \\
0 & \longrightarrow & A & \xrightarrow{g} & B & \xrightarrow{f} & C & & \\
 & & \downarrow w & & \downarrow z & & \| & & \\
0 & \longrightarrow & A' & \xrightarrow{g'} & B' & \xrightarrow{f'} & C & &
\end{array}$$

such that ωu and zv are isomorphisms. Consequently f is right X-determined if and only if f' is right X-determined and moreover $\mathrm{Im}(X, f) = \mathrm{Im}(X, f')$.

b) If $0 \longrightarrow A'' \xrightarrow{g'} B'' \xrightarrow{f''} C$ is exact with f'' right minimal such that there exists morphisms $s\colon B'' \longrightarrow B$ and $t\colon B \longrightarrow B''$ such that $fs = f''$ and $f''t = f$, then there exists a commutative diagram

$$\begin{array}{ccccccc}
0 & \longrightarrow & A'' & \xrightarrow{g''} & B'' & \xrightarrow{f''} & C \\
 & & \downarrow & & \downarrow & & \| \\
0 & \longrightarrow & A' & \xrightarrow{g'} & B' & \xrightarrow{f'} & C
\end{array}$$

with the vertical morphisms isomorphisms.

Similarly, we have the following.

<u>Proposition 2.18</u>: Let $A \xrightarrow{g} B \xrightarrow{f} C \longrightarrow 0$ be exact in $\mathrm{mod}\,\Lambda$.

a) There exists an exact sequence $A \xrightarrow{g'} B' \xrightarrow{f'} C' \longrightarrow 0$ with g' left minimal such that there is a commutative diagram

$$\begin{array}{ccccccc}
A & \xrightarrow{g'} & B' & \xrightarrow{f'} & C' & \longrightarrow & 0 \\
\| & & \downarrow s & & \downarrow t & & \\
A & \xrightarrow{g} & B & \xrightarrow{f} & C & \longrightarrow & 0 \\
\| & & \downarrow q & & \downarrow r & & \\
A & \xrightarrow{g'} & B' & \xrightarrow{f'} & C' & \longrightarrow & 0
\end{array}$$

with qs and rt isomorphisms. Consequently g is left Y-determined if and only if g' is left Y-determined and moreover $\mathrm{Im}(g, Y) = \mathrm{Im}(g, Y')$.

b) If $A \xrightarrow{g''} B'' \xrightarrow{f''} C \longrightarrow 0$ is an exact sequence with g'' minimal such that there exist morphisms $u: B'' \longrightarrow B$ and $v: B \longrightarrow B''$ with $ug'' = g$ and $vg = g''$, then there exists a commutative diagram

$$\begin{array}{ccccccc} A & \xrightarrow{g''} & B'' & \xrightarrow{f''} & C'' & \longrightarrow & 0 \\ \| & & \downarrow & & \downarrow & & \\ A & \xrightarrow{g'} & B' & \xrightarrow{f'} & C' & \longrightarrow & 0 \end{array}$$

with the vertical morphisms isomorphisms.

So far we have been primarily concerned with $(\text{End } X)^{op}$-submodules of (X, C) containing $P(X, C)$. We can generalize some of our previous results in the following way.

Proposition 2.19: Let X and C be in mod Λ. For each $(\text{End } X)^{op}$-submodule H of (X, C) there is an exact sequence $0 \longrightarrow A \xrightarrow{g} B \xrightarrow{f} C$ in mod Λ satisfying the following.

a) f is a minimal right X-determined morphism with $\text{Im}(X, f) = H$.

b) A is in add(DTrX), the full additive subcategory consisting of summands of finite sums of copies of DTrX.

c) f is an epimorphism if and only if $H \supset P(X, C)$.

We also have the following analogue of Proposition 2.19.

Proposition 2.20: Let A and X be modules in mod Λ. For each End X-submodule H of (A, X) there is an exact sequence $A \xrightarrow{g} B \xrightarrow{f} C \longrightarrow 0$ in mod Λ satisfying the following.

a) g is a minimal left X-determined morphism with $\text{Im}(g, X) = H$.

b) C is in $\underline{\underline{\text{add}}}(\text{TrDX})$.

c) g is a monomorphism if and only if $H \supset I(A, X)$.

In order to state the final results of this section it is convenient to have the following definitions.

Two morphisms $f: B \longrightarrow C$ and $f': B' \longrightarrow C$ are said to be <u>isomorphic over</u> C if there is an isomorphism $h: B \longrightarrow B'$ such that $f'h = f$. Two morphisms $g: A \longrightarrow B$ and $g': A \longrightarrow B'$ are said to be <u>isomorphic over</u> A, if there is an isomorphism $h: B \longrightarrow B'$ such that $hg = g'$.

<u>Proposition 2.21</u>: Let X and C be in $\text{mod}\,\Lambda$ and let $0 \longrightarrow A_i \xrightarrow{g_i} B_i \xrightarrow{f_i} C$ be the family of exact sequences in $\text{mod}\,\Lambda$ such that each f is right minimal and each A_i is in $\underline{\underline{\text{add}}}(\text{DTrX})$. Then the map which sends each f_i to (X, f_i) induces a bijection between the over C isomorphism classes of the f_i and the $(\text{End}\,X)^{\text{op}}$-submodules of (X, C).

Similarly, let X and A be in $\text{mod}\,\Lambda$ and let $A \xrightarrow{g_i} B_i \xrightarrow{f_i} C_i \longrightarrow 0$ be the family of exact sequences in $\text{mod}\,\Lambda$ such that each g_i is left minimal and each C_i is in $\underline{\underline{\text{add}}}(\text{TrDX})$. Then the map which sends each g_i to (g_i, X) induces a bijection between the over A isomorphism classes of the g_i and the End X-submodules of (A, X).

<u>§3. Some Applications</u>.

As usual, we assume that R is a commutative artin ring, Λ is an artin R-algebra, and all Λ-modules are finitely generated. In this section we begin exploring some of the implications of the general theory outlined in Section 2.

An important question in representation theory is the following: When is an indecomposable module X isomorphic to a summand of a module M? The notion of morphisms determined by a module gives the following novel formulation of this problem.

Proposition 3 1: Let X be an indecomposable nonprojective Λ-module and let $A = DTrX$. Then the following statements are equivalent for a Λ-module M.

a) X is isomorphic to a summand of M.

b) If $0 \longrightarrow A \xrightarrow{g} B \xrightarrow{f} C \longrightarrow 0$ is a nonsplit exact sequence, then f is right M-determined.

c) There is a nonsplit exact sequence $0 \longrightarrow A \xrightarrow{g} B \xrightarrow{f} C \longrightarrow 0$ such that f is right M-determined.

Proof:

a) implies b). Since X is a nonprojective indecomposable module, we know that TrX is a nonprojective indecomposable Λ^{op}-module (see Proposition 1.3). Hence $A = DTrX$ is a noninjective indecomposable Λ-module. Let $0 \longrightarrow A \xrightarrow{g} B \xrightarrow{f} C \longrightarrow 0$ be a nonsplit exact sequence. Then f is right minimal since A is indecomposable (see Proposition 2.15). Therefore f is right M-determined if and only if $X = TrDA$ is isomorphic to a summand of M (see Proposition 2.15). Hence the assumption that X is isomorphic to a summand of M implies that f is right M-determined.

b) implies c). Since X not being projective implies that $A = DTrX$ is not injective, we know that there is a nonsplit extension $0 \longrightarrow A \xrightarrow{g} B \xrightarrow{f} C \longrightarrow 0$. The hypothesis of b) implies that for this extension f is right M-determined.

c) implies a). Suppose $0 \longrightarrow A \xrightarrow{g} B \xrightarrow{f} C \longrightarrow 0$ is a nonsplit exact sequence with f right M-determined. We have already seen in the proof of a) implies b) that under these circumstances $X = \mathrm{TrDA}$ is isomorphic to a summand of M.

As an immediate consequence of this result we obtain the following characterization of the TrDA for all noninjective indecomposable Λ-modules A.

Corollary 3.2: Let A be a noninjective indecomposable Λ-module. Then the following statements are equivalent for an indecomposable Λ-module X.

a) $X \approx \mathrm{TrDA}$.

b) If $0 \longrightarrow A \xrightarrow{g} B \xrightarrow{f} C \longrightarrow 0$ is a nonsplit exact sequence, then f is right X-determined.

c) There is a nonsplit exact sequence $0 \longrightarrow A \xrightarrow{g} B \xrightarrow{f} C \longrightarrow 0$ with f right X-determined.

Applying the duality $D\colon \mathrm{mod}\,\Lambda \longrightarrow \mathrm{mod}\,\Lambda^{op}$ to the above results we obtain the following results.

Proposition 3.3: Let Y be a noninjective indecomposable Λ-module and let $C = \mathrm{TrDY}$. Then the following statements are equivalent for a module N.

a) Y is isomorphic to a summand of N.

b) If $0 \longrightarrow A \xrightarrow{g} B \xrightarrow{f} C \longrightarrow 0$ is a nonsplit exact sequence. then g is left N-determined.

c) There is a nonsplit exact sequence $0 \longrightarrow A \xrightarrow{g} B \xrightarrow{f} C \longrightarrow 0$ with g left N-determined.

Corollary 3.4: Let C be a nonprojective indecomposable Λ-module. Then the following statements are equivalent for an indecomposable module Y.

a) $Y \approx DTrC$.

b) If $0 \longrightarrow A \xrightarrow{g} B \xrightarrow{f} C \longrightarrow 0$ is a nonsplit exact sequence, then g is left Y-determined.

c) There is a nonsplit exact sequence $0 \longrightarrow A \xrightarrow{g} B \xrightarrow{f} C \longrightarrow 0$ with g left Y-determined.

The following is another general question which is intimately related to our general theory of morphisms determined by modules. Given two exact sequences

$$0 \longrightarrow A' \xrightarrow{g'} B' \xrightarrow{f'} C \longrightarrow 0$$

$$0 \longrightarrow A \xrightarrow{g} B \xrightarrow{f} C \longrightarrow 0$$

in mod Λ, when does there exist a commutative diagram

$$\begin{array}{ccccccccc} 0 & \longrightarrow & A' & \xrightarrow{g'} & B' & \xrightarrow{f'} & C & \longrightarrow & 0 \\ & & \downarrow & & \downarrow & & \| & & \\ 0 & \longrightarrow & A & \xrightarrow{g} & B & \xrightarrow{f} & C & \longrightarrow & 0 \end{array} \quad ?$$

This question is answered in the following result.

Proposition 3.5: Let $0 \longrightarrow A \xrightarrow{g} B \xrightarrow{f} C \longrightarrow 0$ and $0 \longrightarrow A' \xrightarrow{g'} B' \xrightarrow{f'} C \longrightarrow 0$ be exact sequences in mod Λ. There exists a commutative diagram

$$(*) \qquad \begin{array}{ccccccccc} 0 & \longrightarrow & A' & \xrightarrow{g'} & B' & \xrightarrow{f'} & C & \longrightarrow & 0 \\ & & \downarrow & & \downarrow & & \| & & \\ 0 & \longrightarrow & A & \xrightarrow{g} & B & \xrightarrow{f} & C & \longrightarrow & 0 \end{array}$$

if and only if $\mathrm{Im}(\mathrm{TrDA}, f) \supset \mathrm{Im}(\mathrm{TrDA}, f')$.

Proof: Clearly $\mathrm{Im}(\mathrm{TrDA}, f) \supset \mathrm{Im}(\mathrm{TrDA}, f')$ if such a commutative diagram (*) exists.

Suppose now that $\mathrm{Im}(\mathrm{TrDA}, f) \supset \mathrm{Im}(\mathrm{TrDA}, f')$. The fact that f is right TrDA-determined (see Proposition 2.2) means that there is a morphism $g\colon B' \longrightarrow B$ such that $fg = f'$. This implies trivially that a commutative diagram of the type (*) exists.

The dual of this result is the following.

Proposition 3.6: Let $0 \longrightarrow A \xrightarrow{g} B \xrightarrow{f} C \longrightarrow 0$ and $0 \longrightarrow A \xrightarrow{g'} B' \xrightarrow{f'} C' \longrightarrow 0$ be exact sequences in $\mathrm{mod}\ \Lambda$. There exists a commutative diagram

$$\begin{array}{ccccccccc} 0 & \longrightarrow & A & \xrightarrow{g} & B & \xrightarrow{f} & C & \longrightarrow & 0 \\ & & \| & & \downarrow & & \downarrow & & \\ 0 & \longrightarrow & A & \xrightarrow{g'} & B' & \xrightarrow{f'} & C' & \longrightarrow & 0 \end{array}$$

if and only if $\mathrm{Im}(\mathrm{DTrC}, g) \supset \mathrm{Im}(\mathrm{DTrC}, g')$.

A question that is often of interest is the following. Let $0 \longrightarrow A' \xrightarrow{f} A \xrightarrow{g} A'' \longrightarrow 0$ be exact in $\mathrm{mod}\ \Lambda$ and x an element of $\mathrm{Ext}^1_\Lambda(C, A)$. When is there an element x' in $\mathrm{Ext}^1_\Lambda(C, A')$ such that $\mathrm{Ext}^1_\Lambda(C, f)(x') = x$? While we cannot answer this question in this degree of generality, our theory gives an answer in case $A' = \mathrm{Hom}_\Lambda(\Lambda/\underline{a}, A)$ for some two sided ideal $\underline{a}$ of Λ as we now demonstrate. To do this we need the following lemma.

Lemma 3.7: Let $\underline{a}$ be a two sided ideal in Λ and A a $\Lambda/\underline{a}$-module. Let

C be an arbitrary Λ-module and $0 \longrightarrow \Omega C \longrightarrow P \longrightarrow C \longrightarrow 0$ an exact sequence with $P \longrightarrow C$ a projective cover. If $0 \longrightarrow A \xrightarrow{g} B \xrightarrow{f} C \longrightarrow 0$ is exact in mod Λ, then there is a commutative exact diagram

$$\begin{array}{ccccccccc} 0 & \longrightarrow & \Omega C/\underline{a}\Omega C & \longrightarrow & P/\underline{a}\Omega C & \longrightarrow & C & \longrightarrow & 0 \\ & & \downarrow & & \downarrow & & \| & & \\ 0 & \longrightarrow & A & \longrightarrow & B & \longrightarrow & C & \longrightarrow & 0 \end{array}$$

Proof: Since P is projective we know there is an exact commutative diagram

$$\begin{array}{ccccccccc} 0 & \longrightarrow & \Omega C & \longrightarrow & P & \longrightarrow & C & \longrightarrow & 0 \\ & & \downarrow u & & \downarrow v & & \| & & \\ 0 & \longrightarrow & A & \longrightarrow & B & \longrightarrow & C & \longrightarrow & 0 . \end{array}$$

Since $\underline{a}A = 0$, we have that $\underline{a}\Omega C \subset \text{Ker } u$ and hence that $\underline{a}\Omega C$ is also contained in Ker v. The result now follows trivially.

Proposition 3.8: Suppose $\underline{a}$ is a two sided ideal in Λ. Let $0 \longrightarrow \Omega C \longrightarrow P \longrightarrow C \longrightarrow 0$ be an exact sequence in mod Λ with $P \longrightarrow C$ a projective cover and let $0 \longrightarrow \Omega C/\underline{a}\Omega C \longrightarrow P/\underline{a}\Omega C \xrightarrow{t} C \rightarrow 0$ be the induced exact sequence. Then the following statements are equivalent for an exact sequence $x: 0 \longrightarrow A \xrightarrow{g} B \xrightarrow{f} C \longrightarrow 0$ in mod Λ.

a) $\text{Im}(\text{TrDA}, t) \subset \text{Im}(\text{TrDA}, f)$.

b) There is a commutative exact diagram

$$\begin{array}{ccccccccc} 0 & \longrightarrow & A' & \longrightarrow & B' & \longrightarrow & C & \longrightarrow & 0 \\ & & \downarrow & & \downarrow & & \downarrow & & \\ 0 & \longrightarrow & A & \xrightarrow{g} & B & \xrightarrow{f} & C & \longrightarrow & 0 \end{array}$$

with A' in mod $\Lambda/\underline{a}$.

c) If we let $0 \longrightarrow A' \xrightarrow{h} A$ be the inclusion of the submodule $A' = \mathrm{Hom}_\Lambda(\Lambda/\underline{a}, A)$ of A, then there is an x' in $\mathrm{Ext}^1_\Lambda(C, A')$ such that $\mathrm{Ext}^1_\Lambda(C, h)(x') = x$.

Proof:

a) implies b). Trivial consequence of the fact that f is right TrDA-determined.

b) implies c). Trivial.

c) implies a). Follows easily from Lemma 3.7.

For the convenience of the reader we state the dual of Proposition 3.8.

Proposition 3.9: Suppose $\underline{a}$ is a two sided ideal in Λ. Let $0 \longrightarrow A \xrightarrow{u} I \xrightarrow{v} \Omega^{-1}(A) \longrightarrow 0$ be exact in mod Λ with u an injective envelope and let $0 \longrightarrow A \xrightarrow{t} u^{-1}(L) \longrightarrow L \longrightarrow 0$ be the induced exact sequence where $L = \mathrm{Hom}_\Lambda(\Lambda/\underline{a}, \Omega^{-1}A)$. Then the following statements are equivalent for an exact sequence $x: 0 \longrightarrow A \xrightarrow{f} B \xrightarrow{g} C \longrightarrow 0$ in mod Λ.

a) $\mathrm{Im}(t, \mathrm{DTr}C) \subset \mathrm{Im}(f, \mathrm{DTr}C)$.

b) There is a commutative exact diagram

$$\begin{array}{ccccccccc} 0 & \longrightarrow & A & \xrightarrow{f} & B & \xrightarrow{g} & C & \longrightarrow & 0 \\ & & \| & & \downarrow & & \downarrow & & \\ 0 & \longrightarrow & A & \longrightarrow & B'' & \longrightarrow & C'' & \longrightarrow & 0 \end{array}$$

with C'' in mod $\Lambda/\underline{a}$.

c) If $h: C \longrightarrow C/\underline{a}C$ is the canonical epimorphism, then there is an x' in $\text{Ext}^1_\Lambda(C/\underline{a}C, A)$ such that $\text{Ext}^1_\Lambda(h, A)(x') = x$.

Using similar arguments to those used to prove Proposition 3.8, we establish the following.

Proposition 3.10: Let $\underline{a}$ be a two sided ideal in Λ. Let C be a $\Lambda/\underline{a}$-module and $P \xrightarrow{t} C \longrightarrow 0$ a projective cover of C in mod $\Lambda/\underline{a}$. Then the following statements are equivalent for an exact sequence $0 \longrightarrow A \xrightarrow{g} B \xrightarrow{f} C \longrightarrow 0$ in mod Λ.

a) $\text{Im}(\text{TrD}A, f) \supset \text{Im}(\text{TrD}A, t)$.

b) There is a commutative exact diagram

$$\begin{array}{ccccccccc} 0 & \longrightarrow & A' & \longrightarrow & B' & \longrightarrow & C & \longrightarrow & 0 \\ & & \downarrow & & \downarrow & & \| & & \\ 0 & \longrightarrow & A & \longrightarrow & B & \longrightarrow & C & \longrightarrow & 0 \end{array}$$

with A', B', C in mod $\Lambda/\underline{a}$.

c) The sequence of $\Lambda/\underline{a}$-modules

$$0 \longrightarrow (\Lambda/\underline{a}, A) \longrightarrow (\Lambda/\underline{a}, B) \longrightarrow (\Lambda/\underline{a}, C) \longrightarrow 0$$

is exact.

We now state the dual of Proposition 3.10.

Proposition 3.11: Let $\underline{a}$ be a two sided ideal in Λ. Let A be a $\Lambda/\underline{a}$-module and let $0 \longrightarrow A \xrightarrow{t} I$ be an injective envelope of A in mod $\Lambda/\underline{a}$. Then the following statements are equivalent for an exact sequence $0 \longrightarrow A \xrightarrow{f} B \xrightarrow{g} C \longrightarrow 0$ in mod Λ.

a) $\mathrm{Im}(f, \mathrm{DTr}C) \supset \mathrm{Im}(t, \mathrm{DTr}C)$.

b) There is an exact commutative diagram

$$\begin{array}{ccccccccc} 0 & \longrightarrow & A & \longrightarrow & B & \longrightarrow & C & \longrightarrow & 0 \\ & & \| & & \downarrow & & \downarrow & & \\ 0 & \longrightarrow & A & \longrightarrow & B' & \longrightarrow & C' & \longrightarrow & 0 \end{array}$$

with the A, B', C' in $\mathrm{mod}\ \Lambda/\underline{a}$.

c) The sequence of $\Lambda/\underline{a}$-modules

$$0 \longrightarrow \Lambda/\underline{a} \otimes_{\Lambda} A \longrightarrow \Lambda/\underline{a} \otimes_{\Lambda} B \longrightarrow \Lambda/\underline{a} \otimes_{\Lambda} C \longrightarrow 0$$

is exact.

We now turn our attention to the following problem which is obviously related but somewhat different from the problems discussed above. Suppose X is in $\mathrm{mod}\ \Lambda$ and we are given an element x in $\mathrm{Ext}^1_{\Lambda}(C, A)$ with C and A in $\mathrm{mod}\ \Lambda$. When does there exist a morphism $h\colon X' \longrightarrow A$ with X' in $\underline{\underline{\mathrm{add}}}(X)$ such that x is contained in the image of $\mathrm{Ext}^1_{\Lambda}(C, h)\colon \mathrm{Ext}^1_{\Lambda}(C, X') \longrightarrow \mathrm{Ext}^1_{\Lambda}(C, A)$? The answer to this question is based on the following.

<u>Proposition 3.12</u>: The following are equivalent for an exact sequence $0 \longrightarrow A \xrightarrow{g} B \xrightarrow{f} C \longrightarrow 0$.

a) $\mathrm{Im}(\mathrm{TrD}A, f) = P(\mathrm{TrD}A, C)$.

b) If $0 \longrightarrow A' \longrightarrow B' \longrightarrow C \longrightarrow 0$ is exact with A' in $\underline{\underline{\mathrm{add}}}(A)$, then there is an exact commutative diagram

$$\begin{array}{ccccccccc} 0 & \longrightarrow & A & \longrightarrow & B & \longrightarrow & C & \longrightarrow & 0 \\ & & \downarrow & & \downarrow & & \| & & \\ 0 & \longrightarrow & A' & \longrightarrow & B' & \longrightarrow & C & \longrightarrow & 0 \end{array}.$$

c) Let $0 \longrightarrow X \longrightarrow E \longrightarrow C \longrightarrow 0$ be exact. Then there is an exact commutative diagram

$$\begin{array}{ccccccccc} 0 & \longrightarrow & A' & \longrightarrow & B' & \longrightarrow & C & \longrightarrow & 0 \\ & & \downarrow & & \downarrow & & \| & & \\ 0 & \longrightarrow & X & \longrightarrow & E & \longrightarrow & C & \longrightarrow & 0 \end{array}$$

with A' in $\underline{\underline{\text{add}}}\ A$ if and only if there is an exact commutative diagram

$$\begin{array}{ccccccccc} 0 & \longrightarrow & A & \longrightarrow & B & \longrightarrow & C & \longrightarrow & 0 \\ & & \downarrow & & \downarrow & & \downarrow & & \\ 0 & \longrightarrow & X & \longrightarrow & E & \longrightarrow & C & \longrightarrow & 0 . \end{array}$$

Proof:

a) implies b). Let $0 \longrightarrow A' \xrightarrow{g'} B' \xrightarrow{f'} C \longrightarrow 0$ be exact with A' in $\underline{\underline{\text{add}}}\ A$. Then f' is right TrDA'-determined. Since [TrDA'] is a summand of [TrDA], we have that f' is right TrDA-determined. Since f' is an epimorphism we know that $\text{Im}(\text{TrDA}, f') \supset P(\text{TrDA}, C) = \text{Im}(\text{TrDA}, f)$. Therefore the fact that f' is right TrDA-determined implies that there is an exact commutative diagram

$$\begin{array}{ccccccccc} 0 & \longrightarrow & A & \xrightarrow{g} & B & \xrightarrow{f} & C & \longrightarrow & 0 \\ & & \downarrow & & \downarrow & & \| & & \\ 0 & \longrightarrow & A' & \xrightarrow{g'} & B' & \xrightarrow{f'} & C & \longrightarrow & 0 \end{array}$$

b) implies a). We know by our general existence theorem that there is an exact sequence $0 \longrightarrow A' \xrightarrow{g'} B' \xrightarrow{f'} C \longrightarrow 0$ with A' in $\underline{\underline{\text{add}}}\ A$ such that $\text{Im}(\text{TrDA}, f') = P(\text{TrDA}, C)$. Since by the hypothesis of b) there is an exact commutative diagram

$$\begin{array}{ccccccccc} 0 & \longrightarrow & A & \xrightarrow{g} & B & \xrightarrow{f} & C & \longrightarrow & 0 \\ & & \downarrow & & \downarrow & & \| & & \\ 0 & \longrightarrow & A' & \xrightarrow{g'} & B' & \xrightarrow{f'} & C & \longrightarrow & 0 \end{array}$$

we have that $\operatorname{Im}(\mathrm{TrDA}, f) \subset \operatorname{Im}(\mathrm{TrDA}, f') = P(\mathrm{TrDA}, C)$. Hence $\operatorname{Im}(\mathrm{TrDA}, f) = P(\mathrm{TrDA}, C)$, since the fact that f is an epimorphism implies that $\operatorname{Im}(\mathrm{TrDA}, f) \supset P(\mathrm{TrDA}, C)$. Therefore b) implies a).

a) equivalent to c). This is a straightforward consequence of the fact that a) is equivalent to b).

As an easy consequence of this proposition, we have the following result which answers the question posed before Proposition 3.12.

Proposition 3.13: Let $0 \longrightarrow A \xrightarrow{g} B \xrightarrow{f} C \longrightarrow 0$ be exact with f a minimal right TrDA-determined morphism such that $\operatorname{Im}(\mathrm{TrDA}, f) = P(\mathrm{TrDA}, C)$. Then the following are equivalent for an exact sequence $x\colon 0 \longrightarrow X \xrightarrow{j} E \xrightarrow{h} C \longrightarrow 0$ in $\operatorname{Ext}^1_\Lambda(C, X)$.

a) $\operatorname{Im}(\mathrm{TrDX}, f) \subset \operatorname{Im}(\mathrm{TrDX}, h)$.

b) There is a morphism $t\colon A' \longrightarrow X$ with A' in $\underline{\underline{\mathrm{add}}}\, A$ such that x is in $\operatorname{Im} \operatorname{Ext}^1_\Lambda(C, t)$.

We now state the duals of these results. We begin with the dual of Proposition 3.12.

Proposition 3.14: The following statements are equivalent for an exact sequence $0 \longrightarrow A \xrightarrow{g} B \xrightarrow{f} C \longrightarrow 0$.

a) $\operatorname{Im}(g, \mathrm{DTrC}) = I(A, \mathrm{DTrC})$.

b) If $0 \longrightarrow A \xrightarrow{g'} B' \xrightarrow{f'} C' \longrightarrow 0$ is exact with C' in add C, then there is an exact commutative diagram

$$\begin{array}{ccccccccc} 0 & \longrightarrow & A & \xrightarrow{g'} & B' & \xrightarrow{f'} & C' & \longrightarrow & 0 \\ & & \| & & \downarrow & & \downarrow & & \\ 0 & \longrightarrow & A & \longrightarrow & B & \longrightarrow & C & \longrightarrow & 0 \end{array}$$

c) Let $0 \longrightarrow A \longrightarrow E \longrightarrow Y \longrightarrow 0$ be exact. Then there is an exact commutative diagram

$$\begin{array}{ccccccccc} 0 & \longrightarrow & A & \longrightarrow & E & \longrightarrow & Y & \longrightarrow & 0 \\ & & \| & & \downarrow & & \downarrow & & \\ 0 & \longrightarrow & A & \longrightarrow & B' & \longrightarrow & C' & \longrightarrow & 0 \end{array}$$

with C' in add C if and only if there is an exact commutative diagram

$$\begin{array}{ccccccccc} 0 & \longrightarrow & A & \longrightarrow & E & \longrightarrow & Y & \longrightarrow & 0 \\ & & \| & & \downarrow & & \downarrow & & \\ 0 & \longrightarrow & A & \xrightarrow{g} & B & \xrightarrow{f} & C & \longrightarrow & 0 . \end{array}$$

The following is the dual of Proposition 3.13.

Proposition 3.15: Let $0 \longrightarrow A \xrightarrow{g} B \longrightarrow C \longrightarrow 0$ be exact with g a minimal left DTrC-determined morphism such that $\operatorname{Im}(g, \mathrm{DTr}C) = I(A, \mathrm{DTr}C)$. Then the following statements are equivalent for an exact sequence $y\colon 0 \longrightarrow A \xrightarrow{h} E \xrightarrow{j} Y \longrightarrow 0$ in $\operatorname{Ext}^1_\Lambda(Y, A)$.

a) $\operatorname{Im}(h, \mathrm{DTr}Y) \supset \operatorname{Im}(g, \mathrm{DTr}Y)$;

b) There is a morphism $t\colon Y \longrightarrow C'$ with C' in add C such that y is in $\operatorname{Im} \operatorname{Ext}^1(t, A)$.

As an application of Proposition 3.12 and 3.14, we have the following result.

Proposition 3.16: Suppose A and C are in mod Λ.

a) If $P(\mathrm{TrD}A, C) = 0$, then for each submodule C' of C, the induced morphism $\mathrm{Ext}^1_\Lambda(C, A) \longrightarrow \mathrm{Ext}^1_\Lambda(C', A)$ is an epimorphism.

b) If $I(A, \mathrm{DTr}C) = 0$, then for each epimorphism $A \longrightarrow A''$, the induced morphism $\mathrm{Ext}^1_\Lambda(C, A) \longrightarrow \mathrm{Ext}^1_\Lambda(C, A'')$ is an epimorphism.

Proof:

a) Since $P(\mathrm{TrD}A, C) = 0$, we know by our general existence theorem that there is an exact sequence $x\colon 0 \longrightarrow A' \xrightarrow{g} B \xrightarrow{f} C \longrightarrow 0$ with A' in add(A) such that f is right TrDA-determined and $\mathrm{Im}(\mathrm{TrD}A, f) = 0$. Suppose C' is a submodule of C. Then it is easily checked that the exact sequence $y\colon 0 \longrightarrow A' \longrightarrow f^{-1}(C') \xrightarrow{h} C' \longrightarrow 0$ has the property that $\mathrm{Im}(\mathrm{TrD}A, h) = 0$. Since A' is in add A, it follows that h is right TrDA-determined. Moreover if $u\colon C' \longrightarrow C$ is the inclusion we have that $\mathrm{Ext}^1_\Lambda(u, A')(X) = y$.

Suppose now that z is an element of $\mathrm{Ext}^1_\Lambda(C', A')$. Then by Proposition 3.13 we know there is a morphism $t\colon A' \longrightarrow A$ such that the morphism $\mathrm{Ext}^1_\Lambda(C', t)\colon \mathrm{Ext}^1_\Lambda(C', A') \longrightarrow \mathrm{Ext}^1_\Lambda(C', A)$ has the property $\mathrm{Ext}^1_\Lambda(C', t)(y) = z$. From the commutative diagram

$$\begin{array}{ccc}
\mathrm{Ext}^1_\Lambda(C, A') & \xrightarrow{\mathrm{Ext}^1_\Lambda(C,\, t)} & \mathrm{Ext}^1_\Lambda(C, A) \\
{\scriptstyle \mathrm{Ext}^1_\Lambda(u,\, A')}\downarrow & & \downarrow{\scriptstyle \mathrm{Ext}^1_\Lambda(u,\, A)} \\
\mathrm{Ext}^1_\Lambda(C', A') & \xrightarrow[\mathrm{Ext}^1_\Lambda(C',\, t)]{} & \mathrm{Ext}^1_\Lambda(C', A)
\end{array}$$

and the fact that $\mathrm{Ext}^1_\Lambda(u, A')(x) = y$, it follows that $\mathrm{Ext}^1_\Lambda(u, A)(\mathrm{Ext}^1_\Lambda(C, t)(x)) = z$. Thus z is in $\mathrm{Im}\,\mathrm{Ext}^1_\Lambda(u, A)$.

Since this is true for all z in $\mathrm{Ext}_\Lambda^1(C', A)$, it follows that the morphism $\mathrm{Ext}_\Lambda^1(u, A)\colon \mathrm{Ext}_\Lambda^1(C, A) \longrightarrow \mathrm{Ext}_\Lambda^1(C', A)$ is an epimorphism, which finishes the proof of a).

b) Proven similarly to a).

Before going on to the last topic in this section, we point out the following easily verified fact. Let A be an indecomposable Λ-module and suppose $\mathrm{Ext}_\Lambda^1(C, A) \neq 0$. Suppose $x_1, \ldots, x_t$ in $\mathrm{Ext}_\Lambda^1(C, A)$ are a minimal set of generators for $\mathrm{Ext}_\Lambda^1(C, A)$ over the local ring End A. Let $x\colon 0 \longrightarrow A^t \xrightarrow{g} B \xrightarrow{f} C \longrightarrow 0$ represent the element $(x_1, \ldots, x_t)$ in $\mathrm{Ext}_\Lambda^1(C, A)^t = \mathrm{Ext}_\Lambda^1(C, A^t)$. Then f is a right minimal TrDA-determined morphism with $\mathrm{Im}(\mathrm{TrDA}, f) = P(\mathrm{TrDA}, C)$. The obvious dual of this also holds.

The rest of this section is devoted to generalizing Proposition 3.16 by giving a description of the kernel and cokernel of a morphism $\mathrm{Ext}_\Lambda^1(C, A) \longrightarrow \mathrm{Ext}_\Lambda^1(C', A)$ induced by a morphism $f\colon C' \longrightarrow C$. This description is based on the isomorphism

$$(*) \qquad \mathrm{Ext}_\Lambda^1(C, A) \cong \mathrm{Hom}_R(\underline{\mathrm{Hom}}_\Lambda(\mathrm{TrDA}, C), I)$$

which is functorial in C ([2, Chapter I, Proposition 8.7]).

Suppose $f\colon C' \longrightarrow C$ is a morphism. Then f induces a morphism of $(\mathrm{End}\ \mathrm{TrDA})^{op}$-modules $\underline{\mathrm{Hom}}(\mathrm{TrDA}, f)\colon \underline{\mathrm{Hom}}(\mathrm{TrDA}, C') \longrightarrow \underline{\mathrm{Hom}}(\mathrm{TrDA}, C)$. Letting $K = \mathrm{Ker}\ \underline{\mathrm{Hom}}(\mathrm{TrDA}, f)$ and $L = \mathrm{Coker}\ \underline{\mathrm{Hom}}(\mathrm{TrDA}, f)$ we obtain from isomorphism (*) the exact sequence

$$0 \longrightarrow \mathrm{Hom}_R(L, I) \longrightarrow \mathrm{Ext}_\Lambda^1(C, A) \longrightarrow \mathrm{Ext}_\Lambda^1(C', A) \longrightarrow \mathrm{Hom}_R(K, I) \longrightarrow 0.$$

We now give an explicit description of the $(\mathrm{End}\ \mathrm{TrDA})^{op}$-modules K and L. The morphism $f\colon C' \longrightarrow C$ induces the morphism $(\mathrm{TrDA}, f)\colon (\mathrm{TrDA}, C') \longrightarrow (\mathrm{TrDA}, C)$. It then follows that

$K = (\mathrm{TrDA}, f)^{-1}(P(\mathrm{TrDA}, C))/P(\mathrm{TrDA}, C')$ while $L = (\mathrm{TrDA}, C)/(\mathrm{Im}(\mathrm{TrDA}, f), P(\mathrm{TrDA}, C))$. Summarizing this discussion we have:

<u>Proposition 3.17</u>: Let $f: C' \longrightarrow C$ be a morphism in mod Λ and let A be in mod Λ. Then f induces the morphism $h = (\mathrm{TrDA}, f): (\mathrm{TrDA}, C') \longrightarrow (\mathrm{TrDA}, C)$. Set $K = h^{-1}(P(\mathrm{TrDA}, C'))/P(\mathrm{TrDA}, C')$ and $L = (\mathrm{TrDA}, C)/(\mathrm{Im}\ h, P(\mathrm{TrDA}, C))$. Then

a) There is an exact sequence

$$0 \to \mathrm{Hom}_R(L, I) \to \mathrm{Ext}^1_\Lambda(C, A) \xrightarrow{\mathrm{Ext}^1_\Lambda(f, A)} \mathrm{Ext}^1_\Lambda(C', A) \to \mathrm{Hom}_R(K, I) \longrightarrow 0 .$$

b) $\mathrm{Ext}^1_\Lambda(f, A): \mathrm{Ext}^1_\Lambda(C, A) \longrightarrow \mathrm{Ext}^1_\Lambda(C', A)$ is an epimorphism if and only if a morphism $u: \mathrm{TrDA} \longrightarrow C'$ is in $P(\mathrm{TrDA}, C')$ whenever $fu: \mathrm{TrDA} \longrightarrow C$ is in $P(\mathrm{TrDA}, C)$.

c) $\mathrm{Ext}^1_\Lambda(f, A): \mathrm{Ext}^1_\Lambda(C, A) \longrightarrow \mathrm{Ext}^1_\Lambda(C', A)$ is a monomorphism if and only if given any $v: \mathrm{TrDA} \longrightarrow C$ there is a morphism $u: \mathrm{TrDA} \longrightarrow C'$ such that $v = fu + s$ with s in $P(\mathrm{TrDA}, C)$.

As an immediate consequence of this result we have the following.

<u>Corollary 3.18</u>: Suppose A and C are in mod Λ with $P(\mathrm{TrDA}, C) = 0$. Let $f: C' \longrightarrow C$ be a monomorphism. Then the morphism $\mathrm{Ext}^1_\Lambda(f, A): \mathrm{Ext}^1_\Lambda(C, A) \longrightarrow \mathrm{Ext}^1_\Lambda(C', A)$ has the following properties.

a) $\mathrm{Ext}^1_\Lambda(f, A)$ is an epimorphism.

b) $\mathrm{Ext}^1_\Lambda(f, A)$ is an isomorphism if and only if every morphism $\mathrm{TrDA} \longrightarrow C$ has its image in C'.

Similarly (or by duality) one can obtain a description of the kernel and cokernel of a morphism $\mathrm{Ext}^1_\Lambda(C, A) \longrightarrow \mathrm{Ext}^1_\Lambda(C, A')$ induced by a morphism $f: A \longrightarrow A'$. This description is based on the isomorphism

$$\mathrm{Ext}^1_\Lambda(C, A) \longrightarrow \mathrm{Hom}_R(\overline{\mathrm{Hom}}_\Lambda(A, D\mathrm{Tr}C), I)$$

which is functorial in A ([2, Chapter I, Proposition 8.17]) and proceeds exactly as above, an exercise left to the interested reader to carry out.

§4. Almost Split Sequences.

This section is devoted to giving some applications of the theory of morphisms determined by objects to the minimal left and right almost split morphisms and almost split sequences. We start by pointing out some of the elementary connections between these various notions.

Suppose that C is an indecomposable module. Then End C is a local ring and has a unique maximal ideal $\underline{m}$. We know that there is a right minimal C-determined morphism $f: B \longrightarrow C$ such that $\mathrm{Im}(C, f) = \underline{m}$. From this it follows that f has the following properties:

a) f is not a splittable epimorphism since $\mathrm{Im}(C, f) \neq (C, C)$.

b) If $h: B \longrightarrow B$ is such that $fh = f$, then h is an isomorphism

c) If $t: L \longrightarrow C$ is not a splittable epimorphism, then there is an $h: L \longrightarrow B$ such that $fh = t$. For $g: L \longrightarrow C$ is not a splittable epimorphism precisely when $\mathrm{Im}(C, t) \subset \underline{m}$ and if $\mathrm{Im}(C, t) \subset \underline{m} = \mathrm{Im}(C, f)$, then there is an $h: L \longrightarrow B$ such that $fh = t$ since f is right C-determined.

We recall (see [7, Section 2]) that a morphism $f: B \longrightarrow C$ is called a minimal right almost split morphism if it satisfies conditions a), b), and c). On the other hand it is not difficult to show that if a morphism $f: B \longrightarrow C$ is minimal right almost split, i.e. satisfies a), b), c), then C is indecomposable and f is a minimal right C-determined morphism with $\mathrm{Im}(C, f)$ the maximal ideal of End C (see [7, Section 2]).

Similarly, suppose A is an indecomposable module and $\underline{m}$ the maximal ideal of the local ring End A. Then there is a morphism $g: A \longrightarrow B$ which is a minimal left A-determined morphism with $\mathrm{Im}(g, A) = \underline{m}$. From this it follows that g has the following properties.

a') g is not a splittable monomorphism.

b') If $h: B \longrightarrow B$ is a morphism such that $g = hg$, then h is an isomorphism.

c') If $h: A \longrightarrow L$ is not a splittable monomorphism, then there is a morphism $s: B \longrightarrow L$ such that $sg = h$.

We recall that (see [7, Section 2]) that a morphism $g: A \longrightarrow B$ is said to be minimal left almost split if it satisfies a'), b'), and c'). On the other hand it is not difficult to see that if $g: A \longrightarrow B$ is minimal left almost split, than A is indecomposable and g is minimal left A-determined with $\mathrm{Im}(g, A)$ the maximal ideal of End A (see [7, Section 2]).

Suppose C is indecomposable and $f: B \longrightarrow C$ is a minimal right C-determined morphism with $\mathrm{Im}(C, f)$ the maximal ideal of End C. If C is projective, then $P(C, C) = \mathrm{End}\, C$ and so $\mathrm{Im}(C, f)$ does not contain $P(C, C)$. Hence $f: B \longrightarrow C$ is not an epimorphism. In fact it is easily seen that f is isomorphic over C to the inclusion morphism $\underline{r}C \longrightarrow C$, where $\underline{r}$ is the radical of Λ. If C is not projective, the $P(C, C) \subset \mathrm{Im}(C, f)$. Hence letting $\Gamma = \mathrm{End}\, \mathrm{Tr}C$ and E be an injective envelope for the Γ-module $(C, C)/\mathrm{Im}(C, f)$ we have that there is an exact sequence $0 \longrightarrow \mathrm{Hom}_\Gamma(\mathrm{Tr}C, E) \xrightarrow{g} B' \xrightarrow{f'} C \longrightarrow 0$ (see Proposition 2.9) where f' is a minimal right C-determined morphism with $\mathrm{Im}(C, f') = \mathrm{Im}(C, f)$. Since $(C, C)/\mathrm{Im}(C, f)$ is a simple $(\mathrm{End}\, C)^{op}$-module, it is a simple Γ-module and so $E = \mathrm{Hom}_R(\Gamma, I)$. Thus

$\mathrm{Hom}_\Gamma(\mathrm{TrC}, E) \approx \mathrm{Hom}_\Gamma(\mathrm{TrC}, \mathrm{Hom}_R(\Gamma, I)) \approx \mathrm{Hom}_R(\mathrm{TrC}, I) = \mathrm{DTrC}$. Hence $\mathrm{Ker}\, f' \approx \mathrm{DTrC}$. But $\mathrm{Ker} f \approx \mathrm{Ker}\, f'$ since $f: B \longrightarrow C$ and $f': B' \longrightarrow C$ are isomorphic over C because they are both minimal right C-determined morphisms with $\mathrm{Im}(C, f') = \mathrm{Im}(C, f)$.

So we have the exact sequence $0 \longrightarrow \mathrm{DTrC} \xrightarrow{g} B \xrightarrow{f} C \longrightarrow 0$ with f minimal right C-determined and $\mathrm{Im}(C, f)$ the maximal ideal of End C. Since C is a nonprojective indecomposable module, we know that DTrC is indecomposable. Because f is not a splittable epimorphism, we know that $g: \mathrm{DTrC} \longrightarrow B$ is not a splittable monomorphism. Thus $\mathrm{Im}(g, \mathrm{DTrC})$ is contained in the maximal ideal $\underline{m}$ of End DTrC. Suppose t is in $\underline{m}$, i.e. t is an endomorphism of DTrC which is not an isomorphism. Then we have the commutative diagram

$$\begin{array}{ccccccccc} 0 & \longrightarrow & \mathrm{DTrC} & \xrightarrow{g} & B & \xrightarrow{f} & C & \longrightarrow & 0 \\ & & \downarrow t & & \downarrow u & & \| & & \\ 0 & \longrightarrow & \mathrm{DTrC} & \xrightarrow{g'} & B' & \xrightarrow{f'} & C & \longrightarrow & 0 . \end{array}$$

Then f' is right determined by $C = \mathrm{TrD}(\mathrm{DTrC})$. Also $\mathrm{Im}(C, f) \subset \mathrm{Im}(C, f')$. If $\mathrm{Im}(C, f) = \mathrm{Im}(C, f')$, then f' is not a splittable epimorphism and f' is right minimal since DTrC is indecomposable (see Proposition 2.16). But this implies that u and hence t is an isomorphism. This contradicts the fact that t is in $\underline{m}$. Therefore $\mathrm{Im}(C, f')$ contains $\mathrm{Im}(C, f)$ properly and thus $\mathrm{Im}(C, f') = \mathrm{End}(C)$ since $\mathrm{Im}(C, f)$ is the maximal ideal of End C. Hence f' is a splittable epimorphism which implies there is an $h: B \longrightarrow \mathrm{DTrC}$ such that $t = hg$. Therefore $\mathrm{Im}(g, \mathrm{DTrC})$ is the maximal ideal of End(DTrC).

Thus the exact sequence $0 \longrightarrow \mathrm{DTrC} \xrightarrow{g} B \xrightarrow{f} C \longrightarrow 0$ has the following properties.

a) $\mathrm{Im}(g, \mathrm{DTrC})$ is the maximal ideal of End DTrC.

b) g is left minimal since C is indecomposable and $0 \longrightarrow \mathrm{DTrC} \xrightarrow{g} B \xrightarrow{f} C \longrightarrow 0$ does not split.

c) g is left DTrC-determined.

In other words the exact sequence $0 \longrightarrow \mathrm{DTrC} \xrightarrow{g} B \xrightarrow{f} C \longrightarrow 0$ has the property that g is minimal left almost split and f is minimal right almost split. We recall that a sequence $0 \longrightarrow A \xrightarrow{g} B \xrightarrow{f} C \longrightarrow 0$ is said to be almost split if g is minimal left almost split and f is minimal right almost split (see [7, Section 2]). Thus the sequence $0 \longrightarrow \mathrm{DTrC} \xrightarrow{g} B \xrightarrow{f} C \longrightarrow 0$ is almost split.

Having indicated some of the connections between minimal left and right almost split morphisms, we summarize the main features of these notions. The reader is referred to [6] and [7] for proofs not already given.

Proposition 4.1:

a) If P is an indecomposable projective Λ-module, the inclusion $\underline{r}P \longrightarrow P$ is a minimal right almost split morphism.

b) Let C be an indecomposable nonprojective module.

  i) There is an almost split sequence $0 \longrightarrow A \xrightarrow{g} B \xrightarrow{f} C \longrightarrow 0$.

  ii) If $0 \longrightarrow A \xrightarrow{g} B \xrightarrow{f} C \longrightarrow 0$ is almost split, then $A \approx \mathrm{DTrC}$.

  iii) If $0 \longrightarrow A' \xrightarrow{g'} B' \xrightarrow{f'} C \longrightarrow 0$ and $0 \longrightarrow A \xrightarrow{g} B \xrightarrow{f} C \longrightarrow 0$ are almost split, they are isomorphic over C.

c) Let A be an indecomposable noninjective Λ-module.

i) There exists an almost split sequence

$$0 \longrightarrow A \xrightarrow{g} B \xrightarrow{f} C \longrightarrow 0.$$

ii) If $0 \longrightarrow A \xrightarrow{g} B \xrightarrow{f} C \longrightarrow 0$ is almost split, then $C \approx TrDA$.

iii) If $0 \longrightarrow A \xrightarrow{g} B \xrightarrow{f} C \longrightarrow 0$ and $0 \longrightarrow A \xrightarrow{g'} B' \xrightarrow{f'} C \longrightarrow 0$ are almost split, then they are isomorphic over A.

d) If I is an indecomposable injective Λ-module, then the epimorphism $I \longrightarrow I/\mathrm{Soc}\, I$ is minimal left almost split.

e) If $0 \longrightarrow A \xrightarrow{g} B \xrightarrow{f} C \longrightarrow 0$ and $0 \longrightarrow A' \xrightarrow{g'} B' \xrightarrow{f'} C' \longrightarrow 0$ are almost split, then the following are equivalent.

i) $A \approx A'$.

ii) $C \approx C'$.

iii) There exists a commutative diagram

$$\begin{array}{ccccccccc} 0 & \longrightarrow & A & \xrightarrow{g} & B & \xrightarrow{f} & C & \longrightarrow & 0 \\ & & \downarrow & & \downarrow & & \downarrow & & \\ 0 & \longrightarrow & A' & \xrightarrow{g'} & B' & \xrightarrow{f'} & C' & \longrightarrow & 0 \end{array}$$

with the vertical morphisms isomorphisms.

The reader should observe that part e) of the above proposition states that an almost split sequence $0 \longrightarrow A \longrightarrow B \longrightarrow C \longrightarrow 0$ is an invariant of the indecomposable A and the indecomposable module C. Consequently any invariant of the almost split sequences is an invariant of A and C. For example if $B \approx \coprod_{i=1}^{n} B_i$ with the B_i indecomposable, then the following are all numerical invariants of C: the number n, the number of isomorphic B_i, the number of B_i with $L(B_i) > L(C)$, the number of B_i with $L(B_i) < L(C)$. Very little is known at this time about what the module theoretic properties of an almost split sequence $0 \longrightarrow A \longrightarrow B \longrightarrow C \longrightarrow 0$ tells one about A or C. The reader should see [6] and [7] for more details concerning this question.

Before going on with our applications, it is convenient to recall the notion of an irreducible morphism (see [7, Section 2]).

A morphism $f: B \longrightarrow C$ in mod Λ is said to be irreducible if f is not a splittable epimorphism or a splittable monomorphism and if f is the composition $B \xrightarrow{g} X \xrightarrow{h} C$, then either h is a splittable epimorphism or g is a splittable monomorphism. The following are some of the basic facts concerning irreducible morphisms (see [7, Section 2]).

Proposition 4.2: Let $f: B \longrightarrow C$ be a morphism in mod Λ.

a) If f is irreducible, then f is either a proper epimorphism (i.e. is an epimorphism but not an isomorphism) or a proper monomorphism.

b) If C is indecomposable and B is not zero, then f is irreducible if and only if there is a morphism $f': B' \longrightarrow C$ such that the induced morphism $(f, f'): B \coprod B' \longrightarrow C$ is right minimal almost split.

c) If B is indecomposable, then f is irreducible if and only if there is a morphism $f': B \longrightarrow C'$ such that the induced morphism $B \longrightarrow C \coprod C'$ is minimal left almost split.

We now point out some easily verified consequences of this result.

Proposition 4.3:

a) Suppose C is an indecomposable module. There are only a finite number of nonisomorphic modules B shich have the property that there is an irreducible morphism $B \longrightarrow C$ or an irreducible morphism $C \longrightarrow B$.

b) Let $0 \longrightarrow A \longrightarrow B \longrightarrow C \longrightarrow 0$ be an almost split sequence and let B' be isomorphic to a summand of B. Then $L(B')$ is different from $L(A)$ and $L(C)$ where $L(X)$ denotes the length of X. Hence the indecomposable summands of B are not isomorphic to C or A.

We now point out another useful characterization of irreducible morphisms (see [7, Section 2]).

<u>Proposition 4.4</u>: Let $0 \longrightarrow A \xrightarrow{g} B \xrightarrow{f} C \longrightarrow 0$ be a nonsplit exact sequence in mod Λ.

a) f is irreducible if and only if g has the property that given any $h: A \longrightarrow X$ either there is a morphism $s: B \longrightarrow X$ such that $sg = h$ or there is a $t: X \longrightarrow B$ such that $th = g$.

b) g is irreducible if and only if f has the property that given any $h: X \longrightarrow C$ either there is a morphism $s: X \longrightarrow B$ such that $gs = h$ or there is a $t: B \longrightarrow X$ such that $f = ht$.

In connection with this result it is important to make the following observation. Suppose the nonsplit exact sequence $0 \longrightarrow A \xrightarrow{g} B \xrightarrow{f} C \longrightarrow 0$ representing the element x in $\mathrm{Ext}^1_\Lambda(C, A)$ has the property that f is irreducible. Let $h: A \longrightarrow A'$ be a proper epimorphism. The fact that g is a monomorphism and h has a nontrivial kernel means that there is no morphism $s: A' \longrightarrow B$ such that $sh = g$. Hence by Proposition 4.4 we know there is a morphism $t: B \longrightarrow X$ such that $tg = h$, or equivalently, $\mathrm{Ext}^1_\Lambda(C, h)(x)$ in $\mathrm{Ext}^1_\Lambda(C, X)$ is zero. In other words, the nonzero element x in

$\mathrm{Ext}^1_\Lambda(C, A)$ has the property that a morphism $h: A \longrightarrow X$ is a monomorphism if $\mathrm{Ext}^1_\Lambda(C, h)(x)$ in $\mathrm{Ext}^1_\Lambda(C, X)$ is not zero. It is easily checked that this property of x implies that A is indecomposable. This observation leads to the following notions (see [4, Section 2]).

An element x in $\mathrm{Ext}^1_\Lambda(C, \ )(A)$ is said to be <u>minimal</u> if x is not zero and any morphism $h: A \longrightarrow X$ is a monomorphism if $\mathrm{Ext}^1_\Lambda(C, h)(x)$ in $\mathrm{Ext}^1_\Lambda(C, X)$ is not zero. An element y in $\mathrm{Ext}^1_\Lambda(\ , A)(C)$ is said to be <u>minimal</u> if y is not zero and any morphism $h: Y \longrightarrow C$ is an epimorphism if $\mathrm{Ext}^1_\Lambda(h, A)(y)$ in $\mathrm{Ext}^1_\Lambda(Y, A)$ is not zero. We now summarize some of the basic properties of these special types of extensions.

<u>Proposition 4.5</u>:

a) If x in $\mathrm{Ext}^1_\Lambda(C, \ )(A)$ is minimal, then A is indecomposable. If y in $\mathrm{Ext}^1_\Lambda(\ , A)(C)$ is minimal, then C is indecomposable.

b) Let $x: 0 \longrightarrow A \xrightarrow{g} B \xrightarrow{f} C \longrightarrow 0$ be exact.

   i) If f is irreducible, then x viewed as an element in $\mathrm{Ext}^1_\Lambda(C, \ )(A)$ is minimal and so A is indecomposable.

   ii) If g is irreducible, then x viewed as an element of $\mathrm{Ext}^1_\Lambda(\ , A)(C)$ is minimal and so C is indecomposable.

c) Let x be a nonzero element in $\mathrm{Ext}^1_\Lambda(C, A)$.

   i) There is an epimorphism $h: A \longrightarrow A''$ such that the element $\mathrm{Ext}^1_\Lambda(C, h)(x)$ in $\mathrm{Ext}^1_\Lambda(C, \ )(A'')$ is minimal.

   ii) There is a monomorphism $h: C' \longrightarrow C$ such that the element $\mathrm{Ext}^1_\Lambda(h, A)(x)$ in $\mathrm{Ext}^1_\Lambda(\ , A)(C')$ is minimal.

It should be observed that the last part of the above proposition shows that there are lots of exact sequences $x: 0 \longrightarrow A \longrightarrow B \longrightarrow C \longrightarrow 0$ with x a minimal element of $\mathrm{Ext}^1_\Lambda(C, \ )(A)$, or in $\mathrm{Ext}^1_\Lambda(\ ,A)(C)$, or in both $\operatorname{Ext}^1_\Lambda(C, \ )(A)$ and $\operatorname{Ext}^1_\Lambda(\ , A)(C)$.

The applications in the rest of this section are based on the following formulation of Proposition 2.3 which is particularly appropriate for our present purposes.

Proposition 4.6: Suppose $0 \longrightarrow A \xrightarrow{g} B \xrightarrow{f} C \longrightarrow 0$ is an exact sequence in mod Λ representing the element x in $\operatorname{Ext}^1_\Lambda(C, A)$. Further, suppose X is in mod Λ. Then the following statements are equivalent.

a) $(X, f)\colon (X, B) \longrightarrow (X, C) \longrightarrow 0$ is exact.

b) For each $h\colon X \longrightarrow C$ we have that $\operatorname{Ext}^1_\Lambda(h, A)(x)$ in $\operatorname{Ext}^1_\Lambda(X, A)$ is zero.

c) $(g, DTrX)\colon (B, DTrX) \longrightarrow (A, DTrX) \longrightarrow 0$ is exact.

d) For each $h\colon A \longrightarrow DTrX$ we have that $\operatorname{Ext}^1_\Lambda(C, h)(x)$ in $\operatorname{Ext}^1_\Lambda(C, DTrX)$ is zero.

As our first application of Proposition 4.6 we have the following.

Proposition 4.7: Suppose the exact sequence $x\colon 0 \longrightarrow A \xrightarrow{g} B \xrightarrow{f} C \longrightarrow 0$ is a minimal element in $\operatorname{Ext}^1_\Lambda(C, \ )(A)$. Let X be in mod Λ. Then

a) $(X, f)\colon (X, B) \longrightarrow (X, C) \longrightarrow 0$ is exact if and only if $\operatorname{Im}((B, DTrX) \xrightarrow{(g, DTrX)} (A, DTrX))$ contains all monomorphisms $A \longrightarrow DTrX$.

b) Suppose $\operatorname{Im}(X, f) \neq (X, C)$.

i) Then there is a monomorphism $h\colon A \longrightarrow DTrX$ such that $\operatorname{Ext}^1_\Lambda(C, h)(x)$ in $\operatorname{Ext}^1_\Lambda(C, DTrX)$ is not zero.

ii) Soc A is isomorphic to a summand of $\Omega(X)/r\Omega(X)$.

iii) If $f\colon B \longrightarrow C$ is irreducible, then the monomorphism

$0 \longrightarrow A \xrightarrow{g} B$ can be written as a composition $A \xrightarrow{h} \mathrm{DTrX} \longrightarrow B$. In particular, $(\mathrm{DTrX}, B) \neq 0$.

Proof:

a) Suppose $h: A \longrightarrow \mathrm{DTrX}$ is not a monomorphism. Then because x in $\mathrm{Ext}^1_\Lambda(C, \ )(A)$ is minimal we know that $\mathrm{Ext}^1_\Lambda(C, h)(x)$ in $\mathrm{Ext}^1_\Lambda(C, \mathrm{DTrX})$ is zero. But $\mathrm{Ext}^1_\Lambda(C, h)(x)$ being zero is equivalent to there being a morphism $t: B \longrightarrow \mathrm{DTrX}$ such that $tg = h$. Thus $\mathrm{Im}((B, \mathrm{DTrX}) \xrightarrow{(g,\ \mathrm{DTrX})} (A, \mathrm{DTrX}))$ contains all nonmonomorphisms $A \longrightarrow \mathrm{DTrX}$. Thus if $\mathrm{Im}(g, \mathrm{DTrX})$ also contains all monomorphisms, then $\mathrm{Im}(g, \mathrm{DTrX}) = (A, \mathrm{DTrX})$, which finishes the proof of a).

b) i) Since $\mathrm{Im}(X, f) \neq (X, C)$, there is a monomorphism $h: A \longrightarrow \mathrm{DTrX}$ such that h is not in $\mathrm{Im}((B, \mathrm{DTrX}) \longrightarrow (A, \mathrm{DTrX}))$. Then h has the property that $\mathrm{Ext}^1_\Lambda(C, h)(x)$ in $\mathrm{Ext}^1_\Lambda(C, \mathrm{DTrX})$ is not zero.

ii) The monomorphism $h: A \longrightarrow \mathrm{DTrX}$ given in part i) induces a monomorphism $\mathrm{Soc}\, A \longrightarrow \mathrm{Soc}\, \mathrm{DTrX}$. Since $\mathrm{Soc}\, \mathrm{DTrX}$ is semisimple, $\mathrm{Soc}\, A$ is isomorphic to a summand of $\mathrm{Soc}\, \mathrm{DTrX}$. But we know by Proposition 1.4 that $\mathrm{Soc}\, \mathrm{DTrX} \approx \Omega(X)/\underline{r}\Omega(X)$. Hence $\mathrm{Soc}\, A$ is isomorphic to a summand of $\Omega(X)/\underline{r}\Omega(X)$.

iii) Let

$$\begin{array}{ccccccccc}
0 & \longrightarrow & A & \xrightarrow{g} & B & \xrightarrow{f} & C & \longrightarrow & 0 \\
 & & \downarrow h & & \downarrow & & \| & & \\
0 & \longrightarrow & \mathrm{DTrX} & \longrightarrow & B \overset{A}{\times} \mathrm{DTrX} & \longrightarrow & C & \longrightarrow & 0
\end{array}$$

be an exact commutative pushout diagram. Then by i),

the exact pushout sequence

$0 \longrightarrow DTrX \longrightarrow B \underset{A}{\times} DTrX \longrightarrow C \longrightarrow 0$ does not split.

Hence there is no morphism $t: B \longrightarrow DTrX$ such that $tg = h$. Therefore there must be a morphism $t: DTrX \longrightarrow B$ such that $g = th$ since we are assuming that f is irreducible (see Proposition 4.4).

This finishes the proof of the proposition.

As an immediate consequence of this result, we have the following.

Corollary 4.8: Let $x: 0 \longrightarrow A \xrightarrow{g} B \xrightarrow{f} C \longrightarrow 0$ be exact with x a minimal element of $\mathrm{Ext}^1_\Lambda(C, \ )(A)$. Suppose $t: X \longrightarrow C$ is not in $\mathrm{Im}(X, f)$ and the composition $X \xrightarrow{u} Y \xrightarrow{v} C$ is t, for instance $Y = C$ and $v = 1_C$. Then

a) v is not in $\mathrm{Im}(Y, f)$.

b) i) There is a monomorphism $h: A \longrightarrow DTrY$ such that $\mathrm{Ext}^1_\Lambda(C, h)(x)$ in $\mathrm{Ext}^1_\Lambda(C, DTrY)$ is not zero.

ii) Soc A is isomorphic to a summand of $\Omega Y/\underline{r}\Omega Y$.

iii) If f is irreducible, the monomorphism $A \xrightarrow{g} B$ can be written as a composition $A \xrightarrow{h} DTrY \longrightarrow B$. In particular $(DTrY, B) \neq 0$.

Proof:

a) Trivial consequence of the fact that $t: X \longrightarrow C$ is not in $\mathrm{Im}(X, f)$.

b) Follows trivially from a) and Proposition 4.7.

We now explore some consequences of Corollary 4.8.

Proposition 4.9: Suppose $x: 0 \longrightarrow A \xrightarrow{g} B \xrightarrow{f} C \longrightarrow 0$ is minimal in $\mathrm{Ext}^1_\Lambda(C, \ )(A)$ and $f: B \longrightarrow C$ does not generate (B, C) over $(\mathrm{End}\ B)^{op}$.

a) There are monomorphisms $0 \longrightarrow A \longrightarrow DTrB$ and $0 \longrightarrow A \longrightarrow DTrC$.

b) Soc A is a summand of $\Omega(C)/\underline{r}\Omega(C)$ and $\Omega(B)/\underline{r}\Omega(B)$.

c) If f is irreducible, then there is a factorization $A \longrightarrow DTrB \longrightarrow B$ of the monomorphism $g: A \longrightarrow B$. In particualar $(DTrB, B) \neq 0$.

Proof: This is an immediate consequence of Corollary 4.9, once one observes that if $f: B \longrightarrow C$ does not generate (B, C) over $(\mathrm{End}\ B)^{op}$, then $\mathrm{Im}(B, f) \neq (B, C)$.

As our final example of how these ideas can be applied, we have the following.

Proposition 4.10: Let $0 \longrightarrow A \xrightarrow{g} B \xrightarrow{f} C \longrightarrow 0$ be exact with f irreducible and $g \notin P(A, B)$. Further, suppose X is in mod Λ such that $\mathrm{Im}(X, f) \neq (X, C)$. Then

a) $\mathrm{Ext}^1_\Lambda(DTrX, DTrA) \neq 0$;

b) $\mathrm{Ext}^1_\Lambda(B, (DTr)^2X) \neq 0$;

c) DTrX is not projective.

Proof:

a) Since f is irreducible and $\mathrm{Im}(X, f) \neq (X, C)$, we know that $g: A \longrightarrow B$ can be written as a composition $A \xrightarrow{h} DTrX \xrightarrow{j} B$. Since $g \notin P(A, B)$, it follows that $h: A \longrightarrow DTrX$ is not in $P(A, DTrX)$ and $j: DTrX \longrightarrow B$ is not in $P(DTrX, B)$. Hence $\underline{\mathrm{Hom}}\ (A, DTrX) \neq 0$ and $\underline{\mathrm{Hom}}\ (DTrX, B) \neq 0$. Since

$\operatorname{Ext}^1_\Lambda(\mathrm{DTrX}, \mathrm{DTrA}) \cong \operatorname{Hom}_R(\underline{\operatorname{Hom}}(A, \mathrm{DTrX}), I)$ and

$\operatorname{Ext}^1_\Lambda(B, (\mathrm{DTr})^2(X)) \cong \operatorname{Hom}_R(\underline{\operatorname{Hom}}(\mathrm{DTrX}, B), I)$, a) and b) are established

Part c) is an obvious consequence of a).

Some examples of when the hypothesis of Proposition 4.10 is satisfied are the following.

a) Λ is hereditary and A is not projective.

b) Λ is selfinjective and B has no nonzero injective summands.

Propositions 4.7 through 4.10 have been concerned with exact sequence $x: 0 \longrightarrow A \xrightarrow{g} B \xrightarrow{f} C \longrightarrow 0$ with x minimal in $\operatorname{Ext}^1_\Lambda(C, \ )(A)$. Dual results hold for exact sequences with x minimal in $\operatorname{Ext}^1_\Lambda(\ , A)(C)$. The statements and proofs of these results are left to the reader to consider.

§5. Brauer-Thrall $1^1/_2$.

As usual Λ is an artin R-algebra. We denote by ind Λ the full subcategory of mod Λ whose objects are the indecomposable objects in mod Λ. Further, for each integer $i \geq 0$ we denote by $\operatorname{ind}_i \Lambda$ the full subcategory of ind Λ consisting of those indecomposable Λ-modules A with $L(A) = i$, where $L(A)$ denotes the length of A over Λ. Also for each integer $n \geq 0$ we denote the full subcategory $\bigcup_{i \geq n} \operatorname{ind}_i \Lambda$ of ind Λ by $\underline{\underline{C}}_n$. Finally, for each full subcategory $\underline{\underline{C}}$ of ind Λ we denote by $|\underline{\underline{C}}|$ the cardinality of the set of isomorphism classes of objects in $\underline{\underline{C}}$. With this notation in mind, we can state the main result we prove in this section.

Brauer-Thrall $1^1/2$. Λ is not of finite representation type (i.e. $|\text{ind}\,\Lambda| \geq \aleph_0$) if and only if $|\underline{\underline{C}}_n| = |\text{ind}\,\Lambda|$ for all $n \geq 0$.

Clearly this result is stronger than Brauer-Thrall 1 which states that Λ is not of finite representation type if and only if $\text{ind}_i\Lambda$ is not empty for an infinite number of i. On the other hand, it is weaker than Brauer-Thrall 2 which states that if $\text{card}\,\Lambda \geq \aleph_0$, then there are an infinite number of integers i such that $|\text{ind}_i\Lambda| = \text{card}\,\Lambda$. Hence the name Brauer-Thrall $1^1/2$. While Brauer-Thrall 1 was first proven for artin algebras by Roiter [14] and for arbitrary artin rings by Auslander [4], the situation for Brauer-Thrall 2 is not quite so complete.

For finite dimensional algebras over a field k Brauer-Thrall 2 was first proven by Nazarova-Roiter in the case k is perfect [12] and extended by Ringel (unpublished) to k an arbitrary field. To the best of my knowledge there is no proof of Brauer-Thrall 2 for arbitrary artin algebras. Unlike the situation for Brauer-Thrall 1, Brauer-Thrall 2 does not hold for arbitrary artin rings, as was shown by Ringel [13].

Finally, before beginning our proof of Brauer-Thrall $1^1/2$, we would like to say that Smalø has recently given a short elegant proof (unpublished) of the following result. If Λ is an artin algebra, then for each $i \geq 0$ there is a $j > i$ such that $|\text{ind}_i\Lambda| \leq |\text{ind}_j\Lambda|$, which is a generalization to artin algebras of earlier results of Nazarova-Roiter and Ringel. This clearly implies Brauer-Thrall $1^1/2$. In view of all these developments, our proof of Brauer-Thrall $1^1/2$ is being given mainly to illustrate some ideas and techniques which are of interest in their own right and have proven useful in other contexts. It would also be interesting to know if Brauer-Thrall $1^1/2$ holds for arbitrary artin rings.

For the convenience of the reader, we start by recalling some basic facts concerning the radical and socle of finitely presented functors from mod Λ to abelian groups. The reader is referred to [10] for details. We recall that a functor $F: \text{mod } \Lambda \longrightarrow \text{Ab}$ is said to be finitely presented if there is an exact sequence $(A, \ ) \longrightarrow (B, \ ) \longrightarrow F \longrightarrow 0$ with A and B in mod Λ. We denote by f.p. (mod Λ, Ab) the full subcategory of (mod Λ, Ab), the category of all additive functors from mod Λ to abelian groups, whose objects are the finitely presented functors. Finally the radical of an arbitrary functor $F: \text{mod } \Lambda \longrightarrow \text{Ab}$, which we denote by $\underline{r}F$, is the intersection of the maximal subfunctors of F.

Proposition 5.1: Let F be in f.p. (mod Λ, Ab). Then $\underline{r}F$ has the following properties:

a) $\underline{r}F$ is a finitely presented fucntor.

b) $\underline{r}F = F$ if and only if $F = 0$.

c) $F/\underline{r}F$ is a semisimple functor of finite length.

d) If F' is a subfunctor of F, then F/F' is semisimple if and only if $F' \supset \underline{r}F$.

e) If $f: F \longrightarrow G$ is a morphism of finitely presented functors, then $f(\underline{r}F) \subset \underline{r}G$ and $f(\underline{r}F) = \underline{r}G$ if f is an epimorphism.

Proof: See [10, Section 3].

We define $\underline{r}^iF$ for all nonnegative integers i by induction on i. Define $\underline{r}^0F = F$ and $\underline{r}^{i+1}F = \underline{r}(\underline{r}^iF)$. Also we define $\underline{r}^\infty F$, the infinite radical of F, by $\underline{r}^\infty F = \bigcap_{i=0}^{\infty} \underline{r}^iF$. We now list some of the basic properties of the descending chain of subfunctors.

$$F \supset \underline{r}F \supset \underline{r}^2F \supset \dots \supset \underline{r}^iF \supset \dots$$

in the case F is finitely presented. In this connection we recall that the support of a functor F, Supp F, is the full subcategory of ind Λ consisting of all A in ind Λ such that $F(A) \neq 0$. Also if F is finitely generated, then $|\text{Supp } F| < \aleph_0$ if and only if $L(F) < \infty$, where $L(F)$ is the length of F (see [3] and [4]).

Proposition 5.2: Suppose F is in f. p. (mod Λ, Ab).

a) For each nonnegative integer i, we have that $\underline{r}^i F$ is in f. p. (mod Λ, Ab) and $\underline{r}^i F / \underline{r}^{i+1} F$ is semisimple of finite length.

b) If $F' \subset F$, then $L(F/F') < \infty$ if and only if $F' \supset \underline{r}^i F$ for some $i < \infty$.

c) The following statements are equivalent for F.

i) $L(F) < \infty$.

ii) $\underline{r}^i F = 0$ for some $i < \infty$.

iii) $L(F/\underline{r}^\infty F) < \infty$.

d) For A in ind Λ, there is an integer i such that $((A, \ ), \underline{r}^i F) = ((A, \ ), \underline{r}^\infty F)$.

e) $|\text{Supp } (F(\underline{r}^\infty F)| \underline{\subset} \leq \aleph_0$ and $|\text{Supp } (F/\underline{r}^\infty F)| < \infty$ if and only if $L(F) < \infty$.

f) If $f\colon F \longrightarrow G$ is a morphism of finitely presented functors, then $f(\underline{r}^\infty F) \subset \underline{r}^\infty G$ and so f induces a morphism $F/(\underline{r}^\infty F) \longrightarrow G/\underline{r}^\infty G$.

Proof: a), b), and c) follow easily from Proposition 5.1.

d) Since F is finitely presented we know that F(A) is a finitely generated End A-module and is therefore an artin End A-module. Hence the descending chain

$$F(A) \supset \underline{r}F(A) \supset \dots \supset \underline{r}^t F(A) \supset \dots$$

has the property that there is an i such that $\underline{r}^i F(A) = \underline{r}^j F(A)$ for all $j \geq i$. Thus $\underline{r}^i F(A) = \underline{r}^\infty F(A)$ or equivalently $((A,), \underline{r}^i F) = ((A,), \underline{r}^\infty F)$.

e) From d) it follows that $\operatorname{Supp}(F/\underline{r}^\infty F) = \bigcup_{i=0} \operatorname{Supp}(\underline{r}^i F/\underline{r}^{i+1} F)$. Since each $\underline{r}^i F/\underline{r}^{i+1} F$ is of finite length we have that $|\operatorname{Supp}(\underline{r}^i F/\underline{r}^{i+1} F)| < \aleph_0$ and so $|\operatorname{Supp}(F/\underline{r}^\infty F)| \leq \aleph_0$.

Since $F/\underline{r}^\infty F$ is a finitely generated functor, we know that $L(F/\underline{r}^\infty F) < \infty$ if and only if $|\operatorname{Supp}(F/\underline{r}^\infty F)| < \infty$.

f) Follows from Proposition 5.1.

For a functor $F: \operatorname{mod} \Lambda \longrightarrow \operatorname{Ab}$ define $\operatorname{Soc} F$, the <u>socle</u> of F, to be the subfunctor of F generated by the simple subfunctors of F. We now recall some basic facts concerning the socle of finitely presented functors, which are essentially the "duals" of those given for the radical of finitely presented functors (see [10, Section 3]).

<u>Proposition 5.3</u>: Let F be in f. p. $(\operatorname{mod} \Lambda, \operatorname{Ab})$.

a) $\operatorname{Soc} F$ is semisimple of finite length so that $F/\operatorname{Soc} F$ is in f. p. $(\operatorname{mod} \Lambda, \operatorname{Ab})$.

b) F is an essential extension of $\operatorname{Soc} F$ so that $F = 0$ if and only if $\operatorname{Soc} F = 0$.

c) A subfunctor F' of F is semisimple if and only if $F' \subset \operatorname{Soc} F$.

Let F be a finitely presented functor. Define $\operatorname{Soc}^i F$ for all nonnegative integers i by induction as follows: $\operatorname{Soc}^0 F = F$ and $\operatorname{Soc}^{i+1} F$ is the kernel of the natural epimorphism $F \longrightarrow F/\operatorname{Soc}^i F/\operatorname{Soc}(F/\operatorname{Soc}^i F)$. Also define $\operatorname{Soc}^\infty F$, the <u>infinite socle</u> of F, by $\operatorname{Soc}^\infty F = \bigcup_{i=0}^{\infty} \operatorname{Soc}^i F$. We now list some of the basic properties of the ascending chain of subfunctors.

$$\text{Soc } F \subset \text{Soc}^2 F \subset \ldots \subset \text{Soc}^i F \subset \ldots \ (i < \infty).$$

Since there are essentially dual to those given for the radical series in Proposition 5.2, we omit proofs.

<u>Proposition 5.4</u>: Let F be in f. p. (mod Λ, Ab).

a) For all nonnegative i we have that $\text{Soc}^i F$ is finitely presented and $\text{Soc}^{i+1} F/\text{Soc}^i F$ is semisimple of finite length.

b) If $F' \subset F$, then $L(F') < \infty$ if and only if $F' \subset \text{Soc}^i F$ for some $i < \infty$.

c) The following are equivalent:

i) $L(F) < \infty$.

ii) $\text{Soc}^i F = F$ for some i.

iii) $\text{Soc}^\infty F = F$.

d) If A is in ind Λ, then there is an integer i such that $((A, \ ), \text{Soc}^i F) = ((A, \ ), \text{Soc}^\infty F)$.

e) $|\text{Supp Soc}^\infty F| \leq X_0$ and $|\text{Supp Soc}^\infty F| < X_0$ if and only if $L(F) < \infty$.

f) If $f: F \longrightarrow G$ is a morphism of finitely presented functors, then $f(\text{Soc}^\infty F) \subset \text{Soc}^\infty G$ and so f induces a morphism $F/\text{Soc}^\infty F \longrightarrow G/\text{Soc}^\infty G$.

As a consequence of the above we have:

<u>Proposition 5.5</u>: Let $\underline{\underline{C}}$ be the full subcategory of ind Λ consisting of those A in ind Λ such that $L(A, \ ) < \infty$. Then $\underline{\underline{C}} \subset \text{Supp Soc}^\infty(\Lambda, \ )$ and so $|\underline{\underline{C}}| \leq X_0$.

Proof: Let $\Lambda^n \longrightarrow A \longrightarrow 0$ be surjective. Then $0 \longrightarrow (A, \ ) \longrightarrow (\Lambda^n, \ ) = \coprod(\Lambda, \ )$. If $L((A, \ )) < \infty$, then $(A, \ ) \subset \mathrm{Soc}^\infty(\coprod_n(\Lambda, \ )) = \coprod(\mathrm{Soc}^\infty(\Lambda, \ ))$. Soc $\mathrm{Soc}^\infty(\Lambda, \ )(A) \neq 0$. Therefore it follows that $\underline{\underline{C}} \subset \mathrm{Supp}(\mathrm{Soc}^\infty(\Lambda, \ ))$ and so $|\underline{\underline{C}}| \leq \aleph_0$.

Before proceeding to proving the main results of this section, we need need the following very useful fact.

Lemma 5.6: Suppose A and B are in ind Λ and $f\colon A \longrightarrow B$ is not an isomorphism. Then $f = \sum_{i=1}^{n} f_i + h$ where

a) Each f_i is a finite nonzero composition of irreducible morphisms $A = A_0^{(i)} \longrightarrow A_2^{(i)} \longrightarrow \cdots \longrightarrow A_{n_i}^{(i)} = B$ with the $A_j^{(i)}$ indecomposable modules.

b) h is in $\underline{r}^\infty(A, \ )(B)$.

Proof: See [9, Proposition 1.7].

As a consequence of this lemma we have the following.

Proposition 5.7: Let A and B be in ind Λ.

a) Suppose we are given a morphism $t\colon (B, \ )/\underline{r}^\infty(B, \ ) \longrightarrow (A, \ )/\underline{r}^\infty(A, \ )$ which is not an isomorphism. Then there is an $f\colon A \longrightarrow B$ such that $f = \sum f_i$ where each f_i is a finite composition of irreducible morphisms between indecomposable modules such that the diagram

$$\begin{array}{ccc} (B, \ ) & \xrightarrow{(f, \ )} & (A, \ ) \\ \downarrow & & \\ (B, \ )/\underline{r}^\infty(B, \ ) & \xrightarrow{t} & (A, \ )/\underline{r}^\infty(A, \ ) \end{array}$$

commutes.

b) If $((B, \ )/\underline{r}^\infty(B, \ ), (A, \ )/\underline{r}^\infty(A, \ )) \neq 0$ and $A \not\approx B$ then there must be a chain of irreducible morphisms $A = A_0 \xrightarrow{f_0} A_1 \longrightarrow \cdots \xrightarrow{f_{n-1}} A_n = B$ with the A_i indecomposable such that $f_{n-1} \cdots f_0 \neq 0$.

c) If $\underline{\underline{U}}(B)$ is the full subcategory of ind Λ consisting of those A in ind Λ such that $((B, \ )/\underline{r}^{\infty}(B, \ ), (A, \ )/\underline{r}^{\infty}(A, \ )) \neq 0$ then $|\underline{\underline{U}}(B)| \leq \aleph_0$.

Proof:

a) Since $(B, \ )$ is projective, there is a morphism $f'\colon A \longrightarrow B$ such that the diagram

$$\begin{array}{ccc} (B, \ ) & \xrightarrow{(f', \ )} & (A, \ ) \\ \downarrow & & \searrow \\ (B, \ )/\underline{r}^{\infty}(B, \ ) & \xrightarrow{t} & (A, \ )/\underline{r}^{\infty}(A, \ ) \end{array}$$

commutes. Furthermore $f'\colon A \longrightarrow B$ is not an isomorphism since t is not. Thus by Lemma 5.6 we know that $f' = f + h$ where h is in $\underline{r}^{\infty}(A, \ )(B)$ and $f = \sum_{i=1}^{n} f_i$ where each f_i is a finite composition of irreducible morphisms between indecomposable modules. Since h is in $\underline{r}^{\infty}(A, \ )(B)$, the morphism $h\colon A \longrightarrow B$ has the property that $(h, \ )((B, \ )) \subset \underline{r}^{\infty}(A, \ )$. Therefore

$$\begin{array}{ccc} (B, \ ) & \xrightarrow{(f, \ )} & (A, \ ) \\ \downarrow & & \searrow \\ (B, \ )/\mathrm{rad}^{\infty}(B, \ ) & \xrightarrow{t} & (A, \ )/\mathrm{rad}^{\infty}(A, \ ) \end{array}$$

commutes, which completes the proof of a).

b) Easy consequence of a).

c) Given any indecomposable module X we know (see Proposition 4.3) that there is only a finite number of nonisomorphic X_i in ind Λ such that there is an irreducible morphism $X_i \longrightarrow X$. Thus for each integer there are only a finite number of nonisomorphic X_i in ind Λ such that there is a chain of irreducible morphisms $X_i = Y_0 \xrightarrow{f_0} Y_1 \longrightarrow \cdots \xrightarrow{f_{n-1}} Y_{n-1} = B$ with the Y_j in ind Λ. Hence there are at most a denumerable number of nonisomorphic A in ind Λ such that there is a finite chain of irreducible morphisms $A = A_0 \xrightarrow{f_0} A_1 \longrightarrow \cdots \xrightarrow{f_{n-1}} A_{n-1} = B$ with each A_i in ind Λ. Since by part b), for each A in $\underline{\underline{U}}(B)$ with $A \not\approx B$, there is a finite chain of irreducible morphisms in ind Λ $A = A_0 \xrightarrow{f_0} A_1 \xrightarrow{f_1} \cdots \xrightarrow{f_{n-1}} A_{n-1} = B$ with the A_i in ind Λ, we have that $|\underline{\underline{U}}(B)| \le \aleph_0$.

For each integer $i \geq 0$ let $\text{ind}_i\Lambda$ be the full subcategory of ind Λ consisting of those A in ind Λ such that $L(A) = i$. For each n let $\underline{\underline{C}}_n = \bigcup_{i \geq n} \text{ind}_i\Lambda$. Also for each integer n let $\underline{\underline{A}}_n$ be the full subcategory of ind Λ consisting of all A in ind Λ such that $((B, \ )/\underline{r}^{\infty}(B, \ ), (A, \ )/\underline{r}^{\infty}(A, \ )) = 0$ for all B in $\underline{\underline{C}}_n$.

<u>Proposition 5.8</u>: Let A be in ind Λ, then A is in $\underline{\underline{A}}_n$ for some n if and only if $L(A, \ ) < \infty$.

<u>Proof</u>: If $L(A, \ ) < \infty$, then $|\text{Supp}\ (A, \ )| < \aleph_0$. Hence there is some integer n such that $(A, B) = 0$ for all B in $\underline{\underline{C}}_n$. Thus $((B, \ )/\underline{r}^{\infty}(B, \ ), (A, \ )) = 0$ for all B in $\underline{\underline{C}}_n$. But $(A, \ ) = (A, \ )/\underline{r}^{\infty}(A, \ )$ since $L(A, \ ) < \infty$. Hence we have that A is in $\underline{\underline{A}}_n$.

Suppose A is in $\underline{\underline{A}}_n$. We want to show that $L(A, \ ) < \infty$. We do this by showing that $L((A, \ )/\underline{r}^{\infty}(A, \ )) < \infty$ and so $L(A, \ ) < \infty$.

We first observe that the functor $F = (A, \ )/\underline{r}^{\infty}(A, \ )$ has the following properties:

a) $F(X) \neq 0$ for some X in $\underline{\underline{A}}_n$ (namely $X = A$).

b) If $X \xrightarrow{f} Y$ is a morphism such that $f \in \underline{r}^{\infty}(Y, \ )(X)$, then

$F(f) = 0$.

a) is obvious.

b) Suppose $f: X \longrightarrow Y$ is a morphism in $\underline{r}^{\infty}(X, \ )(Y)$. Then the composition $(Y, \ ) \xrightarrow{(f, \ )} (X, \ ) \longrightarrow (X, \ )/\underline{r}^{\infty}(X, \ )$ is zero. Thus the composition

$$\begin{array}{ccc} ((X, \ )/\underline{r}^{\infty}(X, \ ), (A, \ )/\underline{r}^{\infty}(A, \ )) & \xrightarrow{\sim} & ((X, \ ), (A, \ )/\underline{r}^{\infty}(A, \ )) \\ & & \downarrow \\ & & ((Y, \ ), (A, \ )/\underline{r}^{\infty}(A, \ )) \end{array}$$

is zero. Hence the vertical morphism is the zero morphism. But we have the commutative diagram

$$\begin{array}{ccc} ((X, \ ), (A, \ )/\underline{r}^{\infty}(A, \ )) & \xrightarrow{(f, \ )} & ((Y, \ ), (A, \ )/\underline{r}^{\infty}(A, \ )) \\ \downarrow\wr & & \downarrow\wr \\ (A, \ )/\underline{r}^{\infty}(A, \ )(X) & \xrightarrow[(A, \ )/\underline{r}^{\infty}(A, \ )(f)]{} & (A, \ )/\underline{r}^{\infty}(A, \ )(Y) \end{array}$$

so $(A, \ )/\underline{r}^{\infty}(A, \ )(f) = 0$.

We now show that if $F: \text{mod } \Lambda \longrightarrow Ab$ satisfies a) and b), then F must have a simple subfunctor. Suppose F has no simple subfunctor. Let X be in $\underline{\underline{A}}_n$ such that $F(X) \neq 0$. Then we know that there is a chain of morphisms

$$X \xrightarrow{f} X_o \xrightarrow{f_o} X_1 \longrightarrow \cdots \xrightarrow{f_{n-1}} X_n \longrightarrow \cdots$$

between indecomposable modules with the f_i proper monomorphic and f an epimorphism such that $F(f_j \ldots f_o f) \neq 0$ for each j (see [3, Theorem 1.5]). Since the $L(X_i)$ get arbitrarily large $X_i \in \underline{\underline{C}}_n$ for some i. From the fact that $F(f_j \ldots F_o f) \neq 0$ we know that $X \xrightarrow{f_j \ldots f_o f} X_i$ is not in $\underline{r}^{\infty}(X, \ )(X_i)$. Hence $((X_i)/\underline{r}^{\infty}(X_i, \ ), (X, \ )/\underline{r}^{\infty}(X, \ ))$ is not zero. This contradicts the fact that $X_i \in \underline{\underline{C}}_n$. Thus F must have a simple subfunctor.

Returning to $(A, \ )/\underline{r}^{\infty}(A, \ )$ we know that every proper factor of $(A, \ )/\underline{r}^{\infty}(A, \ )$ has a simple subfunctor since it satisfies a) and b) above. Thus $(A, \ )/\underline{r}^{\infty}(A, \ )$ is locally finite. Since it is finitely generated, it must be of finite length. Therefore $L(A, \ ) < \infty$ which is our desired result.

As an easy consequence of Proposition 5.7, we have the following.

Corollary 5.9: Suppose A is in ind Λ and $L(A, \ ) = \infty$. Then given any integer $n \geq 0$ there is a nonzero composition of irreducible morphisms $A \xrightarrow{f_o} A_1 \longrightarrow \dots \xrightarrow{f_{t-1}} A_t$ with the $A_1, \dots, A_t$ indecomposable modules such that $L(A_t) \geq n$.

Proof: Suppose we are given $n \geq 0$ and A in ind Λ such that $L(A, \ ) = \infty$. Since $L(A) = \infty$ we know by Proposition 5.8 that A is not in $\underline{\underline{A}}_k$ where $k > \max(n, L(A))$. Hence we know there is a B in $\underline{\underline{C}}_k$ such that $((B, \ )/\underline{r}^{\infty}(B, \ ), (A, \ )/\underline{r}^{\infty}(A, \ )) \neq 0$. Since $B \not\approx A$, it follows from Proposition 5.7 that there is a nonzero composition of irreducible morphisms $A \longrightarrow A_1 \longrightarrow \dots A_t = B$ with the $A_1, \dots, A_t$ indecomposable modules.

As another application of Proposition 5.8 we prove Brauer-Thrall $1^1/_2$.

Theorem 5.10: Suppose Λ is not of finite representation type. Then $|\underline{\underline{C}}_n| = |\text{ind } \Lambda|$ for all n.

Proof: From Brauer-Thrall 1 we know that $X_o \leq |\underline{\underline{C}}_n|$. So we have

$$X_o \leq |\underline{\underline{C}}_n| \leq |\text{ind } \Lambda| \leq \sum_{B \in \underline{\underline{C}}_n} |\underline{\underline{U}}(B)| + |\underline{\underline{A}}_n| = \sum_{B \in \underline{\underline{C}}_n} |\underline{\underline{U}}(B)| + X_o \leq X_o \cdot |\underline{\underline{C}}_n| + X_o$$

since $|\underline{\underline{U}}(B)| \leq X_o$. From the fact that $|\underline{\underline{C}}_n| \geq X_o$ we have that $X_o \cdot |\underline{\underline{C}}_n| + X_o = |\underline{\underline{C}}_n|$ and so $|\mathrm{ind}\ \Lambda| = |\underline{\underline{C}}_n|$.

As an easy consequence of Theorem 5.10 we have:

<u>Corollary 5.11</u>: Suppose $|\mathrm{ind}\,\Lambda| > X_o$. Then there are an infinite number of n such that $|\mathrm{ind}_n\Lambda| > X_o$.

§6. A Partition of ind Λ.

As usual we assume that Λ is an artin R-algebra. Our purpose in this section is to introduce a certain equivalence relation in the category ind Λ and point out some of its properties.

In Section 5, we dealt only with covariant additive functors from mod Λ to abelian groups. Since the duality D: mod $\Lambda \longrightarrow$ mod Λ^{op} gives an equivalence $(\mathrm{mod}\ \Lambda)^{op} \longrightarrow \mathrm{mod}\ \Lambda^{op}$, the category of contravariant additive functors from mod Λ to abelian groups $((\mathrm{mod}\ \Lambda)^{op}, \mathrm{Ab})$ is equivalent to $(\mathrm{mod}\ \Lambda^{op}, \mathrm{Ab})$ which shows that the results in Section 5 for covariant functors have exact analogues for contravariant functors. We will use these analogues freely, without proof.

Our immediate concern is to show that the relations R_1, R_2, and R_3 which we are about to define on ind Λ generate the same equivalence on ind Λ.

Definition of R_1. Given A and B in ind Λ, we say that $A\ R_1\ B$ if one of the following holds: i) $A \approx B$; ii) there is an irreducible morphism $A \longrightarrow B$; iii) there is an irreducible morphism $B \longrightarrow A$.

Definition of R_2. Given A and B in ind Λ we say that A R_2 B if and only if either $((\ ,A)/\underline{r}^{\infty}(\ ,A), (\ ,B)/\underline{r}^{\infty}(\ ,B)) \neq 0$ or $(((\ ,B)/\underline{r}^{\infty}(\ ,B)), ((\ ,A)/\underline{r}^{\infty}(\ ,A))) \neq 0$.

Definition of R_3. Given A and B in ind Λ we say that A R_3 B if either $((A,\ )/\underline{r}^{\infty}(A,\ ), (B,\ )/\underline{r}^{\infty}(B,\ )) \neq 0$ or $((B,\ )/\underline{r}^{\infty}(B,\ ), (A,\ )/\underline{r}^{\infty}(A,\ )) \neq 0$.

Since all three of these relations obviously have the properties A R_i A and A R_i B implies B R_i A for all A and B in ind Λ, they each generate an equivalence on ind Λ.

Proposition 6.1: The relations R_1, R_2, and R_3 generate the same equivalence relation on ind Λ.

Proof: We show that R_1 and R_2 generate the same equivalence. A similar proof shows that R_1 and R_3 generate the same equivalence relation.

First we observe that if A R_1 B, then A R_2 B. Obviously, if $A \approx B$, then A R_2 B. Suppose there is an irreducible morphism $f: A \longrightarrow B$. Then $(\ ,f): (\ ,A) \longrightarrow (\ ,B)$ has the property that the composition $(\ ,A) \xrightarrow{(\ ,f)} (\ ,B) \longrightarrow (\ ,B)/\underline{r}^2(\ ,B)$ is not zero. Since $\underline{r}^{\infty}(\ ,B) \subset \underline{r}^2(\ ,B)$, we have that the composition $(\ ,A) \xrightarrow{(\ ,f)} (\ ,B) \longrightarrow (\ ,B)/\underline{r}^{\infty}(\ ,B)$ is not zero. Since this composition has $\underline{r}^{\infty}(\ ,A)$ in its kernel, it induces a nonzero morphism $(\ ,A)/\underline{r}^{\infty}(\ ,A) \longrightarrow (\ ,B)/\underline{r}^{\infty}(\ ,B)$. Thus A R_1 B implies A R_2 B.

Suppose A R_2 B, i.e. $((\ ,A)/\underline{r}^{\infty}(\ ,A), (\ ,B)/\underline{r}^{\infty}(\ ,B)) \neq 0$. If $A \approx B$, then A R_1 B and we are done. Suppose $A \not\approx B$. Then by Proposition 5.7, we know there is a nonzero morphism $f: A \longrightarrow B$ which is a finite composition of irreducible morphisms

$A = A_0 \xrightarrow{f_0} A_i \longrightarrow \dots \xrightarrow{f_{n-1}} A_{n-1} = B$ with the A_i indecomposable modules. Thus $A\ R_1\ A_1$, $A_2\ R_1\ A_3$, ..., $A_{n-2}\ R_1\ B$ and so A and B are in the same equivalence class of the equivalence relation generated by R_1, if $A\ R_2\ B$. Therefore R_1 and R_2 generate the same equivalence relation on $\text{ind}\ \Lambda$.

We denote the common equivalence relation on $\text{ind}\ \Lambda$ generated by R_1, R_2, and R_3 by R. For each A in $\text{ind}\ \Lambda$ we denote by $[A]$ the full subcategory of $\text{ind}\ \Lambda$ whose objects are the equivalence class of A under the equivalence relation R. We denote by $\{\underline{\underline{E}}_i\}_{i \in J}$ the family of full subcategories of $\text{ind}\ \Lambda$ where the objects of each $\underline{\underline{E}}_i$ consists of one equivalence class under R.

Proposition 6.2:

a) $|\underline{\underline{E}}_i| \leq \aleph_0$ for all i in I.

b) If $A \in \underline{\underline{C}}_i$ and $B \in \underline{\underline{C}}_j$ with $i \neq j$, then

$$((\ ,A)/\underline{r}^{\infty}(\ ,A), (\ ,B)/\underline{r}^{\infty}(\ ,B)) = 0 = ((\ ,B)/\underline{r}^{\infty}(\ ,B), (\ ,A)/\underline{r}^{\infty}(\ ,A))$$

and

$$((A,\ )/\underline{r}^{\infty}(A,\ ), (B,\ )/\underline{r}^{\infty}(B,\ )) = 0 = ((B,\ )/\underline{r}^{\infty}(B,\ ), (A,\ )/\underline{r}^{\infty}(A,\ )).$$

c) For each $\underline{\underline{E}}_i$ denote by $D(\underline{\underline{E}}_i)$ the full subcategory of $\text{mod}\ \Lambda^{op}$ whose objects are of the form $D(X)$ with X in $\underline{\underline{E}}_i$. Then the $D(\underline{\underline{E}}_i)$ are the full subcategories of $\text{mod}\ \Lambda^{op}$ whose objects constitute an equivalence class for the relation R on $\text{mod}\ \Lambda^{op}$.

Proof:

a) Follows readily from the fact that for each A in $\text{ind}\ \Lambda$, there are only a finite number of nonisomorphic indecomposable modules which have irreducible morphisms to A or from A.

b) Just a restatement of the definition of R.

c) This is an easy consequence of the fact that a morphism $f: A \longrightarrow B$ in ind Λ is irreducible if and only if $D(f): D(B) \longrightarrow D(A)$ in ind Λ^{op} is irreducible.

As our first nonformal result concerning the partition $\{\underline{\underline{E}}_i\}_{i\varepsilon I}$ of mod Λ, we have the following.

<u>Proposition 6.3</u>: The following statements are equivalent for an $\underline{\underline{E}}_i$.

a) $|\underline{\underline{E}}_i| < \aleph_o$.

b) There is an integer n such that $L(A) < n$ for all A in $\underline{\underline{E}}_i$.

c) $L(A, \ ) < \infty$ and $L(\ , A) < \infty$ for all A in $\underline{\underline{E}}_i$.

<u>Proof</u>:

a) implies b). Obvious.

b) implies c). Suppose $L(A) \leq n$ for all A in $\underline{\underline{E}}_i$ and let A be in $\underline{\underline{E}}_i$. If $B \varepsilon \underline{\underline{C}}_{n+1}$, i.e. $L(B) \geq n + 1$, then B is not in $\underline{\underline{E}}_i$. Hence by Proposition 6.2 we have that $((B, \ )/\underline{r}^{\infty}(B, \ ), (A, \ )/\underline{r}^{\infty}(A, \ )) = 0$. Since this is true for all B in $\underline{\underline{C}}_{n+1}$, we have by Proposition 5.8 that $L(A, \ ) < \infty$. Applying the same argument to $D(\underline{\underline{E}}_i)$, we deduce that $L(DA, \ ) < \infty$ or equivalently $L(\ , A) < \infty$. Thus b) implies c).

c) implies a). Suppose A is in $\underline{\underline{E}}_i$ and let $f: A \longrightarrow S$ be an epimorphism with S simple. Since $L(A, \ ) < \infty$, we have that $\underline{r}^{\infty}(A, \ ) = 0$. Hence by Lemma 5.6 we have that $f = \sum_{i=1}^{j} f_i$ with the f_i a finite composition of irreducible morphisms between indecomposable modules. Thus S is in $\underline{\underline{E}}_i$.

Let $S_1, \ldots, S_t$ be the simple modules in $\underline{\underline{E}}_i$. Then we have that $L(\ , S_i) < \infty$ for each i. Thus $|\mathrm{Supp}\ (\ , \coprod_{i=1}^{t} S_i)| < \infty$. But our previous argument shows that

$\underline{\underline{E}}_i \subseteq \mathrm{Supp}\ (\ , \coprod_{i=1}^{t} S_i)$ and so $|\underline{\underline{E}}_i| < \aleph_0$.

As a consequence of this result, we have the following.

Proposition 6.4: Suppose $|\underline{\underline{E}}_i| < \aleph_0$. Let $\tilde{\underline{\underline{E}}}_i$ be the additive category generated by $\underline{\underline{E}}_i$ and let $\underline{\underline{J}}$ be the additive category generated by $\bigcup_{j \neq i} \underline{\underline{E}}_j$. Then

a) The category $\mathrm{mod}\ \Lambda$ is isomorphic to the product of categories $\tilde{\underline{\underline{E}}}_i \times \underline{\underline{J}}$.

b) Λ is isomorphic to a product $\Lambda_1 \times \Lambda_2$ of artin R-algebras such that

i) Λ_1 is an indecomposable R-algebra of finite type with $\mathrm{mod}\ \Lambda_1$ equivalent to $\tilde{\underline{\underline{E}}}_i$ and

ii) $\mathrm{mod}\ \Lambda_2$ is equivalent to $\underline{\underline{J}}$.

Proof:

a) We first show that if A is in $\tilde{\underline{\underline{E}}}_i$ and B is in $\underline{\underline{J}}$, then $(A, B) = 0 = (B, A)$. It clearly suffices to do this in the case A and B are indecomposable. Since $L(A, \ ) < \infty$ we have that $(A, \ )/\underline{r}^{\infty}(A, \ ) = (A, \ )$. Hence

$$(A, B) = ((B, \ ), (A, \ )) = ((B, \ ), (A, \ )/\underline{r}^{\infty}(A, \ )) =$$
$$= ((B, \ )/\underline{r}^{\infty}(B, \ ), (A, \ )/\underline{r}^{\infty}(A, \ )).$$

But $((B, \ )/\underline{r}^{\infty}(B, \ ), (A, \ )/\underline{r}^{\infty}(A, \ )) = 0$ since B is not in $\underline{\underline{E}}_i$. Thus $(A, B) = 0$. The fact that $(B, A) = 0$ can be established similarly. Combining this with the fact that every object of mod Λ is isomorphic to a sum $A \coprod B$ with A in $\tilde{\underline{\underline{E}}}_i$ and B in $\underline{\underline{J}}$ we obtain our desired result, that mod Λ is equivalent to $\tilde{\underline{\underline{E}}} \times \underline{\underline{J}}$.

b) Corresponding to the decomposition $\tilde{\underline{\underline{E}}}_i \times \underline{\underline{J}}$ of mod Λ is a decomposition of Λ into a product $\Lambda_1 \times \Lambda_2$ with mod Λ_1 equivalent to $\tilde{\underline{\underline{E}}}$ and mod Λ_2 is equivalent to mod Λ_2. Since $|\underline{\underline{E}}_i| < X_o$, it follows that mod Λ_1 is of finite type. Finally, it follows easily from the definition of $\underline{\underline{E}}_i$, that $\underline{\underline{E}}_i$, and hence $\tilde{\underline{\underline{E}}}_i$ is an indecomposable category, which means that Λ_1 is an indecomposable R-algebra.

As an easily verified consequence of this result we have the following.

<u>Theorem 6.5</u>: Let $\{\underline{\underline{E}}_i\}_{i \varepsilon I}$ be the decomposition of ind Λ given by the equivalence relation R on ind Λ.

a) K, the set of all i such that $|\underline{\underline{E}}_i| < X_o$, is finite.

b) For each i in K, let $\tilde{\underline{\underline{E}}}_i$ be the additive category generated by $\underline{\underline{E}}_i$ and let $\underline{\underline{J}}$ be the additive category generated by the $\underline{\underline{E}}_j$ with j not in K. Then

i) mod Λ is equivalent to the product $\prod_{i \varepsilon K} \tilde{\underline{\underline{E}}}_i \times J$.

ii) Λ is isomorphic to a product of artin R-algebras $\prod_{j \varepsilon K} \Lambda_j \times \Sigma$ with the properties:

1) Λ_j is an indecomposable artin algebra of finite type with mod Λ_j equivalent to $\tilde{\underline{\underline{E}}}_j$ for each j in K;

2) mod Σ is equivalent to $\underline{\underline{J}}$;

3) if $\Sigma_1 \times \ldots \times \Sigma_t$ is the decomposition of Σ into indecomposable algebras, then no Σ_i is of finite type.

c) Λ is of finite type if and only if $|\underline{\underline{E}}_i| < \aleph_0$ for all i in I.

d) If Λ is an indecomposable algebra, then Λ is of finite type if and only if $|\underline{\underline{E}}_i| < \aleph_0$ for some i.

In [3] it was shown that an artin algebra Λ is not of finite representation type if and only if there are denumerably, nonfinitely generated indecomposable Λ-modules. We now show how the results of this section can be used to obtain somewhat stronger existence theorems for denumerably, nonfinitely generated indecomposable Λ-modules. We begin with the following preliminary result.

Proposition 6.6: Let Λ be an indecomposable artin R-algebra of infinite representation type. Suppose $\underline{\underline{E}}_i$ has the property that $L(A, \ ) < \infty$ for all A in $\underline{\underline{E}}_i$.

a) If $f: A \longrightarrow B$ is a nonzero morphism in ind Λ and A is in $\underline{\underline{E}}_i$, then B is in $\underline{\underline{E}}_i$.

b) If $0 \longrightarrow A \longrightarrow \underline{\underline{I}}_0 \longrightarrow I_1 \longrightarrow \ldots \longrightarrow I_n \longrightarrow \ldots$ is a minimal injective resolution for A and A is in $\underline{\underline{E}}_i$, then I_j is in $\tilde{\underline{\underline{E}}}_i$ for all $j = 0, 1, \ldots$

c) If $0 \longrightarrow A \xrightarrow{f} B \xrightarrow{g} C \longrightarrow 0$ is exact with A and C in $\tilde{\underline{\underline{E}}}_i$, then B is in $\tilde{\underline{\underline{E}}}_i$.

Proof:

a) Let $f: A \longrightarrow B$ be a nonzero morphism in ind Λ with A in $\underline{\underline{E}}_i$. Since $\underline{r}^{\infty}(A, \ ) = 0$, we know that $f = \Sigma f_i$ when the f_i are finite compositions of irreducible morphisms between indecomposable modules. Then $f_i \neq 0$ for some i since $f \neq 0$. This implies that B is in $\underline{\underline{E}}_i$.

b) Let $0 \longrightarrow A \xrightarrow{g} I_o$ be an injective envelope of A with A in $\underline{\underline{E}}_i$. Let $p: I_o \longrightarrow I_j'$ be a split epimorphism with I_j' indecomposable. Then the composition $pg: A \longrightarrow I_j'$ is not zero, since $g: A \longrightarrow I$ is essential. So I_j is in $\underline{\underline{E}}_i$ by part a). Since this is true for all projections $p: I_o \longrightarrow I_j'$, we have that I_o is in $\tilde{\underline{\underline{E}}}$. Using this fact, together with part a), it is easy to establish b).

c) Suppose $0 \longrightarrow A \xrightarrow{g} B \xrightarrow{f} C \longrightarrow 0$ is exact with A and C in $\tilde{\underline{\underline{E}}}_i$. Let $p: B \longrightarrow B'$ be a splittable epimorphism with B' indecomposable. If the composition $pg = 0$, then $\text{Img} \subset \text{Ker}\ \underline{p}$ and so B' is isomorphic to a summand of C and is therefore in $\underline{\underline{E}}_i$. If $pg \neq 0$, then $pg(A') \neq 0$ for some indecomposable summand of A. Since A' is in $\underline{\underline{E}}_i$, it follows that B' is in $\underline{\underline{E}}_i$. Therefore every indecomposable summand of B is in $\underline{\underline{E}}_i$ and so B is in $\tilde{\underline{\underline{E}}}_i$.

Since there are only a finite number of nonisomorphic injective modules in $\text{ind}\,\Lambda$, we have the following as an immediate consequence of Proposition 6.6.

Corollary 6.7: Let Λ be an indecomposable artin algebra which is not of finite representation type. Then there are at most a finite number of i such that $L(A,\ ) < \infty$ for all A in $\underline{\underline{E}}_i$.

Proposition 6.8: Let Λ be an indecomposable artin algebra of infinite representation type. Let $\{\underline{\underline{E}}_i\}_{i\varepsilon I}$ be the usual decomposition of $\text{mod}\,\Lambda$. Let K be the finite subset of I consisting of all i such that $L(A,\ ) < \infty$ for all A in $\underline{\underline{E}}_i$.

a) $I - K$ is not empty.

b) For each j in $I - K$ there exists a denumerably generated (not finitely generated) indecomposable Λ-module X_j with the following properties:

   i) For each j, there is a proper ascending chain of submodules of X_j

$$A_0 \subset A_1 \subset \dots \subset A_n \subset \dots$$

   with the A_k in $\underline{\underline{E}}_j$ such that $\bigcup_{k=0}^{\infty} A_k = X_j$.

   ii) X_j and $X_{j'}$ are not isomorphic if $j \neq j'$.

<u>Proof</u>:

a) If $K = I$, then $L(A, \ ) < \infty$ for all A in ind Λ. But this means that Λ is of finite representation type (see [4, Theorem 3.1]), which is a contradiction. Hence $K \neq I$.

b) Let j be in $I - K$. Then there is an A in $\underline{\underline{E}}_j$ such that $L(A, \ )$ is not finite. Hence $L((A, \ )/\underline{r}^{\infty}(A, \ ))$ is also not finite (see Proposition 5.2). Therefore if we let $F = G/\text{l.f.}G$, where $\text{l.f.}G$ is the maximal locally finite subfunctor of G, we have that $F \neq 0$ and F has no simple subfunctors (see [4, Section 1]). Therefore we know by [3, Theorem 1.5] that there is an infinite sequence of morphisms

$A \xrightarrow{f} A_0 \xrightarrow{f_0} A_1 \longrightarrow \dots \xrightarrow{f_{n-1}} A_n \xrightarrow{f_n} \dots$ having the following properties:

   i) For each $i = 0, 1, \dots$ there are nonzero x_i in $F(A_i)$ such that $F(f_i)(x_i) = x_{i+1}$.

   ii) Each f_i is a proper monomorphism and each A_i is indecomposable.

   iii) $\varinjlim A_i = X_j$ is indecomposable but not finitely generated and $F(X_j) \neq 0$.

Since $F(A_i) \neq 0$ it follows that $(A, \ )/\underline{r}^{\infty}(A, \ )(A_i) \neq 0$, or equivalently, $((A_i, \ )/\underline{r}^{\infty}(A_i), (A, \ )/\underline{r}^{\infty}(A, \ )) \neq 0$. Thus each A_i is in $\underline{\underline{E}}_j$. Consequently the first part of b) is established.

Suppose $j \neq j'$ are in $I - K$. Then by the above construction there is an A in $\underline{\underline{E}}_j$ such that $(A, \ )/\underline{r}^{\infty}(A, \ )(X_j) \neq 0$. Since $\underline{\underline{E}}_j \neq \underline{\underline{E}}_{j'}$, we know that $(A, \ )/\underline{r}^{\infty}(A, \ )(Y) = 0$ for all Y in $\underline{\underline{E}}_{j'}$. Since $X_{j'} = \varinjlim A_i'$ with the A_i' in $\underline{\underline{E}}_{j'}$, it follows that $(A, \ )/\underline{r}^{\infty}(A, \ )(X_{j'}) = 0$ since

$$(A, \ )/\underline{r}^{\infty}(A, \ )(X_{j'}) = \varinjlim (A, \ )/\underline{r}^{\infty}(A, \ )(A_i) = 0$$

(see [3, Pages 5, 6]). Because $(A, \ )/\underline{r}^{\infty}(A, \ )(X_j) \neq 0$ and $(A, \ )/\underline{r}^{\infty}(A, \ )(X_{j'}) = 0$, it follows that X_j is not isomorphic to $X_{j'}$.

Corollary 6.9: Let Λ be an indecomposable finite dimensional algebra over a field k of infinite representation type and suppose that $\text{card } k > \aleph_0$. Then the cardinality of the isomorphism classes of denumerably generated but not finitely generated indecomposable Λ-modules is $\text{card } k$.

Proof: By Brauer-Thrall 2, we know that $|\text{ind } \Lambda| = \text{card } k$. Hence if $\{\underline{\underline{E}}_i\}_{i \in I}$ is the usual partition of $\text{ind } \Lambda$, we have that $\text{card } I = \text{card } k$ since $|\underline{\underline{E}}_i| = \aleph_0$ for all i and $\text{card } k > \aleph_0$. Thus the cardinality of the set of all i with the property that $\underline{\underline{E}}_i$ contains an A with $L(A, \ ) = \infty$ is $\text{card } k$. The corollary now follows immediately from Proposition 6.8.

REFERENCES

[1] Auslander, M., "Comments on the functor Ext," Topology, (1969), 151-166.

[2] ___________., "Functors and morphisms determined by objects," these Proceedings.

[3] ___________., "Large modules over Artin algebras," Algebra, Topology and Categories, Academic Press, 1976.

[4] ___________., "Representation theory of Artin algebras, II," Comm. in Algebra (1974), 269-310.

[5] Auslander, M. and Bridger, I., "Stable module theory," Memoirs of the American Mathematical Society, No. 94, 1969.

[6] Auslander, M. and Reiten, I., "Representation theory of Artin algebras, III," Comm. in Algebra, (1975), 239-294.

[7] _______________________., "Representation theory of Artin algebras, IV," Comm. in Algebra, (1977), 443-518.

[8] _______________________., "Representation theory of Artin algebras, V," Comm. in Algebra, (1977) 519-554.

[9] _______________________., "Representation theory of Artin algebras, VI," Comm. in Algebra, (to appear).

[10] _______________________., "Stable equivalence of dualizing R-varieties," Advances in Math., (1974), 300-366.

[11] Cartan, H. and Eilenberg, S., Homological Algebra, Princeton University Press, Princeton, New Jersey, 1956.

[12] Nazarova, L. A. and Roiter, A. V., "Categorical matrix problems and the Brauer-Thrall conjecture," Preprint, Inst. Math. Acad. Sci., Kiev, 1973.

[13] Ringel, C. M., "Representations of K-species and bimodules," J. of Algebra, (1976), 269-302.

[14] Roiter, A. V., "Unboundness of the dimension of the indecomposable representations of an algebra which has infinitely many indecomposable representations," Izv. Akad. SSR, Sev. Math. 32, (1968) 1275-1282.

THE REPRESENTATIONS OF TAME HEREDITARY ALGEBRAS

V. Dlab
Carleton University
Ottawa, Canada

C. M. Ringel
Universität Bonn
Bonn, Federal Republic of Germany

§1. INTRODUCTION

An (associative, not necessarily commutative) ring R (with 1) is called (right) *hereditary* if each right ideal of R is projective or, equivalently, if the functor $Ext^1(X, -)$ is right exact for any (right) R-module X. Important examples of hereditary rings are the *tensor rings* defined in the following way: Let S be an (artinian) semisimple ring, and ${}_SM_S$ an S-S-bimodule; we denote inductively $M^{i+1} = M^i \otimes_S M$, with $M^0 = S$; then, inducing the multiplication by the tensor product $\otimes$ the direct sum $T(M) = \bigoplus_{i \geq 0} M^i$ becomes a ring. In this paper, we shall consider only *semiprimary* rings; for these rings, the properties of being right hereditary and left hereditary coincide. It is easy to see that the tensor ring T(M) is semiprimary if and only if $M^i = 0$ for some i. Note that there exist examples of hereditary semiprimary rings which are not tensor rings [10]; later in this paper, we shall consider one particular class of such rings in more detail. Also, we should mention the following class of tensor rings: Let F, G be (not necessarily commutative) fields, and ${}_FM_G$ an F-G-bimodule. Writing $S = F \times G$, M can be considered as an S-S-bimodule, and we denote by $R({}_FM_G)$ the tensor algebra $T({}_SM_S)$ which, of course, is just the matrix ring $\begin{pmatrix} F & M \\ 0 & G \end{pmatrix}$. For such a bimodule ${}_F M_G$, the invariant $d({}_F M_G) = (\dim {}_F M) \cdot (\dim M_G)$ is known to be of importance (here, dim denotes the ordinary vector space dimension).

An indecomposable R-module X is a (right) R-module having no non-zero submodules Y_1 and Y_2 such that $X = Y_1 \oplus Y_2$. The semiprimary ring R is said to be of *finite representation type* if there is only a finite number of (isomorphism classes of) indecomposable R-modules. In this case, the indecomposable modules are of finite length, and every (arbitrarily large) module is the direct sum of indecomposable ones [9]. We say that a hereditary semiprimary ring R is of *wild representation type* provided that there are fields F and G and a bimodule $M = {}_F M_G$ with $d(M) > 4$ such that the category of all R(M)-modules of finite length can be embedded as a full and exact subcategory into the category $\mathcal{M}_R$ of R-modules of finite length. (For arbitrary, not necessarily hereditary semiprimary rings, this definition would be too special; in addition to the full exact subcategories one has to consider also those which are representation equivalent to them). This rather technical condition has the following interpretation: If R is a finite dimensional algebra over a commutative field, then the condition is equivalent to the fact that there exists a commutative field k such that for any finite dimensional k-algebra A , the category of all A-modules of finite length can be embedded as a full and exact subcategory into $\mathcal{M}_R$. Thus, it is unreasonable to expect a complete classification of all R-modules of finite length in this case. Also in the general case, the category $\mathcal{M}_R$ with R of wild representation type seems to contain a rather large amount of indecomposable modules of finite length. For example, for such a ring R and any natural number n , there exists ${}_F M_G$ with $d(M) > n$ such that $\mathcal{M}_{R(M)}$ can be embedded as a full and exact subcategory into $\mathcal{M}_R$. If R is neither of finite nor of wild representation type, then R is called *tame*.

The hereditary semiprimary rings of finite representation type, together with all their modules, have been completely described in [4]; a more conceptual account has appeared in [5]. In particular, it has been shown that they are always tensor rings. The tame tensor rings which satisfy some duality condition (which is, for example, satisfied in the case of finite dimensional algebras) have been described in [5]; also all but a certain class of their modules (the "homogeneous" ones) were exhibited in detail there. However, the following construction will show that there are tame hereditary semiprimary rings which are not tensor rings.

Let F be a field, ε an automorphism of F, and δ an $(\varepsilon,1)$-derivation of F (that is, $\delta : F \longrightarrow F$ is additive and satisfies $\delta(f_1f_2) = \varepsilon(f_1)\cdot\delta(f_2) + \delta(f_1)\cdot f_2$; see [2]). This information leads to an F-F-bimodule $M = M(\varepsilon,\delta)$ which, as a left F-space is just ${}_FF \oplus {}_FF$, whereas the right F-action is given by

$$(a,b)\cdot f = (af + b\cdot\delta(f), b\cdot\varepsilon(f)) \quad \text{for } a,b,f \in F .$$

Note that the F-F-submodule $F \oplus 0$ of M and ${}_FF_F$ are canonically isomorphic, and we shall identify them. Now we can define the ring $\tilde{A}_n(\varepsilon,\delta)$ as the $(n+1) \times (n+1)$-matrix ring

$$\begin{pmatrix} F & & & & \\ F & F & & 0 & \\ \vdots & \vdots & \ddots & & \\ F & F & \cdots & F & \\ M & F & \cdots & F & F \end{pmatrix}$$

(it contains, as a subring, the ring of all lower triangular $(n+1) \times (n+1)$ matrices over F) . It is obvious that $\tilde{A}_n(\varepsilon,\delta)$ is hereditary and semiprimary. For $n = 1$, $\tilde{A}_1(\varepsilon,\delta) = R({}_FM_F)$; thus we get a tensor ring. However, for $n > 1$, $\tilde{A}_n(\varepsilon,\delta)$ is a tensor ring only in the case that the $(\varepsilon,1)$-derivation δ is inner [2]. Indeed, $M(\varepsilon,\delta)$

decomposes as a bimodule if and only if $\delta(f) = \varepsilon(f)c - cf$ for a suitable $c \in F$.

Theorem 1. *A tame hereditary semiprimary ring is Morita equivalent to the product of a tensor ring and a finite number of rings of the form* $\tilde{A}_n(\varepsilon,\delta)$ *for a suitable choice of* n's, ε's *and* δ's.

Next we are going to consider the question whether it is possible to describe the category of all $\tilde{A}_n(\varepsilon,\delta)$-modules of finite length. For $\delta = 0$, this has been done in [5] (where the $\tilde{A}_n(\varepsilon,0)$-modules were considered as representations of the extended Dynkin diagram $\tilde{\mathbf{A}}_n$ with respect to a modulation using ε, and a suitable orientation; note that this explains our notation $\tilde{A}_n(\varepsilon,\delta)$). On the other hand, for $n = 1$, the problem was solved in [7] (where the $\tilde{A}_1(\varepsilon,\delta)$-modules were called representations of a non-simple affine bimodule). The general case consists in a combination of these results, as the next theorem reveals. We denote by $F[z;\ \varepsilon,\ \delta]$ the twisted polynomial ring consisting of all finite formal sums $\sum_{t=0} f_t z^t$ with $f_t \in F$, and multiplication defined by $zf = \varepsilon(f)z + \delta(f)$. Evidently, $\tilde{A}_n(\varepsilon,\delta)$ has precisely $n+1$ simple modules $S_0,\ldots, S_n$, which are ordered in such a way that $\mathrm{Ext}^1(S_i, S_{i-1}) \neq 0$ for $1 \leq i \leq n$; then we have also $\mathrm{Ext}^1(S_n, S_0) \neq 0$.

In what follows, always $n > 1$. A sequence $(r_1,\ldots,r_d)$ of integers satisfying $0 \leq r_t \leq n$ is called *regular* if $r_1 \neq n$, $r_d \neq 0$ and $r_{t+1} = \pi(r_t)$ for all t, where π is the cyclic permutation $(1\ 2 \ldots n\text{-}1\ 0\ n)$. Furthermore, $(r_1, r_2,\ldots, r_d)$ is called *preprojective* if $r_1 = 0$, $r_2 = 1$ and $(r_2,\ldots,r_d)$ is regular. Finally, $(r_1,\ldots, r_{d-1}, r_d)$ is called *preinjective* if $r_{d-1} = n-1$, $r_d = n$ and $(r_1,\ldots, r_{d-1})$ is regular.

Theorem 2. *Let* $n > 1$. *The indecomposable* $\tilde{A}_n(\varepsilon,\delta)$*-modules are of the following types: For every regular, preprojective, or preinjective sequence* $(r_1, \ldots, r_d)$, *there is a unique indecomposable module* X *with a composition series*

$$0 = X_o \subset X_1 \subset \ldots . \subset X_d = X ,$$

such that $X_t/X_{t-1} \approx S_{r_t}$ *for all* t . *The direct sums of the remaining indecomposable modules of finite length form an abelian exact subcategory which is equivalent to the category of* $F[z;\ \varepsilon,\ \delta]$*-modules of finite length.*

The equivalence of the categories is given in the following way: The $F[z;\ \varepsilon,\ \delta]$-module W is associated with the $\tilde{A}_n(\varepsilon,\delta)$-module $\underbrace{W \oplus \ldots \oplus W}_{n+1}$ on which $\tilde{A}_n(\varepsilon,\delta)$ operates by the ordinary matrix operation from the right, with the additional condition that, for $(a,b) \in M$ and $w \in W$, we define $w(a,b) = wa + wbz$. The above theorem has several consequences:

Corollary 1. *The representation type of a hereditary semiprimary ring* A *is not determined by the quotient* $A/N(A)^2$, *where* $N(A)$ *denotes the radical of* A .

In fact, taking for F the field $\mathbb{C}(z)$ of rational functions in one variable over the complex numbers, for ε the identity automorphism and for δ the ordinary differentiation of functions, one can use, as in [7], the results of [6] to show that the ring $A = \tilde{A}_n(1,\delta)$ is wild, while the ring $B = \tilde{A}_n(1,0)$ is tame. But for $n > 1$, $A/N(A)^2 = B/N(B)^2$. Furthermore, using the results of [3], the ring $\tilde{A}_n(1,\delta)$, where F is a differentially closed field with a differential δ , will have only finitely many indecomposable modules of any given finite length. In contrast, $\tilde{A}_n(1,0)$ with any infinite field F has infinitely many non-isomorphic indecomposable modules of length d for an infinite number of integers d .

Now, if we assume that F contains a central subfield k of finite index such that the restriction of ε to k is the identity and δ is trivial on k , then (and only then) $\tilde{A}_n(\varepsilon,\delta)$ becomes a finite dimensional k-algebra. Moreover, the $F[z, \varepsilon, \delta]$-modules of finite length are uniserial, so that $\tilde{A}_n(\varepsilon,\delta)$ cannot be wild. Thus, in this case, the two theorems yield the following result.

Corollary 2. *Let* k *be a commutative field and* A *a hereditary finite dimensional* k*-algebra. Then* A *is tame if and only if it is Morita equivalent to the product of a tame tensor algebra and a finite number of algebras of the form* $\tilde{A}_n(\varepsilon,\delta)$.

Let us remark that an indecomposable tame tensor algebra is determined by a modulation of an extended Dynkin diagram.

The paper is divided into 3 sections followed by a remark. In §2, two endofunctor are defined in the category $\mathcal{M}_A$ of all $\tilde{A}_n(\varepsilon,\delta)$-modules of finite length, and are used then in §3 to prove Theorem 2. The proof of Theorem 1 is completed in §4. The final remark refers to a recent paper [1].

§2. CONSTRUCTION OF FUNCTORS

In the representation theory of tensor rings, certain functors play a decisive role; see [5]. One constructs first elementary functors whose behaviour imitates that of the basic reflections in the Weyl group, and then composes them to the Coxeter functors, which correspond to the Coxeter transformations in the Weyl group. The aim of this section is to construct similar functors for the rings of the form $A = \tilde{A}_n(\varepsilon,\delta)$. It turns out that the situation is much easier in this case: The elementary functors which will be denoted by Γ^+ and Γ^- are endofunctors, and one may work with them in the same way as one usually does with the Coxeter functors (which are here the (n+1)-powers $\Gamma^{+(n+1)}$ and $\Gamma^{-(n+1)}$ of the elementary functors).

We denote by M_A the category of all A-modules of finite length. If $S_0, \ldots, S_n$ is the canonical ordering of the simple A-modules introduced in §1, then for every A-module X of finite length, we denote by $\underline{\dim}\, X = (x_0, \ldots, x_n)$ the (n+1)-tuple in which x_i is the number of the composition factors of X isomorphic to S_i ; we shall consider $\underline{\dim}\, X$ to be an element of the rational vector space Q^{n+1}. Observe that A has precisely one projective simple module, namely S_0 , and precisely one injective simple module S_n . Let us reiterate that, since the case n=1 is known by [5] and [7], we shall always assume $n > 1$.

Proposition. *Let* $n > 1$. *For the ring* $A = \tilde{A}_n(\varepsilon,\delta)$, *there exist functors* $\Gamma^+ : M_A \to M_A$ *and* $\Gamma^- : M_A \to M_A$ *with the following properties:*

(i) $\Gamma^+ S_0 = 0$, *while for any other indecomposable* A-*module* X *(with* $\underline{\dim}\, X = (x_0, x_1, \ldots, x_n)$) , *there is a canonical isomorphism* $\Gamma^-\Gamma^+ X \simeq X$ *and*

$$\underline{\dim}\, \Gamma^+ X = (x_1, \ldots, x_n, -x_0 + x_1 + x_n).$$

(ii) $\Gamma^- S_n = 0$, *while for any other indecomposable* A-*module* Y *(with* $\underline{\dim}\, Y = (y_0, \ldots, y_{n-1}, y_n)$, *there is a canonical isomorphism* $\Gamma^+\Gamma^- Y \simeq Y$ *and*

$$\underline{\dim}\, \Gamma^- Y = (-y_n + y_0 + y_{n-1}, y_0, \ldots, y_{n-1}).$$

(iii) Γ^+ *is a right adjoint for* Γ^- .

As a consequence, one gets again the usual properties (compare [5]: Γ^+ is left exact; Γ^- is right exact; if $X \neq S_0$ is indecomposable, then $\mathrm{End}(X) \simeq \mathrm{End}(\Gamma^+ X)$ and $\mathrm{Ext}^1(Y,X) \simeq \mathrm{Ext}^1(\Gamma^+ Y, \Gamma^+ X)$ for all Y etc. [On the other hand, note that the dim-formulas are different from the usual ones. The dim-change is the composition of the expected reflection $(x_0, \ldots, x_n) \longmapsto (-x_0 + x_1 + x_n, x_1, \ldots, x_n)$ and a cyclic permutation. The reason for the appearance of the cyclic permutation lies in the fact that we use a fixed (internally defined) ordering of the simple modules S_i's.]

In order to facilitate the work with the A-modules we shall interpret them as the representations of the species where $F_i = F$ for all i.

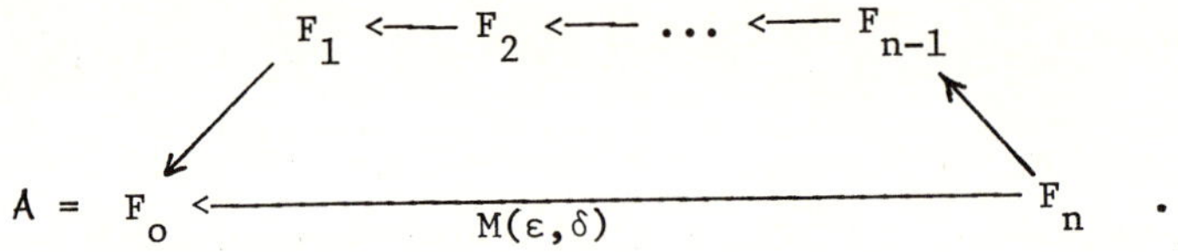

This seems to be the easiest way to get a better understanding of the internal structure of the modules and to have some graphical methods for illustration available. Recall that there is a canonical copy of ${}_F F_F$ embedded into $M(\varepsilon,\delta)$; we denote the embedding by ι . The representation $\underline{X} = (X_i, {}_j\phi_i)$ of the species A consist of $n+1$ F-vector spaces X_i , n linear maps ${}_{i-1}\phi_i : X_i \longrightarrow X_{i-1}$ $(1 \leq i \leq n)$ and a linear map ${}_0\phi_n : X_n \otimes_F M(\varepsilon,\delta) \longrightarrow X_0$. A map $\underline{\alpha} = (\alpha_i) : (X_i, {}_j\phi_i) \longrightarrow (X_i', {}_j\phi_i')$ is given by $n+1$ linear maps $\alpha_i : X_i \longrightarrow X_i'$ such that

$$\alpha_{i-1}\, {}_{i-1}\phi_i = {}_{i-1}\phi_i'\, \alpha_i \quad (1 \leq i \leq n) \quad \text{and} \quad \alpha_0\, {}_0\phi_n = {}_0\phi_n'(\alpha_n \otimes 1) .$$

The full subcategory of all representations $(X_i, {}_j\phi_i)$ of A such that

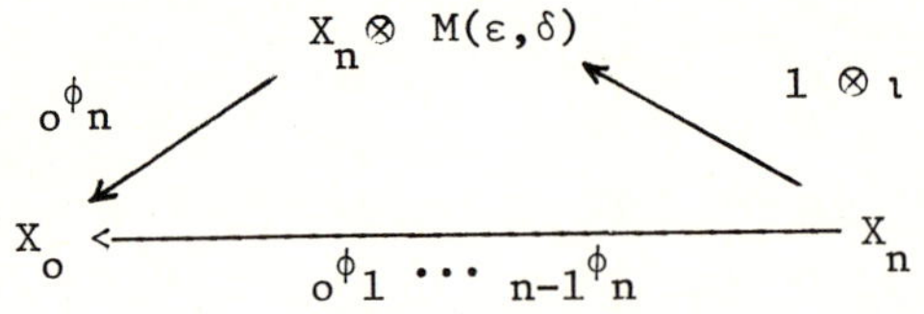

(1)

commutes will be denoted by $\mathcal{Lc}(A)$. This category is equivalent to the category $\mathcal{M}_A$ of all $\tilde{A}_n(\varepsilon,\delta)$-modules of finite length.

It seems to be convenient to have another description of the category $\mathcal{M}_A$ available; in the proof of the Proposition we will use both descriptions simultaneously.

To begin with, we need some additional notation: If X is an F-vector space, and ε an automorphism of the field F, we denote by $X\varepsilon$ the twisted vector space with the scalar multiplication $\cdot$ defined by $x \cdot f = x\,\varepsilon(f)$ for $x \in X$, $f \in F$. If $\alpha : X_F \longrightarrow Y_F$ is F-linear, then the same α, considered as a map $X\varepsilon \longrightarrow Y\varepsilon$, is again F-linear. In particular, considering the F-F-bimodule $F\varepsilon$ (with $f_1 \cdot x \cdot f_2 = f_1 x\,\varepsilon(f_2)$ for $x, f_1, f_2 \in F$), we have $X\varepsilon = X \otimes_F F\varepsilon$. Note that there is an obvious exact sequence of F-F-bimodules

$$0 \longrightarrow F \longrightarrow M(\varepsilon,\delta) \xrightarrow{\pi} F\varepsilon \longrightarrow 0 .$$

Also, recall that for a bimodule ${}_FN_G$ over the fields F and G with finite $\dim {}_FN$, we may define an G-F-bimodule $^*N = \mathrm{Hom}_F({}_FN_G, {}_FF_F)$ (left dual!) such that there is a canonical one-to-one correspondence between the G-linear maps $X_G \longrightarrow Y_F \otimes {}_FN_G$ and the F-linear maps $X_G \otimes {}_G(^*N)_F \longrightarrow Y_F$ for any vector spaces X_G and Y_F. It is obvious that for an automorphism ε of F, one has $^*(F\varepsilon) = F\varepsilon^{-1}$. Also, we will need that $^*(M(\varepsilon,\delta)) = F\varepsilon^{-1} \otimes M(\varepsilon,\delta)$. [Proof [7]: Consider the basis of the left F-space $N = M(\varepsilon,\delta) \otimes F\varepsilon^{-1} \otimes M(\varepsilon,\delta)$ given by the elements $u_{ij} = u_i \otimes 1 \otimes u_j$ $(1 \leq i, j \leq 2)$, where $u_1 = (1,0)$ and $u_2 = (0,1) \in M(\varepsilon,\delta)$. Then,

$$u_{11}\cdot f = \varepsilon^{-1}(f)\cdot u_{11},$$
$$u_{12}\cdot f = \varepsilon^{-1}\delta(f)\cdot u_{11} + f\cdot u_{12},$$
$$u_{21}\cdot f = \delta\varepsilon^{-1}(f)\cdot u_{11} + f\cdot u_{21},$$
$$u_{22}\cdot f = \delta\varepsilon^{-1}\delta(f)\cdot u_{11} + \delta(f)\cdot u_{12} + \delta(f)\cdot u_{21} + \varepsilon(f)\cdot u_{22}.$$

This shows, that $u_{11}, u_{22}, u_{12} + u_{21}$ generate an F-F-bisubmodule U such that ${}_F N_F / {}_F U_F \simeq {}_F F_F$. The canonical map

$$M(\varepsilon,\delta) \otimes (F\varepsilon^{-1} \otimes M(\varepsilon,\delta)) \longrightarrow N/U \simeq {}_F F_F$$

induces the map

$$F\varepsilon^{-1} \otimes M(\varepsilon,\delta) \longrightarrow \mathrm{Hom}_F (M(\varepsilon,\delta), {}_F F_F) \simeq {}^*M(\varepsilon,\delta)$$

which is obviously injective, and is therefore an isomorphism of bimodules.]

We know that the A-modules correspond to those representations of the species

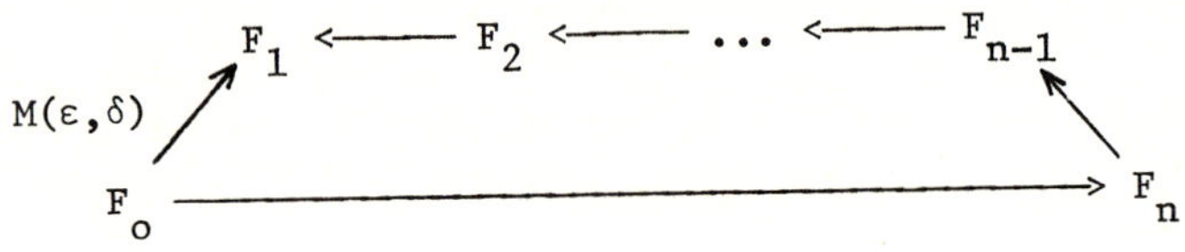

with $F_i = F$ for all i which satisfy a certain commutativity condition (note that we use now a different numbering of the indices!). Equally well, we may consider the species

$\mathcal{B}$ =

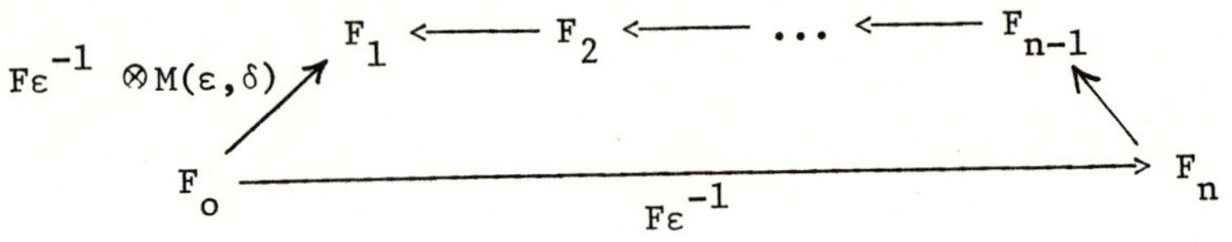

with $F_i = F$ for all i : The categories of the representations are obviously equivalent under the functor $(X_o, \ldots, X_n, {}_j\phi_i) \longmapsto (X_o\varepsilon, X_1, \ldots, X_n, {}_j\phi_i)$. Since $F\varepsilon^{-1} = {}^*(F\varepsilon)$ and $F\varepsilon^{-1} \otimes M(\varepsilon,\delta) = {}^*M(\varepsilon,\delta)$, the representations of $\mathcal{B}$ can be written in the form $(Y_i, {}_j\psi_i)$ with the F-vector spaces Y_i $(0 \leq i \leq n)$ and the linear maps

${}_1\psi_o : Y_o \longrightarrow Y_1 \otimes M(\varepsilon,\delta)$, ${}_{i-1}\psi_i : Y_i \longrightarrow Y_{i-1}$ $(2 \leq i \leq n)$, and ${}_n\psi_o : Y_o \longrightarrow Y_n\varepsilon$.

In this way, the A-modules correspond just to those $(Y_i, {}_j\psi_i)$ for which the diagram

$$\begin{array}{ccccc} & & Y_1 \otimes M(\varepsilon,\delta) & \xrightarrow{1\otimes\pi} & Y_1\varepsilon \\ & {}_1\psi_0 \nearrow & & & \nwarrow {}_1\psi_2 \cdots {}_{n-1}\psi_n \\ Y_0 & & \xrightarrow{\quad {}_n\psi_0 \quad} & & Y_n\varepsilon \end{array} \qquad (2)$$

commutes. The full subcategory of those representations of $\mathcal{B}$ which satisfy the condition (2) will be denoted by $\mathcal{Lc}'(\mathcal{B})$. Obviously, it is equivalent to $\mathcal{M}_A$ and in this way we get an alternative description of the category of all A-modules of finite length.

In the remaining part of this section, we denote the bimodule $M(\varepsilon,\delta)$ simply by M. Recall that we have an exact sequence of bimodules $0 \longrightarrow F \xrightarrow{\iota} M \xrightarrow{\pi} F\varepsilon \longrightarrow 0$.

Let X_1, X_n be F-spaces, and $\xi : X_n \longrightarrow X_1$ a linear map. We claim that the following sequence is exact:

$$0 \longrightarrow X_n \xrightarrow{\begin{pmatrix}\xi \\ 1\otimes\iota\end{pmatrix}} X_1 \oplus (X_n \otimes M) \xrightarrow{\begin{pmatrix}1\otimes\iota & -\xi\otimes 1 \\ 0 & 1\otimes\pi\end{pmatrix}} (X_1 \otimes M) \oplus X_n\varepsilon \xrightarrow{(1\otimes\pi,\xi)} X_1\varepsilon \longrightarrow 0 .$$

For, it is easy to check that the composition of any two consecutive maps is zero, and the following diagram shows that the sequence is the extension of two exact sequences:

$$\begin{array}{ccccccccc} 0 & \longrightarrow & & & X_1 & \xrightarrow{1\otimes\iota} & X_1 \otimes M & \xrightarrow{1\otimes\pi} & X_1\varepsilon \longrightarrow 0 \\ & & & & \downarrow\begin{pmatrix}1\\o\end{pmatrix} & & \downarrow\begin{pmatrix}1\\o\end{pmatrix} & & \\ 0 & \longrightarrow & X_n & \longrightarrow & X_1 \oplus (X_n \otimes M) & \longrightarrow & (X_1 \otimes M) \oplus X_n\varepsilon & \longrightarrow & X_1\varepsilon \longrightarrow 0 \\ & & & & \downarrow (0,1) & & \downarrow (0,1) & & \\ 0 & \longrightarrow & X_n & \xrightarrow{1\otimes\iota} & X_n \otimes M & \xrightarrow{1\otimes\pi} & X_n\varepsilon & \longrightarrow & 0 \end{array}$$

Let Q be the image of the middle map $\begin{pmatrix} 1 \otimes \iota & -\xi \otimes 1 \\ 0 & 1 \otimes \pi \end{pmatrix}$, and let $\eta_1, \eta_n, \mu_1,$ and μ_n be the corresponding canonical maps; hence, in the diagram

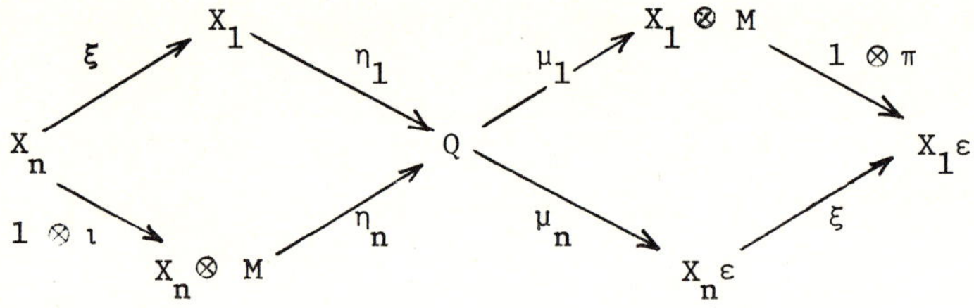

the left square is a pushout and the right square is a pullback. This pair of squares will be of importance in the sequel. The dimension of Q_F is easy to calculate:

$$\dim Q_F = \dim (X_1)_F + \dim (X_n)_F \, .$$

Now, we are going to define the functor

$$\Gamma^+ \; : \; M_A \simeq \mathcal{L}c(A) \longrightarrow \mathcal{L}c'(B) \;\simeq\; M_A$$

of the Proposition. Let $\underline{X} = (X_i, {}_j\phi_i)$ be a representation in $\mathcal{L}c(A)$. Take the map $\xi = {}_1\phi_2 \cdots {}_{n-1}\phi_n : X_n \longrightarrow X_1$ and form the above pair of squares. The commutativity condition (1) shows that we can factor the two maps ${}_0\phi_1$ and ${}_0\phi_n$ through Q and get

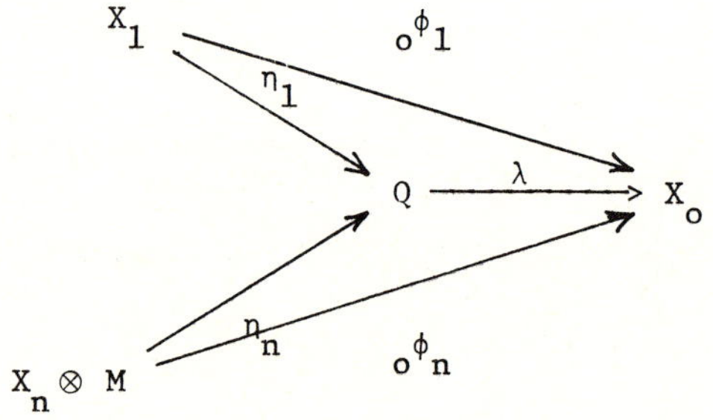

Now, define Y_o to be the kernel of the map λ ; thus we have

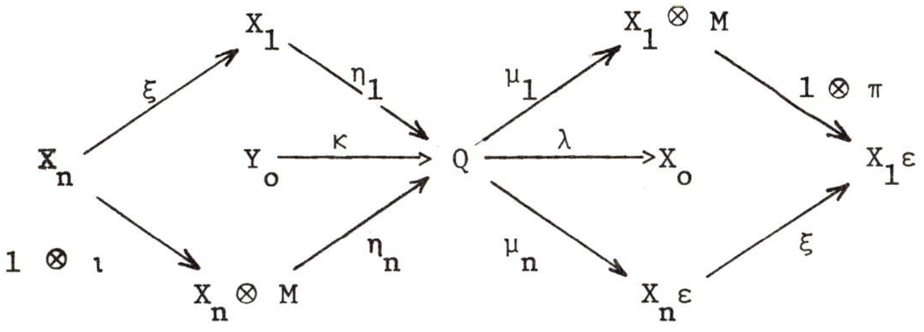

If, in addition, we define $Y_i = X_i$ for $i \neq 0$, ${}_{i-1}\psi_1 = {}_{i-1}\phi_1$ for $2 \leq i \leq n$, and ${}_1\psi_o = \mu_1 k$, ${}_n\psi_o = \mu_n k$, we get a representation $\underline{Y} = (Y_i, {}_j\psi_i)$ in $\mathcal{Lc}'(\mathcal{B})$; put $\Gamma^+\underline{X} = \underline{Y}$. Note that the commutativity condition (2) follows from the commutativity of the right square above. It is obvious that this construction is functorial.

The reverse functor Γ^- is equally easy to construct given a representation $\underline{Y} = (Y_i, {}_j\psi_i)$ in $\mathcal{Lc}'(\mathcal{B})$, we define $\Gamma^-\underline{Y} = \underline{X} = (X_i, {}_j\phi_i)$ as follows. Put $X_i = Y_i$ for $i \neq 0$, and ${}_{i-1}\phi_i = {}_{i-1}\psi_i$ for $\underline{2 < i < n}$. Then denoting the composition ${}_1\phi_2 \cdots {}_{n-1}\phi_n$ of these maps by ξ, we use again for this ξ the pair of squares. This time, we factor the two maps ${}_1\psi_o$ and ${}_n\psi_o$ through Q, and, in this way, we get $k : Y_o \longrightarrow Q$. Denote the cokernel of k by X_o and the cokernel map by $\lambda : Q \longrightarrow X_o$. In order to complete the definition of Γ^- we set ${}_o\phi_1 = \lambda\eta_1$ and ${}_o\phi_n = \lambda\eta_n$.

Next, assume that $\underline{X} \in \mathcal{Lc}(\mathcal{A})$ is given. Then either the map $\lambda : Q \longrightarrow X_o$ (defined above) is surjective, in which case we obviously have $\Gamma^-\Gamma^+\underline{X} = \underline{X}$, and also $\dim (Y_o)_F = \dim Q_F - \dim (X_o)_F$. Or, λ is not surjective, and then a copy of the representation

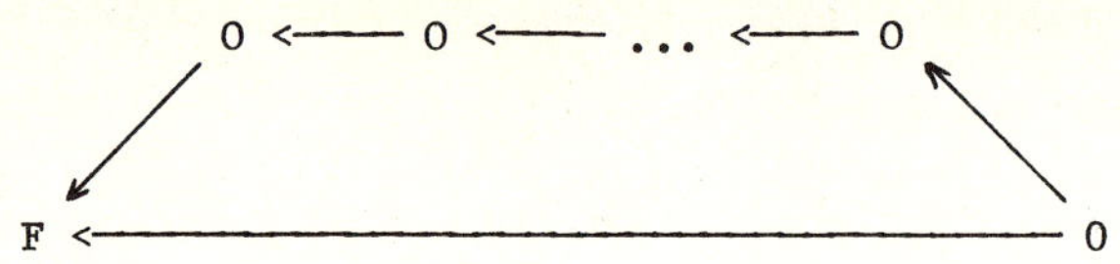

splits off $\underline{X}$. Thus, if $\underline{X}$ is indecomposable, then it is simple projective (of course, conversely, this is the only simple projective representation in $Lc(A)$). Similar arguments apply for $\underline{Y} \in Lc'(B)$ and establish that either $\Gamma^+\Gamma^-\underline{Y} = \underline{Y}$, or that a simple injective representation is a direct summand of $\underline{Y}$.

The dimension $\underline{\dim}\, X$ of an A-module X was defined in terms of the canonical ordering of the simple modules S_i introduced in §1 . Thus, if the A-module Y corresponds to the representation $\underline{Y} = (Y_i, {}_j\psi_i)$ in $Lc'(B)$ and if $\dim (Y_i)_F = y_i$, then $\underline{\dim}\, Y = (y_1, \ldots, y_n, y_o)$. This immediately yields the $\underline{\dim}$-formula, since for an indecomposable representation $\underline{X} = (X_i, {}_j\phi_i)$ in $Lc(A)$ which is not simple projective and which satisfies $\Gamma^+\underline{X} = \underline{Y}$, we have

$$\dim Y_o = \dim Q - \dim X_o = \dim X_1 + \dim X_n - \dim X_o .$$

To complete the proof of the Proposition, it remains to show that the functors Γ^+ and Γ^- are adjoint. Let $\underline{X} \in Lc(A)$, $\underline{Y} \in Lc'(B)$ and $\underline{\alpha} = (\alpha_i) : \underline{Y} \longrightarrow \Gamma^+\underline{X}$. In order to define the corresponding map $\Gamma^-\underline{Y} \longrightarrow \underline{X}$, we only replace $\alpha_o : Y_o \longrightarrow (\Gamma^+\underline{X})_o$ by a suitable map $\beta : (\Gamma^-\underline{Y})_o \longrightarrow X_o$ defined by the following diagram

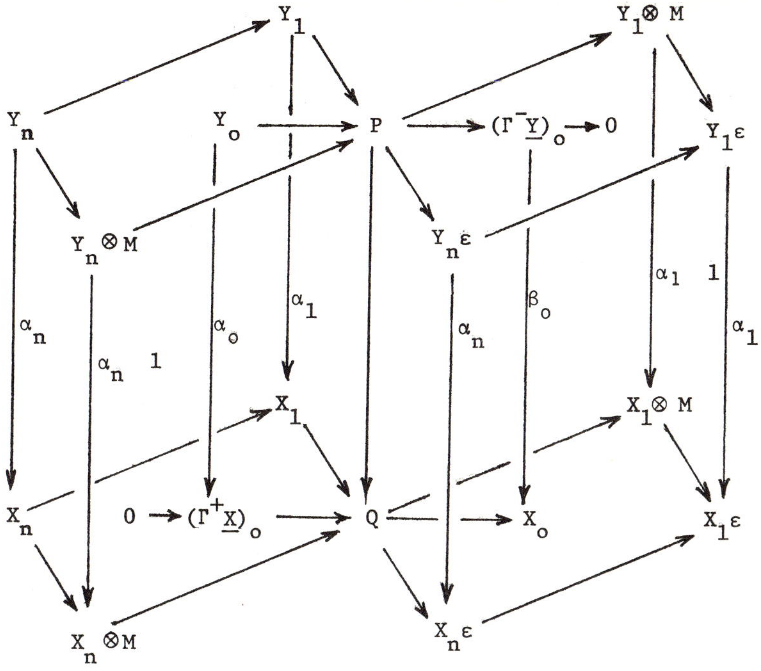

It is clear that, in this way, we get a bijection between Hom ($\underline{Y}$, $\Gamma^+\underline{X}$) and Hom ($\Gamma^-\underline{Y}$, $\underline{X}$) which is natural in both arguments.

§3. THE INDECOMPOSABLE $\tilde{A}_n(\varepsilon,\delta)$-MODULES

Our aim is to prove Theorem 2 of §1. Hence, we deal with the modules over a ring $A = \tilde{A}_n(\varepsilon,\delta)$ for some fixed $n > 1$, ε and δ. As in the case of tame tensor rings, we introduce the notion of *defect* of a module. Thus, let X be an A-module of finite length with $\underline{\dim}\, X = (x_0,\ldots, x_n)$. The defect ∂X of X is defined to be the difference between the number of simple injective and the number of simple projective composition factors of X:

$$\partial X = - x_0 + x_n \quad .$$

First, let us formulate the following (rather trivial) assertions.

Lemma. *Let* X *be an indecomposable* A-*module. If* $X \neq S_o$, *then* $\partial X = \partial\Gamma^{+}X$. *If* $\Gamma^{+n}X \neq 0$, *then the lengths of* $\Gamma^{+n}X$ *and* X *satisfy the formula*

$$\text{length } \Gamma^{+n}X = \text{length } X + (n+1)\ \partial X \quad .$$

Proof. The first assertion follows from

$$\partial\Gamma^{+}X = -\ x_1 - x_o + x_1 + x_n = -\ x_o + x_n = \partial X \ .$$

To verify the second formula, observe that

$$\underline{\dim}\ \Gamma^{+}X = (y_o, y_1, \ldots, y_n)\ , \text{ where } \ y_i = -x_o + x_i + x_n \quad (0 \leq i \leq n) \ .$$

Thus the length of $\Gamma^{+n}X$ equals to

$$\sum_{i=o}^{n} y_i = (n+1)(-x_o + x_n) + \sum_{i=0}^{n} x_i = (n+1)\ \partial X + \text{length } X \ ,$$

as required.

Now, the procedure to describe all indecomposable A-modules of finite length is rather clear. Since it follows precisely the arguments used in the case of a tame tensor ring, we shall outline only the main steps; for further details, one is referred to [5] and [7].

If X is an indecomposable A-module of negative defect, then, for some r, $\Gamma^{+r}X = 0$. For, otherwise one could apply Γ^{+n} inductively and would get non-zero modules $\Gamma^{+nt}X$ for all $t \in \mathbb{N}$. The length of $\Gamma^{+nt}X$ equals length $X + t(n+1)\ \partial X$, which has to become negative for large t ; a contradiction. If r is the least number with $\Gamma^{+(r+1)}X = 0$, then $\Gamma^{+r}X \simeq S_o$ (the simple projective module) and therefore $X \simeq \Gamma^{-r}S_o$. Consequently, the dimension type of X is of the form

$(\underbrace{x, \ldots, x}_{S}, x-1, \ldots, x-1)$ with $1 \leq S \leq n$. These modules are called *preprojective;* they are uniquely determined by their dimension types, and are all of defect -1. Similarly, the indecomposable modules of positive defect are called *preinjective.* They are of the form $\Gamma^{+r}S_n$, the dimension type is of the form $(\underbrace{x, \ldots, x}_{S}, x+1, \ldots, x+1)$ for some S with $1 \leq S \leq n$, and their defect is $+1$.

Now, the direct sums of indecomposable A-modules of zero defect form an abelian, exact, extension closed subcategory $\mathcal{R}$ of $\mathcal{M}_A$; the objects of $\mathcal{R}$ will be called *regular.* Some of the simple objects of $\mathcal{R}$ (simple in the category $\mathcal{R}$, not necessarily simple A-modules) can be easily listed: $S_1, \ldots, S_{n-1}$ (these modules are actually simple A-modules) and the indecomposable A-module T of dimension type $\underline{\dim}\, T = (1,0,\ldots,0,1)$ corresponding to the non-zero elements in $\mathrm{Ext}^1(S_n, S_o)$. The A-modules from $\mathcal{R}$ whose composition factors (in $\mathcal{R}$) are all of the forms $S_1, \ldots, S_{n-1}$, and T, form a serial subcategory $\mathcal{U}$ of global dimension 1. The indecomposable A-modules which belong to $\mathcal{U}$ are uniquely determined by their lowest composition factor and their length (in $\mathcal{R}$). Note that $\mathcal{U}$ is stable under Γ^+ and Γ^-; for example, $\Gamma^+T = S_{n-1}$, $\Gamma^+S_i = S_{i-1}$ (for $2 \leq i \leq n-1$), and $\Gamma^+S_1 = T$. Also, $\mathcal{R}$ is the product category of $\mathcal{U}$ and of the category $\mathcal{H}$ of all homogeneous A-modules; here, an object of $\mathcal{R}$ is called *homogeneous* if none of its composition factors (in $\mathcal{R}$) equals S_i $(1 \leq i \leq n-1)$ or T. Equivalently, an A-module X is homogeneous if the maps ${}_{i-1}\phi_i$ $(1 \leq i \leq n)$ of $\underline{X} = (X_i, {}_j\phi_i)$ are all isomorphisms. Here, of course, we consider X as a representation $\underline{X}$ of the species A. [In order to see that $\mathcal{R}$ is the product category of $\mathcal{U}$ and $\mathcal{H}$, one shows that $\mathrm{Ext}^1(H, S_i) = 0 = \mathrm{Ext}^1(S_i, H)$ for all simple homogeneous objects H and $1 \leq i \leq n-1$. Indeed, given an exact sequence

$$0 \longrightarrow H \longrightarrow E \longrightarrow S_i \longrightarrow 0 \quad ,$$

one can embed S_i into E using $\ker {}_{i-1}\phi_i$. Similarly, given an exact sequence

$$0 \longrightarrow S_i \longrightarrow E \longrightarrow H \longrightarrow 0 \quad ,$$

one can embed H into E using the image of ${}_i\phi_{i+1}$. Now applying the functor Γ^+ to H and S_1, we get also that $\mathrm{Ext}^1(H,T) = 0 = \mathrm{Ext}^1(T,H)$ for all simple homogeneous objects H.]

The category $\mathcal{H}$ is easily seen to be equivalent to the category of all $F[z;\varepsilon,\delta]$-modules of finite length. The equivalence functor $\mathcal{M}_{F[z;\varepsilon,\delta]} \longrightarrow \mathcal{M}_A$ is defined by $W \longmapsto (X_o,\dots, X_n, {}_j\phi_i)$, where $X_i = W$ for all i, ${}_{i-1}\phi_i$ is the identity map for $1 \leq i \leq n$, and ${}_o\phi_n : W \otimes M \longrightarrow W$ is given by ${}_o\phi_n\ (w \otimes (a,b)) = wa + wbz$. [Here again, we have identified $\mathcal{M}_A$ with $\mathcal{L}c(A)$. The description of the functor in terms of actual A-modules is given in the introduction!] For, in the case that $\underline{X} = (X_i, {}_j\phi_i)$ is homogeneous, we may identify the different vector spaces X_i via the maps ${}_{i-1}\phi_i$, so that these maps become the identity maps. Thus, the only map of interest is ${}_o\phi_n : X_n \otimes M \longrightarrow X_o$. However, according to (1), the restriction of ${}_o\phi_n$ to $X_n \otimes F$ is the identity map, too. Consequently, the only invariant are the values ${}_o\phi_n(w \otimes (o,1))$ for $w \in X_n = X_o$. If we define $w \cdot z = {}_o\phi_n(w \otimes (o,1))$, then we get an $F[z;\varepsilon,\delta]$-module structure on X_n.

To complete the proof of Theorem 2, it remains to show that the indecomposable preprojective, or preinjective A-modules, as well as the indecomposable A-modules in $\mathcal{U}$ are uniquely characterized by the existence of a composition series

$$0 = X_o \subset X_1 \subset \dots \subset X_d = X$$

with $X_t/X_{t-1} \simeq S_{r_t}$ for a preprojective, preinjective, or regular sequence $(r_1, \ldots, r_d)$, respectively. Consider first the case of an indecomposable module X in $\mathcal{U}$. Then X has a unique composition series in $\mathcal{U}$ the factors are S_i $(1 \leq i \leq n-1)$ and T. There is a unique refinement of this series to a composition series of A-modules, and the indices of the composition factors obviously form a regular sequence. Conversely, if an indecomposable A-module X has a composition series which corresponds to a regular sequence, then (calculating the defect) X has to be regular. Now, since we can embed either one of the S_i's $(1 \leq i \leq n-1)$ or T into X, X cannot belong to $\mathcal{H}$ and therefore has to belong to $\mathcal{U}$. Next, consider the preprojective modules. Note that there is an inclusion

$$S_0 \hookrightarrow \Gamma^- S_0 \hookrightarrow \Gamma^{-2} S_0 \hookrightarrow \ldots \hookrightarrow \Gamma^{-k} S_0$$

with the factors $\Gamma^{-(k+1)} S_0 / \Gamma^{-k} S_0 \simeq \Gamma^{-k} S_1$. For, consider the sequence

$$0 \longrightarrow S_0 \longrightarrow \Gamma^- S_0 \longrightarrow S_1 \longrightarrow 0 ,$$

and apply Γ^{-k}. If we consider this inclusion series for some $X = \Gamma^{-k} S_0$, we see that there is a unique refinement to a composition series of A-modules and that the indices of the composition factors of this series form a preprojective sequence. A similar argument works in the case of preinjective modules.

Note that the methods and results of this section are not restricted to the A-modules of finite length only, but that they can be used to deal with certain classes of A-modules of arbitrary length. For example, one can show, as in [8], that every union X of a chain

$$X^{(0)} \subset X^{(1)} \subset \ldots \subset X^{(t)} \subset \ldots$$

of indecomposable A-mofules of finite length is again indecomposable, and that either every non-zero endomorphism of X is a monomorphism or an epimorphism, so that, in particular, the endomorphism ring $\mathrm{End}(X)$ of X has no zero-divisors.

§4. WILD RINGS

Let R be a semiprimary ring. Since we are interested only in the representation type of R, we may suppose that R is indecomposable (that is, R cannot be written as the product of two rings), and basic (every simple factor ring is a field). Let N be the radical of R.

Lemma. *Let* f, g *be orthogonal primitive idempotents such that* $F = fRf$ *and* $G = gRg$ *are fields. Then there is a full and exact embedding of the category of* $R({}_F fNg_G)$*-modules of finite length into* mod_R.

Proof. Let $h = 1-f-g$; thus f,g,h form a complete set of orthogonal idempotents. Hence R can be written in the form

$$R = \begin{pmatrix} F & fNg & fNh \\ gNf & G & gNh \\ hNf & hNg & hRh \end{pmatrix}.$$

and any module M_R can be decomposed into the direct sum of abelian group $M = Mf \oplus Mg \oplus Mh$, on which those matrices operate. Let $(X_F, Y_G, \phi : X_F \otimes {}_F fNg_G \longrightarrow Y_G)$ be an $R({}_F fNg_G)$-module. We define an R-module M in the following way: Let $Mf = X$, $Mg = Y$, and $Mh = X \otimes fNh$, and let the scalar multiplication be given by the maps

$$Mf \otimes fNg = X \otimes fNg \xrightarrow{\phi} Y = Mg ,$$

$$Mf \otimes fNh = X \otimes fNh \xrightarrow{id} Mh ,$$

$$Mh \otimes hNg = X \otimes fNh \otimes hNg \xrightarrow{1 \otimes mult} X \otimes fNg \xrightarrow{\phi} Y = Mg ;$$

all the other maps be zero. It is easy to verify that, in this way, we get an R-module and that the respective functor is a full and exact embedding.

This shows, in particular, that in the case that R is tame we have $d({}_F fNg_G) \leq 4$ for all such idempotents f and g.

Assume now, in addition, that R is hereditary. Then for any primitive idempotent e, eRe is a field. Let $e_1, \ldots, e_m$ be a complete set of orthogonal primitive idempotents. The product ΠF_i of the fields $F_i = e_i R e_i$ is a subring of R which complements the radical: $N \oplus \Pi F_i = R$. If ${}_R M_R$ is an R-R-bimodule, we denote by ${}_i M_j$ the submodule ${}_i M_j = e_i M e_j$. Obviously, $M = \oplus\, {}_i M_j$. Usually, we will consider ${}_i M_j$ as an F_i-F_j-bimodule. Assuming that R is tame, we know that $d({}_i N_j) \leq 4$.

On the other hand, if $d({}_i N_j) \leq 3$ for all i,j, then R has to be a tensor ring. Namely, the only way R can fail to be a tensor ring is that for some i,j the extension

$$0 \longrightarrow {}_i(N^2)_j \longrightarrow {}_i N_j \longrightarrow {}_i(N/N^2)_j \longrightarrow 0 \qquad (3)$$

does not split as a sequence of F_i-F_j-bimodules. But then necessarily $\dim_{F_i}({}_i N_j) \geq 2$, and $\dim ({}_i N_j)_{F_j} \geq 2$; thus $d({}_i N_j) \geq 4$. Therefore, we may assume that there exists a pair i,j with $d({}_i N_j) = 4$ such that the corresponding sequence (3) does not split. Necessarily, we have $(d_i(N^2)_j) = 1$, and $d({}_i(N/N^2)_j) = 1$. Let $i = i_o, i_1, \ldots, i_n = j$ be a sequence of maximal length such that ${}_{i_{t-1}}N_{i_t} \neq 0$ for $1 \leq t \leq n$. changing the indices (replacing i_t by t), we are in the situation

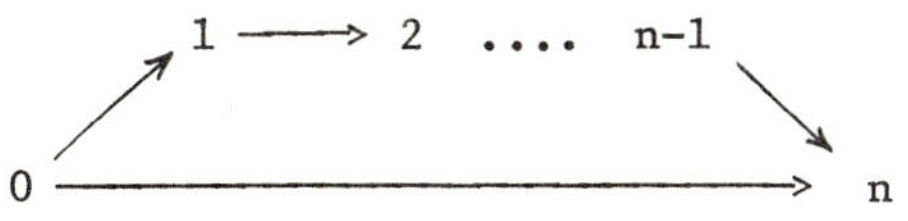

such that for all bimodules ${}_{i-1}N_i$, we have $d({}_{i-1}N_i) = 1$. (For, using the multiplication map, the tensor product ${}_oN_1 \otimes \cdots \otimes {}_{n-1}N_n$ is canonically embedded into ${}_o(N^2)_n$.) In the case that the idempotents $e_o, \ldots, e_n$ form a complete set of orthogonal idempotents, we are just in the situation of a ring of the form $A_n(\varepsilon,\delta)$. For, $d({}_oN_n) = 4$ and the fact that ${}_oN_n$ is not a simple bimodule, imply readily that ${}_oN_n$ is of the form $M(\varepsilon,\delta)$ for some ε, δ [7].

Thus, we may assume that there is e_{n+1} with ${}_iN_{n+1} \neq 0$ for some i . (The case ${}_{n+1}N_i \neq 0$ can be treated similarly.) Let i be the largest possible number with ${}_iN_{n+1} \neq 0$. In case $i = n$, we are in the situation

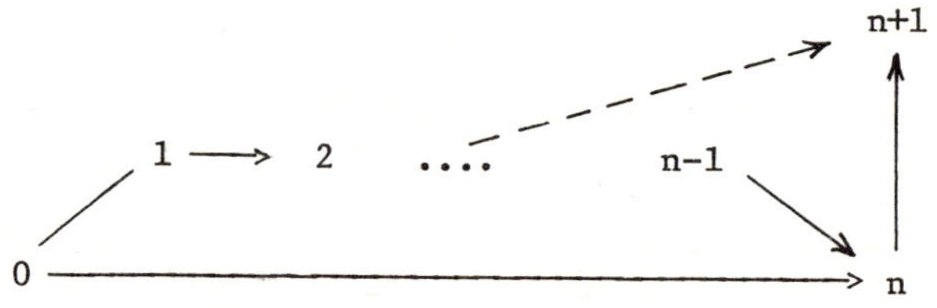

possibly with some additional arrows (indicated by --->). It is easy to see that there is a full and exact embedding of the category of representations of the species C

$$F_o \xrightarrow{{}_oN_n} F_n \xrightarrow{{}_nN_{n+1}} F_{n+1} ,$$

(taking for the maps ${}_i\phi_{i-1}$ $(1 \leq i \leq n-1)$ the identity maps, and for the additional arrows ----> the zero maps). But C is wild: Consider the (unique) indecomposable representations $\underline{X}$ with dimension type (t, t+1, 0) and $\underline{Y}$ with dimension type (t+1, t, tb) , where $b = \dim ({}_nN_{n+1})_{F_{n+1}}$ for some fixed t . Then it is obvious that there are no homomorphisms $\underline{X} \to \underline{Y}$ or $\underline{Y} \to \underline{X}$ (except zero), that the rings of endomorphisms of both $\underline{X}$ and $\underline{Y}$ are fields, and that $\mathrm{Ext}^1(\underline{Y}, \underline{X})$ is arbitrary large (depending on t). It is well-known (see for example [7], lemma 1.5) that there exists a full and exact embedding of $L(\mathrm{Ext}^1(Y,X)^L)$ into $L(C)$.

$\mathcal{D}$ =

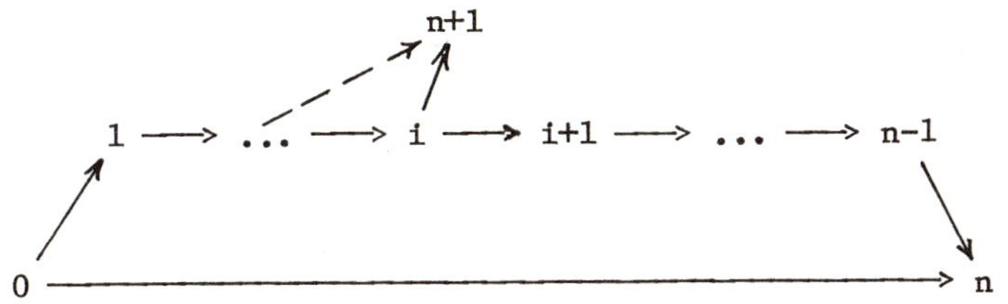

can be reduced to the previous situation using the functor $\Gamma^{-(n-i)}$. Namely, define a functor from the category of representations of

$\mathcal{D}'$ =

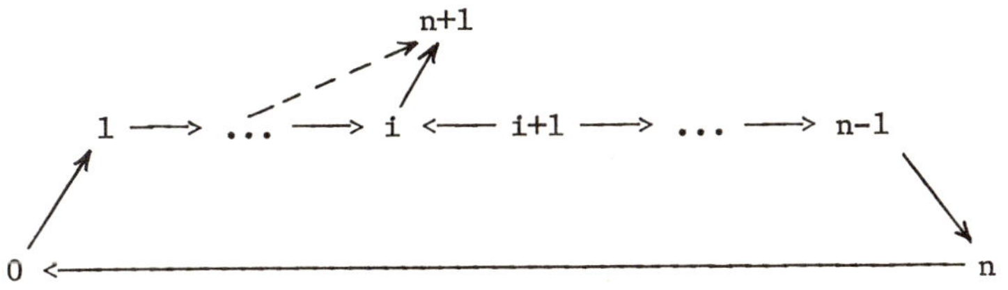

into the category of representations of $\mathcal{D}$, which coincides on the circuit with $\Gamma^{-(n-i)}$ and is the identity elsewhere. This functor kills only a certain number of injective modules, but the full and exact subcategory which is embedded into $L(\mathcal{D}')$ is mapped bijectively into $L(\mathcal{D})$.

In combination with Theorem 2, this completes the proof of Theorem 1.

§5. REMARK

Since this paper deals in some detail with the relationship between hereditary semiprimary rings and tensor rings, a particular class of tensor rings should be mentioned which attracted some interest lately. In a recent paper [1], M. Auslander and M. I. Platzek develop parts of a general representation theory of hereditary artin algebras, and they stress the fact that the techniques used in their paper are applicable for all hereditary artin algebras, not just for those associated with a k-species (those called, in this paper, tensor rings). However, at the beginning of the proof of one of the main theorems (4.1), the authors introduce the following property for a hereditary artin algebra R

(P) If S_o, S_1, S_2 are non-isomorphic simple modules such that $\mathrm{Ext}^1(S_1, S_o) \neq 0$ and $\mathrm{Ext}^1(S_2, S_o) \neq 0$, then $\mathrm{Hom}(P(S_1),P(S_2))=0$, where $P(S_i)$ denotes the projective cover of S_i,

and work with it throughout the proof. In fact, this property is equivalent to the property that the ring is a tensor ring whose diagram does not contain any circuit of the form

Indeed, the diagram of R is constructed in the following way: The points correspond to the simple R-modules, and the arrow $i \longrightarrow j$ means that $\mathrm{Ext}^1(S_i, S_j) \neq 0$. Of course, the condition (P) just excludes circuits of the form mentioned above. However, if this type of circuits is excluded, then the ring R is in fact a tensor ring: Let e_i be pairwise orthogonal idempotents with $P(S_i) = e_iR$. Then $\mathrm{Ext}^1(S_i, S_j) \neq 0$ is equivalent to the fact that $e_i(N/N^2)e_j \neq 0$, and $\mathrm{Hom}(P(S_j), P(S_i)) \neq 0$ is equivalent to the fact that $e_iRe_j \neq 0$ (or, to $e_iNe_j \neq 0$ if $S_i \neq S_j$). The condition (P) therefore can be rephrased as follows:

(P') For any i,j, only one of $e_i(N^2)e_j$ and $e_i(N/N^2)e_j$ can be non-zero.

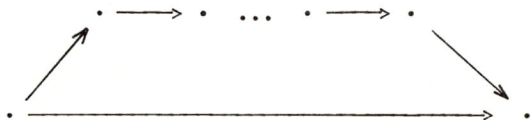

This, of course, immediately implies that for any i,j , the exact sequence

$$0 \longrightarrow e_i(N^2)e_j \longrightarrow e_iNe_j \longrightarrow e_i(N/N^2)e_j \longrightarrow 0$$

splits, and therefore R is a tensor ring.

REFERENCES

1. M. Auslander and I. M. Platzek, Representation Theory of Hereditary Artin Algebras. Preprint 1976.

2. P. M. Cohn, Free Rings and their Relations, Academic Press, New York, 1971.

3. J. H. Cozzens, Homological Properties of the Ring of Differential Polynomials, Bull. Amer. Math. Soc. 76(1970), 75-79.

4. V. Dlab and C. M. Ringel, On Algebras of Finite Representation Type, J. Algebra 33(1975), 306-394.

5. V. Dlab and C. M. Ringel, Indecomposable Representations of Graphs and Algebras, Memoirs Amer. Math. Soc, No.173, Providence, 1976.

6. J. C. MacConnell and J. C. Robson, Homomorphisms and Extensions of Modules over certain Differential Polynomial Rings, J. Algebra 26(1973), 319-342.

7. C. M. Ringel, Representations of K-species and Bimodules, J. Algebra 40 (1976).

8. C. M. Ringel, Unions of Chains of Indecomposable Modules, Comm. Algebra 3(1975), 1121-1144.

9. C. M. Ringel and H. Tachikawa, QF-3 Rings, J. Reine und Angew. Math. 272(1975), 49-72.

10. A. Zaks, A Note on Semi-primary Hereditary Rings, Pacific J. Math. 23(1967), 627-628.

THE BILINEAR INVARIANTS OF A 2-GROUP

K. L. Fields
University of California, San Diego
San Diego, California

At the 1950 International Congress of Mathematicians, Richard Brauer announced that the Schur index of an irreducible representation of a p-group is either 1 or 2. It was not until 1958, however, that a proof of this result was published [1]. Roquette derived it from the following extension of Clifford's Theorem: *If a finite group* G *acts irreducibly on the* K-*vector space* M *and* $N \lhd G$ *then either (1)* ${}_N M$ *is a direct sum of isomorphic irreducible* N-*modules or (2)* M *is induced from an irreducible* KH-*module* M_1 *such that* $N \subseteq H \subsetneq G$ *and* $\mathrm{End}_{KG}(M) \cong \mathrm{End}_{KH}(M_1)$. Thus, to prove Brauer's announced result one may assume that abelian normal subgroups are cyclic, and hence (P. Hall) that the p-group is cyclic or, if $p = 2$, either dihedral, quasi-dihedral, or generalized quaternion. Since the representations of these groups are well known, the result follows.

We wish to make use of Roquette's observation here to prove:

Theorem. If a 2-*group* G *acts absolutely irreducibly on a real* K-*vector space of dimension* ≥ 2 *then* G *leaves invariant a bilinear form equivalent over* K *to (a constant times) a direct sum of copies of*

$$\begin{bmatrix} 2+(\xi + \xi^{-1}) & 0 \\ 0 & 2-(\xi + \xi^{-1}) \end{bmatrix}$$

where ξ *is some* 2^k-*root of* 1 *such that* $\xi + \xi^{-1} \in K$.

Recall that any finite group G (necessarily of even order) which acts absolutely irreducibly on a real vector space leaves invariant a definite bilinear form B which is uniquely determined up to a constant multiple. (By a real field K we mean a subfield of the reals.)

Recall also that if T is a representation of a subgroup $H \subseteq G$ then, for any $g \in G$, the induced representation $T^G(g)$ is of the form $P \otimes T(h)$ for some $h \in H$, where P is a permutation matrix and $\otimes$ denotes the tensor product of matrices. Hence $T^G(g)^t\, T^G(g)$ is a direct sum of copies of $T(h)^t\, T(h)$. Now B is a multiple of $\Sigma\, g^t g$ where here g denotes the matrix representation of an element of G. Hence if this representation is induced from one of H, then B is a direct sum of copies of a form left invariant by H.

Now let G be as in our theorem. By Roquette and the above argument, we may assume that the abelian normal subgroups of G are cyclic, so that G is either cyclic, generalized quaternion, dihedral, or quasi-dihedral. The first two are not real possibilities since they have no absolutely irreducible actions* on a real vector space of dimension ≥ 2. Any real irreducible K-representation of a dihedral 2-group is generated by two matrices of the form

$$\begin{bmatrix} \xi + \xi^{-1} & -1 \\ 1 & 0 \end{bmatrix}, \quad \begin{bmatrix} 0 & 1 \\ 1 & 0 \end{bmatrix}$$

where ξ is a 2^k-root of 1 such that $\xi + \xi^{-1} \in K$ (cf. Curtis and Reiner, p. 339). One verifies that the bilinear form

$$\begin{bmatrix} 2 & \xi + \xi^{-1} \\ \xi + \xi^{-1} & 2 \end{bmatrix}$$

is preserved. This form is equivalent over K to the diagonal form

* *We may and do assume that all of our representations are faithful.*

$$\frac{1}{2}\begin{bmatrix} 2+(\xi + \xi^{-1}) & 0 \\ 0 & 2-(\xi + \xi^{-1}) \end{bmatrix}$$

Finally, any irreducible representation of a quasi-dihedral 2-group must preserve such a form since it contains a dihedral subgroup of index 2 which must also act irreducibly.

BIBLIOGRAPHY

[1] P. Roquette, Realisierung von Darstellungen endlicher nilpotenter Gruppen, Archiv. der Math. 9 (1958) 241-250.

ON A GENERALIZATION OF SERIAL RINGS

Kent R. Fuller
The University of Iowa
Iowa City, Iowa

Recent results of several authors can be combined to show that every left module over a ring R is a direct sum of modules with distributive submodule lattices if and only if R is an artinian ring with finitely many finitely generated indecomposable left modules each of which has a distributive submodule lattice. (See [8, Remark 7(2)].) We shall say that a ring satisfying these conditions is of <u>left distributive module type</u>.

A module with a distributive submodule lattice is called a <u>distributive module</u>. Distributive modules have been studied in [3], [4], and [12]. Two facts that give insight into their structure are their characterization [3, Theorem 1] as those modules whose factor modules all have square-free socles (so that linearly independent submodules of distributive modules are homologically independent), and their property that submodules are stable under endomorphisms provided that almost any minimum or maximum conditions are satisfied [12, p. 305]. Also recall the well-known/easy-to-prove facts that a distributive module of finite composition length has a finite lattice of submodules; and that over an artinian ring whose simple factor rings have infinite centers the converse is true.

Skornjakov [11] and (assuming minimum condition) Fuller [7] proved that if every left module over a ring is a direct sum of uniserial modules then the ring is one of Nakayama's serial (or generalized uniserial) rings. Our purpose here is to prove (Theorem 5) that a ring of left distributive module type has left and right ideal structure that is just one step away from that of a serial ring. Both the left and the right regular representations of a ring of left distributive module type are direct sums of (indecomposable projective) modules whose lattices have Hasse diagrams of one of the following types:

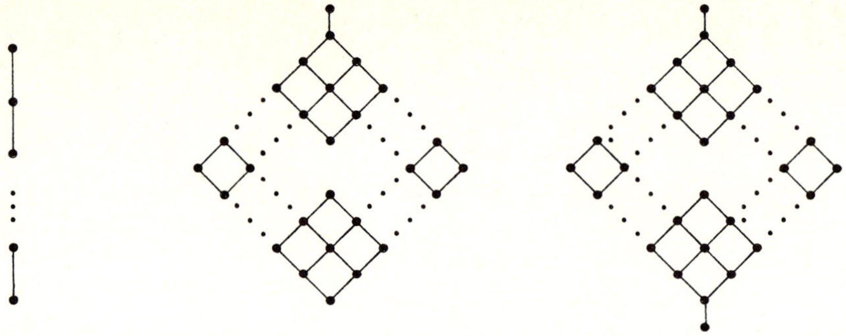

This is, however, in marked contrast to the serial case, for there the indecomposable modules have lattices with Hasse diagrams of the same type as the indecomposable projective modules, while a ring of distributive module type can have indecomposable modules (e.g., ones containing four minimal submodules (see Corollary 7)) whose submodule lattices are virtually impossible to draw. Also the theorem allows us to characterize one-sided serial rings and special K-algebras with radical-square-zero of distributive module type, and it strongly suggests that one-sided distributive module type may imply the other.

Throughout this paper R is an associative ring with identity element 1 and radical J = J(R). We denote the injective envelope of a module M by E(M), the nth lower Loewy factor of M by $Soc_n(M)$ and the Jordan-Holder length of M by c(M).

We shall need the following lemmas. The first was essentially proved in [8].

Lemma 1 (1) If R is a left artinian ring over which every indecomposable left module has square free top (e.g., if R is a ring of left distributive module type) then, for every primitive idempotent e in R, eR is distributive.

(2) If, for each primitive idempotent e in R, Re and eR are distributive then there are anti-isomorphisms between the submodule lattices of Re and E(eR/eJ) and between those of eR and E(Re/Je), and every indecomposable R-module with a simple socle (top) is injective (projective) over some factor ring of R.

Proof. See [8]--especially Remark 7(2) and Lemma 5.

The second lemma is an interlacing lemma that we shall use to build nondistributive modules.

Lemma 2 Let R be a left artinian ring with primitive idempotents $e_1, \cdots, e_\ell$ such that $Re_1, \cdots, Re_\ell$ are pairwise nonisomorphic. Suppose that there are minimal left ideals $S_i \subseteq Re_i$ such that $S_1 \cong \cdots \cong S_\ell$ and $S_i \subseteq Soc(R_R)$ $(i=1,\cdots,\ell)$. Then there is a minimal left ideal $L \subseteq S_1 \oplus \cdots \oplus S_\ell$ such that $(Re_1 \oplus \cdots \oplus Re_\ell)/L$ is indecomposable.

Proof. Assume that $e_1, \cdots, e_\ell$ are orthogonal and let $e = e_1 + \cdots + e_\ell$. Let f be a primitive idempotent in R such that $Rf/Jf \cong S_i$ $(i=1,\cdots,\ell)$. Then there are elements $0 \neq x_i = fx_ie_i \in S_i$ $(i=1,\cdots,\ell)$. Let

$$L = R(x_1 + \cdots + x_\ell) \leq Re = Re_1 \oplus \cdots \oplus Re_\ell.$$

Then with an argument similar to one given in [6] we see that for each $\gamma \in End(Re/L)$ there is an element $eae \in eRe \cong End(Re)$ such that

$$(x_1 + \cdots + x_\ell)eae \in L$$

and such that

$$\gamma \text{ is nilpotent or invertible iff } eae \text{ is.}$$

Since the $x_i \in Re_i \cap Soc(R_R)$ and $e_iae_j \in J$ whenever $i \neq j$, the first condition means that there is an $r \in R$ with

$$fx_ie_iae_i = rfx_i \qquad (i=1,\cdots,\ell).$$

But then, since $rfx_i = 0$ iff $rf \in Jf$ and $fx_ie_iae_i = 0$ iff $e_iae_i \in e_iJe_i$, these equations imply that either each e_iae_i is invertible in e_iRe_i or else they all are nilpotent. Thus

$$eae = \Sigma e_iae_i + \Sigma_{i \neq j}\, e_iae_j$$

is either invertible in eRe or nilpotent, so by the relationship between γ and eae, $End(Re/L)$ is local.

Next we apply Lemma 2 to obtain a lemma that has the dual conclusion to and a proof suggested by that of [14, Proposition 2.5]. It is at this point that we seem to need the hypothesis that indecomposable modules are distributive, rather than that they merely have square free tops and socles.

Lemma 3 If R is a ring of left distributive module type and $\mathrm{Soc}_2({}_R M)$ is uniserial then M is uniserial.

Proof. Suppose that $\mathrm{Soc}_2(M)$ is uniserial, but that M is not, so there is a smallest integer k such that $\mathrm{Soc}_k(M)/\mathrm{Soc}_{k-1}(M)$ is not simple. Then $k > 2$ and there exist primitive idempotents e_1 and e_2 and elements $x_i = e_i x_i \in \mathrm{Soc}_k(M)/\mathrm{Soc}_{k-1}(M)$ $(i=1,2)$ that span distinct simple submodules of $\mathrm{Soc}_k(M)/\mathrm{Soc}_{k-1}(M)$. Then Re_1x_1 and Re_2x_2 both have Loewy length k, so $Je_1x_1 = \mathrm{Soc}_{k-1}(M) = Je_2x_2$ and, since M is indecomposable (hence distributive), $Re_1 \not\cong Re_2$. Since $k > 2$, Je_1/J^2e_2 and Je_2/J^2e_2 both have a submodule isomorphic to $Je_1x_1/J^2e_1x_1 = Je_2x_2/J^2e_2x_2$, and similarly J^2e_1/J^3e_2 and J^2e_2/J^3e_2 have isomorphic simple submodules. Now we can apply Lemma 2 to Re_1/J^3e_1 and Re_2/J^3e_2 over R/J^3 to obtain a nondistributive indecomposable module, contrary to hypothesis.

Finally we need the following lemma about distributive modules. Although we don't need it here, the proof is valid for transfinite lower Loewy series.

Lemma 4 Let M be a distributive module with submodules $K,L \leq M$. Then

$$\mathrm{Soc}_n(K+L) = \mathrm{Soc}_n(K) + \mathrm{Soc}_n(L) \qquad (n=1,2,\cdots).$$

Proof. A simple inductive argument shows that in any module $M \geq K$, $\mathrm{Soc}_n(K) = \mathrm{Soc}_n(M) \cap K$. Thus, letting $N = K + L$ we have, by distributivity,

$$\begin{aligned}\mathrm{Soc}_n(N) &= \mathrm{Soc}_n(N) \cap (K+L) \\ &= (\mathrm{Soc}_n(N) \cap K) + (\mathrm{Soc}_n(N) \cap L) \\ &= \mathrm{Soc}_n(K) + \mathrm{Soc}_n(L).\end{aligned}$$

Now we are ready to prove the main result.

Theorem 5 Let R be a ring of left distributive module type. Then each left and each right indecomposable projective R-module P is distributive, and if P is not uniserial then Rad(P) is the sum of two uniserial modules whose intersection is either zero or simple.

Proof. The first conclusion is by Lemma 1.

For the rest, we shall first prove that if $\mathrm{Soc}({}_RQ)$ is simple and Q/JQ is not, then $Q = K_1 + K_2$ with K_1 and K_2 uniserial and $K_1 \cap K_2 = \mathrm{Soc}\, Q$. Since $\mathrm{Soc}({}_RQ)$ is simple, we know by Lemma 1 that Q is injective modulo its annihilator. Thus we may assume that $Q = E(Rf/Jf)$ for some primitive idempotent f and also that Q is faithful, so that every minimal left ideal of R is isomorphic to Rf/Jf. Let $Re_1 \oplus \cdots \oplus Re_\ell \to Q \to 0$ be a projective cover. Then, since, by Lemma 1, Q is distributive, the Re_i are pairwise nonisomorphic. By [7, Theorem 2.4] each e_iR/e_iJ embeds in fR. Thus we have simple modules

$$Rf/Jf \cong S_i \le Re_i \quad \text{and} \quad e_iR/e_iJ \cong T_i \le fR$$

$(i=1,\cdots,\ell)$. Since Re_i is distributive, the left fRf-module fRe_i is uniserial [8, Lemma 4]. Its unique minimal submodule must be fS_i. Since fR is distributive, T_i is stable under endomorphisms [12], so $0 \ne fT_ie_i \le_{fRf} fRe_i$. Thus $0 \ne fS_i \subseteq fT_ie_i \subseteq \mathrm{Soc}(R_R)$, so since S_i is simple, $S_i \subseteq \mathrm{Soc}(R_R)$ $(i=1,\cdots,\ell)$. Hence, by Lemma 2, we have an indecomposable module $(Re_1 \oplus \cdots \oplus Re_\ell)/L$ whose socle contains a direct sum of $\ell-1$ copies of Rf/Jf. Thus $\ell = 2$ and Q has a projective cover $Re_1 \oplus Re_2 \to Q \to 0$. Now, since Re_1 and Re_2 are distributive, our assumption that Q is faithful implies that they embed in Q. By what we have just proved $c(\mathrm{Soc}_2(Q)/\mathrm{Soc}(Q)) \le 2$, so either $\mathrm{Soc}_2(Re_1)$ and $\mathrm{Soc}_2(Re_2)$ are uniserial or Re_1/S_1 and Re_2/S_2 have a composition factor in common. The latter would yield a nondistributive indecomposable module $(Re_1 \oplus Re_2)/L$. Thus $\mathrm{Soc}_2(Re_i)$, and hence, by Lemma 3, Re_i is uniserial $(i=1,2)$. Thus we see that Q is spanned by two uniserial submodules $Q = K_1 + K_2$. By Lemmas 3 and 4, $\mathrm{Soc}_2(Q) = \mathrm{Soc}_2(K_1) + \mathrm{Soc}_2(K_2)$ is not uniserial, so it follows that $K_1 \cap K_2 = \mathrm{Soc}(Q)$.

Returning to the general setting in which R is any ring of left distributive module type, since R is artinian and the indecomposable injective left R modules are finitely generated, there is a ring S (necessarily of right distributive module type) whose category of finitely generated right modules is Morita dual to the category of finitely generated left R-modules [10, Theorem 6.3]. Denoting the duality by ( )*, we recall that there is an anti-isomorphism between the lattices of submodules of M and (M)* (e.g., see [1, 24.5]). In particular, by Lemma 3 applied to right S-modules, if M/J^2M is uniserial then ${}_RM$ is uniserial. Now let e be a primitive idempotent in R. If Soc(Re) is simple and Re is not uniserial then Je has simple socle and non-simple top. So in this case, let ${}_RQ = Je$. If Soc(Re) is not simple let $Q_S = (Re)^*$ to obtain submodules of Re that have zero intersection, maximal (= Je) sum and are uniserial. This proves that the left indecomposable projective R-modules satisfy the conclusion of the theorem. Similarly, so do the right ones over S. But, by Lemma 1, submodule lattice of eR is anti-isomorphic to that of E = E(Re/Je) which in turn is anti-isomorphic to the submodule lattice of the indecomposable projective right S-module (E)*, so the proof is complete.

An artinian ring is of <u>left invariant module type</u> [5] in case each of its indecomposable left modules is quasi-injective. In [8, Theorem 6] we characterized these rings as the ones such that for all primitive idempotents e and f, (i) Re is uniserial, (ii) $Je/J^3e \cong Jf/J^3f$ and $J^2e \neq 0$ implies $Re \cong Rf$, (iii) $c(eJ/eJ^2) \leq 2$, and (iv) eR is distributive; and we observed that they are of left and right distributive module type. Theorem 5 allows us to prove for one-sided serial rings that left distributive module type is equivalent to right, and to give a lattice theoretic characterization (cf. [14, Proposition 4.4]) of rings of left invariant module type.

<u>Corollary 6</u> Let R be a left serial ring. Then the following statements are equivalent:

(a) R is of left distributive module type;

(b) R is of left invariant module type;

(c) For each primitive idempotent $e \in R$, eR is distributive and eJ is the sum of at most two uniserial right ideals;

(d) Each indecomposable right projective R-module is either uniserial or its radical is the direct sum of two uniserial right ideals with no composition factors in common;

(e) R is of right distributive module type.

Proof. (a) $\Rightarrow$ (d). Assume (a) and let P_R be indecomposable and projective. Then by the theorem, either PJ is the direct sum of two uniserial modules or $Soc(P)$ is simple. By Lemma 1 with $E(P) = E(eR/eJ)$, since R is left serial, the latter implies that P is uniserial.

(d) $\Rightarrow$ (c). This is almost obvious.

(c) $\Rightarrow$ (b). Clearly condition (c) (together with Lemma 1) implies that every right (left) R-module M with M/MJ^2 ($Soc_2(M)$) uniserial is itself uniserial. Now suppose that $Je/J^3e \cong Jf/J^3f$ and both are of length 2. Let $E = E(Re/J^3e) = E(Rf/J^3f)$, let $K \cong Re/J^3e$ and $L \cong Rf/J^3f$ with $K,L \leq E$. By Lemma 1, E is distributive and, by Lemma 4, $Soc_2(K+L) = Soc_2(K) + Soc_2(L)$. Since $Soc_2(K) \cong Soc_2(L)$, the distributivity of E forces $Soc_2(K) = Soc_2(L)$. But then, by Lemma 3, $K + L$ is uniserial, so $K = L$. This proves that (c) implies condition (ii) of [8, Theorem 6]; and the remaining conditions are immediate from (c).

(b) $\Rightarrow$ (a) and (e). This is [8, Corollary 3].

(e) $\Rightarrow$ (c). By Theorem 5.

In conclusion we shall use the theorem and the results of Gabriel [9] as set forth in [2] to determine which special K-algebras with radical-square-zero are of distributive module type. To accomplish this we also need the following lemma which clearly has other implications for rings of distributive module type.

Lemma 7 Let $e_1,\cdots,e_\ell$ be orthogonal primitive idempotents in R such that $Re_1,\cdots,Re_\ell$ are pairwise nonisomorphic. Suppose that there are left ideals $S_i \oplus T_i \leq Je_i$ $(i=1,\cdots,\ell)$ such that $T_i \cong S_{i+1}$ and both are minimal and contained in $Soc(R_R)$ $(i=1,\cdots,\ell-1)$ while $S_i \not\cong T_i$ $(i=2,\cdots,\ell-1)$. Then there is an indecomposable left R-module M such that $M = N_1 + \cdots + N_\ell$ with the N_i independent modulo JM and $N_i \cong Re_i$ $(i=1,\cdots,\ell)$. Moreover M contains a direct sum $V_1 \oplus \cdots \oplus V_{\ell+1}$ such that $S_i \oplus T_i \cong V_i \oplus V_{i+1} \leq N_i$ $(i=1,\cdots,\ell)$.

Proof. Let $f_1,\cdots,f_{\ell-1}$ be primitive idempotents such that $Rf_i/Jf_i \cong T_i \cong S_{i+1}$ $(i=1,\cdots,\ell-1)$. Let $0 \neq f_{i-1}s_ie_i = s_i \in S_i$ $(i=2,\cdots,\ell)$ and $0 \neq f_it_ie_i = t_i \in T_i$ $(i-1,\cdots,\ell-1)$. Let $L = R(t_1+s_2) + \cdots + R(t_{\ell-1}+s_\ell)$ and let $M = (Re_1 \oplus \cdots \oplus Re_\ell)/L$. Then the proof that M is indecomposable is similar to that of Lemma 2. The N_i are the canonical images of the Re_i in M.

In order to construct the left diagram associated with an artinian ring R with $J^2 = 0$ and basic set of idempotents $e_1, \cdots, e_n$, one lists two rows of n points each

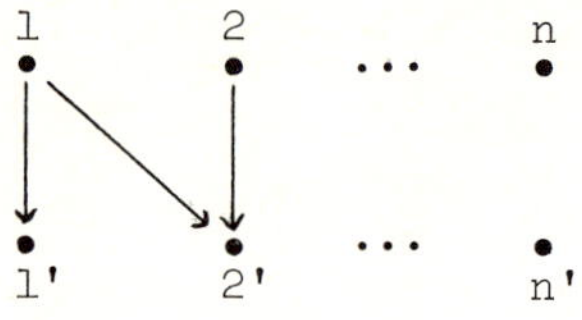

and draws k_{ij} arrows from point i to point j' whenever the Re_j/Je_j-homogeneous component of Je_i has length k_{ij}. Gabriel [9] proved that if R is a special K-algebra then R is of finite module type if and only if (ignoring arrowheads and labels) the connected components of the diagram form Dynkin diagrams of type A_m (a line with m points), D_m, E_6, E_7, or E_8. Moreover (see [2]), each component contributes a certain number ($(m-1)m/2$ for A_m) of indecomposable left R-modules of length ≥ 2, the totality of which, together with the simple left R-modules, consists of one copy of each indecomposable left R-module. From these facts, Theorem 5, and Lemma 7 we get

Corollary 8 Let R be an artinian ring with $J^2 = 0$. If R is of distributive module type then the diagram associated with R is a disjoint union of diagrams of type A_m with the property that if the points i and i' both appear in a component then there is an arrow from i to i' (i.e., Re_i/Je_i embeds in Je_i). If R is a special K-algebra, the converse is also true.

Proof. From Theorem 5 we see that if R is of distributive module type then at most two arrows begin at any point i or end at any point i'. An examination of the module M of Lemma 7 shows that all such M are distributive if and only if the diagrams are as described. The first statement follows.

For the converse, if the condition holds, each component A_m comes from $Re_1, \cdots, Re_\ell$ (top row) which in turn yield a distributive module M like the one in Lemma 7 that has $(m-1)m/2$ indecomposable modules of length ≥ 2 among the submodules of its factor modules.

Corollary 9 Let R be an hereditary special K-algebra with

$J^2 = 0$. Then R is of distributive module type if and only if the diagram associated with R is the disjoint union of diagrams of type A_m.

Proof. The hereditary condition prevents i from being connected to i' in the construction of the diagram for R.

REFERENCES

1. F. W. Anderson and K. R. Fuller, Rings and Categories of Modules, Springer-Verlag, New York-Heidelberg-Berlin, 1974.
2. M. Auslander and I. Reiten, On the Representation Type of Triangular Matrix Rings, J. London Math. Soc., Vol. 12, No. 2 (1976), pp 371-382.
3. V. P. Camillo, Distributive Modules, J. Algebra, Vol. 36 (1975), pp 16-25
4. P. M. Cohn, Free Rings and Their Relations, Academic Press, New York, 1971.
5. S. E. Dickson and K. R. Fuller, Algebras for Which Every Indecomposable Right Module Is Invariant in Its Injective Envelope, Pacific J. Math., Vol. 31 (1969), pp 655-658.
6. S. E. Dickson and K. R. Fuller, Commutative QF-1 Artinian Rings Are QF, Proc. Amer. Math. Soc., Vol. 24 (1970), pp 667-670.
7. K. R. Fuller, On Indecomposable Injectives Over Artinian Rings, Pacific J. Math., Vol. 29 (1969), pp 115-135.
8. K. R. Fuller, Rings of Left Invariant Module Type, to appear.
9. P. Gabriel, Indecomposable Representations I, Manuscripta Math., Vol. 6 (1972), pp 71-103.
10. K. Morita, Duality of Modules and Its Applications to the Theory of Rings with Minimum Condition, Sci. Rep. Tokyo Kyoiku Diagaku, Vol. 6 (1958), pp 85-142.
11. L. A. Skornjakov, When Are All Modules Semi-Chained?, Mat. Zametki, Vol. 5 (1969), pp 173-182.
12. W. Stephenson, Modules Whose Lattice of Submodules Is Distributive, Proc. London Math. Soc., Vol. 28, No. 3 (1974), pp 291-310.
13. H. Tachikawa, On Rings for Which Every Indecomposable Right Module Has a Unique Maximal Submodule, Math. Z., Vol. 71 (1959), pp 200-222.
14. H. Tachikawa, On Algebras of Which Every Indecomposable Representation Has an Irreducible One As the Top or the Bottom Loewy Constituent, Math. Z., Vol. 75 (1961), pp 215-227.

ON THE STRUCTURE OF INDECOMPOSABLE MODULES*

Edward L. Green
University of Illinois
Champaign-Urbana, Illinois

§0. INTRODUCTION

Algebraists have long been engaged in the study of indecomposable modules. In the last few years there have been many new insights concerning such modules over Artinian rings, leading to renewed interest in the subject [e.g. see 1,2,4,5,7,8 and 9 for a fuller bibliography]. Recently a class of indecomposable modules, called modules with cores, was defined and studied in detail [9]. The indecomposability of modules with cores is defined in terms of its submodules - see section 1. It is this "internal" characterization of indecomposability that has led to the definitions of s-indecomposable modules and ∞-indecomposable modules. In section 1 we define this hierarchy of classes of indecomposable modules and derive some of the basic properties about them. It is hoped that a more detailed study of the concepts introduced here will lead to a fuller understanding of the structure of indecomposable modules. More precisely, the study of how the submodules of a module overlap hopefully will yield interesting new results.

A second purpose of this paper is to announce a number of new results of the author and R. Gordon. To achieve this, in section 2 we restrict our attention to modules over Artin rings and recall the notion of pasting, introduced in [3]. Section 3 summarizes the results on pasting and modules with cores found in [9].

Examples, showing that the classes of indecomposable modules introduced in section 1 are in fact distinct, are given in section 4. We end with a number of open questions in section 5.

---

*This research was partially supported by a grant from the N.S.F.

All rings have a unit and all modules are left modules unless otherwise stated.

§1

We begin by recalling a concept due to Bass [6]. A submodule N of a module M is called <u>nonsuperfluous</u> if there exists a proper submodule N' of M such that $N + N' = M$. It should be noted that the concept of a nonsuperfluous submodule is dual to the notion of a nonessential submodule of a module (in the same way as the radical of a module is dual to the socle of a module). Moreover, all the following definitions have dual definitions which, in general, lead to a distinct hierarchy of classes of indecomposable modules. Such dualizations are left to the reader except for the definition of the cocore of a module which is given in the next section.

<u>Definition.</u> Let R be a ring and let M be an R-module. Let s be a positive integer. We say that M is <u>s-indecomposable</u> if given $s+1$ nonsuperfluous submodules $N_1, \ldots, N_{s+1}$ such that $\sum_{i=1}^{s+1} N_i = M$ then $\bigcap_{i=1}^{s+1} N_i \neq (0)$. We say that M is <u>∞-indecomposable</u> if M is s-indecomposable for all $s \geq 1$. Finally, we say <u>M has a core</u> if the intersection of all the nonsuperfluous submodules of M is not zero. This intersection, whether zero or not, will be called the <u>core of M</u> and will be denoted $C(M)$.

The next proposition shows how these concepts are interrelated.

<u>Proposition 1.1</u>. Let R be a ring. Let M be an R-module. Then

i) M is indecomposable <=> M is 1-indecomposable.

ii) If M is (s+1)-indecomposable then M is s-indecomposable, for $s \geq 1$.

iii) If M has a core then M is ∞-indecomposable.

The proof follows easily from the definitions and is omitted. We should remark that given s, we provide examples of modules which are s-indecomposable but not (s+1)-indecomposable. Later in this section we provide examples of modules which are ∞-indecomposable but have zero cores. These examples must have infinite length as shown by the next result.

Proposition 1.2. Let R be a ring and let M be an R-module. Assume that M has an essential submodule N of finite length; say, the length of N is n. Then the following statements are equivalent:

(1). M is n-indecomposable

(2). M is ∞-indecomposable

(3). M has a core.

Proof. We have (3) => (2) and (2) => (1) by 1.1. We show that (1) => (3). Suppose that $C(M) = 0$. Then $N \cap C(M) = 0$. Since the length of N is n and since C(M) is the intersection of all nonsuperfluous submodules of M, it is easy to see that one may find n nonsuperfluous submodules $N_1,\dots,N_n$ such that $N \cap (\bigcap_{i=1}^{n} N_i) = 0$. Thus, since N is essential, $\bigcap_{i=1}^{n} N_i = (0)$. Since $\sum_{i=1}^{n} N_i + M = M$, letting $N_{n+1} = M$ we see that $N_1,\dots,N_{n+1}$ are nonsuperfluous submodules such that $\sum_{i=1}^{n+1} N_i = M$ and $\bigcap_{i=1}^{n+1} N_i = (0)$. Thus M is not n-indecomposable and we are done. □

An easy consequence of this proposition is that if a module has finite length then it is ∞-indecomposable if and only if it has a core. In general, ∞-indecomposable modules seem to be of more interest than modules with cores. For example, commutative domains are always ∞-indecomposable when viewed as modules over themselves, but they rarely have cores since the core of a module is contained in the Jacobson radical of the module.

Before proceeding, we introduce some more notation. The top of a module M, top(M), is just M/rad(M), where rad(M) = Jacobson radical of M. If M' is a submodule of M we say that M' is a covering submodule of M if M' is a nonsuperfluous submodule of M and the map top(M') → top(M), induced from the inclusion map, is a monomorphism. A module M will be called local if top(M) is a simple module.

The next result shows that the condition on a module that it be s-indecomposable for some $s > 1$, is a strong condition in that it implies that all covering submodules of the module are indecomposable.

Proposition 1.3. Let R be a ring. Let $s > 1$. Suppose that M is an s-indecomposable R-module. Then every covering submodule of M is (s-1)-indecomposable.

Proof. Let M' be a covering submodule of M. We may assume that $M' \neq M$. Let M'' be a proper submodule of M such that $M' + M'' = M$. Note that M'' is a nonsuperfluous submodule of M. Suppose that $N_1, \ldots, N_s$ are nonsuperfluous submodules of M' such that $\sum_{i=1}^{s} N_i = M'$. Since M' is a nonsuperfluous submodule of M, each N_i is a nonsuperfluous submodule of M. Letting $N_{s+1} = M''$, we see that since M is s-indecomposable, $\bigcap_{i=1}^{s+1} N_i \neq 0$ and hence $\bigcap_{i=1}^{s} N_i \neq 0$. Thus M' is (s-1)-indecomposable. □

It should be mentioned that an example of a module such that every covering submodule is 1-indecomposable but which is not 2-indecomposable is provided in section 4.

§2

We now restrict our attention to Artin rings. Unless otherwise stated, R will denote a left Artin ring with radical $\underline{r}$. It is easy to see that all of the following types of R-modules have nonzero cores: modules with simple tops, modules with simple socles and modules with waists. (Recall that a module M has a waist if there is a proper nonzero submodule N of M which contains or is contained in every other submodule of M.) The class R-modules with cores in general contains more indecomposable modules than these (see [9]). Since all finitely generated R-modules are sums of nonsuperfluous submodules with cores, (in fact, they are sums of local submodules), one method of studying the module category of R would be first to analyze R-modules with cores and then study amalgamations (i.e. pushouts) of these modules. This program is started in [9], where modules with cores are studied in detail and amalgamations of local modules and other types of amalgamations are also considered. A number of the results of this investigation have been announced [10,11,12]. Other results are reported on here; namely, those dealing with the "pasting of modules".

Recall that if S is a ring and $a:A \to B$ is an S-homomorphism then we say that a is pastable if there is a module X extending A such that the triangle

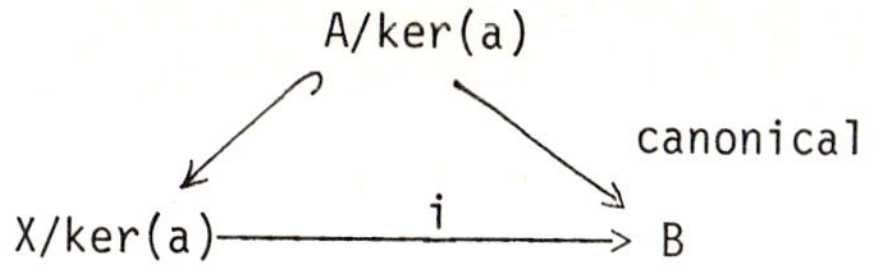

commutes for some isomorphism i. We refer to X as a pasted module of a. For more details see [3].

In the next section we give a method, involving pasting, of constructing new modules with cores from old. Secondly, we show that one may reduce the study of modules with cores to a study of a particular type of pasting, together with the study of modules with semi-simple cores.

We end this section by noting that most Artin rings have indecomposable modules with zero cores. The next results make this explicit. For the sake of completeness we define the dual concept of a core. The cocore of a module M, $C^0(M)$, is the sum of the nonessential submodules of M. We say that M has a cocore if $C^0(M) \neq M$.

Proposition 2.1. [9] Let R be a left Artin ring such that there is an R-module with a simple top containing at least three nonisomorphic simple submodules. Then there is an indecomposable finitely generated R-module M with $C(M) - (0)$ and $C^0(M) = M$.

It is worth noting that the proof is constructive and can be improved if R is a radical squared zero Artin algebra--see Corollary 2.4 below. The next result uses the concept of Auslander and Reiten of "almost split sequence" [4].

Proposition 2.2. Let R be an Artin algebra. Let (*) $0 \to S \to B \to M \to 0$ be an almost split sequence with S a simple R-module. Suppose that B has at least three nonzero summands in a direct sum decomposition. Then $C(M) = (0)$ and $C^0(M) = M$.

Proof. Assume that $B = B_1 \oplus B_2 \oplus B_3$ with $B_i \neq 0$. Since (*) is an almost split exact sequence, the B_i's have length different from the length of S and hence the B_i's are not simple modules. Furthermore the induced maps $S \to B_i$ are not split. Since M is indecomposable, for each $i = 1,2,3$, $S \to \coprod_{i \neq j} B_j$ is a monomorphism whose image is nonessential. From this it is easy to see that the image of each B_i in M is nonessential. Since the images of the B_i generate M, we see that $C^o(M) = M$.

To see that $C(M) = 0$, we note that $B_1 \oplus B_2$ maps monomorphically into M and that the images of B_1 and B_2 are covering submodules of M. Thus the intersections of the images of B_1 and B_2 are zero in M and we get our conclusion. □

<u>Corollary 2.3</u>. If R is an Artin algebra with $\underline{r}^2 = 0$ and if S is a simple R-module such that the length of the injective envelope of S is greater than or equal to 4 then there is a finitely generated indecomposable R-module M with $C(M) = 0$ and $C^o(M) = 0$.

<u>Proof</u>. By [4], if $0 \to S \to B \to M \to 0$ is the almost split sequence for S, then B is the projective cover of the top of the injective envelope of S. Thus B has n summands, where $n+1$ is the length of the injective envelope of S. The result now follows from 2.2. □

<u>Corollary 2.4</u>. If R is an Artin algebra with $\underline{r}^2 = 0$ and there is a local R-module of length ≥ 4, then there is a finitely generated indecomposable R-module M with $C(M) = 0$ and $C^o(M) = M$.

<u>Proof</u>. Let X be a local module of length ≥ 4. Let $S = \text{top}(X)$. Then the projective cover of S has length ≥ 4. Let $S = \text{top}(X)$. Then the projective cover of S has length ≥ 4. Dualizing, we may apply 2.3 to get a right R-module M^* such that $C(M^*) = 0$, $C^o(M^*) = M^*$ and M^* is a finitely generated and indecomposable. Letting M = dual of M^* works. □

§3

The following result enables one to construct new modules with cores from given ones.

<u>Proposition 3.1</u>. [9] Let R be a left Artin ring with radical $\underline{r}$. Let $f: M \to N$ be an R-homomorphism such that $\ker(f) = \underline{r}M$ and $\text{im}(f) = \text{soc}(N)$. Then $C(M)$ is contained in the core of every pasted module of f.

Application of the above result is dependent on the existence of pasted modules of f. We now give a method of constructing smaller modules with cores from given ones that is applicable provided that the given module does not have a semi-simple core. Before giving the construction, we note that if A is a submodule of B then $C(B)/(C(B) \cap A) \subseteq C(B/A)$.

Let M be an R-module. Let $S_0(M) = soc(M)$. Let $S_1(M) = p^{-1}(soc(M/S_0(M)))$, where $p:M \to M/S_0(M)$ is the canonical surjection. Now if $C(M)$ is not semi-simple, then $0 \neq C(M)/S_0(M) \cap C(M) \subseteq C(M/S_0(M))$. Now let $X = S_1(M)$, $N = M/S_0(M)$ and $f:X \to N$ via $\underset{\underset{X}{\parallel}}{S_1(M)} \to S_1(M)/S_0(M) \to \underset{\underset{N}{\parallel}}{M/S_0(M)}$. It is easy to see that M is a pasted module for f. Note that $\ell\ell(M/S_0(M)) = \ell\ell(M) - 1$ where $\ell\ell(A)$ denoted the Loewy length of A. Thus, starting with a module M, such that $C(M)$ is not semi-simple we may view M as a pasted module of a morphism $f:X \to N$ where $\ell\ell(X) = 2$, $\ell\ell(N) = \ell\ell(M) - 1$ and $C(N) \neq 0$.

If $C(N)$ is not semi-simple we may continue until we eventually get a module with a nonzero semi-simple core. In this way we see that the general study of modules with cores may be reduced to those with semi-simple cores and pasting.

In [9], the technique is refined so that M is a pasted module for some $f:X \to N$ such that

(1). $\ell\ell(X) = 2$ and $soc(X) = \underline{r}x$

(2). $ker(f) = \underline{r}X$

(3). $im(f) \cap r^n N \neq 0$, where $\ell\ell(N) = n+1$.

Note that (1) is equivalent to X being of Loewy length 2 and having no semi-simple summands. Furthermore, given (1), (3) is equivalent to the Loewy length of every pasted module being $n+2$. This refinement is achieved by taking X to be an appropriate summand of $S_1(M)$. We now investigate when one may reverse the above process; that is, the following result determines the conditions on $f:X \to N$ so that if one can paste, one gets a module M so that (1)-(3) hold and $C(M) \neq 0$.

Theorem 3.2. [9] Let R be a left Artin ring with radical $\underline{r}$. Let $f:X \to N$ be an R-homomorphism such that

i) $\ell\ell(X) = 2$ and X has no semi-simple summands.

ii) $ker(f) = \underline{r}X$.

iii) $\operatorname{im}(f) \cap r^n N \neq 0$, where $n + 1 = \ell\ell(N)$.
iv) $\operatorname{im}(f) \cap C(N) \neq 0$.

Then every pasted module of f has a nonzero core.

Note that condition iv) does not appear in the reduction discussed before the theorem. In [9] there is an example which shows that there is an Artin algebra R and an R-morphism $f:X \to N$ such that $C(N) \neq 0$ and conditions (i)-(iii) are satisfied, yet there are two pasted modules of f--one with zero core and the other with nonzero core.

§4

This section is devoted to the construction of examples of s-indecomposable modules. We use the notation and definitions of [9, appendix]. Let s be given and let k be a fixed field. Consider the quiver $\mathcal{Q}$ with $2s+4$ vertices, numbered $1,\ldots,s+2$, $1',\ldots,(s+2)'$ having exactly one arrow from vertex i to vertex j' provided that $j \neq i$. For example, for $s = 1$, is

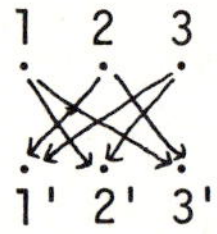

. Let T be the associated tensor k-algebra for $\mathcal{Q}$. Note that T is a hereditary Artin algebra with radical squared zero. Let P_i (respectively P_i') be the projective module associated to the ith (resp. i'^{th}) vertex. Let $M_1 = P_1$. Now $P_3' \ \ldots \ P_{s+2}'$ maps by canonical inclusions into M_1 and P_2. Let M_2 be the pushout of the maps. Note that $\underline{r}M_2 \cong \amalg_1^{s+2} P_i'$, and $C(M_2) \cong \amalg_3^{s+2} P_i'$. Let M_3 be the pushout of the canonical injections of $\amalg_{i\neq 3} P_i'$ into M_3 and P_3. Continuing in this fashion one finally constructs M_{s+2} which we will denote as M. It is not hard to check that diagrammatically M corresponds to the representation of $\mathcal{Q}$ given by putting the same one dimensional vector space at each vertex and letting each morphism be the identity.

We claim that M is s-indecomposable but not s+1-indecomposable. First note that there are only a finite number of local covering modules, $N_1,\ldots,N_{s+2}$ of M, where N_i is isomorphic to P_i. Furthermore, $N_i \cap N_j \cong \amalg_{\ell\neq i,j} P'$. Now $\sum_{i=1}^{s+2} N_i = M$ and each N_i is nonsuperfluous.

Since $\bigcap N_i = (0)$ we see that M is not s+1-indecomposable.

If $i \neq j$, $N_i + N_j$ contains $\underline{r}M$. It follows that there are only a finite number of nonsuperfluous submodules of M, namely finite sums of the N_i. Using the fact that $\underline{r}M$ is a submodule of all the nonlocal nonsuperfluous submodules and that $\underline{r}N_i \cong \amalg_{j\neq i} P_j'$, it is not hard to show that M is s-indecomposable.

Finally, returning to the case of $s = 1$, from the description of the nonsuperfluous submodules it is easily verified that every nonsuperfluous submodule is 1-indecomposable (and in fact, every proper nonsuperfluous submodule has a core) yet the module is not 2-indecomposable.

§5

In section 3 a method of reduction was given showing that one may study all modules with cores by studying those modules with nonzero semi-simple cores and pastings.

Question 1. Suppose that R is a left Artin ring and there is a nonlocal R-module with semi-simple core of length ≥ 2. Is R of infinite representation type? If so, can one explicitly construct indecomposable modules of arbitrarily large length from the given nonlocal module having a core?

Remarks. If R is an Artin algebra with radical squared zero, the answer is yes [9,13]. In general, one may construct from such modules, a module with zero cocore by amalgamating certain submodules [9].

Question 2. Can one classify or find special properties or techniques for dealing with modules with semi-simple nonzero cores?

We now introduce a class of modules which have all finitely generated modules with cores as factor modules. We begin with the construction of these modules. Let R be a left Artin ring. Let E be an indecomposable injective R-module with $S = soc(E)$. Let V be a finitely generated submodule of E. Assume for simplicity of notation that V is not local. Let $f : \amalg_{i=1}^{n} P_i \longrightarrow V$ be a projective cover of V with the P_i's indecomposable projective R-modules. Let $g:P \to S$ be a projective cover of S. Let $f_i = f|_{P_i}$. Since $S \subseteq f_i(P_i)$, for all i, there exists maps $a_i:P \to P_i$ such that $f_i a_i = g$ for all i. Let $M(V,f_i,a_i)$ denote the cokernel of the following map:

$$\amalg^{n-1} P \xrightarrow{\begin{pmatrix} a_1 & 0 & & 0 \\ -a_2 & a_2 & \ddots & \\ 0 & -a_3 & \ddots & a_{n-1} \\ & 0 & & a_n \end{pmatrix}} \amalg_{i=1}^{n} P_i \quad .$$

We now describe how, given a finitely generated R-module X with nonzero core, to find $M(V,f_i,a_i)$ such that there is a surjection $p:M(V,f_i,a_i) \longrightarrow X$ having the property that the induced map $top(M(V,f_i,a_i)) \longrightarrow top(X)$ is an isomorphism. Again for simplicity we assume that X is not local. Let X be given with $C(X) \neq 0$. Choose a simple submodule S contained in $C(X)$. Let $i:S \to E$ be an injective envelope for S. Then there exists $h:X \to E$ such that h restricted to S coincides with $i:S \to E$. Let $V = h(X)$. In [9] it is shown that since S is contained in the core of X, the map from $top(X)$ to $top(V)$ induced from h is an isomorphism. Since X is not local, it follows that V is not local. Let $f^*: \amalg_{i=1}^{n} P_i \longrightarrow X$ be a projective cover of X (with the P_i indecomposable projective R-modules). Then $hf^*: \amalg P_i \longrightarrow V$ is a projective cover of V. Now use hf^* for the projective cover of V in the construction of $M(V,f_i,a_i)$ above; i.e., $f_i = hf^*|_{P_i}$, a_i arbitrary. It is not hard to show that the following diagram commutes:

$$\begin{array}{ccccccccc} & & \amalg^{n-1} P & \longrightarrow & \amalg_{i=1}^{n} P_i & \longrightarrow & M(V,f_i,a_i) & \longrightarrow & 0 \\ & & \downarrow 0 & & \| & & & & \\ 0 & \longrightarrow & \ker(F^*) & \longrightarrow & \amalg_{i=1}^{n} P_i & \xrightarrow{f^*} & X & \longrightarrow & 0 \end{array}$$

Thus there is a surjective map $M(V,f_i,a_i) \longrightarrow X$ making the diagram commute. It follows that the induced map from $top(M(V,f_i,a_i)$ to $top(X)$ is an isomorphism. Thus the class of $M(V,f_i,a_i)$ has the class of finitely generated R-modules with nonzero cores as factor modules. (One may take care of the local case by defining $M(V,f_i,a_i)$ as the projective cover of V.)

R. Gordon has shown that in general the $M(V,f_i,a_i)$ depend on the choice of the f_i and a_i. This differs from the radical square zero finite representation type case. He has also shown that for all choices of f_i and a_i the $M(V,f_i,a_i)$ need not have a nonzero core. But this still leaves the following questions:

Question 3. Is there a choice of the f_i,a_i for which $M(V,f_i,a_i)$ is indecomposable? It should be remarked that Gordon's examples show that for some choice of the a_i,f_i, $M(V,f_i,a_i)$ can even decompose. All his examples occur over rings of infinite respresentation type.

Question 4. If R is of finite representation type,

(a) do the $M(V,f_i,a_i)$ depend on the f_i and a_i?

(b) is $M(V,f_i,a_i)$ indecomposable for some choice of f_i and a_i?

(c) does $M(V,f_i,a_i)$ have a core for some choice of f_i and a_i?

An affirmative answer to these questions would show that for appropriate choices of the f_i and a_i, we would have a set of indecomposable modules, (namely, the corresponding $M(V,f_i,a_i)$) such that each module with a core is a factor of one these.

REFERENCES

1. Auslander, M.; Representation theory of artin algebras I, Comm. in Algebra (1974), 177-268.

2. Auslander, M.; Representation theory of artin algebras II, Comm. in Algebra, (1974), 269-310.

3. Auslander, M., Green, E. L., Reiten, I,; Modules with waists, Ill. J. Math. 19 (1975), 467-478.

4. Auslander, M, and Reiten, I.; Representation theory of artin algebras III, Comm. in Algebra (1975), 239-294.

5. Auslander, M. and Reiten, I.; Stable equivalence of dualizing R-varieties, Advances in Math. 12 (1974), 306-366.

6. Bass, H.; Finistic dimension and a homological generalization of semi-primary rings, Trans. Amer. Math. Soc. 95 (1960), 466-488.

7. Dlab, V. and Ringel, C.; On algebras of finite representation type, J. Algebra 33 (1975), 306-394.

8. Gabriel, P.; Indecomposable representations II, in "Symposia Mathematica," Vol. XI, Academic Press, New York/San Francisco/London (1973), 81-104.

9. Gordon, R. and Green, E. L.; Modules with cores and amalgamations of indecomposable modules, Memoirs of the Am. Math. Soc., to appear.

10. Gordon, R. and Green, E. L.; Indecomposable modules: Modules with cores, Bull. Amer. Math. Soc., 82 (1976), 590-592.

11. Gordon, R. and Green, E. L.; Indecomposable modules: Amalgamations, Bull. Amer. Math. Soc., to appear.

12. Green, E. L.; Diagrammatic techniques in the study of indecomposable modules, Oklahoma Conf. on Ring Theory, 1976, to appear.

13. Müller, W.; Unzerlegbare moduln über aritnschen Ringen, Math. Zeitschrift 137 (1974), 197-226.

AUTOMORPHISM GROUPS OF SIMPLE ALGEBRAS AND GROUP ALGEBRAS

Gerald J. Janusz
University of Illinois
Urbana, Illinois

INTRODUCTION

In this paper we study the group Aut (A) of all automorphisms of a simple algebra A having finite dimension over its center K. Since the automorphisms which act as the identity on K are known to form the subgroup In Aut (A) of inner automorphisms, the interest lies in determining which automorphisms of K can be extended to A. In the first section we treat this problem with no restriction on K. In the second section we restrict K to be an algebraic number field and a fairly complete answer is given to the question. In the third section we apply our results to give information about the automorphism group of the group algebra Q(G) of a finite group G over the rational field Q. We give some evidence to support the conjecture that the equality Aut Q(G) = In Aut Q(G) can hold only in the trivial case when G has order one.

§1. EXTENSIONS OF CENTRAL AUTOMORPHISMS

Let K be a field and Aut (K) its group of automorphisms. For an element σ in Aut (K) let K_σ denote the K - K bimodule Kz_σ on which K acts according to the rule $z_\sigma x = \sigma(x)z_\sigma$ for x in K.

Let A denote a central simple K-algebra. That is A is a simple ring which is finite dimensional over its center K. The class of A in the Brauer group $\mathcal{B}(K)$ is denoted by [A]. For each automorphism σ of K we may form the vector space $A_\sigma = K_\sigma \otimes A$ (tensor products are taken over K unless otherwise indicated). A_σ becomes a ring with multiplication induced by defining

$$(z_\sigma x \otimes a)(z_\sigma y \otimes b) = z_\sigma xy \otimes ab.$$

It is easy to see that $e_\sigma = z_\sigma \otimes 1$ is the identity element and A_σ is a K-algebra with center Ke_σ when the left action of K is on the first factor. Thus A_σ is a central simple K-algebra.

The following properties are easily verified:

(1) $K_\sigma \otimes K_\tau \cong K_{\sigma\tau}$ as K-K bimodules;

(2) $M_n(A_\sigma) \cong M_n(A)_\sigma$ as K-algebras.

Here $M_n(A)$ denotes the ring of $n \times n$ matrices over A.

From these two properties one deduces that $[A] \longrightarrow [A_\sigma]$ defines an action of the group Aut(K) upon $\mathbf{B}(K)$.

Let Aut (K;A) denote the subgroup of Aut (K) which fixes the class [A]. Thus σ is in Aut (K;A) if and only if $A \cong A_\sigma$ as K-algebras. We shall prove that the automorphisms of K which can be extended to an automorphism of A are precisely those in Aut (K;A).

<u>Theorem 1</u>. Let $\underset{\sim}{r}$ denote the restriction map which carries an automorphism of the central simple K-algebra A to the induced automorphism on K. Then the following sequence is exact:

$$(*) \qquad 1 \longrightarrow \text{In Aut } (A) \longrightarrow \text{Aut } (A) \xrightarrow{\underset{\sim}{r}} \text{Aut } (K;A) \longrightarrow 1.$$

Proof. If τ is an automorphism of A and its restriction to K, $\underset{\sim}{r}(\tau)$, is the identity, then the Skolem-Noether Theorem implies τ is an inner automorphism. Hence the first part of the exactness holds. For an arbitrary automorphism τ of A, we show now $\underset{\sim}{r}(\tau)$ lies in Aut(K;A). Let $\underset{\sim}{r}(\tau)$ be denoted by σ. Define a mapping

$$f : A_\sigma = K_\sigma \otimes A \longrightarrow A$$

by the formula

$$f(z_\sigma x \otimes a) = \tau(xa).$$

One easily sees f is well-defined and is a ring isomorphism. Moreover for x in K,

$$f(xe_\sigma) = f(z_\sigma \sigma^{-1}(x) \otimes 1) = \tau\sigma^{-1}(x) = x$$

so that f is a K-algebra isomorphism between A_σ and A. Thus $[A] = [A_\sigma]$ and σ lies in Aut (K;A).

Conversely suppose σ lies in Aut (K;A). Then there is a K-algebra isomorphism f from A_σ to A. The function τ defined on A by $\tau(a) = f(z_\sigma \otimes a)$ is a ring isomorphism (and hence τ is in Aut (A)). Moreover for x in K,

$$\tau(x) = f(z_\sigma \otimes x) = f(\sigma(x)z_\sigma \otimes 1) = \sigma(x)f(e_\sigma) = \sigma(x)$$

and so $\underset{\sim}{r}(\tau) = \sigma$ and the theorem is proved.

This result tells when an automorphism of K is the restriction of an automorphism of A. However an automorphism of order n on K need not be the restriction of an automorphism of order n on A. This is the question of splitting. We say the sequence (*) splits over a subgroup H of Aut (K;A) if there is a homomorphism $\underset{\sim}{s} : H \longrightarrow \text{Aut}(A)$ such that $\underset{\sim}{r}\underset{\sim}{s}$ = identity on H. In order to give a criterion for splitting over H, we shall assume H is finite. The general case seems to involve the Galois theory of simple rings which is not developed to a satisfactory state.

Assume H is a finite subgroup of Aut (K;A). Let F denote the subfield of K consisting of elements fixed by H.

Theorem 2. The sequence (*) splits over H if and only if there is an F-central simple algebra B such that A is isomorphic to $K \otimes_F B$ as K-algebras.

Proof. Suppose A is isomorphic to $K \otimes_F B$ as K-algebras. If we identify these two algebras, then $\underset{\sim}{s} : \sigma \longrightarrow \sigma \otimes 1$ gives an imbedding of H into Aut(A) and $\underset{\sim}{r}\underset{\sim}{s}(\sigma) = \sigma$ so we have the splitting.

Conversely suppose we have a splitting map $\underset{\sim}{s}$ defined on H. We let B denote the set of elements of A fixed by all elements in $\underset{\sim}{s}(H)$. We shall prove B has the desired properties.

Let D denote the trivial crossed product of K by H; that is

$$D = \sum_{\sigma \text{ in } H} Ku_\sigma, \quad u_\sigma u_\tau = u_{\sigma\tau}, \quad u_\sigma x = \sigma(x)u_\sigma.$$

Then D is an F-central simple algebra and in fact is a matrix ring over F. D has a unique simple module which can be identified with K. The action of D on K is determined by $xu_\sigma(y) = x\sigma(y)$. D also acts on A by the rule $xu_\sigma(a) = x\underset{\sim}{s}(\sigma)(a)$. These two actions are compatible with the inclusion of K in A because $\underset{\sim}{r}\underset{\sim}{s}(\sigma) = \sigma$. Since A is a finitely generated D module, it is isomorphic to a direct sum of copies of the unique simple module K. As a K-vector space, such a module is one-dimensional so we have

$$A = Ky_1 \oplus \cdots \oplus Ky_m.$$

We may assume there is a D-module isomorphism between K and Ky_i given by $x \longrightarrow xy_i$. Since $u_\sigma(1) = 1$ it follows that y_i is left fixed by all $\underset{\sim}{s}(\sigma)$; that is y_i lies in B. So we have KB = A.

From the definition of B we see $F \subseteq B$ so the tensor product $K \otimes_F B$ is defined and there is a well defined map from it onto KB = A induced by $x \otimes b \longrightarrow xb$. It remains to show this map is a monomorphism This will follow if we show B is a simple ring with center F; for then $K \otimes_F B$ is also simple.

In the ring $\mathrm{Hom}_F(A,A)$, Let A_L and D_L be the rings of left multiplication induced by elements of A and D respectively. We have the relations

$$(u_\sigma)_L \, a_L = \underset{\sim}{s}(\sigma)(a)_L \, (u_\sigma)_L$$

so the ring generated by A_L and D_L is a "crossed-product" of A_L by $\underset{\sim}{s}(H)$. We have

$$A_L D_L = \sum_\sigma A_L \, u_{\sigma L}$$

and one may check that $A_L D_L$ is a simple ring with center F. The proof requires very little change from the proof in the usual case in which A_L is a field.

Now A is a left $A_L D_L$ module and so the endomorphism ring of A over $A_L D_L$ is a simple ring having center F. (This is true for any module over $A_L D_L$.) However the endomorphism ring of A over $A_L D_L$ is the set of right multiplications, a_R, which commute with every u_σ. This holds precisely for the right multiplications induced by elements of B. Since $B \cong B_R$ we have the result that B is F-central simple and the proof is complete.

§2. THE NUMBER FIELD CASE

Now we assume K is an algebraic number field. The elements of the Brauer group of K are uniquely determined by Hasse invariants. We shall identify Aut $(K;A)$ in these terms. The reader may refer to Reiner's book [3] for a survey of the properties of Hasse invariants. A very brief summary of properties will be mentioned below.

Let v be a valuation of K and K_v the completion at v. Suppose K_v is a p-adic field; that is K_v is a finite extension of Q_p, the p-adic rationals. A central simple K_v algebra can be represented as a crossed product $(W,\phi,\pi^s) = C$ in which W/K_v is an unramified extension, ϕ is the Frobenius automorphism, and π is a prime element of K_v. The Hasse invariant of C is the coset in the additive group Q/Z of the rational number $s/(W:K_v)$.

If K_v is the real field then $\mathbf{B}(K_v)$ has order two; the non-trivial element has invariant $1/2$ in Q/Z. If A is a K-central simple algebra, $\text{inv}_v A$ denotes the Hasse invariant of $K_v \otimes A$. When K_v is the complex field, $\text{inv}_v A = 0$ for all A.

The group Aut (K) acts naturally upon the set of valuations; v^σ is defined by $v^\sigma(x) = v(\sigma(x))$. For a central simple K-algebra A, let $I(A)$ denote the subgroup of Aut (K) consisting of those σ which satisfy $\text{inv}_v A = \text{inv}_{v^\sigma} A$ for all valuations v of K. We shall prove $I(A)$ coincides with Aut $(K;A)$. First we need a preliminary result.

<u>Lemma</u>. For $i = 1,2$ let L_i be a finite extension of the p-adic rationals Q_p and A_i a central simple L_i-algebra. If A_1 and A_2 are isomorphic as rings, then $\text{inv}\, A_1 = \text{inv}\, A_2$.

Proof. Let $f : A_1 \longrightarrow A_2$ be the ring isomorphism. Then $f(L_1) = L_2$ since the centers must correspond. Express A_1 as a crossed product (W_1,ϕ_1,π_1^s) with W_1/L_1 unramified, ϕ_1 the Frobenius automorphism and π_1 a prime element of L_1. Then $\text{inv}\, A_1 \equiv s/(W_1;L_1) \bmod Z$. Now $A_2 = (f(W_1), f\phi f^{-1}, f(\pi_1)^s)$ and $f(W_1)/L_2$ is unramified, $f\phi f^{-1}$ is the Frobenius automorphism and $f(\pi_1)$ is a prime element of L_2. Hence $\text{inv}\, A_2 \equiv s/(f(W_1):L_2) \equiv \text{inv}\, A_1$.

Remark: The lemma is trivially true also in the case when L_i is isomorphic to either the real or complex field.

Theorem 3. For any K-central simple algebra A we have Aut $(K;A) = I(A)$.

Proof: Let σ be an element in Aut $(K;A)$. Then there is an automorphism σ' of A which extends σ. Also there is an isomorphism σ'' from K_v to K_{v^σ} which extends σ for each valuation v. Hence $\sigma'' \otimes \sigma'$ is a well defined ring isomorphism between $K_v \otimes A$ and $K_{v^\sigma} \otimes A$. By the lemma we have $\mathrm{inv}_v A = \mathrm{inv}_{v^\sigma} A$ and so σ is in $I(A)$.

Conversely let σ be an element of $I(A)$. Let σ'' be the isomorphism from K_v to K_{v^σ} and let σ' be the map from A to A_σ defined by $\sigma'(a) = 1 \otimes a$. Then $\sigma'' \otimes \sigma'$ is a ring isomorphism between $K_v \otimes A$ and $K_{v^\sigma} \otimes A_\sigma$. By the lemma we have $\mathrm{inv}_v A = \mathrm{inv}_{v^\sigma} A_\sigma$. However the choice of σ insures $\mathrm{inv}_v A = \mathrm{inv}_{v^\sigma} A$. Thus A and A_σ have the same invariant at every valuation v. It follows $[A] = [A_\sigma]$ and so σ is in Aut $(K;A)$.

§3. GROUP ALGEBRAS

We use the results of sections one and two to prove results about Aut $(Q(G))$ when G is a finite group. Fix the following notation:

$$Q(G) = A_1 \oplus \cdots \oplus A_n, \quad A_i \text{ simple},$$

$$\text{center} = K_1 \oplus \cdots \oplus K_n.$$

Let m_i denote the index of A_i. Thus A_i is a central simple K_i-algebra which is a matrix ring of some size over a division ring which has dimension $(m_i)^2$ over K_i. We shall require two results about simple summands of group algebras.

Theorem [1] [2]. (a) A primitive m_i'th root of unity, ε_{m_i} lies in K_i.

(b) If σ is in Aut (K_i) and $\sigma(\varepsilon_{m_i}) = (\varepsilon_{m_i})^r$, then $\mathrm{inv}_v A_i = r \cdot \mathrm{inv}_{v^\sigma} A_i$ for all valuations v.

From this theorem and theorems 1 and 3 we obtain the following.

Corollary 1. An automorphism of K_i lifts to an automorphism of A_i if and only if it fixes a primitive m_i'th root of unity; that is $\text{Aut}(K_i;A_i) = \text{Gal}(K_i/Q(\varepsilon_{m_i}))$.

Corollary 2. Let N be the subgroup of $\text{Aut}(Q(G))$ consisting of all the automorphisms which send each A_i into itself. Then there is a normal series

$$\text{Aut}(Q(G)) > N > \text{In Aut}(Q(G)).$$

(a) $\text{Aut}(Q(G))/N$ is a finite group acting as permutations of the A_i;

(b) $N/\text{In Aut}(Q(G)) = \text{Gal}(K_1/Q(\varepsilon_{m_1})) \times \cdots \times \text{Gal}(K_n/Q(\varepsilon_{m_n}))$.

We use the information contained in this corollary to give evidence for the following conjecture.

Conjecture. If G is a finite group of order $\neq 1$, then $Q(G)$ has an outer automorphism.

If this conjecture is false, let G be a counterexample of least possible order. We prove some facts about G.

Fact 1. G is a non-abelian simple group.

Proof: Suppose $H \triangleleft G$. Since $Q(G)$ is semi-simple, the natural homomorphism $Q(G) \longrightarrow Q(G/H)$ splits and every automorphism of $Q(G/H)$ extends to an automorphism of $Q(G)$. In particular all automorphisms of $Q(G/H)$ are also inner. Thus $H = G$ or $H = 1$ and so G is simple.

Suppose G is abelian of order p. Then one of the simple summands of $Q(G)$ is the field $Q(\varepsilon_p)$ which has $p-2$ non-inner automorphisms all of which extend to non-inner automorphisms of $Q(G)$. Hence $p=2$ and $Q(G) \cong Q \oplus Q$ which has an outer automorphism exchanging the two summands. This conflict proves fact 1.

Fact 2. If X is an absolutely irreducible character of G having Schur index m, then the field of values $Q(X)$ equals $Q(\varepsilon_m)$. In particular if X has Schur index one or two, then X is rational valued.

Proof: If $X(A_i) \neq 0$, then each element of $Gal(K_i/Q(\varepsilon_{m_i}))$ corresponds to an outer automorphism. Hence $K_i = Q(\varepsilon_{m_i})$. But it is well-known that K_i is isomorphic to the field $Q(X)$.

Fact 3. If X_1 and X_2 are two rational valued absolutely irreducible characters of G, each having Schur index one, then $X_1(1) \neq X_2(1)$.

Proof: If all automorphisms of $Q(G)$ are inner, then no two simple summands of $Q(G)$ can be isomorphic . The contrary case would allow the two summands to be exchanged, giving an outer automorphism. If X is rational valued of Schur index one, then the simple summand of $Q(G)$ corresponding to X is $M_n(Q)$ with $n = X(1)$. We cannot have two summands isomorphic to $M_n(Q)$ so the result follows as distinct rational characters correspond to distinct summands.

An examination of the character tables available to me at this time has failed to produce any possible G which satisfies the condition described by these three facts. There is no example known of a simple group with a character of Schur index ≥ 3. All the examples checked have at least two rational characters of Schur index one and the same degree, or else a character for which Fact 2 does not hold.

REFERENCES

1. M. Bernard, The Schur subgroup, I, J. of Alg. 22 (1972), 374-377.

2. M. Bernard and M. Schacher, The Schur subgroup, II, J. of Alg. 22 (1972), 378-385.

3. I. Reiner, Maximal Orders, Academic Press, New York, 1975.

REPRESENTATION THEORY OF HEREDITARY ARTIN ALGEBRAS

María Inés Platzeck
Brandeis University
Waltham, Massachusetts

Maurice Auslander
Brandeis University
Waltham, Massachusetts

INTRODUCTION

Throughout this paper we assume that Λ is an artin algebra, that is an artin ring that is a finitely generated modulo over its center C, which is also an artin ring. We denote by mod (Λ) the category of finitely generated (left) Λ-modules. Let D: mod $(\Lambda) \longrightarrow$ mod (Λ^{op}) be the usual duality given by $X \longrightarrow \mathrm{Hom}_C(X,I)$, where I is the injective envelope over C of $C/_{\mathrm{rad}\ C}$ and let Tr: $\underline{\mathrm{mod}}$ $(\Lambda) \longrightarrow \underline{\mathrm{mod}}$ (Λ^{op}) be the duality between the category $\underline{\mathrm{mod}}$ (Λ) of finitely generated modules modulo projectives over Λ and $\underline{\mathrm{mod}}$ (Λ^{op}), the category of finitely generated modules modulo projectives over Λ^{op} given by the transpose (see [3]).

We assume in all that follows that Λ is an hereditary artin algebra. All the modules considered are finitely generated left modules. We prove that the following conditions are equivalent for an indecomposable Λ-module M.

a) There exists an integer $n \geq 0$ such that $(D\mathrm{Tr})^n M$ is projective.

b) There are only a finite number of nonisomorphic indecomposable modules X such that $\mathrm{Hom}_\Lambda(X,M) \neq 0$.

We prove that a) and b) are equivalent using the notion of almost split sequences and irreducible maps (see [4]). More precisely, we prove that a) and b) are equivalent to: c) There is a projective module C_0 and a chain of irreducible maps of indecomposable Λ-modules

$$C_0 \xrightarrow{f_1} C_1 \longrightarrow \cdots \xrightarrow{f_k} C_k = M.$$

It is known that Λ is of finite representation type if and only if all the indecomposable Λ modules verify b) or, what is equivalent, if and only if the simple Λ modules verify b) (see[2]). So we obtain, in particular, that Λ is of finite representation type if and only if for every indecomposable module M there is some $n \geq 0$ such that $(DTr)^n M$ is projective, or what is equivalent, if and only if for every simple Λ module S there is $n \geq 0$ such that $(DTr)^n S$ is projective.

We study properties of the indecomposable modules M verifying the equivalent conditions a), b) and c). For example, we prove for such a module M that $\mathrm{End}_\Lambda(M)$ is a division ring, and that $\mathrm{Ext}^1_\Lambda(M,M) = 0$. The properties proven for these modules will hold for all the indecomposable modules when the ring is of finite representation type.

Let $Gr(\Lambda)$ denote the Grothendieck group of Λ and let $[M]$ denote the element of $Gr(\Lambda)$ determined by the Λ-module M. We define a group isomorphism $c\colon Gr(\Lambda) \longrightarrow Gr(\Lambda)$ such that $C([M]) = [DTrM]$ for a non-projective indecomposable Λ-module M. This isomorphism is an important tool in the study of the representation theory of the ring. For example, using it we prove that if M and N are indecomposable Λ-modules and M verifies the property b), then $[M] = [N]$ implies $M \cong N$. In particular, if Λ is of finite representation type the indecomposable modules are determined by their composition factors.

We also define a quadratic form q from $Gr(\Lambda)$ into the field Q of rational numbers. To do it, we may assume that Λ is an algebra over a field k; then $q([M]) = \dim_k \mathrm{End}(M) - \dim_k \mathrm{Ext}^1_\Lambda(M,M)$, and we prove that the following conditions are equivalent:

a) Λ is of finite representation type.

b) There is some integer $n > 0$ such that $c^n = \mathrm{Id}_{Gr(\Lambda)}$.

c) q is positive definite.

Using these results an explicit description of the set of indecomposable Λ-modules can be given when the ring Λ is of finite representation type.

Using diagramatic techniques many of the results obtained here for arbitrary hereditary artin algebras have already been proven for algebras associated to quivers by Bernstein, Gelfand, Ponomarev, and Gabriel, (see [8] and [12]) and, in a more general context, for the hereditary algebra associated to a K-species by V. Dlab and C. Ringel in [11]. However, the basic ideas and techniques of proof used here are quite different from the diagramatic techniques used by the previous named authors and are applicable to all hereditary artin algebras, not just those associated with a K-species. The connections between these various points of view in case one is dealing with an algebra associated to a K-species are best illustrated by the fact that the functors DTr and C^+ as well as the functors TrD and C^- are isomorphic, where C^+ and C^- are the Coxeter functors (see [11]) (S. Brenner and M. Butler) and the fact that the Coxeter transformation corresponding to the bilinear form associated to a K-species in [11] is precisely the isomorphism $c: \mathrm{Gr}(\Lambda) \longrightarrow \mathrm{Gr}(\Lambda)$ mentioned above.

In another paper we are going to extend these results to artin algebras stably equivalent to an hereditary artin algebra. We recall that two algebras are stably equivalent if their categories of finitely generated modulos modulo projectives are equivalent (see [7]). Since artin algebras of radical square zero are stably equivalent to an hereditary algebra, this provides a unified approach to the hereditary and radical square zero cases, and also includes other algebras not studied previously.

§1. MODULES M SUCH THAT THE FUNCTOR $\mathrm{Hom}_\Lambda(\ ,M)$ HAS FINITE LENGTH.

It is known (see [2] Theorem 3.1) that an artin algebra Λ is of finite representation type if and only if the representable functors $\mathrm{Hom}_\Lambda(\ ,M)$ have finite length, for every Λ module M. We recall that the functor $\mathrm{Hom}_\Lambda(\ ,M)$ has finite length if and only if the number of non-isomorphic indecomposable modules X such that $\mathrm{Hom}_\Lambda(X,M) \neq 0$ is finite. To study the representation theory of the hereditary artin algebra Λ we are going to determine when for a given indecomposable Λ module M the length of the representable functor $\mathrm{Hom}_\Lambda(\ ,M)$ is finite. This will give us, in particular, a criterion to decide whether Λ is of finite representation type or not. We will also study properties of the indecomposable modules M such that the length of $\mathrm{Hom}_\Lambda(\ ,M)$ is finite. These properties will hold for all the modules when the ring Λ is of finite representation type.

We will denote the representable functor $\mathrm{Hom}_\Lambda(\ ,M)$ by $(\ ,M)$. Let $\mathrm{mod}_P\Lambda$ be the full subcategory of $\mathrm{mod}\ \Lambda$ whose objects are the Λ modules with no nonzero projective summands.

We recall some definitions and results from ([3], [4], and [5]) that will be needed later:

A map $g: B \longrightarrow C$ in an arbitrary additive category $\underline{\underline{C}}$ is said to be *right almost split* if it is not a splittable epimorphism, C is indecomposable and given any map $f: X \longrightarrow C$ which is not a splittable epimorphism there is a map $h: X \longrightarrow B$ such that $gh = f$. The map $g: B \longrightarrow C$ is said to be *right minimal* if for any commutative diagram

$$\begin{array}{ccc} B & \xrightarrow{g} & C \\ {\scriptstyle h}\downarrow & \nearrow_{g} & \\ B & & \end{array},$$

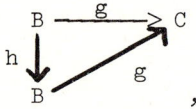

h is an isomorphism. The map $g: B \longrightarrow C$ is said to be *minimal right almost split* if it is right minimal and right almost split. There are analogous definitions by replacing right by left (see [4]).

A non-split exact sequence of finitely generated Λ-modules $0 \longrightarrow A \xrightarrow{f} B \xrightarrow{g} C \longrightarrow 0$ is almost split if A and C are indecomposable and given any morphism $h: X \longrightarrow C$ which is not a splittable epimorphism, there is some $s: X \longrightarrow B$ such that $gs = h$ (see [3]). Given a nonprojective indecomposable Λ module C (or a noninjective Λ-module A) then there exists a uniquely determined almost split sequence $0 \longrightarrow A \longrightarrow B \longrightarrow C \longrightarrow 0$, and $A \cong DTrC$ $(C \cong TrDA)$ (see [3]). Let C in mod Λ be indecomposable. Then if C is not projective a map $g: B \longrightarrow C$ is minimal right almost split if and only if $0 \longrightarrow Ker(g) \longrightarrow B \longrightarrow C \longrightarrow 0$ is an almost split sequence. If C is projective then $g: B \longrightarrow C$ is minimal right almost split if and only if g is a monomorphism and $g(B) = \underline{r}C$.

We recall also that a map $g: B \longrightarrow C$ is said to be irreducible if g is neither a split monomorphism nor a split epimorphism and for any commutative diagram

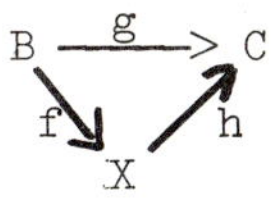

f is a splittable monomorphism or h is a splittable epimorphism. If C in mod (Λ) is indecomposable then a map $g: B \longrightarrow C$ where B is non-zero is irreducible if and only if there is some map $g': B' \longrightarrow C$ such that $(g,g'): B \amalg B' \longrightarrow C$ is minimal right almost split (see [4]).

There are analogous results for left almost split maps.

Let now M be an indecomposable module and assume that $f: P \longrightarrow M$ is an irreducible map with P indecomposable projective. Then there is a Λ-module K and a map $g: K \longrightarrow M$ such that $(f,g): P \amalg K \longrightarrow M$ is a minimal right almost split map; so

$0 \longrightarrow DTrM \longrightarrow P \amalg K \longrightarrow M \longrightarrow 0$ is the almost split sequence. Then $DTrM \longrightarrow P \amalg K$ is minimal left almost split, so the induced map $DTrM \longrightarrow P$ is irreducible. But we know that the inclusion $\underline{r}P \longrightarrow P$ is minimal left almost split, so DTrM is a direct summand of $\underline{r}P$. We have then

Proposition 1.1: If M is an indecomposable nonprojective module and there is an irreducible map $f\colon P \longrightarrow M$ with P indecomposable projective, then DTrM is a direct summand of $\underline{r}P$. If there is an irreducible map $M \longrightarrow P$ with P indecomposable projective, then M is a direct summand of $\underline{r}P$.

We denote by $\underline{\mathrm{Hom}}_\Lambda(M,N)$ and $\overline{\mathrm{Hom}}_\Lambda(M,N)$ the group of morphisms from M to N in $\underline{\mathrm{mod}}\ \Lambda$ and $\overline{\mathrm{mod}}\ \Lambda$ respectively. For a Λ-module X we denote $X^* = \mathrm{Hom}_\Lambda(X,\Lambda)$. We recall that if M is in $\mathrm{mod}_P\Lambda$ and $P_1 \xrightarrow{f} P_0 \longrightarrow M \longrightarrow 0$ is a minimal projective presentation for M, then the transpose of M, TrM, is the cokernel of the map $f^*\colon P_0^* \longrightarrow P_1^*$. Tr induces a duality $\mathrm{Tr}\colon \underline{\mathrm{mod}}\ \Lambda \longrightarrow \underline{\mathrm{mod}}\ \Lambda^{op}$, and the compositions $\mathrm{DTr}\colon \underline{\mathrm{mod}}\ \Lambda \longrightarrow \overline{\mathrm{mod}}\ \Lambda$ and $\mathrm{TrD}\colon \overline{\mathrm{mod}}\ \Lambda \longrightarrow \underline{\mathrm{mod}}\ \Lambda$ are inverse equivalences that have the following property (see [5], Proposition 1.2). Let $\mathrm{mod}_I\Lambda$ be the full subcategory of mod Λ of the modules with no nonzero injective summands.

Proposition 1.2: Let $f\colon M \longrightarrow N$ be an irreducible map. Then:

a) If M,N are in $\mathrm{mod}_P\Lambda$, $\mathrm{DTrf}\colon \mathrm{DTrM} \longrightarrow \mathrm{DTrN}$ is an irreducible map.

b) If M,M are in $\mathrm{mod}_I\Lambda$, $\mathrm{TrDf}\colon \mathrm{TrDM} \longrightarrow \mathrm{TrDN}$ is an irreducible map.

We assume in all that follows that Λ is an hereditary artin algebra.

We observe that if M and N are in $\mathrm{mod}_P\Lambda$ and $f\colon M \longrightarrow N$ is a map that factors through a projective then since Λ is **hereditary** $f = 0$. Som $\mathrm{Hom}_\Lambda(M,N) = \underline{\mathrm{Hom}}_\Lambda(M,N)$ and therefore $\mathrm{mod}_P\Lambda = \underline{\mathrm{mod}}_P\Lambda$. Similarly,

$\mathrm{mod}_I\Lambda = \overline{\mathrm{mod}}_I\Lambda$. Then in this case the transpose $\mathrm{Tr}: \underline{\mathrm{mod}}\ \Lambda \longrightarrow \underline{\mathrm{mod}}\ \Lambda^{op}$ is in fact a functor from $\mathrm{mod}_P\Lambda$ to $\mathrm{mod}_P\Lambda^{op}$. If M is in $\mathrm{mod}_P\Lambda$ and $P_1 \xrightarrow{f} P_0 \longrightarrow M \longrightarrow 0$ is a minimal projective presentation for M then, since Λ is hereditary, f is a monomorphism, so we have an exact sequence $0 \longrightarrow M^* \longrightarrow P_0^* \xrightarrow{f^*} P_1^* \longrightarrow \mathrm{Ext}^1_\Lambda(M,\Lambda) \longrightarrow 0$. So $\mathrm{Tr}M = \mathrm{Ext}^1_\Lambda(M,\Lambda)$ and Tr is therefore the restriction of $\mathrm{Ext}^1_\Lambda(\ ,\Lambda)$ to $\mathrm{mod}_P\Lambda$.

For an indecomposable module M in $\mathrm{mod}\ \Lambda$, if we write $D\mathrm{Tr}M$ it is implicitely assumed that M is not projective. We define $(D\mathrm{Tr})^nM$, which will play an important role for studying when the representable functor $(\ ,M)$ has finite length, by $(D\mathrm{Tr})^0M = M$, $(D\mathrm{Tr})^nM = D\mathrm{Tr}(D\mathrm{Tr})^{n-1}M$, if $n \geq 1$.

Lemma 1.3: Let M, N be indecomposable Λ-modules. If for some $n \geq 0$ $(D\mathrm{Tr})^nM$ is projective and $(D\mathrm{Tr})^jN$ is not projective for all $j \leq n$, then $\mathrm{Hom}_\Lambda(N,M) = 0$.

Proof: We observe first that if P is an indecomposable projective Λ-module and L is in $\mathrm{mod}_P(\Lambda)$ then $\mathrm{Hom}_\Lambda(L,P) = 0$. Suppose now that M,N are indecomposable modules such that $(D\mathrm{Tr})^nM$ is projective and $(D\mathrm{Tr})^jN$ is not projective if $j \leq n$. Then

$$\mathrm{Hom}_\Lambda(N,M) \simeq \mathrm{Hom}_\Lambda(D\mathrm{Tr}N, D\mathrm{Tr}M) \simeq \ldots \simeq \mathrm{Hom}_\Lambda((D\mathrm{Tr})^nN,(D\mathrm{Tr})^nM)) = 0$$

because $(D\mathrm{Tr})^nM$ is projective and $(D\mathrm{Tr})^nN$ is in $\mathrm{mod}_P(\Lambda)$.

Proposition 1.4: Let M be an indecomposable Λ-module such that $(D\mathrm{Tr})^nM$ is projective for some $n \geq 0$. Then the functor $(\ ,M)$ has finite length. In particular $(\ ,P)$ has finite length for any finitely generated projective module P.

Proof: Let X be an indecomposable module and $f: X \longrightarrow M$ a non-zero map. We know by the lemma that there is $j \leq n$ such that $(D\mathrm{Tr})^jX$ is projective. So the indecomposable modules X such that

$(X,M) \neq 0$ are all of the form $(TrD)^j P$, for some indecomposable projective Λ-module P and some j between 0 and n; they are only finite in number and consequently $(,M)$ has finite length.

Proposition 1.5: Let M be an indecomposable Λ-module in $mod_P(\Lambda)$. The following conditions are equivalent:

a) There is a chain of irreducible maps of indecomposable Λ-modules $P = C_0 \xrightarrow{f_1} C_1 \longrightarrow \cdots \xrightarrow{f_k} C_k = M$ with P projective.

b) There is an integer $n \geq 0$ such that $(DTr)^n M$ is projective.

c) $(,M)$ has finite length.

d) There is a chain of irreducible maps of indecomposable Λ-modules $P = C_0 \xrightarrow{f_1} C_1 \longrightarrow \cdots \xrightarrow{f_k} C_k = M$ with P projective and such that the composition $f_k \cdots f_1$ is not zero. If any of these conditions is verified, then the integer of n of b) is smaller than or equal to the length of any chain of irreducible maps of indecomposable modules $P = C_0 \xrightarrow{f_1} C_1 \longrightarrow \cdots \xrightarrow{f_k} C_k = M$ with P projective.

Proof: Proposition 1.4 proves that b) ===> c). Obviously, d) ===> a). a) ===> b). Let now $P = C_0 \xrightarrow{f_1} C_1 \longrightarrow \cdots \xrightarrow{f_k} C_k$ be a chain of irreducible maps of indecomposable Λ-modules with P projective. To finish the proof of the proposition we only need to prove that there is a nonnegative integer $n \leq k$ such that $(DTr)^n C_k$ is projective. We prove it by induction on k. If $k = 1$ we have an irreducible map $f_1 : P \longrightarrow M$ with P indecomposable projective; we know then by Proposition 1.1 that $DTrM$ is a direct summand of $\underline{r}P$ and is, therefore, projective.

We assume now that the assertion is true for $k \leq j$. Let $P = C_0 \xrightarrow{f_1} \cdots C_j \xrightarrow{f_{j+1}} C_{j+1}$ be a chain of irreducible maps of indecomposable Λ-modules with P projective. By the induction hypothesis we know that $(DTr)^s(C_j)$ is projective for some nonnegative integer $s \leq j$. If $(DTr)^r(C_{j+1})$ is projective for some $r \leq s$ then the assertion is true, so we may assume that $(DTr)^r(C_{j+1})$ is not projective for every $r \leq s$; we know then by repeated application of Proposition 1.2 that the map $(DTr)^s(f_{j+1})\colon (DTr)^s(C_j) \longrightarrow (DTr)^s(C_{j+1})$ is irreducible. But $(DTr)^s(C_j)$ is projective, so $DTr((DTr)^s(C_{j+1})) = (DTr)^{s+1}(C_{j+1})$ is projective; this ends the proof because $s+1 \leq j+1$. c) ===> d) It is proven in [6] that if M is indecomposable, (,M) has finite length and there is a nonzero map from an indecomposable module N to M, then there is a chain of irreducible maps of indecomposable Λ-modules $N = C_0 \xrightarrow{f_1} C_1 \longrightarrow \cdots \xrightarrow{f_k} C_k = M$ such that the composition $f_k \cdots f_1 \neq 0$. c) ===> d) follows now observing that there is always a nonzero map from an indecomposable projecti e P to M.

<u>Corollary 1.6</u>: There is only a denumerable number of indecomposable modules M such that (,M) has finite length.

<u>Observation</u>: Using other techniques it is possible to prove that Corollary 1.6 holds for all artin algebras, not only for those that are hereditary.

As an artin algebra is of finite representation type if and only if the length of (,M) is finite for every module M, or, what is equivalent, the length of (,S) is finite for all simple modules S (see [2], theorem 3.1), we now summarize these remarks in the following theorem.

<u>Theorem 1.7</u>: The following conditions are equivalent:

a) Λ is of finite representation type.

b) For every indecomposable module M in mod (Λ) there exists $n \geq 0$ such that $(DTr)^n M$ is projective.

c) For every simple Λ-module S there exists $n \geq 0$ such that $(DTr)^n S$ is projective.

d) For every indecomposable nonprojective Λ-module M there is a chain of irreducible maps of indecomposable Λ-modules $P = C_0 \xrightarrow{f_1} C_1 \longrightarrow \cdots \xrightarrow{f_k} C_k = M$ with P projective.

d') For every indecomposable nonprojective Λ-module M there is a chain of irreducible maps of indecomposable Λ-modules $P = C_0 \xrightarrow{f_1} C_1 \longrightarrow \cdots \xrightarrow{f_k} C_k = M$ with P projective such that the composition $f_k \ldots f_1$ is not zero.

e) For every simple nonprojective Λ-module S there is a chain of irreducible maps of indecomposable Λ-modules $P = C_0 \xrightarrow{f_1} C_1 \longrightarrow \cdots \xrightarrow{f_k} C_k = S$ with P projective.

e') For every simple nonprojective Λ-module S there is a chain of maps as indicated in e) verifying also that the composition $f_k \ldots f_1$ is not zero.

This theorem allows us to give a concrete description of the indecomposable Λ-modules when the ring is of finite representation type.

<u>Corollary 1.8</u>: Let $I_1, \ldots, I_n$ be a complete set of nonisomorphic indecomposable injective Λ-modules and assume that Λ is of finite representation type. Let $n_i \geq 0$ be such that $(DTr)^{n_i}(I_i)$ is projective. Then

$\mathcal{D} = \{(DTr)^j(I_i), 0 \leq j \leq n_i\}$ is a complete set of nonisomorphic indecomposable Λ-modules.

<u>Proof</u>: Assume that $M = (DTr)^r(I_i)$ and $N = (DTr)^t(I_j)$ are isomorphic, for some $i, j, r \leq n_i, t \leq n_j$. We will see that then $r = t$ and $i = j$.

We may assume that $r \geq t$. From $M = (DTr)^r(I_i) \simeq (DTr)^t(I_j) = N$ we get $(TrD)^t(DTr)^r(I_i) \simeq (TrD)^t(DTr)^t(I_j)$, so $(DTr)^{r-t}(I_i) \simeq I_j$. Since I_j is injective $r - t$ must be zero because the dual of the transpose of a module is in $\text{mod}_I(\Lambda)$; then

$I_i \simeq I_j$, so $i=j$ and $r=t$.

Then the modules $P_i = (DTr)^{n_i}(I_i)$ are pairwise nonisomorphic; as their number is the same as the number of indecomposable injective Λ-modules, we have that $\{P_1,\ldots,P_n\}$ is a complete set of nonisomorphic indecomposable projective Λ-modules.

Let now M be an indecomposable Λ-module. Since Λ is of finite representation type there is some $m \geq 0$ such that $(DTr)^m(M)$ is projective, and is therefore isomorphic to one of the P_i, $i=1,\ldots,n$. Assume $(DTr)^m(M) \cong P_i$. Then $(DTr)^m(M) \simeq (DTr)^{n_i}(I_i)$; if $m > n_i$ we have $(DTr)^{m-n_i}(M) \simeq I_i$, so $DTr((DTr)^{m-n_i}M) = I_i$, contradiction, because I_i is injective. Therefore $m \leq n_i$ and $M \simeq (DTr)^{n_i-m}(I_i)$, i.e., M is isomorphic to one object in $\mathcal{D}$. This finishes the proof of the corollary.

Let P be an indecomposable projective Λ-module and $f: P \longrightarrow P$ be a nonzero homomorphism. Then Im(f) is projective, so $P \longrightarrow \mathrm{Im}(f)$ is a splittable epimorphism; as f is nonzero and P is indecomposable, $\mathrm{Ker}(P \longrightarrow \mathrm{Im}(f)) = 0$ so f is a monomorphism; therefore, P and Im(f) have the same length, so $P = \mathrm{Im}(f)$ and f is then an isomorphism. This proves that $\mathrm{End}_\Lambda(P)$ is a division ring. Using Proposition 1.5 we can prove now the following generalization.

Corollary 1.9: Let M,N be indecomposable modules in mod (Λ) such that $(\,,M)$ has finite length and assume that there exist nonzero maps $f: M \longrightarrow N$ and $g: N \longrightarrow M$. Then f and g are isomorphisms. In particular, $\mathrm{End}_\Lambda(M)$ is a division ring for every indecomposable M in mod (Λ) such that the length of $(\,,M)$ is finite. So, if Λ is of finite representation type $\mathrm{End}_\Lambda(M)$ is a division ring for every finitely generated indecomposable Λ-module M.

Proof: We assume first that M is projective. Then Im(g) is also projective, so the map $N \longrightarrow \mathrm{Im}(g)$ is a splittable epimorphism; as

N is indecomposable and $g \neq 0$, g must be a monomorphism and then N is projective and length $(N) \leq$ length (M). Applying now the same argument to the projective module N and to the map f we obtain that f is a monomorphism, so length $(M) \leq$ length (N). Therefore M and N have the same length and the morphisms f and g are isomorphisms.

Assume now that $(\,,M)$ has finite length; let $n \geq 0$ be such that $(DTr)^n(M)$ is projective. We may assume that $(DTr)^j(N)$ is not projective if $j < n$. Then

$0 \neq Hom_\Lambda(M,N) \simeq Hom_\Lambda(DTrM,DTrN) \simeq \ldots \simeq Hom_\Lambda((DTr)^n M,(DTr)^n N)$ and

$0 \neq Hom_\Lambda(N,M) \simeq Hom_\Lambda(DTrN,DTrM) \simeq \ldots \simeq Hom_\Lambda((DTr)^n N,(DTr)^n M)$.

Here $(DTr)^n M$ is projective and we know by the first part of the proof that $(DTr)^n(f)$, $(DTr)^n(g)$ are isomorphisms, so f and g are isomorphisms (remember that $DTr\colon mod_P\Lambda \longrightarrow mod_I\Lambda$ is an equivalence of categories).

We now use Lemma 1.3 and the previous result to study some properties of the modules $Ext^1_\Lambda(M,N)$, with M and N indecomposable. We have

Proposition 1.10: Let M,N be indecomposable Λ-modules and assume that for some $n \geq 0$ $(DTr)^n M$ is projective and $(DTr)^j N$ is not projective if $j < n$. Then $Ext^1_\Lambda(M,N) = 0$. Therefore, for every indecomposable Λ-module M such that $(\,,M)$ has finite length $Ext^1_\Lambda(M,M) = 0$. In particular, if Λ is of finite representation type $Ext^1_\Lambda(M,M) = 0$ for every indecomposable Λ-module M.

Proof: We know by ([3], Proposition 2.2) that for every A in $mod_P(\Lambda)$, B in $mod(\Lambda)$ there is an isomorphism

$\underline{Hom}(A,B) \simeq Tor_1^\Lambda(TrA,B)$. For A in $mod(\Lambda)$ and B in $mod(\Lambda^{op})$ $Ext^1_\Lambda(A,DB) \simeq D(Tor_1^\Lambda(B,A))$ (see [9], p.p. 119). Thus $Ext^1_\Lambda(M,N) \simeq D\underline{Hom}(TrDN,M)$, for every N in $mod_I(\Lambda)$.

Then, since $\underline{\mathrm{Hom}}(\mathrm{TrDN},M) = \mathrm{Hom}(\mathrm{TrDN},M)$ we have $\mathrm{Ext}^1_\Lambda(M,N) \cong D(\mathrm{Hom}_\Lambda(\mathrm{TrDN},M))$.

$(\mathrm{DTr})^n M$ is projective, but $(\mathrm{DTr})^j(\mathrm{TrDN}) = (\mathrm{DTr})^{j-1}N$ is not projective if $j - 1 < n$, i.e., if $j \leq n$. Then, by Lemma 1.3, $\mathrm{Hom}_\Lambda(\mathrm{TrDN},M) = 0$. Thus $\mathrm{Ext}^1_\Lambda(M,N) = 0$.

Another case where $\mathrm{Ext}^1_\Lambda(M,N)$ is zero is considered in the following.

Proposition 1.11: Let M be an indecomposable nonprojective Λ-module such that $(\mathrm{DTr})^n M$ is projective for some $n \geq 0$. If $0 \longrightarrow \mathrm{DTrM} \longrightarrow E \longrightarrow M \longrightarrow 0$ is the almost split sequence then $\mathrm{Ext}^1_\Lambda(M,E) = 0$.

Proof: We know that $D(\mathrm{Ext}^1_\Lambda(M,E)) \cong \underline{\mathrm{Hom}}(\mathrm{TrDE},M)$. Since DTr induces an isomorphism $\underline{\mathrm{Hom}}_\Lambda(\mathrm{TrDE},M) \cong \overline{\mathrm{Hom}}_\Lambda(E,\mathrm{DTrM})$, what we want to prove is that $\overline{\mathrm{Hom}}_\Lambda(E,\mathrm{DTrM}) = 0$. It is enough to prove that if E_1 is an indecomposable summand of E then $\overline{\mathrm{Hom}}_\Lambda(E_1,\mathrm{DTrM}) = 0$. Let $\beta: E_1 \longrightarrow \mathrm{DTrM}$. Since $0 \longrightarrow \mathrm{DTrM} \longrightarrow E \longrightarrow M \longrightarrow 0$ is an almost split sequence and E_1 is a summand of E there is an irreducible map $g: \mathrm{DTrM} \longrightarrow E_1$. If $\beta \neq 0$ and $g \neq 0$ then we know by Proposition 1.9 that β and g are isomorphisms. This contradicts the fact that g is irreducible. Thus $\beta = 0$ or $g = 0$. But $g \neq 0$; so $\beta = 0$.

§2. THE ISOMORPHISM c ASSOCIATED TO AN HEREDITARY ARTIN ALGEBRA.

We assume throughout this section that Λ is an hereditary artin algebra. We recall that the Grothendieck group $\mathrm{Gr}(\Lambda)$ is the abelian group F/R, where F is the free abelian group on the set of isomorphism classes $[M]$ of Λ-modules M and R is subgroup generated by all $[M] - [M'] - [M'']$ whenever there is an exact sequence $0 \longrightarrow M' \longrightarrow M \longrightarrow M'' \longrightarrow 0$ of Λ-modules. We will define a group isomorphism $c: \mathrm{Gr}(\Lambda) \longrightarrow \mathrm{Gr}(\Lambda)$ with the property that Λ is of finite representation type if and only if c^m is the identity map $Gr(\Lambda) \rightarrowtail Gr(\Lambda)$, for some positive integer m. Using c we will prove that if Λ is a ring of finite representation type then two indecomposable modules with the same composition factors are isomorphic.

Let $S_1, \ldots, S_n$ be a complete set of nonisomorphic simple Λ-modules. Since Λ is an artin ring $Gr(\Lambda)$ is a free abelian group having as a basis the set $\beta = \{[S_1], \ldots, [S_n]\}$. If M is a Λ-module, let $P_o(M)$ denote the projective cover of M and $I_o(M)$ the injective envelope of M.

<u>Lemma 2.1</u>: The sets $\{[P_o(S_1)], \ldots, [P_o(S_n)]\}$ and $\{[I_o(S_1)], \ldots, [I_o(S_n)]\}$ are bases for $Gr(\Lambda)$.

<u>Proof</u>: For any simple Λ-module S_i we have a sequence $0 \longrightarrow \underline{r}P_o(S_i) \longrightarrow P_o(S_i) \longrightarrow S_i \longrightarrow 0$. As Λ is an hereditary ring $\underline{r}P_o(S_i)$ is projective, so $\underline{r}P_o(S_i) = \sum_{j=1}^{n} n_j\, [P_o(S_j)]$ for some integers n_j and then $[S_i] = [P_o(S_i)] - \sum_{j=1}^{n} n_j\, [P_o(S_j)]$, so $\{[P_o(S_1)], \ldots, [P_o(S_n)]\}$ is a set of generators of $Gr(\Lambda)$ and is, therefore, a basis of $Gr(\Lambda)$.

The rest of the proof follows in an analogous way.

We define now a group homomorphism $c\colon Gr(\Lambda) \longrightarrow Gr(\Lambda)$ by $c([P_o(S_i)]) = -[I_o(S_i)]$. Obviously c is an isomorphism, since it carries a basis to a basis.

We fix the basis $\beta = \{[S_1], \ldots, [S_n]\}$ of $Gr(\Lambda)$. We shall say that an element in $Gr(\Lambda)$ is <u>positive</u> if all its coordinates in β are non-negative. An element in $Gr(\Lambda)$ is <u>negative</u> if all its coordinates in β are non-positive. We have:

<u>Proposition 2.2</u>:

a) If M is in $\mathrm{mod}(\Lambda)$, $c([M]) = [D(\mathrm{Ext}^1_\Lambda(M,\Lambda))] - [D(\mathrm{Hom}_\Lambda(M,\Lambda)]$.

b) If M is an indecomposable nonprojective Λ-module then $c([M]) = [DTrM]$.

c) If M is indecomposable then M is projective if and only if $c[M]$ is negative.

d) If M is indecomposable then $c[M]$ is positive or negative.

Proof:

a) Let M be in $\mathrm{mod}(\Lambda)$ and let $0 \longrightarrow P_1 \longrightarrow P_0 \longrightarrow M \longrightarrow 0$ be a minimal projective presentation for M. Then we have

$0 \longrightarrow M^* \longrightarrow P_0^* \longrightarrow P_1^* \longrightarrow \mathrm{Ext}^1_\Lambda(M,\Lambda) \longrightarrow 0$, so

$0 \longrightarrow D(\mathrm{Ext}^1_\Lambda(M,\Lambda)) \longrightarrow D(P_1^*) \longrightarrow D(P_0^*) \longrightarrow D(M^*) \longrightarrow 0$.

Then $[D(\mathrm{Ext}^1_\Lambda(M,\Lambda))] - [D(M^*)] = [D(P_1^*)] - [D(P_0^*)]$. It is well known that if $P_0(S)$ is the projective cover of the simple S, then $(P_0(S))^* \simeq P_0(D(S))$, so $D(P_0(S)^*) \simeq I_0(D(D(S))) \simeq I_0(S)$. Therefore $[D(P_0(S)^*] = [I_0(S)] = -c([P_0(S)])$. Then, for every projective module Q, $c([Q]) = -[D(Q^*)]$, so we have

$$[D(\mathrm{Ext}^1_\Lambda(M,\Lambda))] - [D(M^*)] = -c([P_1]) + c([P_0]) = c([P_0] - [P_1]) = c([M]).$$

This ends the proof of a).

b) We only need to observe that if M is in $\mathrm{mod}_P(\Lambda)$ then $M^* = 0$.

c) is a trivial consequence of a) and b), and d) follows from b) and c).

Corollary 2.3: If M and N are indecomposable Λ-modules with $[M] = [N]$ then

a) M is projective if and only if N is projective.

b) If M is projective, then $M \simeq N$.

c) If M is not projective, then $[DTrM] = [DTrN]$.

Proof:

a) From $[M] = [N]$ we get $c([M]) = c([N])$. By d) of Proposition 2.2, M is projective if and only if $c([M]) = c([N])$ is negative, and this is equivalent to say that N is projective.

b) Let M be a projective module; we know by a) that N is also projective. If $S = M/\underline{r}M$, M is the projective cover of S; as $[M] = [N]$ we know that S is a composition factor of N, so there is a nonzero homomorphism $f\colon P_0(S) \longrightarrow N$; for the same reason, as N is projective there is a nonzero homomorphism $g: N \longrightarrow M$; by Corollary 1.9, f and g are isomorphisms, so b) holds.

c) If M is not projective we know by a) that N is not projective. By c) of Proposition 2.2, $c([M]) = [DTrM]$, $c([N]) = [DTrN]$. But $[M] = [N]$ implies $c([M]) = c([N])$, i.e., $[DTrM] = [DTrN]$.

Proposition 2.4: Let M and N be finitely generated indecomposable Λ-modules and assume that $(\,,M)$ has finite length. If $[M] = [N]$ then M is isomorphic to N. In particular, if Λ is of finite representation type two indecomposable modules with the same composition factors are isomorphic.

Proof: As the length of $(\,,M)$ is finite we know by Proposition 1.6 that there is an integer $n \geq 0$ such that $(DTr)^n M$ is projective. If $(DTr)^{n-1} N$ is not projective we have, by repeated application of Corollary 2.3 c) that $[(DTr)^n M] = [(DTr)^n N]$. The module $(DTr)^n M$ is projective, so, by Corollary 2.3 b), $(DTr)^n N \simeq (DTr)^n M$ and therefore $N \simeq M$.

If $(DTr)^r N$ is projective for some $r \leq n$ then the functor $(\,,N)$ has finite length and we can apply the same argument.

Proposition 2.5: If c^m is the identity map of $Gr(\Lambda)$ for some integer $m > 0$ then for every injective indecomposable Λ-module I there is a non-negative integer $n < m$ such that $(DTr)^n(I)$ is projective.

Proof: If $(DTr)^{m-1}(I)$ is not projective then we have, by Proposition 2.2 that $c^m([I]) = [(DTr)^m I]$. But $c^m = Id$, so $[(DTr)^m I] = [I]$. Then, in $Gr(\Lambda^{op})$, $[D(I)] = [D(DTr)^m I]$; but $D(I)$ is a projective Λ-module, so, by Corollary 2.3 b) $D(I) \simeq D(DTr)^m I$, i.e. $I \simeq (DTr)^m I$; contradiction, because $(DTr)^m I$ is in $mod_I(\Lambda)$. Therefore, there is $r \leq m - 1$ such that $(DTr)^r I$ is projective.

We can prove now the following criterion for Λ to be of finite representation type:

Theorem 2.6: Λ is of finite representation type if and only if there exists an integer $m > 0$ such that $c^m = \mathrm{Id}$.

Proof: If $c^m = \mathrm{Id}$ we know by the preceding proposition, combined with Proposition 1.5, that the length of $(\ ,I)$ is finite for every injective indecomposable module I. Let S be a simple Λ-module and let $I_0(S)$ be its injective envelope; then there is a monomorphism $0 \longrightarrow (\ ,S) \longrightarrow (\ ,I_0(S))$, where $(\ ,I_0(S))$ has finite length. Therefore the length of $(\ ,S)$ is finite for every simple Λ-module S, thus Λ is of finite representation type.

Conversely, assume now that Λ is of finite representation type. Let M be an indecomposable Λ-module; $c([M]) = [\mathrm{DTrM}]$ if M is not projective and $c([M]) = -[D(M^*)]$ otherwise. So c transforms the finite set $\{\pm[N]$, N is an indecomposable Λ-module$\}$ into itself. Therefore, there is an integer $m > 0$ such that $c^m = \mathrm{Id}$, since c is a permutation of a finite set.

As an application of this we give an estimate of the number of indecomposable modules, which is not an equality, in general.

Corollary 2.7: If $c^m = \mathrm{Id}$ for an integer $m > 0$, n is the number of simple Λ-modules and r is the number of injective projective modules then the number of indecomposable Λ-modules is smaller than or equal to $(n-r)\cdot m + r$.

Proof: Let $I_1,\ldots,I_n$ be a complete set of nonisomorphic indecomposable injective Λ-modules. By Proposition 2.5, for every i there is a nonnegative integer $n_i < m$ such that $(\mathrm{DTr})^{n_i} I_i$ is projective. Therefore the set $\mathcal{D} = \{(\mathrm{DTr})^j(I_i),\ 0 \leq j \leq n_i\}$ has at most $(n-r)\cdot m + r$ elements. By Corollary 1.8 it is a complete set of nonisomorphic indecomposable Λ-modules.

§3. THE QUADRATIC FORM ASSOCIATED TO AN HEREDITARY ARTIN ALGEBRA

We assume in this section that Λ is an hereditary artin algebra. We consider the free abelian group $Gr(\Lambda)$, the Grothendieck group of Λ, as a module over the ring Z of integers. We will associate to Λ a quadratic form $q: Gr(\Lambda) \longrightarrow Q$, the field of rational numbers. When Λ is the hereditary algebra associated to a k-species the quadratic form that we define here is the quadratic form associated to the species (see[10], [12], and [13]). For this type of hereditary algebras, many of the results proven in this section can be obtained from [11] or are directly proven there.

Our aim is to prove in this section that the isomorphism $c: Gr(\Lambda) \longrightarrow Gr(\Lambda)$ defined in the preceding section is the Coxeter transformation associated to the quadratic form $\underline{q}$. We begin by defining $\underline{q}$ and recalling the definition of the Coxeter transformation.

To define q we may assume that the ring Λ is indecomposable, because, if $\Lambda = \Lambda_1 x \ldots x \Lambda_n$ is a product of indecomposable rings Λ_i, then the category of finitely generated Λ-modules is the product category $mod(\Lambda_1) x \ldots x\, mod(\Lambda_n)$, so $Gr(\Lambda) = Gr(\Lambda_1) x \ldots x\, Gr(\Lambda_n)$ and we define the quadratic form associated to Λ as the orthogonal sum of the quadratic forms associated to the rings Λ_i.

Therefore we may assume in what follows that Λ is indecomposable. Then its center $Z(\Lambda)$ is a local ring; let $0 \neq c \in Z(\Lambda)$ and let $L_c: \Lambda \longrightarrow \Lambda$ be the multiplication by c. Since $L_c(1) \neq 0$ there exists an indecomposable projective Λ-module P such that $L_c|P \neq 0$; but $End_\Lambda(P)$ is a division ring, so $L_c: P \longrightarrow P$ is an isomorphism and consequently c is not nilpotent. Thus c has an inverse in the local ring $Z(\Lambda)$; so we have proven:

<u>Proposition 3.1</u>: The center of an indecomposable hereditary artin

algebra is a field.

Therefore we assume in the rest of this section that Λ is an hereditary indecomposable artin algebra over the field K.

If $0 \longrightarrow M' \longrightarrow M \longrightarrow M'' \longrightarrow 0$ is an exact sequence in mod (Λ), then, for every N in mod(Λ), $0 \longrightarrow (N,M') \longrightarrow (N,M) \longrightarrow (N,M'') \longrightarrow \mathrm{Ext}^1_\Lambda(N,M') \longrightarrow \mathrm{Ext}^1_\Lambda(N,M) \longrightarrow \mathrm{Ext}^1_\Lambda(N,M'') \longrightarrow 0$ is exact and in $\mathrm{Gr}(\Lambda)$ we have $[(N,M)] - [\mathrm{Ext}^1_\Lambda(N,M)] = [(N,M')] - [\mathrm{Ext}^1_\Lambda(N,M')] + [(N,M'')] - \mathrm{Ext}^1_\Lambda(N,M'')]$.

In analogous way we get

$[(M,N)] - [\mathrm{Ext}^1_\Lambda(M,N)] = [(M',N)] - [\mathrm{Ext}^1_\Lambda(M',N)] + [(M'',N)] - [\mathrm{Ext}^1_\Lambda(M'',N)]$.

Therefore the map $B_1\colon \mathrm{Gr}(\Lambda) \times \mathrm{Gr}(\Lambda) \longrightarrow Z$ defined by $B_1([M],[N]) = \dim_K(M,N) - \dim_K \mathrm{Ext}^1_\Lambda(M,N)$, for M,N in mod$(\Lambda)$ is a bilinear form. Let B from $\mathrm{Gr}(\Lambda) \times \mathrm{Gr}(\Lambda)$ to the field Q of rational numbers be the associated symmetric bilinear form, i.e., for x, y in $\mathrm{Gr}(\Lambda)$, $B(x,y) = 1/2.(B_1(x,y) + B_1(y,x))$ and let q be the corresponding quadratic form, defined by $q(x) - B(x,x)$. (See [13].)

We observe that if Λ is an algebra over another field K' and q' is the quadratic form corresponding to the K'-algebra Λ, then $q = a.q'$, for some nonzero rational number a.

Since Λ is hereditary, if $f\colon P \longrightarrow Q$ is a nonzero morphism between indecomposable projective Λ-modules, then f is a monomorphism. Therefore we can define an order in the set of indecomposable projective Λ-modules by setting $P \leq Q$ if and only if $(P,Q) \neq 0$, where P, Q are arbitrary indecomposable projectives. This order induces an order in the set of simple Λ-modules by setting $S \leq S'$ if and only if $P_0(S) \leq P_0(S')$ for any pair of simple Λ-modules S and S'.

Let $\mathcal{S}$ be a complete set of nonisomorphic simple Λ-modules. An <u>admissible indexing</u> of $\mathcal{S}$ is an order preserving bijection $a\colon \mathcal{S} \longrightarrow \{1, 2, \ldots, n\}$. Then a is admissible if and only if, calling

$a^{-1}(i) = S_i$, $S_i \leq S_j$ implies $i \leq j$, or what is equivalent, $i > j$ implies $S_i \not\leq S_j$, i.e., $(P_o(S_i), P_o(S_j)) = 0$.

Admissible indexings always exist: we choose S_1 minimal with respect to the order established above between the simples, i.e., if $S \neq S_1$, then $S_1 \not\geq S$. We choose S_i such that $S_i \not\geq S$ for $S \neq S_1,\ldots,S_i$.

Lemma 3.2: Assume that the indexing $S_1,\ldots,S_n$ of a complete set of nonisomorphic simple Λ-modules is admissible. Then:

a) $B_1(S_i, S_i) = \dim_K(S_i, S_i)$, for $i = 1,\ldots,n$.

b) $\mathrm{Ext}^1_\Lambda(S_i, S_j) \simeq (\underline{r}P_o(S_i)/\underline{r}^2P_o(S_i), S_j)$, for every $i, j = 1,\ldots,n$.

Thus

$$B_1(S_i, S_j) = \begin{cases} 0 \text{ if } i < j \\ \dim_K \mathrm{Ext}^1_\Lambda(S_i, S_j) = \dim_K(\underline{r}P_o(S_i)/\underline{r}^2P_o(S_i), S_j), \text{ if } i > j. \end{cases}$$

c) $(P_o(S_i), P_o(S_j)) = 0$ if and only if $(I_o(S_i), I_o(S_j)) = 0$.

Therefore, we can define an admissible indexing

$$a\colon \{D(S_1),\ldots,D(S_n)\} \longrightarrow \{1, 2,\ldots,n\}$$

in the set of simples of Λ^{op} by writing $a(D(S_i)) = n + 1 - i$, for $i = 1,\ldots,n$.

Proof: From $0 \longrightarrow \underline{r}P_o(S_i) \longrightarrow P_o(S_i) \longrightarrow S_i \longrightarrow 0$ we have $\ldots \longrightarrow (P_o(S_i), S_j) \longrightarrow (\underline{r}P_o(S_i), S_j) \longrightarrow \mathrm{Ext}^1_\Lambda(S_i, S_j) \longrightarrow 0$. If $i \leq j$ then $(\underline{r}P_o(S_i), S_j) = 0$ so $\mathrm{Ext}^1_\Lambda(S_i, S_j) = 0$. If $i \neq j$ then $(P_o(S_i), S_j) = 0$ and, since $(\underline{r}P_o(S_i), S_j) \simeq (\underline{r}P_o(S_i)/\underline{r}^2P_o(S_i), S_j)$, then $\mathrm{Ext}^1_\Lambda(S_i, S_j) \simeq (\underline{r}P_o(S_i)/\underline{r}^2P_o(S_i), S_j)$. This proves a) and b).

c) Suppose that there is a nonzero homomorphism $\alpha\colon P_o(S_i) \longrightarrow P_o(S_j)$. Then α is a nomomorphism and therefore there is $\beta\colon P_o(S_j) \longrightarrow I_o(S_i)$ such that the diagram

$$\begin{array}{ccccc} 0 & \longrightarrow & P_o(S_i) & \xrightarrow{\alpha} & P_o(S_j) \\ & & \pi\downarrow & \swarrow \beta & \\ & & S_i & & \\ & & i\downarrow & & \\ & & I_o(S_i) & & \end{array}$$

commutes, where π is the canonical epimorphism and i the inclusion. β is nonzero and induces a monomorphism $P_o(S_j)/\mathrm{Ker}(\beta) \longrightarrow I_o(S_i)$, so there is a morphism $I_o(S_i) \xrightarrow{\delta} I_o(S_j)$ such that the diagram

$$\begin{array}{ccccc} 0 & \longrightarrow & P_o(S_j)/\mathrm{Ker}(\beta) & \xrightarrow{\beta} & I_o(S_i) \\ & & \pi'\downarrow & \swarrow \delta & \\ & & S_j & & \\ & & i'\downarrow & & \\ & & I_o(S_j) & & \end{array}$$

commutes, where π' is the canonical projection and i' the inclusion. Then $\delta \neq 0$; so $(I_o(S_i), I_o(S_j)) \neq 0$. This proves that $(P_o(S_i), P_o(S_j)) \neq 0$ implies that $(I_o(S_i), I_o(S_j)) \neq 0$. The converse follows by duality. Thus $(P_o(S_i), P_o(S_j)) = 0$ if and only if $(P_o(D(S_j)), P_o(D(S_i))) = 0$.
This implies c).

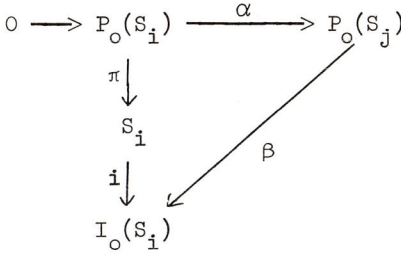

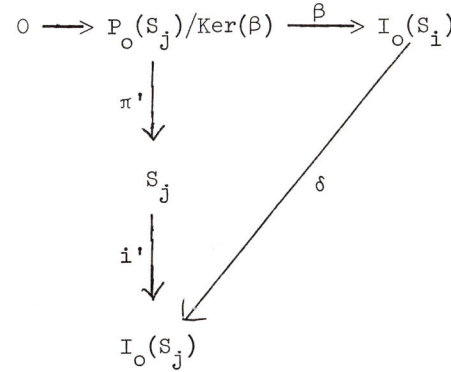

Let $S_1,\ldots,S_n$ be a complete set of nonisomorphic simple Λ-modules. We identify $Gr(\Lambda)$ with Z^n, by means of the isomorphism that transforms the basis $\{[S_1],\ldots,[S_n]\}$ of $Gr(\Lambda)$ to the canonical basis of Z^n; let Q be the field of rational numbers. We shall denote also by B the extension of $B: Gr(\Lambda) \times Gr(\Lambda) \longrightarrow Q$ to a bilinear form $Q^n \times Q^n \longrightarrow Q$ and by c the extension of the map $c: Gr(\Lambda) \longrightarrow Gr(\Lambda)$ defined in §2 to a map: $Q^n \longrightarrow Q^n$.

For every $i = 1,\ldots,n$, $B([S_i],[S_i]) \neq 0$; let $\sigma_i : Q^n \longrightarrow Q^n$ denote the symmetry with respect to the vector $[S_i]$, i.e., for M in mod(Λ),

$$\sigma_i([M]) = [M] - 2 \cdot \frac{B([M],[S_i])}{B([S_i],[S_i])} \cdot [S_i]$$

and let $C = \sigma_n \cdots \sigma_1$. The subgroup W of the general linear group $Gl(n,Q)$ generated by the symmetries $\sigma_1 \ldots,\sigma_n$ is called the Weyl group and C is the Coxeter transformation corresponding to the basis $\{[S_1],\ldots,[S_n]\}$ of Q^n (see [11]). The Coxeter transformation depends on the indexing of the elements of the basis, but for two admissible indexings of the simples of Λ the corresponding Coxeter transformations are the same.

We assume in all that follows that the indexing of the simples $S_1,\ldots,S_n$ of Λ is admissible; we shall prove that the morphism c defined in the preceding section is the Coxeter transformation associated to the basis $\{[S_1],\ldots,[S_n]\}$ of $Gr(\Lambda)$.

For M in mod(Λ) and $i = 1,\ldots,n$ we denote by $[M]_i$ the multiplicity of the simple S_i as a composition factor of M, i.e., in $Gr(\Lambda)$, $[M] = \sum_{i=1}^{n} [M]_i [S_i]$.

Lemma 3.3:

a) $S_{j+1},\ldots,S_n$ are not composition factors of $P_0(S_j)$, for $j=1,\ldots,n-1$.

If M is in mod(Λ) then

b) Let $1 \leq t \leq n$. If S_j is not a composition factor of M for $1 \leq j < t$ then $[M]_t = [socM]_t$, where socM is the socle of the Λ-module M.

c) Let $1 \leq t \leq n$. If S_j is not a composition factor of M for $t < j \leq n$

then $[M]_t = [M/\underline{r}M]_t$.

Proof:

a) If S_k is a composition factor of $P_o(S_j)$ then there is a nonzero morphism $P_o(S_k) \longrightarrow P_o(S_j)$ and therefore $k \leq j$.

b) Assume that for some $i > 1$ the simple S_t is isomorphic to a direct summand of $soc^i M/soc^{i-1}M$. Then there is an epimorphism $\lambda: soc^i M \longrightarrow S_t$.

Let $\pi: P_o(S_t) \longrightarrow S_t$ be a projective cover and let $\alpha: P_o(S_t) \longrightarrow soc^i M$ be such that $\lambda\alpha = \pi$. Since, by a), the only possible composition factors of $\underline{r}P_o(S_t)$ are $S_1, \ldots, S_{t-1}$ and these are not composition factors of M, then $\alpha|\underline{r}P_o(S_t) = 0$. Thus $S_t \simeq Im(\alpha)$ and then $Im(\alpha)$ is simple and contained in $soc^i M \subseteq M$, so $Im(\alpha) \subseteq socM$. Therefore $\lambda\alpha = 0$, because $Ker(\lambda) \supseteq soc^{i-1}M \supseteq socM$, since $i > 1$. Then $\pi = \lambda\alpha = 0$, contradiction. Therefore S_t is not isomorphic to a summand of $soc^i M/soc^{i-1}M$ for $i > 1$. Then $[M]_t = [socM]_t$. c) can be obtained from b) by duality, using Lemma 3.2 c).

For each $i = 1, \ldots, n$ we consider the modules $P_{i,k}$, $k = 0, \ldots, i - 1$ defined in the following way: If L and N are Λ-modules, let $\tau_L(N)$ denote the trace of L in N, that is, the submodule of N generated by the images of the morphisms from L to N. We write $P_{i,o} = P_o(S_i)$; if $k > 0$, $P_{i,k} = P_{i,k-1}/\tau_{S_k}(P_{i,k-1})$. By b) of the previous lemma, S_1 is not a composition factor of $P_{i,1}$. It follows easily by induction on k and using b) that $S_1, \ldots, S_k$ are not composition factors of $P_{i,k}$, and therefore $[P_i]_k = [P_{i,k-1}]_k$, so we have

Lemma 3.4: For $i = 1, \ldots, n$, $j \leq i$

a) $\dim_K(S_j, P_{i,j-1}) = [P_o(S_i)]_j \cdot \dim_K(S_j, S_j)$

b) $[P_{i,j-1}] = [P_o(S_i)] - \sum_{k<j} [P_o(S_i)]_k [S_k]$

We now prove two formulas which reduce the proof that c is the

Coxeter transformation to an easy computation.

Lemma 3.5:

a) For $i = 2,\dots,n$ and $j < i$, $B([S_j], [P_o(S_i)] - \sum_{k<j} [P_o(S_i)]_k \cdot [S_k])$
$= \frac{1}{2} [P_o(S_i)]_j \dim(S_j,S_j)$.

b) For $i = 1,\dots,n-1$ and $j > i$,
$B([S_j],[I_o(S_i)] - \sum_{k>j}[I_o(S_i)]_k[S_k]) = \frac{1}{2} [I_o(S_i)]_j \cdot \dim_K(S_j,S_j)$.

Proof:

a) Let $1 < i \leq n$ and $j < i$. By b) of the previous lemma we have to prove that $B([S_j],[P_{i,j-1}]) = \frac{1}{2}[P_o(S_i)]_j \cdot \dim_K(S_j,S_j)$.

Since $(P_{i,j-1},S_j) = 0$, $B([S_j],[P_{i,j-1}] =$
$\frac{1}{2}(\dim_K(S_j,P_{i,j-1}) - \dim_K \mathrm{Ext}^1_\Lambda(S_j,P_{i,j-1}) - \dim_K \mathrm{Ext}^1_\Lambda(P_{i,j-1}, S_j))$.

From Lemma 3.3 we know that

$\dim_K(S_j,P_{i,j-1}) = [P_o(S_i)]_j \cdot \dim_K(S_j,S_j)$, so it is enough to prove that $\mathrm{Ext}^1_\Lambda(S_j,P_{i,j-1}) = \mathrm{Ext}^1_\Lambda(P_{i,j-1},S_j) = 0$. From the exact sequence $0 \longrightarrow \underline{r}P_o(S_j) \longrightarrow P_o(S_j) \longrightarrow S_j \longrightarrow 0$ we get

$\dots \longrightarrow (\underline{r}\, P_o S_j), P_{i,j-1}) \longrightarrow \mathrm{Ext}^1_\Lambda(S_j,P_{i,j-1}) \longrightarrow 0$, but the only possible composition factors of $\underline{r}P_o(S_j)$ are $S_1,\dots,S_{j-1}$ that are not composition factors of $P_{i,j-1}$. So $(\underline{r}P_o(S_j), P_{i,j-1}) = 0$, hence $\mathrm{Ext}^1_\Lambda(S_j,P_{i,j-1}) = 0$.

We want to see now that $\mathrm{Ext}^1_\Lambda(P_{i,j-1},S_j) = 0$. We shall prove, more generally, that $\mathrm{Ext}^1_\Lambda(P_{i,j-1},S_t) = 0$ for all $t \geq j$. This is true for $j = 1$ because $P_{i,o} = P_o(S_i)$ is projective. We assume that it is true for $j \leq k$. Let $k_i = [P_o(S_i)]_k$. From $0 \longrightarrow \coprod_1^{k_i} S_k \longrightarrow P_{i,k-1} \longrightarrow P_{i,k} \longrightarrow 0$ we get for every $s \geq k+1$ the exact sequence

$\dots \longrightarrow (\coprod_1^{k_i} S_k,S_s) \longrightarrow \mathrm{Ext}^1_\Lambda(P_{i,k},S_s) \longrightarrow \mathrm{Ext}^1_\Lambda(P_{i,k-1},S_s) \longrightarrow 0$.

By the induction hypothesis $\mathrm{Ext}^1_\Lambda(P_{i,k-1},S_s) = 0$ and, since $(S_k,S_s) = 0$ because $s > k$, we have that $\mathrm{Ext}^1_\Lambda(P_{i,k},S_s) = 0$.

Theorem 3.6: With the previous notations, $C = c$.

Proof: The set $\{[P_o(S_1)],\dots,[P_o(S_n)]\}$ is a basis of Q^n and $c([P_o(S_i)]) = -[I_o(S_i)]$. So to prove that $C = c$ it is enough to prove that $C([P_o(S_i)]) = -[I_o(S_i)]$. Thus we compute $\sigma_n\cdots\sigma_1([P_o(S_i)])$.

$\sigma_1([P_o(S_i)]) = [P_o(S_i)] - 2\dfrac{B([P_o(S_i)],[S_1])}{B([S_1],[S_1])}\cdot[S_1] = [P_o(S_i)] - [P_o(S_i)]_1\cdot[S_1] = [P_{i,1}]$.

We assume by induction that $\sigma_j\cdots\sigma_1([P_o(S_i)]) = [P_{i,j}]$, for $1 \leq j < i - 1$. Then $\sigma_{j+1}\cdot\sigma_j\cdots\sigma_1([P_o(S_i)]) = \sigma_{j+1}[P_{i,j}] = [P_{i,j}] - [P_o(S_j)]_{j+1}[S_{j+1}] = [P_{i,j+1}]$ because, by Lemma 3.5,

$$2\,\frac{B([P_{i,j}],\ [S_{j+1}])}{B([S_{j+1}],\ [S_{j+1}])} = [P_o(S_i)]_{j+1}$$

Thus $\sigma_{i-1}\cdots\sigma_1([P_o(S_i)]) = [P_{i,i-1}] = [S_i]$

Using b) of the previous lemma one can prove in a similar form that $\sigma_{i+1}\cdots\sigma_n([I_o(S_i)]) = [S_i]$. Hence $\sigma_n\cdots\sigma_{i+1}([S_i]) = [I_o(S_i)]$. Then $\sigma_n\cdots\sigma_1([P_o(S_i)]) = \sigma_n\cdots\sigma_{i+1}\sigma_i([S_i]) = \sigma_n\cdots\sigma_{i+1}(-[S_i]) = -[I_o(S_i)]$, i.e., $C([P_o(S_i)]) = -[I_o(S_i)]$. Therefore $C = c$.

§4. OTHER PROPERTIES OF THE QUADRATIC FORM q

Let Λ be an hereditary artin algebra. We devote this section to prove:

Theorem 4.1: The following conditions are equivalent

a) Λ is of finite representation type.

b) There is some $m > 0$ such that $c^m = Id_{Gr(\Lambda)}$.

c) $\underline{q}$ is positive definite.

The equivalence of a) and b) has already been proven in §2, Theorem 2.6. On the other hand, it is well known and not hard to prove that when B is positive definite the Weyl group W is finite, so b) ===> c); we transcribe from [11] a proof of this fact, for the sake of completeness.

Let f be a natural number such that $B([S_i],[S_i]) = \dim_K End(S_i) \leq f$, for all $i \leq n$. Let M be the set of all integral vectors $x \in Q^n$ such that $B(x,x) \leq f$. Let P be a regular matrix which transforms B into diagonal form. Then $B(x,x) = \sum_{i=1}^{n} c_i y_i^2$, where $y = P.x$ and $c_i \in Q$; $c_i > 0$ for every i because B is positive definite. If h is the common denominator of the entries in P, then hP is an integral matrix. If $x \in M$, $hy = hP.x$ is integral and, since $B(x,x) \leq f$, $|y| \leq \sqrt{f/c_i}$, with $1 \leq i \leq n$. Then, under the linear transformation P,M is mapped into a finite set, so M is finite, because P is regular. Since $B(\sigma x,\sigma x) = B(x,x)$ for every $\sigma \in W$, then W transforms M into itself. The canonical basis of Q^n is contained in M; therefore, W can be embedded into the symmetric group of M and is, therefore, finite.

Therefore, a) $\Longleftrightarrow$ b) and c) $\Longrightarrow$ b). Thus we only need to prove that a) and b) imply c).

We will prove first that if the ring Λ is of finite representation type, then Λ verifies the property:

(P) Given three nonisomorphic simple Λ-modules S, S', and S" such that $\mathrm{Ext}^1_\Lambda(S', S) \neq 0$ and $P_o(S') \subseteq \underline{r}P_o(S")$, then $\mathrm{Ext}^1_\Lambda(S",S) = 0$.

Therefore, to prove that a) and b) imply c) we only need to consider rings that verify the property (P). So we prove now

<u>Proposition 4.2</u>: If Λ does not verify (P) then Λ is of infinite representation type.

<u>Proof</u>: Since Λ does note verify the condition (P) there are three nonisomorphic simple Λ-modules S_1, S_2, S_3 such that $\mathrm{Ext}^1_\Lambda(S_3, S_1) \neq 0$, $\mathrm{Ext}^1_\Lambda(S_2, S_1) \neq 0$ and $P_o(S_2) \subseteq \underline{r}P_o(S_3)$. Let $P = P_o(S_1) \coprod P_o(S_2) \coprod P_o(S_3)$. Then the ring $\Lambda_P = \mathrm{End}_\Lambda(P)$ is also an hereditary artin algebra that does not verify (P) and has only three nonisomorphic simple Λ-modules. Also, if Λ_P is of infinite representation type then, by [1], Proposition 6.5, Λ is of infinite representation type.

Therefore we may assume that S_1, S_2, S_3 are a complete set of nonisomorphic simple Λ-modules. Since $\mathrm{Ext}^1_\Lambda(S_2, S_1) \neq 0$, then $P_o(S_1) \subseteq \underline{r}P_o(S_2)$. Also, $P_o(S_2) \subseteq \underline{r}P_o(S_3)$. If $\mathrm{Ext}^1_\Lambda(S_3, S_2) = 0$ then $P_o(S_2) \subseteq \underline{r}^2P_o(S_3)$, so

$P_o(S_1) \subseteq \underline{r}^3 P_o(S_3)$. Since Λ has only three simples then $\underline{r}^3 P_o(S_3) = 0$ and therefore $P_o(S_1) = 0$; contradiction. Thus $\mathrm{Ext}^1_\Lambda(S_3, S_2) \neq 0$.

Let now $f_i = \dim_K(S_i, S_i)$. Since, for $i > j = 1,2,3$, $\mathrm{Ext}^1_\Lambda(S_i, S_j)$ is a nonzero vector space over the division ring $\mathrm{End}_\Lambda(S_i)^{op}$ and over $\mathrm{End}_\Lambda(S_j)$, we have that $\dim_K \mathrm{Ext}^1_\Lambda(S_i, S_j) \geq \max\{f_i, f_j\}$. Therefore, if we call $T = S_1 \amalg S_2 \amalg S_3$, then $B([T],[T]) = \dim_K(T, T) - \dim_K \mathrm{Ext}^1_\Lambda(T,T) \leq 0$.

Since $\mathrm{Ext}^1_\Lambda(S_3, S_2)$ and $\mathrm{Ext}^1_\Lambda(S_3, S_1)$ are nonzero, then S_2 and S_1 are summands of $\underline{r}P_o(S_3)/\underline{r}^2 P_o(S_3)$, and consequently, there is a factor M of $P_o(S_3)$ such that $[M] = [S_1] + [S_2] + [S_3] = [T]$. The module M is indecomposable and $0 \geq B([M],[M]) = \dim_K(M,M) - \dim_K \mathrm{Ext}^1_\Lambda(M,M)$. Hence $\dim_K \mathrm{Ext}^1_\Lambda(M,M) \neq 0$ and we know then by Proposition 1.10 that the functor $(\ ,M)$ has infinite length. Therefore Λ is of infinite representation type.

Let $\underline{n} = (n_1,\dots,n_n) \in Z^n$, with $n_i \geq 0$. Let $\mathcal{S}_{\underline{n}}$ be the full subcategory of mod Λ whose objects are the modules M such that $[M] = \Sigma\, n_i[S_i]$ in $\mathrm{Gr}(\Lambda)$, i.e., such that M has composition factors $S_1,\dots,S_n$ with multiplicities $n_1,\dots,n_n$ respectively.

Let N be a Λ-module. Since Λ is a K-algebra, we can consider N as a K-vector space. So $\mathrm{Hom}_K(N,N)$ can be identified with $M_m(K)$, the ring of $m \times m$ matrices over K, with $m = \dim_K N$. Then $\mathrm{Hom}_\Lambda(N,N)$ is a K-subspace of $M_m(K)$ and therefore the group of Λ automorphisms of N, $H = \mathrm{Aut}_\Lambda(N)$, is an algebraic subgroup of the algebraic group $\mathrm{Gl}(m,K)$.

Let $V = \mathrm{Ext}^1_\Lambda(N,N)$. V is a K-vector space and therefore an affine K-variety. We define $\varphi_i\colon H \longrightarrow \mathrm{End}_K(V)$, $i = 1, 2$, by $\varphi_1(f) = \mathrm{Ext}^1_\Lambda(f,N)$ $\varphi_2(f) = \mathrm{Ext}^1_\Lambda(N,f)$, for $f \in H$. Then φ_1 and φ_2 are linear maps. Let $\Theta\colon H \longrightarrow H \times H$ be the map defined by $\Theta(h) = (h^{-1}, h)$, for $h \in H$. Then the composition $\chi\colon H \xrightarrow{\Theta} H \times H \xrightarrow{\varphi_1 \times \varphi_2} \mathrm{Aut}_K(V) \times \mathrm{Aut}_K(V) \longrightarrow \mathrm{Aut}_K(V)$ is a map of algebraic groups.

We consider now the algebraic group H acting on V by $h.v = \chi(h)(v)$, for $h \in H$, $v \in V$. The restriction of this action to K^*, the subgroup of H consisting of multiplications by nonzero elements in K, is trivial. So we consider H/K^* acting on V.

To study modules in $\underline{\underline{\mathcal{S}}}_{\underline{n}}$ with $\underline{n} = (n_1,\ldots,n_n)$, $n_i \geq 0$, we consider the semisimple module $N = \coprod_{i=1}^{n} S_i^{n_i}$. We will prove that when the ring Λ verifies the property (P) we can define a surjective correspondence φ between the objects in $\underline{\underline{\mathcal{S}}}_{\underline{n}}$ and the elements in $V = \operatorname{Ext}^1_\Lambda(N,N)$ such that elements of V corresponding to isomorphic modules belong to the same orbit under the action of H/K^*.

Let us see now how we can deduce from this that conditions a) and b) of Theorem 4.1 imply condition c) in the case when K is an infinite field. Let $\underline{n} = (n_1,\ldots,n_n)$ with $n_i \geq 0$ be nonzero and such that $B(\underline{n},\underline{n}) \leq 0$. Since $[N] = \sum_{i=1}^{n} n_i\ [S_i]$, then $B(\underline{n},\underline{n}) = \dim_K \operatorname{End}_\Lambda(N) - \dim_K \operatorname{End}^1_\Lambda(N,N)$. But $\dim_K \operatorname{End}_\Lambda(N)$ is equal to the dimension of the algebraic variety $H = \operatorname{Aut}_\Lambda(N)$ so, since $B(\underline{n},\underline{n}) \leq 0$ we have that $\dim_K H - \dim_K V \leq 0$, so $\dim_K V > \dim_K(H/K^*)$. When the field K is infinite this implies that the number of orbits of V under the action of H/K^* is infinite and, therefore, that there are an infinite number of nonisomorphic modules in $\underline{\underline{\mathcal{S}}}_{\underline{n}}$. Thus Λ is of infinite representation type.

What follows is devoted to defining the correspondence φ that we mentioned above, between modules in $\underline{\underline{\mathcal{S}}}_{\underline{n}}$ and elements in $\operatorname{Ext}^1_\Lambda(N,N)$. We state first some results that will be needed.

<u>Lemma 4.3</u>: Let S be s simple Λ-module and let M be a module such that S is not a composition factor of M. If $\operatorname{Ext}^1_\Lambda(S',S) = 0$ for every composition factor S' of M, then $\operatorname{Ext}^1_\Lambda(M,S) = 0$.

<u>Proof</u>: Follows by induction on the length of M.

Lemma 4.4: Let M be a Λ-module. If a simple S is a composition factor of M that is not contained in the socle of M, then there is a simple $S' \subseteq \text{soc } M$ such that $P_o(S') \subseteq \underline{r}\, P_o(S)$.

Proof: Let $i > 1$ by such that $S \subseteq \text{soc}^i M/_{\text{soc}^{i-1} M}$. Let $\alpha: P_o(\text{soc}^i M) \longrightarrow \text{soc}^i M$ be a projective cover of $\text{soc}^i M$. Let $\rho: \text{soc}^i M \longrightarrow \text{soc}^i M/_{\text{soc}^{i-1} M}$ be the canonical epimorphism. Then $\alpha^{-1}\rho^{-1}(S) \simeq P_o(S)$. Let $S_1 \subseteq \text{soc}(\rho^{-1}(S))$. We have

$$\begin{array}{ccc} \alpha^{-1}\rho^{-1}(S) & \xrightarrow{\alpha} & \rho^{-1}(S) \\ & & \cup|\, i \\ P_o(S_1) & \xrightarrow{\pi} & S_1 \longrightarrow 0 \end{array}$$

There is $\theta: P_o(S_1) \longrightarrow \alpha^{-1}\rho^{-1}(S)$ such that $\alpha\theta = i\pi$. If θ is an isomorphism then $\alpha\theta = i\pi$ is an epimorphism and therefore i is an epimorphism; so $S_1 = \rho^{-1}(S)$. Then $\rho^{-1}(S)$ is simple and therefore $\rho: \rho^{-1}(S) \longrightarrow S$ is an isomorphism. So $\rho^{-1}(S) \subseteq \text{soc } M$. Thus $\rho(\rho^{-1}(S)) - 0$, contradiction. This proves that θ is not an isomorphism, so $\theta(P_o(S_1)) \subseteq \underline{r}(\alpha^{-1}\rho^{-1}(S)) \simeq \underline{r}\, P_o(S)$. Since Λ is hereditary θ is a monomorphism. This finishes the proof of the lemma.

Another result that we will need is the following:

Lemma 4.5: Assume Λ verifies (P). Let S be a simple and let M be a Λ-module such that $\text{Ext}^1_\Lambda(S',S) \neq 0$ for every $S' \subseteq \text{soc } M$. Then $\text{Ext}^1_\Lambda(S'',S) = 0$ for every composition factor S'' of $M/\text{soc } M$.

Proof: Let S'' be a composition factor of $M/\text{soc } M$. Then, by Lemma 4.4, there is a simple $S' \subseteq \text{soc } M$ such that $P_o(S') \subseteq rP_o(S'')$. But $S' \subseteq \text{soc} M$ implies, by hypothesis, that $\text{Ext}^1_\Lambda(S',S) \neq 0$. Since Λ verifies (P) then $\text{Ext}^1_\Lambda(S'',S) = 0$.

We shall use the following notations. Let S_i be a simple Λ-module and let M be a Λ-module. Let $S_i(M)$ denote the maximal submodule of

M with the property that $\mathrm{Ext}^1_\Lambda(S, S_i) = 0$ for every composition factor S of $S_i(M)$. If $\lambda : M \longrightarrow M'$ is a morphism, then $\lambda(S_i(M)) \subseteq S_i(M')$, so λ induces $\bar{\lambda} : M/S_i(M) \longrightarrow M'/S_i(M')$. We denote by F_i the functor from $\mathrm{mod}(\Lambda)$ to itself defined by $F_i(M) = M/S_i(M)$, $F_i(\lambda) = \bar{\lambda}$. The canonical epimorphism $M \longrightarrow M/S_i(M) = F_i(M)$ induces a morphism of functors $\alpha_i : \mathrm{Ext}^1_\Lambda(\ ,S)F_i \longrightarrow \mathrm{Ext}^1_\Lambda(\ ,S)$, where $\mathrm{Ext}^1_\Lambda(\ ,S)F_i$ is the composition $\mathrm{mod}\ \Lambda \xrightarrow{F_i} \mathrm{mod}\ \Lambda \xrightarrow{\mathrm{Ext}(\ ,S)} \mathrm{Mod}\ K$. On the other hand, the inclusion $\mathrm{soc}F_i(M) \longrightarrow F_i(M)$ induces a functorial morphism $\beta_i(M) : \mathrm{Ext}^1_\Lambda(F_i(M), S_i) \longrightarrow \mathrm{Ext}^1_\Lambda(\mathrm{soc}F_i(M), S_i)$.

Proposition 4.6: Let M be a Λ-module and let S_i be a simple that is not a composition factor of M. Then:

(a) $\alpha_i(M)$ is an isomorphism.

(b) If Λ verifies (P) and S is a composition factor of M, then $\mathrm{Ext}^1_\Lambda(S, S_i) \neq 0$ if and only if $S \subseteq \mathrm{soc}F_i(M)$.

(c) $\beta_i(M)$ is an isomorphism.

Proof:

(a) From the exact sequence $0 \longrightarrow S_i(M) \longrightarrow M \longrightarrow M/S_i(M) \longrightarrow 0$ we obtain the exact sequence $\ldots \longrightarrow (S_i(M), S_i) \longrightarrow \mathrm{Ext}^1_\Lambda(M/S_i(M), S_i) \xrightarrow{\alpha_i(M)} \mathrm{Ext}^1_\Lambda(M, S_i) \longrightarrow \mathrm{Ext}^1_\Lambda(S_i(M), S_i) \longrightarrow \ldots$ Since S_i is not a composition factor of M and $S_i(M) \subseteq M$, then $(S_i(M), S_i) = 0$. Since $\mathrm{Ext}^1_\Lambda(S, S_i) = 0$ for every composition factor S of $S_i(M)$ then, by Lemma 4.3 $\mathrm{Ext}^1_\Lambda(S_i(M), S_i) = 0$. Therefore $\alpha_i(M)$ is an isomorphism.

b) Let $S \subseteq \mathrm{soc}\ F_i(M)$ and consider $0 \longrightarrow S_i(M) \longrightarrow M \xrightarrow{\pi} F_i(M) \longrightarrow 0$. Then $\pi^{-1}(S) \supsetneq S_i(M)$. So, by the maximality of $S_i(M)$ there is a composition factor S' of $\pi^{-1}(S)$ such that $\mathrm{Ext}^1_\Lambda(S', S_i) \neq 0$. Since the composition factors of $S_i(M)$ don't have this property then, necessarily, $\mathrm{Ext}^1_\Lambda(S, S_i) \neq 0$. So, for every $S \subseteq \mathrm{soc}\, F_i(M)$, $\mathrm{Ext}^1_\Lambda(S, S_i) \neq 0$. Since Λ verifies (P) then, applying Lemma 4.5, we have that $\mathrm{Ext}^1_\Lambda(S'', S_i) = 0$ for every composition factor S'' of $F_i(M)/_{\mathrm{soc}F_i(M)}$. This proves b).

c) By Lemma 4.3 we have that $\mathrm{Ext}^1_\Lambda(F_i(M)/_{\mathrm{soc}F_i(M)}, S_i) = 0$. Then, from $0 \to \mathrm{soc}\, F_i(M) \to F_i(M) \to F_i(M)/_{\mathrm{soc}F_i(M)} \to 0$ we obtain $0 \to \mathrm{Ext}^1_\Lambda(F_i(M), S_i) \xrightarrow{\beta_i(M)} \mathrm{Ext}^1_\Lambda(\mathrm{soc}F_i(M), S_i) \to 0$ i.e., $\beta_i(M)$ is an isomorphism.

Let now $\underline{n} = (n_1, \ldots, n_n)$ and let M be in $\mathcal{E}_{\underline{n}}$, i.e., M is a Λ-module such that $[M] = \sum_{i=1}^{n} n_i[S_i]$ in $\mathrm{Gr}(\Lambda)$. We will associate to M a family of factor modules $H_i(M)$, $i = 0, \ldots, n$, defined in the following way:

$$H_0(M) = M \ , \ H_i(M) = H_{i-1}(M)/_{\tau_{S_i}(H_{i-1}(M))}, \ i = 1, \ldots, n \ .$$

Then, by Lemma 3.3 we know that $H_{i-1}(M)$ does not contain $S_1, \ldots, S_{i-1}$ as composition factors and that $\tau_{S_i}(H_{i-1}(M)) \cong S_i^{n_i}$. We fix an isomorphism $\Theta_{i,M}: S_i^{n_i} \to \tau_{S_i}(H_{i-1}(M))$, for every $i = 1, \ldots, n$. If $\lambda: M \to M'$ is a morphism in $\mathcal{E}_{\underline{n}}$, then λ induces a morphism $H_{i-1}(\lambda): H_{i-1}(M) \to H_{i-1}(M')$. We define $\lambda^{(i)}: S_i^{n_i} \to S_i^{n_i}$ to be the map such that the diagram

$$\begin{array}{ccc}
\tau_{S_i}(H_{i-1}(M)) & \xrightarrow{H_{i-1}(\lambda)|\tau_{S_i}(H_{i-1}(M))} & \tau_{S_i}(H_{i-1}(M')) \\
\Theta_{i,M}\downarrow & & \downarrow\Theta_{i,M'} \\
S_i^{n_i} & \xrightarrow{\lambda^{(i)}} & S_i^{n_i}
\end{array}$$

commutes.

We observe that H_i is a functor from $\mathcal{E}_{(n_1, \ldots, n_n)}$ to $\mathcal{E}_{(0,\ldots,0,\, n_{i+1},\ldots,n_n)}$. Obviously, if λ is an isomorphism, then $\coprod_{i=1} \lambda^{(i)} \in \mathrm{Aut}_\Lambda (\coprod_{i=1} S_i^{n_i})$.

We assume now in that follows that Λ verifies (P). For $i=1,\dots,n$, let $\mathcal{S}_i = \{j: \mathrm{Ext}^1_\Lambda(S_j,S_i) \neq 0\}$. If M is in $\mathcal{L}_n$, then, by Proposition 4.6 b), there is an isomorphism $\epsilon_{i,M}$: soc $F_i(H_i(M)) \longrightarrow \coprod_{j\in \mathcal{S}_i} S_j^{n_j}$

By means of this identification the isomorphism

$\beta_{i,H_i(M)}\alpha^{-1}_{i,H_i(M)}: \mathrm{Ext}^1_\Lambda(H_i(M), S_i) \longrightarrow \mathrm{Ext}^1_\Lambda(\mathrm{soc} F_i(H_i(M)), S_i)$ defines an isomorphism

$$\gamma_{H_i(M)}: \mathrm{Ext}^1_\Lambda(H_i(M), S_i^{n_i}) \longrightarrow \mathrm{Ext}^1_\Lambda(\coprod_{j\in \mathcal{S}_i} S_j^{n_j}, S_i^{n_i}) = \mathrm{Ext}^1_\Lambda(\coprod_{j=1}^{n} S_j^{n_j}, S_i^{n_i})$$

and $\epsilon_{i,M}$ can be chosen in such a way that, if $\lambda: M \longrightarrow M'$ is an isomorphism in $\mathcal{L}_n$ then the diagram

$$\begin{array}{ccc}
\mathrm{Ext}^1_\Lambda(H_i(M), S_i^{n_i}) & \xrightarrow{\gamma_{H_i(M)}} & \mathrm{Ext}^1_\Lambda(\coprod_{j=1}^{n} S_j^{n_j}, S_i^{n_i}) \\
\uparrow {\scriptstyle \mathrm{Ext}^1_\Lambda(H_i(\lambda), S_i^{n_i})} & (1) & \uparrow {\scriptstyle \mathrm{Ext}^1_\Lambda(\coprod \lambda^{(j)}, S_i^{n_i})} \\
\mathrm{Ext}^1_\Lambda(H_i(M'), S_i^{n_i}) & \xrightarrow{\gamma_{H_i(M')}} & \mathrm{Ext}^1_\Lambda(\coprod_{j=1}^{n} S_j^{n_j}, S_i^{n_i})
\end{array}$$

commutes.

A way to choose $\epsilon_{i,M}$ with the above property is the following: let $j \in \mathcal{S}_i$ and let $N_j(M)$ be the maximal submodule of M that does not contain S_j as a composition factor. The modules $H_{j-1}(M)$ and $F_i(H_i(M))$ are both factors of M by a submodule that does not contain S_j as a composition factor, since $\tau_{S_j}(H_{j-1}(M)) \simeq S_j^{n_j}$ and $\tau_{S_j}(F_i(H_i(M)) \simeq S_j^{n_j}$. Then there are canonical epimorphisms $H_{j-1}(M) \longrightarrow M/N_j(M)$ and $F_i(H_i(M)) \longrightarrow M/N_j(M)$ that induce isomorphisms $\rho: \tau_{S_j}(H_{j-1}(M)) \longrightarrow \tau_{S_j}(M/N_j(M))$, $\theta: \tau_{S_j}(F_i(H_i(M)) \longrightarrow \tau_{S_j}(M/N_j(M))$.

We have fixed an isomorphism $\theta_{j,M}: S_j^{n_j} \longrightarrow \tau_{S_j}(H_{j-1}(M))$. Let $\epsilon_{i,j,M} = \theta^{-1}_{j,M}\, \rho^{-1}\, \theta: \tau_{S_j}(F_i(H_i(M)) \longrightarrow S_j^{n_j}$. Then the isomorphism

$\epsilon_{i,M} = \sum_{j \in S} \epsilon_{i,j,M}$ verifies the required property.

We can state now:

Theorem 4.7: Assume that Λ verifies (P). Let $\underline{n} = (n_1,\ldots,n_n)$ and let $N = \coprod_{i=1}^{n} S_i^{n_i}$. Then there is a surjective correspondence φ between modules in $\mathcal{S}_{\underline{n}}$ and elements of $V = \mathrm{Ext}^1_\Lambda(N,N)$ such that

a) If M is in $\mathcal{S}_{\underline{n}}$ and $x_i \in \mathrm{Ext}^1_\Lambda(H_i(M), S_i^{n_i})$ is the element determined by the sequence $0 \longrightarrow S_i^{n_i} \longrightarrow H_{i-1}(M) \longrightarrow H_i(M) \longrightarrow 0$, then $\varphi(M) = \sum_{i=1}^{n} \gamma_{H_i(M)}(x_i)$.

b) We consider as above V as an algebraic variety and $H = \mathrm{Aut}_\Lambda(N)$ as an algebraic group acting on V. If M and M' in $\mathcal{S}_{\underline{n}}$ are isomorphic, then $\varphi(M)$ and $\varphi(M')$ belong to the same orbit of V under the action of H.

Proof: The surjectivity of φ defined by a) follows easily by induction on the biggest k such that $n_k \neq 0$.

b) If $\lambda: M \longrightarrow M'$ is an isomorphism and $\coprod \lambda^{(i)} \in \mathrm{Aut}_\Lambda(N)$ is the corresponding element, then one can easily check using that the diagram (1) commutes that $\varphi(M') = \coprod_{i=1}^{n} \lambda^{(i)} \cdot \varphi(M)$.

With this we have proven

Corollary 4.8: Assume that Λ verifies (P) and that K is an infinite field. Then if Λ is of finite representation type the bilinear form B is positive definite.

Proof of Theorem 4.1: We observed that we only have to prove that a) and b) imply c). This follows from Corollary 4.8 and Proposition 4.2 when K is an infinite field. So we assume now that K is finite. Let $K(t)$ be the quotient field of the polynomial ring $K(t)$. Let $\Lambda' = \Lambda \otimes_K K(t)$. Since K is perfect and $\Lambda/\underline{r}$ is semisimple, then $\Lambda/\underline{r} \otimes_K K(t)$ is semisimple. On the other hand, let P be an indecomposable projective Λ-module. Then $P \otimes_K K(t)$ is a projective Λ'-module.

We will prove that it is indecomposable. We have that $End_{\Lambda\otimes_K K(t)}(P \otimes_K K(t)) \simeq End_\Lambda(P) \otimes_K K(t)$. Since Λ is hereditary, $End_\Lambda(P)$ is a division ring, that is a finite extension of K and is, therefore, a field, that we denote by L. Then $L \otimes_K K(t)$ is a domain and is a finite extension of the field $K(t)$. Therefore $L \otimes_K K(t)$ is a field, so $End_\Lambda(P \otimes K(t))$ is a field, hence $P \otimes_K K(t)$ is indecomposable.

Combining this with the fact that $\Lambda/\underline{r} \otimes_K K(t)$ is semisimple we have that $S \otimes_K K(t)$ is a simple Λ'-module, for every simple Λ-module S. Therefore, if P is indecomposable projective and $\underline{r}P = \coprod P_i^{\alpha_i}$, with P_i indecomposable projective, then $\underline{r}(P \otimes_K K(t)) \simeq \underline{r}P \otimes_K K(t) \simeq \coprod (P_i \otimes_K K(t))^{\alpha_i}$

We identify $Gr(\Lambda)$ with $Gr(\Lambda')$ by means of the isomorphism that transforms the class [M] of the module M into $[M \otimes_K K(t)]$. Then, by the previous observation, the bilinear forms B_Λ and $B_{\Lambda'}$ associated to the rings Λ and Λ' are the same. Hence the corresponding Coxeter transformations C_Λ and $C_{\Lambda'}$ are equal.

Assume now that there is $n > 0$ such that $c^n = 1$. Since Λ' is an hereditary algebra over the infinite field $K(t)$ and $c_\Lambda^n = c_{\Lambda'}^n = Id$, then Λ' is of finite representation type. Therefore $B_\Lambda = B_{\Lambda'}$ is positive definite. This ends the proof of b) ===>c) when K is a finite field.

We indicate now an alternative proof of the statement b) ===>c) of Theorem 4.1, using results concerning representations of K-species proven by V. Dlab and C. Ringel in [10] and [11].

Assume that Λ is of finite representation type; we may also assume that Λ is basic. We know then by ([11], Proposition 10.2) that Λ is isomorphic to the tensor algebra of its K-species Q. Since Λ is of finite representation type, there are only a finite number of nonisomorphic indecomposable representations over the K-species Q. We have then (see [10]) that the quadratic form associated to Q is positive definite. But this quadratic form is precisely the quadratic form associated to the tensor algebra corresponding to Q that is, as we have just seen, the algebra Λ. This proves that if Λ is of finite representation type then the quadratic form $\underline{q}$ is positive definite.

REFERENCES

[1] M. Auslander, Representation theory of artin algebras I, Communications in Algebra 1(3), 1974, Pages 177-268.

[2] M. Auslander, Representation theory of artin algebras II, Communications in Algebra 1(4), 1974, Pages 269-310.

[3] M. Auslander and I. Reiten, Representation theory of artin algebras III, Communications in Algebra 3(3), 1975, Pages 239-294.

[4] M. Auslander and I. Reiten, Representation theory of artin algebras IV, To appear in Communications in Algebra.

[5] M. Auslander and I. Reiten, Representation theory of artin algebras V, To appear in Communications in Algebra.

[6] M. Auslander and I. Reiten, Representation theory of artin algebras VI, To appear in Communications in Algebra.

[7] M. Auslander and I. Reiten, Stable equivalence of artin algebras, Conference in Orders, Group Rings and related topics, Springer Verlag No. 353, 1973.

[8] I N. Bernstein, I. M. Gelfand, V. A. Ponomarev, Coxeter functors and a theorem of Gabriel. Uspechi Math. Nauk 28, 1973.

[9] H. Cartan, S. Eilenberg, Homological Algebra, Princeton University Press, 1956.

[10] V. Dlab and C. M. Ringel, On algebras of finite representation type, Carleton Lecture Notes No. 2, 1973

[11] V. Dlab and C. M. Ringel, Representation of graphs and algebras, Carleton Lecture Notes No. 8, 1974.

[12] P. Gabriel, Réprésentations indecomposables, Séminaire Bourbaki, 26e année. 1973/74, No. 444.

[13] C. M. Ringel, Representations of K-species and bimodules. Sonderforschungsbereich 40---Theoretische Mathematik, Univ. Bonn.

INDECOMPOSABLE INTEGRAL REPRESENTATIONS OF CYCLIC p-GROUPS*

Irving Reiner
University of Illinois
Urbana, Illinois

§1. INTRODUCTION

On of the oldest problems in the theory of integral representations is to classify, relative to unimodular similarity, all integral matrices which satisfy the equation $X^n = I$. Here, two such matrices X, X' are called unimodularly similar if $X' = PXP^{-1}$ for some matrix P which is unimodular over the ring of rational integers Z.

Reformulating this problem in terms of modules, we start with a cyclic group $G = \langle x : x^n = 1 \rangle$, and let ZG denote its integral group ring. By a ZG-lattice we mean a left ZG-module which is finitely generated and torsionfree as Z-module. Such lattices have finite free Z-bases, and give rise to integral matrices X such that $X^n = I$. Unimodular similarity of matrices corresponds to isomorphism of ZG-lattices. The basic problem then becomes: determine all isomorphism classes of ZG-lattices. Since every lattice is expressible as a direct sum of indecomposable lattices, the problem splits into two parts:

Question I Find all isomorphism classes of indecomposable ZG-lattices, and

Question II Determine when two direct sums of indecomposable lattices are isomorphic.

Let $n(ZG)$ be the number of indecomposable classes of ZG-lattices. As part of Question I, we want to know whether $n(ZG)$ is finite or infinite. As we shall see below, this weaker question can be answered completely, for arbitrary finite groups G.

*Research partially supported by the National Science Foundation.

When G is cyclic of prime order p, Question I was answered by Diederichsen [4], a student of Zassenhaus. Let ω_1 denote a primitive p-th root of 1 over the rational field Q, and set

$$K = Q(\omega_1), \quad R = Z[\omega_1] = \text{alg. int. } \{K\}, \tag{1}$$

so R is the ring of all algebraic integers in K. Let h_R be the ideal class number of R, that is, the number of isomorphism classes of fractional R-ideals in K. Diederichsen showed that

$$n(ZG) = 2h_R + 1,$$

and gave a full list of non-isomorphic indecomposable ZG-lattices. The list is

$$Z\,;\, B_1,\cdots,B_h\,;\, E_1,\cdots,E_h, \tag{2}$$

where $h = h_R$. Here, G acts trivially on Z. The $\{B_i\}$ are ideals of R, one for each ideal class, with the generator x of G acting on each B_i by multiplication by ω_1. Finally, for each i $(1 \leq i \leq h)$, E_i denotes a non-split extension

$$0 \to Z \to E_i \to B_i \to 0$$

of ZG-lattices. (It turns out that for fixed i, all such non-split extensions are mutually isomorphic.)

Diederichsen also considered briefly the case where G is cyclic of order 4, and "proved" that $n(ZG)$ is infinite in this case. The error in his proof was discovered independently by Troy [15] and Roiter [13] in 1960; they found that $n(ZG) = 9$ in this case. Soon thereafter, Heller and Reiner [7] proved that $n(ZG)$ is finite for G cyclic of order p^2, where p is any prime. They showed also that $n(ZG)$ is infinite if G is a non-cyclic p-group, or a cyclic p-group of order greater than p^2. These results were independently obtained by Berman and Gudivok [1].

Jones [8] used the above results to prove

<u>Theorem</u> The number of non-isomorphic indecomposable ZG-lattices is finite if and only if for each prime p dividing $|G|$, the p-Sylow subgroups of G are cyclic of order p or p^2.

Indeed, Jones established a more general result, as follows: let Λ be a Z-order in a semisimple Q-algebra A, and let $\underline{S}(\Lambda)$ be the set of all rational primes p at which the p-adic completion Λ_p is not a maximal Z_p-order in A_p. (The set $\underline{S}(\Lambda)$ is necessarily finite; in the case where $\Lambda = ZG$, the set $\underline{S}(\Lambda)$ consists of all prime divisors of $|G|$.) Then $n(\Lambda)$ is finite if and only if $n(\Lambda_p)$ is finite for each $p \in \underline{S}(\Lambda)$. Furthermore, for fixed p, let H be a p-Sylow subgroup of the group G; then $n(Z_pG)$ is finite if and only if $n(Z_pH)$ is finite.

Despite these results, very little progress has been made in asnwering Questions I and II, even for groups G for which $n(ZG)$ is finite. Reiner [9] answered Question II for the case where G is cyclic of order p, and gave a new proof of Diederichsen's solution of Question I. For G cyclic of order p^2, Reiner [11] recently gave a full set of non-isomorphic indecomposable ZG-lattices and determined $n(ZG)$ explicitly for all primes p, except those which are improperly irregular. Recall the following definitions: the prime p is <u>regular</u> if $p \nmid h_R$, where R is as in (1). Next, let h_0 denote the ideal class number of $Z[\omega_1 + \omega_1^{-1}]$, the ring of algebraic integers in the maximal real subfield of K. Call p <u>properly regular</u> if $p \mid h_R$ but $p \nmid h_0$. If $p \mid h_R$ and $p \mid h_0$, then p is <u>improperly irregular</u>. We remark that in all cases known so far, every prime is either regular or improperly irregular (see Borevich and Shafarevich [2]).

Let us put

$$L = Q(\omega_2), \quad S = Z[\omega_2] = \text{alg. int. } \{L\}, \tag{3}$$

where ω_2 is a primitive p^2-th root of 1, and let h_S denote the ideal class number of S. Let G be cyclic of order p^2, where p is a regular prime. Reiner [11] proved that

$$n(ZG) = 1 + 2h_R + 2h_S + [4N_1 + (\varepsilon_p - 1)(N_1 - p^{[(p-3)/2]})]h_R h_S, \tag{4}$$

where

$$N_1 = \sum_{r=0}^{p-2} p^{[(p-3-r)/2]}, \quad \varepsilon_p = \begin{cases} 2, & p \equiv 1 \pmod 4, \\ 1, & \text{otherwise.} \end{cases} \tag{5}$$

Here, the brackets in the exponent denote "greatest integer function", except that $p^{[s]}$ is to be interpreted as 1 whenever $s \le 0$. These formulas give $n(ZG) = 9,13,40$ for $p = 2,3,5$, respectively, and clearly $n(ZG)$ increases rapidly with p.

Corresponding formulas for $n(ZG)$ are available [11] when the prime p is properly irregular. Let $\delta(k)$ denote the number of Bernoulli numbers $B_1,\cdots,B_k$ whose numerators are multiples of p. (We remark that an odd prime p is regular if and only if $\delta((p-3)/2) = 0$.) Let G be cyclic of order p^2, where p is properly irregular. Then $n(ZG)$ is given by a formula exactly like that in (4), except that N_1 is replaced by N_1', where

$$N_1' = \sum_{r=0}^{p-2} p^{[(p-3-r)/2] + \delta[(p-2-r)/2]}.$$

In this article, we shall outline some of the calculations which lead to formula (4). We shall also give a partial solution to Question II, for the case where G is cyclic of order p^2. This partial solution already shows that "cancellation" need not hold true for direct sums of ZG-lattices; in other words, there may exist lattices L,M,N with $L \oplus M \cong L \oplus N$ but where M is not isomorphic to N.

§2. INDECOMPOSABILITY

Let p be prime, and let the subscript p indicate p-adic completion. For each ZG-lattice M, its completion M_p is a Z_pG-lattice. We now prove (see Heller and Reiner [7, I, (1.5)])

Proposition 1 Let G be a cyclic p-group, and M any ZG-lattice. Then M is indecomposable if and only if M_p is indecomposable.

Proof Let $A = QG$, $\Lambda = ZG$, $A_p = Q_pG$, where Q_p is the p-adic completion of Q. Since G is a cyclic p-group, the decomposition of A into simple components is given by

$$A \cong F_1 \oplus \cdots \oplus F_r,$$

where each F_i is a cyclotomic extension field of the form $Q(\omega)$, with ω some p-power root of 1. As is well known, the p-adic valuation on Q extends uniquely to a P_i-adic valuation on F_i, where P_i is the unique maximal ideal in alg. int.$\{F_i\}$ containing p (indeed, $P_i=(1-\omega)Z[\omega]$). Thus $(F_i)_p$ is a field, namely, the P_i-adic completion of F_i. The decomposition of A_p into simple components is therefore given by

$$\Lambda_p \cong (F_1)_p \oplus \cdots \oplus (F_r)_p .$$

Hence every simple left A_p-module is the completion of a simple A-module. Since every finitely generated A_p-module W is a finite direct sum of simple A_p-modules, it follows that each such W is the completion of some finitely generated A-module. We shall use this fact below.

Now let M be any Λ-lattice. If M is decomposable, then obviously so is M_p. Assume conversely that there is a non-trivial decomposition

$$M_p = X_1 \oplus X_2, \quad X_i = \Lambda_p\text{-lattice}.$$

Then

$$(QM)_p = Q_pX_1 \oplus Q_pX_2, \quad Q_pX_i = A_p\text{-module} .$$

The preceding discussion shows that there exist finitely generated left A-modules V_1, V_2 such that $Q_pX_i = Q_pV_i$, $i = 1,2$. We now set

$$Y_i = X_i \cap V_i \quad (\text{intersection in } Q_pX_i), \quad i = 1,2 .$$

Then (see [10, (5.2)]) each X_i is a $\tilde{\Lambda}$-lattice, where $\tilde{\Lambda}$ is the <u>localization</u> of Λ at the prime ideal pZ of Z, that is,

$$\tilde{\Lambda} = \{\lambda/s : \lambda \in \Lambda,\ s \in Z,\ (s,p) = 1\} .$$

Furthermore, $(Y_i)_p = X_i$, $QY_i = V_i$. Thus we have a pair of $\tilde{\Lambda}$-lattices, namely $Y_1 \oplus Y_2$ and $\tilde{M}$, whose p-adic completions coincide. It follows at once (see [10, (5.2)]) that $\tilde{M} = Y_1 \oplus Y_2$.

We hope to use this decomposition of $\tilde{M}$ to obtain a decomposition of M. Set $N = M \cap Y_1$, a Λ-sublattice of M; we shall show that N is a direct summand of M, and to begin with, let us verify that $\tilde{N} = Y_1$. Since $N \subset Y_1$, it is clear that $\tilde{N} \subset Y_1$. On the other hand, each $y \in Y_1$ lies in $\tilde{M}$, so we may write $y = m/s$ for some $m \in M$, $s \in Z$, where $(s,p) = 1$. Then $m = sy \in Y_1$, so $m \in N$, and thus $y = m/s \in \tilde{N}$. This establishes the fact that $\tilde{N} = Y_1$.

Let $L = M/N$, a left Λ-module. We show that L is a Λ-lattice by proving that L is Z-torsionfree. Indeed, suppose that $sm \in N$, where $s \in Z$, $s \neq 0$, and $m \in M$. Then $sm \in \tilde{N}$; but $\tilde{M}/\tilde{N} = \tilde{M}/Y_1 \cong Y_2$, so $\tilde{M}/\tilde{N}$ is Z-torsionfree. Thus we have $m \in \tilde{N}$, that is, $m \in Y_1$.

Therefore $m \in M \cap Y_1$, so $m \in N$ as desired. Hence L is a lattice, and we have an exact sequence of Λ-lattices

$$0 \to N \to M \to L \to 0 . \tag{6}$$

The localization of (6) at p is split, since the inclusion $N \subset M$ was constructed from the direct sum decomposition $\tilde{M} = \tilde{N} \oplus Y_2$. On the other hand, let q be any prime different from p. Since q is relatively prime to the order of the p-group G, it follows (see [10, (41.1)]) that Λ_q is a maximal order in A_q. But then the Λ-lattice L_q is projective (see [(10, (21.5)]), and thus the completion of sequence (6) at q is split. Thus the sequence (6) must split globally, since it splits locally at every prime (see [10, §3]). Therefore $M \cong N \oplus L$ as Λ-lattices. This decomposition is non-trivial, because $\tilde{N} = Y_1$, and the proof of the Proposition is finished.

<u>Remark</u> If G is an arbitrary p-group, where p is an odd prime, then Proposition 1 remains valid. By Feit [5, (14.5)], QG is a direct sum of full matrix algebras $M_n(F_i)$, where each F_i is a cyclotomic field of the form $Q(\omega)$, with ω a p-power root of 1. The discussion in the proof of Proposition 1 is easily modified to show that, also in this case, every finitely generated Q_pG-module is the completion of some finitely generated QG-module. The rest of the proof of Proposition 1 carries over unchanged.

Returning to the case where G is a cyclic p-group, we see that the problem of determining all indecomposable ZG-lattices breaks up into two parts:

<u>Question Ia</u> Find all indecomposable Z_pG-lattices, and

<u>Question Ib</u> Given an indecomposable Z_pG-lattice X, find a full set of non-isomorphic ZG-lattices M such that $M_p \cong X$.

Let us recall the definition of genus (see [10, §27]). For an arbitrary finite group G, two ZG-lattices M,N are in the same <u>genus</u> (notation: $M \vee N$) if $M_q \cong N_q$ as Z_qG-lattices for each prime q. Further, it suffices to consider just those primes which divide $|G|$. Hence, for the case where G is a p-group, Question Ib is the same as trying to determine all isomorphism classes of ZG-lattices belonging to a given indecomposable genus.

For the case where G is cyclic of order p^2, Question Ia was settled by Heller and Reiner [7], and also by Berman and Gudivok [1]. Question Ib was answered by Reiner [11] for all primes p, except those which are improperly irregular (if any such primes exist!). In the next section, we shall discuss these solutions, with special emphasis on one of the subcases which arises in the solutions.

§3. CYCLIC GROUPS OF ORDER p^2

Throughout this section, let G be a cyclic group with generator x of order p^2. Let $\Phi_j(x)$ denote the cyclotomic polynomial of order p^j and degree $\phi(p^j)$. Then we may write

$$x^{p^2} - 1 = \prod_{j=0}^{2} \Phi_j(x), \qquad ZG \cong Z[x]/(\Phi_0(x)\Phi_1(x)\Phi_2(x)) .$$

Further, for R and S as defined in (1) and (3), we have

$$R \cong Z[x]/(\Phi_1(x)), \qquad S \cong Z[x]/(\Phi_2(x)).$$

Now let M be any ZG-lattice, and define

$$M_0 = \{m \quad M : (x^p-1)m = 0\}, \qquad M_1 = M/M_0 .$$

It is easily verified that M_0 and M_1 are Z-torsionfree, so we have an exact sequence of ZG-lattices

$$0 \to M_0 \to M \to M_1 \to 0 . \tag{7}$$

Both M_0 and M_1 are uniquely determined (up to isomorphism) by M, and in fact M_0 is a characteristic submodule of M.

Since $(x^p-1)M_0 = 0$, we may view M_0 as a ZH-lattice, where H is a cyclic group of order p, with generator x. By the results of Diederichsen and Reiner described in §1, we know the structure of M_0 explicitly, namely, M_0 is isomorphic to a direct sum of copies of the $2h_R + 1$ modules listed in (2). Furthermore, if we write

$$M_0 \cong Z^{(r)} \oplus \bigoplus_{i=1}^{h} \{B_i^{(s_i)} + E_i^{(t_i)}\} , \tag{8}$$

where the superscripts denote multiplicities, then the invariants of M_0, which determine M_0 up to isomorphism, are the three non-negative integers r, $\sum_{i=1}^{h} s_i$, $\sum_{i=1}^{h} t_i$, and the R-ideal class of the product $\prod_{i=1}^{h} B_i^{s_i+t_i}$. (This ideal class is often called the Steinitz class of M_0.) Thus, the module M_0 occurring in (7) is known quite explicitly.

On the other hand, we have

$$(x^p-1)\ \Phi_2(x)M = (x^{p^2}-1)M = 0\ ,$$

and so

$$\Phi_2(x)M \subset M_0, \quad \text{that is,} \quad \Phi_2(x)M_1 = 0\ .$$

Thus M_1 may be viewed as an S-module, and is clearly finitely generated. Let us verify that M_1 is S-torsionfree. There is a multiplicative norm map $N_{L/Q} : L \to Q$. For each nonzero $s \in S$, its norm $N_{L/Q}(s)$ is a nonzero rational integer, which is expressible in the form $s's$ for some $s' \in S$. Suppose now that $sm_1 = 0$, where $m_1 \in M_1$. Then also $s'sm_1 = 0$, and hence $m_1 = 0$ because M_1 is Z-torsionfree. Thus M_1 is an S-lattice, so by Steinitz's Theorem (see [10, (4.13)]) we may write

$$M_1 \cong C_1 \dotplus \cdots \dotplus C_c\ , \tag{9}$$

an external direct sum of S-ideals C_i in L. The invariants which determine M_1 up to isomorphism are its rank c, and its Steinitz class (that is, the S-ideal class of the product $C_1 \cdots C_c$ in L).

The problem of classifying all ZG-lattices M is thus reduced to a question on extensions of lattices: given the lattices M_0 and M_1 in (8) and (9), determine a full set of non-isomorphic extensions M as in (7). This same question arises in the local case as well, when we try to find all Z_pG-lattices. The only simplification of the local case is that each of the ideals $(B_i)_p$ $(C_i)_p$ is principal.

We shall now consider in detail the special case where M_0 is itself an R-lattice, say

$$M_0 = B_1 \dotplus \cdots \dotplus B_b. \tag{10}$$

However, we change notation, so that now the $\{B_i\}$ denote arbitrary R-ideals in K, rather than representatives of the ideal classes of R. The invariants, which determine M_0 up to isomorphism, are its rank b

and its Steinitz class (that is, the ideal class of $B_1 \cdots B_b$). This special case contains many interesting features, and gives insight into the general case.

Suppose now that M_0 ia given as in (10), and M_1 as in (9). Since we are interested in M_0 and M_1 only up to isomorphism, we can replace each of the ideals B_i, C_j by other ideals in the same ideal class. Having done so, and changing notation, we may hereafter assume that

$$(B_i)_p = R_p, \quad (C_j)_p = S_p, \quad 1 \leq i \leq b, \quad 1 \leq j \leq c. \tag{11}$$

Let $\Lambda = ZG$, and let us compute $\mathrm{Ext}^1_\Lambda(M_1, M_0)$. Since M_1 is a Λ-lattice, we have $|G| \cdot \mathrm{Ext}^1_\Lambda(M_1, M_0) = 0$ (see Curtis and Reiner [3, (75.10)]). Thus $\mathrm{Ext}^1_\Lambda(M_1, M_0)$ is a finitely generated p-torsion Z-module, and hence is isomorphic to its p-adic completion. This gives

$$\mathrm{Ext}^1_\Lambda(M_1, M_0) \cong \mathrm{Ext}^1_{\Lambda_p}((M_1)_p, (M_0)_p) \cong \mathrm{Ext}^1_{\Lambda_p}(S_p^{(c)}, R_p^{(b)}).$$

Now there is a Λ-exact sequence

$$0 \to \Phi_2(x)\Lambda \to \Lambda \to S \to 0,$$

where x is a generator of G. This gives

$$\begin{aligned}\mathrm{Ext}^1(S,R) &\cong \mathrm{Hom}_\Lambda(\Phi_2(x)\Lambda, R)/\text{image of } \mathrm{Hom}_\Lambda(\Lambda, R)\\ &\cong R/pR = \bar{R}, \quad \text{say}.\end{aligned}$$

Then also

$$\mathrm{Ext}^1_{\Lambda_p}(S_p, R_p) \cong \bar{R},$$

and we obtain

$$\mathrm{Ext}^1_\Lambda(M_1, M_0) \cong \bar{R}^{b \times c}, \tag{12}$$

where the right-hand expression is the group of all $b \times c$ matrices with entries in the ring $\bar{R}$.

Proposition 2 Let $\Lambda = ZG$, and let

$$0 \to M_0 \to M \to M_1 \to 0, \quad 0 \to M_0' \to M' \to M_1' \to 0, \tag{13}$$

be a pair of exact sequences of Λ-lattices, where M_0 and M'_0 are R-lattices, and M_1, M'_1 are S-lattices. Then $M \cong M'$ if and only if there exists a commutative diagram

$$\begin{array}{ccccccccc} 0 & \to & M_0 & \to & M & \to & M_1 & \to & 0 \\ & & \alpha\downarrow & & \phi\downarrow & & \beta\downarrow & & \\ 0 & \to & M'_0 & \to & M' & \to & M'_1 & \to & 0 \end{array} \tag{14}$$

in which α and β are isomorphisms.

Proof Given exact sequences as in (13), it is easily seen that M_0 is the characteristic submodule of M defined immediately preceding (7), and likewise M'_0 is the corresponding module for M'. Any Λ-homomorphism $\phi : M \to M'$ then induces an R-homomorphism $\alpha : M_0 \to M'_0$ and an S-homomorphism $\beta : M_1 \to M'_1$. The assertions in the Proposition are then obviously true.

We may restate the above in terms of extension classes. Given an extension M of M_1 by M_0, let $\xi_M \in \operatorname{Ext}^1_\Lambda(M_1,M_0)$ be the extension class determined by M. Then we have

Corollary Let M, M' be a pair of extensions of M_1 by M_0, and let ξ_M, $\xi_{M'}$ be their extension classes in $\operatorname{Ext}^1(M_1,M_0)$. Then $M \cong M'$ if and only if

$$\alpha\xi_M = \xi_{M'}\beta \quad \text{for some } \alpha \in \operatorname{Aut} M_0, \quad \beta \in \operatorname{Aut} M_1. \tag{15}$$

We shall say that ξ_M and $\xi_{M'}$ are strongly equivalent whenever condition (15) is satisfied. Thus, given M_0 and M_1, the isomorphism classes of ZG-lattices M occurring in (7) are in one-to-one correspondence with the strong equivalence classes of elements in $\operatorname{Ext}^1_{ZG}(M_1,M_0)$. In order to use this fact, we must calculate the actions of $\operatorname{Aut}(M_0)$ and $\operatorname{Aut}(M_1)$ on $\operatorname{Ext}^1(M_1,M_0)$.

For any pair of R-ideals B, B' in K, we have

$$\operatorname{Hom}_\Lambda(B,B') = \operatorname{Hom}_R(B,B') \cong B^{-1}B',$$

where $\Lambda = ZG$ as usual. Hence we may identify the endomorphism ring $\operatorname{End}(M_0)$ with

$$[B_j\, B_i^{-1}]_{1 \le i,j \le b},$$

where M_0 is given by (10). The above symbol denotes the ring of all $b \times b$ matrices whose (i,j)-entry ranges over all elements of $B_j B_i^{-1}$. Further, $\mathrm{Aut}(M_0)$ consists of all such matrices having an inverse of the same form. By virtue of condition (11), which we have imposed without loss of generality, we have $(B_j B_i^{-1})_p = R_p$ for each i and j. Thus we obtain a composition of homomorphisms

$$f_0 : \mathrm{Aut}(M_0) \to \mathrm{Aut}(M_0)_p \cong GL(b,R_p) \to GL(b,\bar{R}) . \tag{16}$$

Identifying $\mathrm{Ext}^1_\Lambda(M_1,M_0)$ with $\bar{R}^{b \times c}$ as in (12), we see that $\mathrm{Aut}(M_0)$ acts via f_0 by left multiplication on $\bar{R}^{b \times c}$.

In order to study the action of $\mathrm{Aut}(M_1)$ on $\mathrm{Ext}(M_1,M_0)$, we show first that there exists a commutative diagram

$$\begin{array}{ccc} S & \xrightarrow{N} & R \\ & \searrow^{\sigma} & \downarrow \rho \\ & & \bar{R} \end{array} \tag{17}$$

in which N is the norm map $N_{L/K}$, and ρ and σ are canonical ring surjections. Here, $\rho : R \to R/pR = \bar{R}$ is canonical. Further we have

$$S \cong Z[x]/(\Phi_2(x)), \qquad \bar{R} = Z[x]/(\Phi_1(x),p) .$$

Since $\Phi_2(x)$ lies in the ideal $(\Phi_1(x),p)$, there is clearly a canonical surjection $\sigma : S \to \bar{R}$.

For $a \in Z$, we have $N(a) = a^p \equiv a \pmod p$, and thus $(\rho \circ N)(a) = \sigma(a)$. Next, the cost of x in S is precisely ω_2, and $N(\omega_2) = (-1)^{p-1}\omega_1$. Thus both $\rho \circ N$ and σ carry the coset of x in S onto the coset of x in $\bar{R}$. Hence in order to prove that diagram (17) is commutative, it suffices to verify that the composite $\rho \circ N$ is a ring homomorphism. Clearly the composite is multiplicative, since both N and ρ are multiplicative. On the other hand, for $s_1, s_2 \in S$ we have

$$N(s_1+s_2) = N(s_1) + N(s_2) + \mathrm{Tr}(s') \quad \text{for some} \quad s' \in S,$$

where <u>Tr</u> denotes the trace from L to K. But $\mathrm{Tr}(s') \in pR$, since $S = R[\omega_2]$ and we have

$$\mathrm{Tr}(R) \subset pR, \quad \mathrm{Tr}(\omega_2^k) = 0, \quad 1 \le k \le p-1 .$$

Therefore $N(s_1+s_2) \equiv N(s_1) + N(s_2) \pmod{pR}$, and so

$$(\rho \circ N)(s_1+s_2) = (\rho \circ N)(s_1) + (\rho \circ N)s_2.$$

This completes the proof of the commutativity of (17). The result is due to Galovich [6].

Now let M_1 be given by (9), and assume without loss of generality that (11) holds true. Then we have

$$\mathrm{End}(M_1) \cong [C_j C_i^{-1}]_{1 \le i,j \le c} ,$$

and as above we obtain a homomorphism f_1 by composition of maps:

$$f_1 : \mathrm{Aut}(M_1) \to \mathrm{Aut}(M_1)_p \cong GL(c,S_p) \to GL(c,\bar{R}). \tag{18}$$

The last map is induced from the ring homomorphism σ, extended to a map $S_p \to \bar{R}$. When we identify $\mathrm{Ext}^1_\Lambda(M_1,M_0)$ with $\bar{R}^{b \times c}$ as in (12), we find at once that $\mathrm{Aut}(M_1)$ acts via f_1 by right multiplication on $\bar{R}^{b \times c}$.

In view of Proposition 2 and its Corollary, it is natural to introduce the following

<u>Definition</u> Let $\xi, \xi' \in \bar{R}^{b \times c}$. We call ξ and ξ' <u>strongly equivalent</u> (notation: $\xi \approx \xi'$) if

$$\xi' = \alpha\xi\beta \quad \text{for some } \alpha \in \mathrm{Aut}(M_0),\ \beta \in \mathrm{Aut}(M_1).$$

Here, α acts as does $f_0(\alpha)$, and β acts as does $f_1(\beta)$.

We wish to determine the strong equivalence classes of $\bar{R}^{b \times c}$; as observed above, such classes are in one-to-one correspondence with extensions of an S-lattice M_1 of rank c by an R-lattice M_0 of rank b. We begin by setting

$$\bar{Z} = Z/pZ, \quad \lambda = 1 - x.$$

Then $\Phi_1(x) \equiv \lambda^{p-1} \bmod p \cdot Z[x]$, so we obtain

$$\bar{R} \cong \bar{Z}[x]/(\Phi_1(x)) \cong \bar{Z}[\lambda]/(\lambda^{p-1}) .$$

Thus $\bar{R}$ is a local principal ideal ring, with ideals $\{\lambda^i\bar{R} : 0 \le i \le p-1\}$. Each $v \in \bar{R}$ is expressible as $v = \lambda^i u$, $0 \le i \le p-1$, where $u \in u(\bar{R})$, the group of units of $\bar{R}$. Of course, v determines the exponent $\underline{i}$ uniquely, and determines u modulo λ^{p-1-i}.

Proposition 3 For $b \leq c$, every $\xi \in \bar{R}^{b \times c}$ is strongly equivalent to a matrix of the form $[D\ 0]$, where

$$D = \operatorname{diag}(\lambda^{k_1} u_1, \cdots, \lambda^{k_b} u_b),\ 0 \leq k_1 \leq \cdots \leq k_b \leq p-1,\ u_i \in u(\bar{R}) \ . \tag{19}$$

An analogous result holds for the case where $b \geq c$.

Proof By an elementary matrix we mean a matrix obtained from an identity matrix by placing a nonzero entry in some off-diagonal position. Elementary row or column operations, by definition, are obtained from left or right multiplications by elementary matrices. If $b \leq c$, then by elementary row and column operations on a given matrix $\xi \in \bar{R}^{b \times c}$, we can bring ξ into the form $[D\ 0]$, with D as in (19).

On the other hand, let f_0, f_1 be the homomorphisms defined in (16) and (18). Since there is a ring surjection $R_p \to \bar{R}$, every elementary matrix in $GL(b,\bar{R})$ is the image of some elementary matrix in $GL(b,R_p)$. But each element $r \in R_p$ is congruent mod p to some $r' \in B_j B_i^{-1}$ (for fixed i,j), by virtue of (11). It follows at once that every elementary matrix in $GL(b,\bar{R})$ lies in the image of f_0. In the same way, each elementary matrix in $GL(c,\bar{R})$ lies in the image of f_1. But then ξ is strongly equivalent to $[D\ 0]$, and the proof is finished.

If ξ is strongly equivalent to $[D\ 0]$, with D as in (19), then of course ξ is also equivalent to $[D\ 0]$ in the usual (weaker) sense, that is, $\xi = \mu \cdot [D\ 0] \cdot \nu$ for some $\mu, \nu \in GL(\bar{R})$. Thus, the set $\{\lambda^{k_1}, \dots, \lambda^{k_b}\}$ is uniquely determined by ξ, and is precisely the set of elementary divisors of ξ, hereafter denoted by el. div. (ξ). We have at once

Corollary Let $\xi \approx \xi'$, where $\xi, \xi' \in \bar{R}^{b \times c}$. Then

$$\text{el. div.}(\xi) = \text{el. div.}(\xi'). \tag{20}$$

Now let M be a ZG-lattice which is an extension of M_1 by M_0, with extension class $\xi_M \in \bar{R}^{b \times c}$. Since ξ_M can be diagonalized under strong equivalence (by Proposition 3), it follows that M must decompose into a direct sum of ideals of R, ideals of S, and indecomposable ZG-lattices $\{L_i\}$, where

$$0 \to B_i \to L_i \to C_i \to 0$$

is exact, and corresponds to the extension class $\lambda^{k_i} u_i \in \overline{R}$. We must still settle the question as to when two indecomposable lattices are isomorphic. First of all, two R-ideals are Λ-isomorphic if and only if they lie in the same ideal class. Thus we obtain h_R non-isomorphic indecomposable Λ-lattices, given by R-ideals representing the h_R ideal classes of R. Likewise, there are h_S non-isomorphic indecomposable Λ-lattices given by S-ideals. It remains to settle the question as to when $L_i \cong L_j$. By Proposition 2, such an isomorphism can exist only when $B_i \cong B_j$, $C_i \cong C_j$. Changing notation, let us start with a pair

$$0 \to B \to L \to C \to 0, \quad 0 \to B \to L' \to C \to 0, \tag{21}$$

corresponding to extension classes $\xi, \xi' \in \overline{R}$, respectively. The earlier discussion shows at once that $L \cong L'$ if and only if $\xi' = \alpha\xi\beta$ for some $\alpha \in u(R)$, $\beta \in u(S)$. But the image of $u(S)$ in $u(\overline{R})$ is contained in the image of $u(R)$ in $u(\overline{R})$, by virtue of the commutative diagram (17). Hence $L \cong L'$ if and only if $\xi' = \alpha\xi$ for some $\alpha \in u(R)$.

Let us write $\xi = \lambda^r u$, where $0 \le r \le p-2$ and $u \in u(\overline{R})$. We may view u as a unit in the ring R', where $R' = \overline{Z}[\lambda]/(\lambda^{p-1-r})$, since ξ determines $u \bmod \lambda^{p-1-r}$. Then for fixed B and C, and for each r, there are as many non-isomorphic extensions L (belonging to the exponent r) as there are orbits of $u(R')$ under the action of $u(R)$. These orbits have been determined by Galovich [6] for the case where the prime p is either regular or properly irregular; for simplicity, let us consider here only the case where p is a regular odd prime. Galovich showed that $u(R')$ is the direct product of a cyclic group of order p-1 and an elementary abelian p-group generated by the units

$$1 + \lambda,\ 1 + \lambda^2,\ 1 + \lambda^3,\ \cdots,\ 1 + \lambda^{p-2-r}.$$

On the other hand, the image of $u(R)$ in $u(R')$ is the product of the cyclic factor of order p-1, and a subgroup generated by units of the form

$$1 + \lambda,\ 1 + \lambda^2 + \alpha_2\lambda^3,\ 1 + \lambda^4 + \alpha_4\lambda^5,\ \cdots,\ 1 + \lambda^k + \alpha_k\lambda^{k+1},$$

where each $\alpha_i \in R'$, and where k is the largest even integer $\le p-2-r$. The "missing" exponents are $3,5,7,\cdots,p-2-r$, where the last term occurs only for even r. There are $[(p-3-r)/2]$ such missing exponents, provided we interpret $[s]$ to be 0 whenever $s \leq 0$. Thus, the number of orbits is $p^{[(p-3-r)/2]}$, and we obtain:

Theorem 1 Let p be a regular odd prime, and let B, C be ideals of R,S, respectively. Let L range over all extensions of C by B corresponding to extension classes $\lambda^r u \in \overline{R}$, where $0 \leq r \leq p-2$ and $u \in u(\overline{R})$. For fixed B, C and r, there are precisely $p^{[(p-3-r)/2]}$ non-isomorphic indecomposable ZG-lattices L of this type. As B, C and r vary, the total number of non-isomorphic L's thus obtained is $h_R h_S N_1$, where N_1 is given in (5).

Remarks i) The result holds also for $p = 2$.

ii) If p is a properly irregular prime, the exponent $[(p-3-r)/2]$ must be increased by $\delta[(p-2-r)/2]$, where δ is the function defined in §1.

iii) The same type of analysis can be carried out for the general case where M_0 is given by (8) and M_1 by (9). There are relatively few choices for which M_p is indecomposable (see [7]), and it turns out that these can occur only when $c \leq 1$, and when $\overline{M}_0$ is one of the modules $0, Z, B_i, E_i, Z \oplus B_i, Z \oplus E_i$. This analysis yields formula (4) and its analogue for properly irregular primes (see [11]).

§4. INVARIANTS OF DIRECT SUMS

Our aim here is to provide partial answer to Question II of §1. Keeping the notation and terminology of §3, we shall deal mainly with ZG-lattices which are extensions of S-lattices by R-lattices, In §3, before Proposition 3, we introduced the concept of strong equivalence of matrices over $\overline{R}$. Starting with an R-lattice M_0 and an S-lattice M_1, we had constructed homomorphisms

$$f_i : \mathrm{Aut}(M_i) \longrightarrow GL(\overline{R}), \quad i = 1,2,$$

defined in (16) and (18). Let $b = \text{rank } M_0$, $c = \text{rank } M_1$, and let $\xi, \xi' \in \overline{R}^{b \times c}$. We called ξ, ξ' strongly equivalent ($\xi \approx \xi'$) if $\xi' = f_0(\alpha) \cdot \xi \cdot f_1(\beta)$ for some $\alpha \in \mathrm{Aut}(M_0)$, $\beta \in \mathrm{Aut}(M_1)$. We showed in Proposition 3 that $\xi \approx [D\ 0]$ (or its transpose), where

$$D = \mathrm{diag}(\lambda^{k_1}u_1, \cdots, \lambda^{k_b}u_b),\ 0 \le k_1 \le \cdots \le k_b \le p-1,\quad u_i \in u(\bar{R}),$$

and we called $\{\lambda^{k_1}, \cdots, \lambda^{k_b}\}$ the elementary divisors of ξ. Let us also define

$$u(\xi) = u_1 \cdots u_b \in u(\bar{R}), \tag{22}$$

although $u(\xi)$ is not uniquely determined by ξ. We are now ready to state the key results, which give necessary and sufficient conditions for strong equivalence:

Theorem 2 Let $\xi, \xi' \in \bar{R}^{b \times c}$, where $b \neq c$. Then $\xi \approx \xi'$ if and only if ξ and ξ' have the same elementary divisors.

Theorem 3 Let $\xi, \xi' \in \bar{R}^{b \times b}$. Then $\xi \approx \xi'$ if and only if

i) ξ and ξ' have the same elementary divisors, and

ii) There exists a unit $w \in u(R)$ such that

$$u(\xi') \equiv w \cdot u(\xi) \bmod \lambda^{p-1-k} \bar{R},$$

where λ^k is the elementary divisor of ξ with largest exponent. (This condition is automatically satisfied if $k = p - 1$.)

We sketch the proofs, which will be given in detail in another article [12]. For each $u \in u(\bar{R})$, the matrix $\mathrm{diag}(u, u^{-1}) \in GL(2, \bar{R})$ is expressible as a product of elementary matrices. Hence

$$\mathrm{diag}(u_1^{-1}, u_1, 1, \cdots, 1) \in GL(b, \bar{R})$$

lies in the image of f_0, and likewise

$$\mathrm{diag}(u_1^{-1}, u_1, 1, \cdots, 1) \in GL(c, \bar{R})$$

lies in the image of f_1. This readily implies that, with ξ and D as above,

$$\xi \approx [D\ 0] \approx [\mathrm{diag}(\lambda^{k_1}, \cdots, \lambda^{k_{b-1}}, \lambda^{k_b}u(\xi))\quad 0].$$

If $b \neq c$, the same procedure enables us to eliminate the factor $u(\xi)$, and so we obtain Theorem 2.

In the case where $b = c$, we are faced with the problem of deciding when two matrices of the form

$$\mathrm{diag}(\lambda^{k_1},\cdots,\lambda^{k_{b-1}},\lambda^{k_b}u(\xi))$$

are strongly equivalent. The sufficiency of the conditions stated in Theorem 3 is easily established, as is the necessity of condition i) of Theorem 3. It is somewhat more difficult to prove the necessity of condition ii), however.

Let us apply Theorem 2 and 3 to the question as to when two direct sums of indecomposable ZG-lattices are isomorphic. To begin with, let $\{B_i : 1 \leq i \leq h_R\}$ be ideals of R representing the h_R ideal classes of R, and let $\{C_j : 1 \leq j \leq h_S\}$ be ideal class representatives for S. For each pair (i,j), let $M(i,j;\lambda^r u)$ denote an extension of C_j by B_i, corresponding to the extension class $\lambda^r u \in \overline{R}$, where $0 \leq r \leq p-2$. Here, $u \in u(\overline{R})$ ranges over a set T_r of representatives of the orbits of $u(\overline{R}/(\lambda^{p-1-r}))$ under the action of $u(R)$. As shown in §3, every extension M of an S-lattice M_1 by an R-lattice M_0 must be a direct sum of modules of the form B_i, C_j, and $M(i,j;\lambda^r u)$.

Suppose now that M is a direct sum of such modules, as follows:

i) B_i occurs with multiplicity b_i, $1 \leq i \leq h_R$,

ii) C_j occurs with multiplicity c_j, $1 \leq j \leq h_S$, and

iii) $M(i,j;\lambda^r u_k)$ occurs with multiplicity $e(i,j,r,k)$ for $1 \leq i \leq h_R$, $1 \leq j \leq h_S$, $0 \leq r \leq p-2$, $u_k \in T_r$.

Given such an M, we can give a full set of invariants which characterize M up to ZG-isomorphism. We note first that

$$\begin{aligned} M_0 &= \{\bigoplus_i B_i^{(b_i)}\} \oplus \{\bigoplus_{i,j,r,k} B_i^{e(i,j,r,k)}\}, \\ M_1 &= \{\bigoplus_j C_j^{(c_j)}\} \oplus \{\bigoplus_{i,j,r,k} C_j^{e(i,j,r,k)}\}. \end{aligned} \tag{23}$$

Then two of the invariants of M are

$$\begin{aligned} b &= R\text{-rank of } M_0 = \sum_i b_i + \sum_{i,j,r,k} e(i,j,r,k), \\ c &= R\text{-rank of } M_1 = \sum_j c_j + \sum_{i,j,r,k} e(i,j,r,k). \end{aligned} \tag{24}$$

Two more invariants of M are the Steinitz classes of M_0 and M_1; the Steinitz class of M_0 is the R-ideal class of the product $\Pi B_i^{b_i} \cdot \Pi B_i^{e(i,j,r,k)}$ in K, and the class of M_1 is obtained analogously.

Next, the extension class $\xi_M \in \overline{R}^{b \times c}$ has the form

$$\xi_M = \begin{pmatrix} D & 0 \\ 0 & 0 \end{pmatrix} ,$$

where D is a diagonal matrix in which, for fixed r $(0 \leq r \leq p-2)$ and fixed $u_k \in T_r$, the diagonal entry $\lambda^r u_k$ occurs with multiplicity $\sum_{i,j} e(i,j,r,k)$. Thus, another invariant of M is given by

$$\text{el. div. } (D) = \{\lambda^r \text{ with multiplicity } \sum_{i,j,k} e(i,j,r,k): 0 \leq r \leq p-2\}. \quad (25)$$

Of course, el. div. (ξ_M) is the union of el. div. (D) together with a set of zeros, the number of which can be calculated at once from the knowledge of b,c and the set el. div. (D). Thus we obtain:

<u>Theorem 4</u> Let M be as above. The invariants of M, which completely determine the genus of M, are the integers b,c in (24) and the set el. div. (D) in (25). Furthermore, if some $b_i \neq 0$ or some $c_j \neq 0$, then the invariants which completely determine the ZG-isomorphism class of M are the genus invariants b,c and el. div. (D), together with the Steinitz classes of M_0 and M_1.

<u>Theorem 5</u> Let M be as above, but where each $b_i = 0$ and each $c_j = 0$. Let r_0 be the largest value of r for which some $e(i,j,r,u) \neq 0$, and set

$$R_0' = \overline{Z}[\lambda]/(\lambda^{p-1-r_0}) .$$

Let $u(M)$ denote the image of

$$\prod_{i,j,r,k} u_k^{e(i,j,r,k)}$$

in $u(R_0')$. Then the orbit of $u(M)$ in $u(R_0')$, under the action of $u(R)$, is an invariant of M. Furthermore, the ZG-isomorphism class of M is completely determined by the genus invariants b,c and el. div. (D), together with the Steinitz classes of M_0 and M_1, and the orbit of

$u(M)$ in $u(R_0')$ under the action of $u(R)$.

Remarks 1. In the preceding theorem, the orbits of $u(R_0')$ under the action of $u(R)$ are in one-to-one correspondence with the elements of the factor group $u(R_0')$/image of $u(R)$.

2. Theorem 4 tells us that in a direct sum of the type considered above, if any B_i or C_j occurs as a summand, then we may replace each u_k by 1 without affecting the isomorphism class of the sum. Thus, "cancellation" does not hold for ZG-lattices which are extensions of an S-lattice by an R-lattice, except for certain very special cases.

3. Even if no B_i or C_j occurs as a summand of M, it follows from Theorem 5 that we may change each u_k modulo λ^{p-1-r_o}, without affecting the isomorphism class of the direct sum. Furthermore, if $r_o \geq p-4$, then by Galovich's results there is only one orbit in $u(R_0')$ under the action of $u(R)$; hence when $r_o \geq p-4$, we may replace each u_k by 1 without affecting the isomorphism class of the direct sum.

4. Most of the results in Theorem 4 and 5 can also be obtained by using the following result of Roiter [14]

Theorem 6 Let Γ be a Z-order in a semisimple Q-algebra and let T be a faithful Γ-lattice. Let M,N be any Γ-lattices in the same genus. Then there exists a Γ-lattice T' in the genus of T, such that

$$M \oplus T \cong N \oplus T'.$$

In our case, we may choose $\Gamma = Z[x]/(\Phi_1(x)\Phi_2(x))$, a factor ring of Λ. Then each $M(i,j;\lambda^r u)$ is a faithful Γ-lattice, and hence by Roiter's Theorem, any direct sum of such lattices is isomorphic to a corresponding sum in which at most one u is different from 1. Indeed, in our special case, Roiter's Theorem can be proved easily, and it can be shown as well that of all the unit factors $\{u_k\}$ can be concentrated at one of the summands. This type of argument yields a somewhat simpler proof of Theorem 4, and of the "sufficiency" part of Theorem 5. It does not seem to establish, however, that the orbit described in Theorem 5 is indeed an invariant of M.

5. It is relatively easy to give analogues of Theorem 4 for the case of arbitrary ZG-lattices, not just those which are extensions of S-lattices by R-lattices. For example, let M be an extension of M_1 by M_0, where M_0 is given by (8) and M_1 by (9). Suppose that M has as direct summand either $Z \oplus E_i$ for some i (see (2)), or else $Z \oplus B_i \oplus C_j$ for some i,j. Then the isomorphism class of M is completely determined by the genus of M, together with the Steinitz classes of M_0 and M_1. See [12] for further details.

1. S. D. Berman and P. M. Gudivok, Indecomposable representations of finite groups over the ring of p-adic integers, Izv. Akad. Nauk SSSR Ser. Mat. 28 (1964), 875-910; English transl., Amer. Math. Soc. Transl. (2) 50 (1966), 77-113.

2. Z. I. Borevich and I. R. Shafarevich, Number Theory, Academic Press, New York, 1966.

3. C. W. Curtis and I. Reiner, Representation theory of finite groups and associative algebras, Pure and Appl. Math., vol. XI, Interscience, New York, 1962; 2nd ed., 1966.

4. F. E. Diederichsen, Über die Ausreduktion ganzzahliger Gruppendarstellungen bei arithmetischer Aquivalenz, Abh. Math. Sem. Univ. Hamburg 14 (1938), 357-412.

5. W. Feit, Characters of Finite Groups, Math. Lecture Notes Series, Benjamin, New York, 1967.

6. S. Galovich, The class group of a cyclic p-group, J. Algebra, Vol. 30 (1974), pp. 368-387.

7. A. Heller and I. Reiner, Representations of cyclic groups in rings of integers. I, II, Ann. of Math. (2) 76 (1962), 73-92; (2) 77 (1963), 318-328.

8. A. Jones, Groups with a finite number of indecomposable integral representations, Michigan Math J. 10 (1963), 257-261.

9. I. Reiner, Integral representations of cyclic groups of prime order, Proc. Amer. Math. Soc. 8 (1957), 142-146.

10. I. Reiner, Maximal Orders, Academic Press, London, 1975.

11. I. Reiner, Integral representations of cyclic groups of order p^2, Proc. Amer. Math. Soc., Vol. 58 (1976), pp. 8-12.

12. I. Reiner, Invariants of integral representations, to appear.

13. A. V. Roiter, On the representations of the cyclic group of fourth order by integral matrices, Vestnik Leningrad. Univ. 15 (1960), no. 19, 65-74. (Russian).

14. A. V. Roiter, On integral representations belonging to a genus, Izv. Akad. Nauk SSSR Ser. Mat. 30 (1966), 1315-1324; English transl. Amer. Math. Soc. Transl. (2) 71 (1968), 49-59.

15. A. Troy, Integral representations of cyclic groups of order p^2, Ph.D. Thesis, University of Illinois, Urbana, Ill., 1961.

NON-UNIQUENESS IN CROSSED PRODUCTS*

Murray Schacher
University of California
Los Angeles, California

Let D be a division ring which is finite-dimensional over its center K. By Wedderburn's theorem [3, Theorem 4.1.2] we have $[D : K] = n^2$ for some integer n called the degree of D. D is said to be a crossed product if there is a maximal subfield L of D which is a Galois extension of K. If $G = Gal(L/K)$, the Galois group of L over K, then G is a group of order n which is associated to D via L. It is tempting to ask to what extent G is an invariant of D. Examples abound in [5] to show that in general G is by no means unique. In this note we will show that the lack of uniqueness is as broad as can be expected; an example will be constructed in which the G that occur exhaust (up to isomorphism) all groups of order n.

Throughout this paper we will abide by the notation and terminology of [5]. Following [5], we refer to a finite-dimensional division ring which is central over K as a K-division ring. A finite-dimensional field extension L of K is called K-adequate if L can be embedded in some K-division ring. It is proved in [5, Proposition 2.2] that, when K is an algebraic number field, L is K-adequate if and only if L is contained in some K-division ring as a <u>maximal</u> <u>subfield</u>. A finite group G is called K-admissible if $G = Gal(L/K)$ for some Galois extension L of K which is K-adequate.

We will have to use freely the number theory associated to algebraic number fields and the description of division rings defined over them in terms of their Hasse invariants. For an exposition of this material we recommend either [1] or [4]. If $L \supset K$ are number fields and π a prime of K we write K_π for the completion of K at π. When the ambiguity is harmless we will also write L_π for the completion of L at any of the primes extending π.

*Research supported in part by NSF Grant MCS 76-06988.

We will prove:

Theorem Let n be any integer ≥ 2. There is an algebraic number field K and a K-division ring D (both depending on n) so that: any group G of order n is isomorphic to $Gal(L/K)$ for some maximal subfield L of D. Thus D can be expressed as a crossed product using any desired group of order n.

Proof Suppose $G_1, G_2, \cdots, G_t$ exhaust, up to isomorphism, all groups of order n. We set $G_0 = G_1 \times G_2 \times \cdots \times G_t$, the direct product of the G_i. Clearly G_0 is a group of order n^t. By [5, Theorem 9.1] there is an algebraic number field K so that G_0 is K-admissible. Using [5, Proposition 2.2] we conclude: there is a K-division ring D_0 of degree n^t over K so that G_0 is the Galois group of some maximal subfield of D_0. Let D be the Wedderburn constituent of $n^{t-1}D$, i.e. D is the (unique) K-division ring determined so that $\underbrace{D_0 \otimes_K D_0 \otimes_K \cdots \otimes_K D_0}_{n^{t-1}\text{ times}}$ is a ring of matrices over D. Since division rings over number fields have exponent equal to index [1, Theorem 33, page 150], D is a K-division ring of degree n. After minor modifications of D_0 we will show that the D constructed in this way is the desired example.

Suppose p is any prime dividing n, so that $n = p^a m$, $(p,m) = 1$. In order for D_0 to be K-adequate we must have by [5, Proposition 2.5]: two primes π_1 and π_2 of K so that $[L_{\pi_i} : K_{\pi_i}]$ is divisible by p^{ta}, $i = 1,2$, and the Hasse invariant of D_0 at π_i is of form a_i/b_i where $(p,a_i) = 1$ and $p^{ta} | b_i$. Using [2, Theorem 8, Chapter 7] we may assume that the invariants of D_0 are prime to p for all other primes π of K; this is the modification alluded to in the last paragraph. From now on we assume D_0 has been so conditioned and that D is constructed from D_0 as outlined above. From [6, §2, Chapter XI] the Hasse invariants of D are not divisible by p except at the primes π_1 and π_2.

We may write L_0 as a composite $L_0 = L_1 L_2 \cdots L_t$ where L_i is normal over K and $Gal(L_i/K) \cong G_i$. It is enough to show that L_i splits D for $i = 1,2,\ldots,t$, for then [1, Theorem 27, page 61] allows us to conclude that L_i is contained in D as a maximal subfield. To simplify notation let $L = L_1$, $G = G_1$; we will show L splits D.

We have $L_0 = LM$ where $Gal(M/K) \cong G_2 \times \cdots \times G_t$. Clearly the local degree of M at π_1 and π_2 cannot be divisible by a power of p greater than $a(t - 1)$. But the local degree of L_0 at these primes is divisible by p^{at}, so we conclude that the local degree of L at π_1 and π_2 is divisible by p^a. As this argument was relevant to any p dividing n, we conclude by [5, Proposition 2.5] that L splits D. This concludes the proof of the theorem.

One might ask what fields are candidates for K in examples as pathological as those in the theorem. Clearly p-adic fields are out of the question, since for them all Galois groups are obliged to be solvable. From [5, Corollary 10.4] we can see that function fields in one variable over a finite field are also out of the question. However, purely transcendental extensions of number fields will also work by inducing up the examples produced in the theorem. A possible classification of these fields would perhaps be a good project for future research.

REFERENCES

1. A. A. Albert, Structure of Algebras, American Mathematical Society, New York, 1939.

2. E. Artin and J. Tate, Class Field Theory, Harvard University Press, Cambridge, Mass., 1961.

3. I. Herstein, Non-Commutative Rings, Carus Monograph, 1968.

4. I. Reiner, Maximal Orders, Academic Press, New York, 1975.

5. M. Schacher, Subfields of Division Rings, J. Algebra, Vol. 9, No. 4, (1968), 451-477.

6. J.-P. Serre, Corps Locaux, In. Act. Sci. Ind. no. 1296, Hermann, Paris, 1962.

LARGE MODULES OVER ARTINIAN RINGS

R. B. Warfield, Jr.
University of Washington
Seattle, Washington

In this paper we discuss methods of constructing infinitely generated modules with various properties over Artinian rings. In the first section we give a smoother proof than the original one of the existence of a representation equivalence between a certain subcategory of R-modules and the category of all modules over a suitable principal ideal domain, where R is any commutative Artinian ring which is not a principal ideal ring. As an application we discuss the number of countably generated indecomposable modules. In the second section these results are generalized to arbitrary left Artinian rings with non-distributive ideal lattice. In the third section we discuss methods essentially due to Corner for constructing modules with specified endomorphism rings. The point is to give a short and simple proof of such a theorem, rather than to get the best possible results. We use this to give examples of the failure of the Krull-Schmidt theorem for modules over Artinian rings and for torsion-free modules of finite rank over a discrete valuation ring.

This is a partly expository paper, but the results of section 2 and some of the applications in sections 1 and 3 are new.

§1. MODULES OVER COMMUTATIVE RINGS

The main result of this section is Theorem 1, which originally appeared in [16]. The proof presented here is considerably more transparent than the original one, and our intention is to emphasize the elementary nature of the construction.

We recall (e.g. from [14]) that if R is a commutative Noetherian ring, and if for every maximal ideal m of R, m/m^2 has dimension at most 1 over R/m, then R is a finite product of Dedekind domains and Artinian principal ideal rings. Therefore, if R is a commutative Artinian ring which is not a principal ideal ring, then R contains an ideal I such that R/I is a local ring with maximal ideal m, such that $m^2 = 0$ and m has R/m-dimension 2. We may assume, therefore, that R is originally a ring of this sort, and that m is generated by the two elements a and b. We let $k = R/m$.

If M is any R-module, multiplication by a and b respectively induce k-linear transformations α and β, from M/mM to mM. We will say that M is a-translatable if α is an isomorphism of M/mM onto mM.

If M is a-translatable, we define a translation (linear transformation) $t : M/mM \to M/mM$ by $t = \alpha^{-1}\beta$. This gives M/mM a natural $k[t]$-module structure. If M and N are a-translatable R-modules, and $f: M \to N$ is a homomorphism, then f induces a $k[t]$-module homomorphism $M/mM \to N/mN$. We therefore have a functor Φ from the category $\underline{T}$ of a-translatable R-modules (a full subcategory of the category of all R-modules) to the category $k[t]$-Mod of all $k[t]$-modules.

Definition. An additive functor $F : \underline{A} \to \underline{B}$ between additive categories is a representation equivalence [1] if it is full and dense and a morphism f of $\underline{A}$ is an isomorphism if $F(f)$ is an isomorphism.

We recall that F is full if the map $\mathrm{Hom}(X,Y) \to \mathrm{Hom}(F(X),F(Y))$ is surjective, and dense if for every object X of $\underline{B}$ there is an object Y of $\underline{A}$ with $F(Y) \cong X$. It is remarked in [1, p.54] that the condition that f is an isomorphism whenever $F(f)$ is an isomorphism is equivalent to the condition that the kernel of the ring homomorphism $\mathrm{End}(X) \to \mathrm{End}(F(X))$ be contained in the Jacobson radical of $\mathrm{End}(X)$ for all objects X of $\underline{A}$, and we will use this description.

Theorem 1. The functor $\Phi : \underline{T} \to \mathrm{Mod}\ k[t]$ described above is a representation equivalence.

Proof. By construction, it is clear that the kernel of the map $\mathrm{End}(X) \to \mathrm{End}(\Phi(X))$ is an ideal whose square is zero, so the last condition in the definition of a representation equivalence is trivially verified. We next show that Φ is full. Let $f : A/mA \to B/mB$ be a $k[t]$-module homomorphism. Let $\phi : A \to A/mA$ and $\chi : B \to B/mB$ be the natural maps, and let X and Y be subsets of A and B respectively such that (i) ϕ and χ take X and Y (respectively) bijectively onto bases of the k-vector spaces A/mA and B/mB, (ii) $\phi(X) \cap \ker(f)$ spans $\ker(f)$, and (iii) $f(\phi(X)) \subseteq \chi(Y)$. We define $\gamma : X \to B$ as follows: If $f(\phi(x)) = 0$, let $\gamma(x) = 0$, and otherwise, let $\gamma(x)$ be the unique

$y \in Y$ such that $\chi(y) = f(\phi(x))$. We now wish to define $g : A \to B$ by the formula $g(\Sigma r_i x_i) = \Sigma r_i \gamma(x_i)$ $(x_i \in X, r_i \in S)$. This will clearly do everything we want if, in fact, it is well defined. To check this it suffices to show that if $\Sigma r_i x_i = 0$ then $\Sigma r_i \gamma(x_i) = 0$, where we may assume that the x_i are distinct elements of X. If $\Sigma r_i x_i = 0$ then clearly $r_i \in m$ since ϕ takes X bijectively onto a basis for A/mA. We write $r_i = au_i - bv_i$, so our equation becomes $a\Sigma u_i x_i = b\Sigma v_i x_i$, or equivalently, $t\Sigma v_i \phi(x_i) = \Sigma u_i \phi(x_i)$. The equation $\Sigma r_i \gamma(x_i) = 0$ similarly translates into $t\Sigma v_i \chi\gamma(x_i) = \Sigma u_i \chi\gamma(x_i)$. Since on X $\chi\gamma = f\phi$, the second equation follows from the first and the fact that f is a $k[t]$-modules homomorphism. This comples the proof that Φ is a full functor.

Finally, we must show that Φ is dense. We first find an $A \in \underline{T}$ such that $\Phi(A) \cong k[t]$. Define A to be F/K where F is a free S-module on the generators x_i $(1 \le i < \omega)$ and K is generated by the elements $ax_{i+1} - bx_i$, $i \ge 1$. We now let M be arbitrary $k[t]$-module and we choose a free resolution $0 \to F_1 \to F_2 \to M$ 0. Taking direct sums of the module A just constructed, we obtain modules G_1 and G_2 in $\underline{T}$ such that $\Phi(G_i) = F_i$ and we obtain a map $f\colon G_1 \to G_2$ corresponding to the map $F_1 \to F_2$. It is easy to see that f is necessarily a monomorphism. One now only has to check that $\Phi(G_2/G_1) \cong M$. It is clear that it is enough to show that G_2/G_1 is a-translatable. If $B = G_2/G_1$, then it is clear that the map $\alpha\colon B/mB \to mB$ is surjective. To prove that it is injective, let $x \in G_2$, and let $\bar{x}$ be the corresponding element of B, and suppose that $a\bar{x} = 0$. It follows that $ax \in G_1$. Since $a^2 = 0$, it follows that $a(ax) = 0$ so $ax \in mG_1$ (since G_1 is a-translatable). Therefore, for some $y \in G_1$, $ay = ax$, so that $a(x-y) = 0$. Since G_2 is a-translatable, this means that $x-y \in mG_2$. Putting this together, we see that $x \in G_1 + mG_2$ so $\bar{x} \in mB$. This shows that B is a-translatable, and completes the proof of the theorem.

Corollary 1.1. If R is a commutative Artinian local ring which is not a principal ideal ring, and k is the residue field of R, then R has $|k|^{\aleph_0}$ nonisomorphic countably generated indecomposable modules.

Proof. By Theorem 1 it suffices to find the right number of countably generated indecomposable modules over $k[t]$. The modules we find are all submodules of the quotient field $k(t)$. If π is any set of irreducible polynomials in $k[t]$, then by $\pi^{-1}k[t]$ we mean the subring of $k(t)$ consisting of rational functions whose denominators are products of irreducible polynomials in π. If π_1 and π_2 are two distinct sets of monic irreducible polynomials in $k[t]$, then $\pi_1^{-1}k[t]$ and $\pi_2^{-1}k[t]$ are not isomorphic as $k[t]$-modules. (For example, if $f \in \pi_1$, $f \notin \pi_2$, then every element in $\pi_1^{-1}k[t]$ is divisible by f, but this is not true for all elements of $\pi_2^{-1}k[t]$.) If k is finite or countable, there are exactly $\aleph_0$ distinct monic irreducible polynomials in $k[t]$, and choosing different sets of these we obtain $2^{\aleph_0} = |k|^{\aleph_0}$ nonisomorphic indecomposable countably generated modules. If k is uncountable, there are exactly $|k|$ distinct monic irreducible polynomials in $k[t]$. If π is any countable set of these, then $\pi^{-1}k[t]$ is a countably generated indecomposable module. There are clearly $|k|^{\aleph_0}$ such subsets.

Remark: Griffith showed in [10] that every commutative Artinian ring which is not a principal ideal ring has indecomposable modules which are not finitely generated. The construction above is essentially that given by the author in [16] though the estimate on the number of indecomposables is not made so precise there. Auslander [2] showed that if R is an Artin algebra which is not of finite module type then it has an indecomposable module which is not finitely generated. The question is still open for Artinian rings. Ringel and Tachikawa show in [13] that if a ring is of finite module type then every module is a direct sum of finitely generated modules. If the truth of the second Brauer-Thrall conjecture can be proved in general (and not just for algebras over perfect fields, which seems to be the state of the theory at this writing), then Auslander's construction would yield as many nonisomorphic indecomposable modules as the cardinality of the field (assuming the field infinite), which would still not be as many as there are in the commutative case. There is still a gap here, therefore, between the commutative and noncommutative theories.

Corollary 1.2. If R is a commutative Noetherian ring such that every countably generated R-module is a direct sum of indecomposable modules, then R is an Artinian principal ideal ring.

Proof. We first show that a discrete valuation ring does not have this property. Let R be a discrete valuation ring with prime p and M the module given by generators $x_i (0 \le i < \infty)$ and relations $px_0 = 0$, $p^n x_n = x_0$. If Q is the quotient field of R, then the module Q/R has elements $z_i (0 \le i < \infty)$ such that $z_0 \ne 0$, $pz_0 = 0$ and

$pz_{n+1} = z_n$ $(n \geq 1)$, from which it follows that in our module M, $x_0 \neq 0$. Clearly x_0 is divisible by all powers of p, while M/Rx_0 is a direct sum of cyclic modules, and hence has no nonzero elements divisible by all powers of p. M cannot be a direct sum of cyclic modules (since it has a nonzero element divisible by all powers of p) and it cannot have a divisible summand (since such a summand could not be contained in Rx_0 and thus it would have a nonzero image in M/Rx_0, which, however, has no nonzero divisible submodule). By [12, Theorem 10] every indecomposable torsion module over a discrete valuation ring is cyclic or divisible, so M cannot be a direct sum of indecomposable modules.

If D is a Dedekind domain which is not a field and R is one of its localizations, then a countably generated torsion R-module is countably generated as a D-module. Hence, we have shown that no Dedekind domain (other than a field) has the property that all of its countably generated modules are direct sums of indecomposable modules. By our remarks at the beginning of this section, we know that if R is any commutative Noetherian ring which is not an Artinian principal ideal ring, then either R has a homomorphic image which is a Dedekind domain which is not a field, or R has a homomorphic image S of the type treated in Theorem 1. In this case, by Theorem 1 there is a category $\underline{T}$ of S-modules which is representation equivalent to Mod k[t] for some field k. Since we have just finished proving that any principal ideal domain has countably generated modules which are not direct sums of indecomposables, the result is proved.

2. RINGS WHOSE IDEAL LATTICE IS NOT DISTRIBUTIVE

In this section we will extend the results of the previous section to left Artinian rings whose lattice of two-sided ideals is not distributive. (In the commutative case, this class coincides with the class of Artinian rings which are not principal ideal rings.) The main result of this section includes Theorem 1 as a special case, but its proof is much more complicated than that of Theorem 1.

The connection between infinite module type and nondistributive ideal lattices was first noticed by J. P. Jans, who used a nondistributive ideal lattice to construct large finite generated indecomposables in [11]. His ideas were generalized from algebras to Artinian rings by Colby [5] and Dickson and Kelly [8]. Dickson and Kelly construct infinitely generated modules, and some of our notation is borrowed from them.

We prove that for any left Artinian ring R whose ideal lattice is not distributive there is a category of R-modules which is representation equivalent to the category of all modules over $D[t]$, where D is a suitable division ring. Starting from this, the methods of the previous and following sections generalize easily to give various "pathological modules" over these rings. We leave details of this to the reader. We remark that this method improves the results of [3] (in effect changing the hypothesis from "I(3)" to "I(2)" in the terminology of that paper).

We may assume that R is a left Artinian ring, and that there are ideals I, I', J, K, and L such that $J + K = J + L = K + L = I$ and $J \cap K = J \cap L = K \cap L = I'$. Without loss of generality we may assume that $I' = 0$ and that if N is the nilradical of R, that $NI = 0 = IN$. We let $\eta : K \to J$ be the projection induced by the bimodule decomposition $I = J \oplus L$. η is an (R,R)-bimodule isomorphism. There is some local idempotent e such that $Ke \neq 0$. η restricts to an (R,eRe)-bimodule isomorphism $Ke \to Je$, which we still call η.

An R-module A will be called η-translatable if it satisfies the following conditions:

(i) The projective cover P of A is a direct sum of copies of Re.

(ii) The induced homomorphism $P/IP \to A/IA$ is an isomorphism.

(iii) The induced map $JP \to JA$ is an isomorphism.

(iv) If $b \in K$, $bA \subseteq \eta(b)A$.

(v) If $x \in A$ and $b_1, b_2 \in K$ and $b_1x = \eta(b_1)y$ then $b_2x = \eta(b_2)y$.

We assume (as we may, by Morita equivalence, which preserves the lattice of two-sided ideals) that R is a basic ring, so that R/N is a product of division rings. In this case, if N' is the annihilator of Re/NRe, then $D = R/N'$ is a division ring, and for any module A of the type described, A/NA may be regarded as a D-vector space. We show that, as before, A/NA can be given the structure of a $D[t]$-module.

If $x \in A-NA$, and $b \in K$, $b \neq 0$ then by (iv) there is a $y \in A$ such that $bx = \eta(b)y$. If $\bar{x}$ and $\bar{y}$ are the corresponding elements of A/NA, we define $t\bar{x} = \bar{y}$. If $x_1 - x \in NA$ then $bx_1 - bx \in INA = 0$ (since $IN = 0$). A similar argument shows that $\bar{y}$ is uniquely determined by $\bar{x}$ and b. This shows that t is well-defined, and (v) implies that the definition is independent of the element $b \in K$ chosen, $(b \neq 0)$.

Because t is defined in terms of the action of elements of the ring, it is easy to verify that if $f : A \to B$ is an R-homomorphism of

η-translatable modules, then $\bar{f} : A/NA \to B/NB$ is a $D[t]$-homomorphism. We therefore have a functor Φ from the category of η-translatable modules to the category of $D[t]$-modules, as before.

Theorem 2. The functor Φ just defined is a representation equivalence between the category $\underline{T}$ of η-translatable modules and the category Mod $D[t]$.

Proof. The proof that $\Phi(f)$ is an isomorphism if and only if f is an isomorphism is (as before) a consequence of the nilpotence of the radical.

To prove that Φ is full, we suppose we are given $f : A/NA \to B/NB$ and choose a set of generators X for A as before, and define $\gamma : X \to A$ and g as before. We must show that $\Sigma r_i x_i = 0$ ($x_i \in X$) implies $\Sigma r_i \gamma(x_i) = 0$, where we may assume that the x_i are distinct elements of X. In this case, it is clear from condition (ii) that $r_i \in R(1-e) + I$ and we may assume that $r_i \in I$. Since $I = J \oplus K$, we write $r_i = a_i - b_i$, $a_i \in J$, $b_i \in K$. We thus must show that

(1) $\Sigma a_i x_i = \Sigma b_i x_i$

implies

(2) $\Sigma a_i \gamma(x_i) = \Sigma b_i \gamma(x_i)$.

If $\eta(b_i) = d_i$, then $t\bar{x}_i = \bar{y}_i$, where the bar denotes reduction modulo N, and

(3) $d_i y_i = b_i x_i$.

We expand

(4) $y_i = \Sigma c_{ij} x_j \qquad (c_{ij} \in Re)$.

This implies that $\Sigma a_i x_i = \Sigma d_i c_{ij} x_j$, and since all coefficients are in Je,

(5) $a_k = \Sigma d_m c_{mk}$

(using condition (iii)).

Since f is a $D[t]$-module homomorphism, and $t\bar{x}_i = \bar{y}_i$, we conclude from (3) and (4) that

$$(6)\qquad t\,\overline{\gamma(x_i)} = \Sigma c_{ij}\,\overline{\gamma(x_j)}$$

whence

$$(7)\qquad b_i\gamma(x_i) = d_i\,\Sigma_j c_{ij}\gamma(x_j).$$

From (5) we also have

$$(8)\qquad a_i\gamma(x_i) = \Sigma_m d_m c_{mi}\gamma(x_i).$$

Combining (7) and (8) we obtain

$$\Sigma_i b_i\gamma(x_i) = \Sigma_{i,j}\, d_i c_{ij}\gamma(x_j) = \Sigma_j(\Sigma_m d_m c_{mj}\gamma(x_j))$$

$$= \Sigma_j a_j\gamma(x_j)$$

which is the desired formula (2).

We finally prove that Φ is dense. We first need to construct an η-translatable module A such that $\Phi(A) \cong D[t]$. We define A by taking a free module on generators $x_1, x_2, \ldots$ and adding the relations $(1-e)x_i = 0$ and $bx_i = \eta(b)x_{i+1}$, $b \in K$. It is routine to verify the conditions (i) - (v). We then consider η-translatable modules A and B and an $f : B \to A$ such that $\bar{f} : B/NB \to A/NA$ is injective. The proof will be completed (as in the proof of Theorem 1) be showing $A/f(B)$ is η-translatable. Choosing suitable projective covers for A and B, we see that (i), (ii), and (iii) are clear, and that $f : B \to A$ is a split monomorphism modulo I. (iv) holds since it holds in A. (v) easily reduces to knowing that if b_1 and b_2 are in K and $\eta(b_1)z \in f(B)$, then $\eta(b_2)z \in f(B)$. The fact that f is a split monomorphism modulo I and condition (iii) applied to A makes it clear that $f(B) \cap JA = f(JB)$. From this it is clear that for some $w \in B$, $z-f(w) \in NA$, whence (since $IN = 0$) one has $\eta(b_2)z = \eta(b_2)f(w)$.

§3. MODULES WITH SPECIFIED ENDOMORPHISM RINGS

Corner showed in [6] that any reduced countable torsion-free ring is the endomorphism ring of some countable torsion-free Abelian group. He used this to prove, for example, that there are countable torsion-free groups with no indecomposable summands. In [16], the author used similar techniques to show that if R is a discrete valuation ring and R^* is its completion, and if R^* has uncountable transcendence degree over R, and if E is any countably generated reduced torsion-free R-algebra, then there is a countably generated torsion-free R-module with endomorphism ring isomorphic to E. As an application, it is shown in [16] that every commutative Artinian ring which is not a principal ideal ring has a countably generated module which has no indecomposable summands. (One uses the discrete valuation ring $k[t]_{(t)}$, where k is the residue field of the local Artinian ring. If k is uncountable, the result of a direct application of Theorem 1 will not be a countably generated module, but it does have countable rank, and a suitable countably generated submodule of it can be found.)

In a subsequent paper [4], Brenner and Ringel prove a generalization of the theorem in [16]. They relax the condition on the transcendence degree of the completion of the discrete valuation ring, and improve the result to give an imbedding of certain categories of modules over R-algebras into the category of R-modules (where R is the discrete valuation ring). In this section, we prove a result which is contained in theirs, but which is proved in a considerably more elementary way. The purpose of this section is to present a very natural and short proof of a theorem of this type, rather than to prove the best possible result.

Theorem 3. Let R be a discrete valuation ring and R* its completion and assume that R* is not algebraic over R. Let Λ be a finitely generated R-algebra and $\underline{C}$ a full subcategory of Mod Λ such that if X is an object of $\underline{C}$ then X is free as an R-module. Then there is a full imbedding of $\underline{C}$ into Mod R.

Proof. Let Λ be generated by $e_1,\ldots,e_n$ with $e_1 = 1$, and let $a_1,\ldots,a_n$ be elements of R* quadratically independent over R. We let p be the prime of R and for every Λ-module X, we let X* be its p-adic completion (as an R-module), which clearly can be given the structure of a Λ-module in a natural way, and is also an R*-module. Let $\tau = \Sigma_{i=1}^{n} a_i e_i$ which we can regard as an element of $R^* \otimes_R \Lambda$. For any $X \in \underline{C}$, let $\Phi(X)$ be the pure R-submodule of X* generated by $X + \tau X$, (that is, all elements $y \in X^*$ such that for some positive integer n, $p^n y = x + \tau z$ for some x and z in X.) This clearly defines a faithful functor $\underline{C} \to$ Mod R since if $f : X \to Y$ is a homomorphism of Λ-modules, then f extends to a unique $f^* : X^* \to Y^*$ which is a $R^* \otimes \Lambda$-homomorphism.

To show that Φ is a full functor, we first show that if $\theta : \Phi(X) \to \Phi(Y)$ is an R-homomorphism then $\theta(X) \subseteq Y$. We let θ^* be the extension of θ to an R^*-homomorphism $X^* \to Y^*$, using the fact that $X^* = \Phi(X)^*$. We calculate

$$(*) \qquad \theta\tau x = \theta\Sigma a_i e_i = \Sigma a_i \theta^* e_i x .$$

The first term of the expression on the right is $a_1\theta x$. For some integer k, $p^k\theta x \in Y + \tau Y$, and $p^k\theta\tau x \in Y + \tau Y$, so $p^k\theta x = y + \tau z$, $(y,z \in Y)$. Substituting this expression in the previous equation, we obtain

$$p^k\theta\tau x = a_1 y + a_1\Sigma a_i e_i z + \cdots$$

Since Y is a free R-module, and the elements $a_i (1 \leq i \leq n)$ are quadratically independent over R, it follows that the submodules Y, a_iY, and a_ia_jY of $Y^*(1 \leq i,j \leq n)$ are independent R-submodules of Y^*. Hence in equation (*), the coefficient of each term a_ia_j is zero. The coefficient of a_1^2 is z, so $z = 0$, which implies that $p^k\theta x \in Y$. Since Y is a pure submodule of Y^*, it follows that $\theta(x) \subseteq Y$.

It is clear that θ is determined by its action on X, so to show that the map $\text{Hom}_\Lambda (X,Y) \to \text{Hom}_R(\Phi(X), \Phi(Y))$ is surjective, we need only show that θ restricted to X is a Λ-homomorphism. If $x \in X$, then since $\theta^* : X^* \to Y^*$ is an R^*-module homomorphism, $\theta^*(a_iX) \subseteq a_iY$, so $\theta\tau x \in \tau Y$. If $\theta\tau x = \tau y$, then

$$(**) \qquad \Sigma a_i\theta^* e_i x = \Sigma a_i e_i y.$$

Again, since the submodules a_iY are independent, we obtain $a_1\theta e_1 x = a_1 e_1 y$, and since $e_1 = 1$, $\theta x = y$. From the other terms of (**), we obtain $\theta e_1 x = e_i y$, and since we just verified that $y = \theta x$, we obtain $\theta e_i x = e_i\theta x$ for all i, $1 \leq i \leq n$. This shows that the restriction of θ to X is a Λ-module homomorphism and completes the proof that the functor is full.

Corollary 3.1. If R is a discrete valuation ring with completion R^* such that R^* is not algebraic over R and such that R is not Hensel, then the Krull-Schmidt theorem fails for torsion-free R-modules of finite rank.

Proof. By a result of E. G. Evans, Jr. [9], since R is not Hensel, there is a finite R-algebra A and a finitely generated A-module M with two direct sum decompositions into indecomposable modules which are not isomorphic. Inspection of Evan's construction shows that A and M are both torsion-free over R. An application of the previous theorem yields a full exact embedding of the category of finitely generated A-modules into the category of torsion-free R-modules of finite rank, thus proving the stated result.

Remark: For the ring $Z_{(p)}$ of integers localized at a prime p, the failure of the Krull-Schmidt theorem for torsion-free modules of finite rank has been folklore for many years, but no examples are in the literature. (The first examples seem to have been obtained about 15 years ago by M. C. R. Butler.) It has been recently proved by E. L. Lady (not yet published) that if R is a Hensel discrete valuation ring, then the torsion-free R-modules of finite rank do satisfy the Krull-Schmidt theorem.

Corollary 3.2. If S is a commutative Artinian local ring which is not a principal ideal ring, m is the maximal ideal of S and k = S/m, then there is an S-module M with endomorphism ring E such that E/J(E) is a finite dimensional k-algebra and such that the Krull-Schmidt theorem does not hold for direct sum decompositons of M.

Proof. It is easily verified that the ring $k[t]_{(t)}$ is not Hensel. We obtain immediately from Theorem 3 a countably generated $k[t]_{(t)}$-module M with the desired properties. (We refer to [17] for a proof that if R is a discrete valuation ring with maximal ideal I and M a reduced torsion-free R-module of finite rank, and E is the endomorphism ring of M, then E/J(E) is a finite dimensional R/I-module.)

The reader who likes his modules countably generated will notice that if the field k is uncountable then the module we just constructed will not be countably generated. The example can be adjusted to give a countably generated module in which the Krull-Schmidt theorem fails, using the methods used in [16], but in the process one will lose the stated property of the endomorphism ring.

One can easily generalize the above results to include noncommutative localizations of rings like D[t] (where D is a noncommutative division ring). Suppose R is a noncommutative discrete valuation ring (a local domain, not a division ring, such that every one sided ideal is a power of the maximal ideal). Suppose that the completion of R contains an element which centralizes R and which is not algebraic. By a finitely generated R-algebra we mean a ring S with a ring homomorphism $R \to S$ taking 1 to 1, such that S is generated over R by a finite number of elements that centralizes R. In this situation all of the arguments of Theorem 3 go through completely. Clearly a localization $D[t]_{(t)}$ satisfies these conditions, so the results of Theorems 2 and 3 can be used to produce modules with various properties over any left Artinian ring whose ideal lattice is not distributive.

The two corollaries above are proved using commutative methods. However, in [16] a construction is given of a countably generated module which violates the Krull-Schmidt and cancellation properties. The proof uses the author's version of Corner's theorem. The R-algebra involved is finitely generated and commutatively is not essential, so this method can be used to yield countably generated indecomposable modules A, B, and C over any left Artinian whose ideal lattice is not distributive, such that B and C are not isomorphic, and such that $A \oplus B \cong A \oplus C$.

In [4], Brenner and Ringel consider the ring $\Lambda_6(R)$ consisting of 6×6 matrices over R which are zero except on the diagonal and top row. Adapting arguments of Corner's from [7], they construct a full exact embedding $\phi : R\text{ Mod} \to \Lambda_6(R)\text{ Mod}$ (for any ring R) such that $\phi({}_R R)$ as an R-module is free on as many generators as one likes, provided that the number of generators is less than the first strongly inaccessible cardinal. Combining this with the other results (when R is a discrete valuation ring) we obtain for any left Artinian ring whose ideal lattice is not distributive a family of very large indecomposable modules. It would be interesting to know if the set theoretic methods which have recently enabled one to cross the inaccessible cardinal barrier in Abelian group theory would also do so here.

REFERENCES

1. M. Auslander, Representation dimension of Artin Algebras, Queen Mary College Lecture Notes, 1971.

2. M. Auslander, Large modules over Artin algebras, in "Algebra, Topology and Category Theory", ed. A. Heller and M. Tierney, Academic Press, New York, 1976.

3. S. Brenner, Some modules with nearly prescribed endomorphism rings, J. Alg. 23 (1972), 250-262.

4. S. Brenner and C. Ringel, Pathological modules over tame rings, preprint.

5. R. R. Colby, On indecomposable modules over rings with minimum condition, Pac. J. Math 19 (1966), 23-33.

6. A. L. S. Corner, Every countable reduced torsion-free ring is an endomorphism ring, Proc. London Math. Soc. (3) 13 (1963), 687-710.

7. A. L. S. Corner, Endomorphism algebras of large modules with distinguished submodules, J. Alg. 11 (1969), 155-185.

8. S. E. Dickson and G. M. Kelly, Interlacing methods and large indecomposables, Bull. Austr. Math. Soc., 3 (1970), 337-348.

9. E. G. Evans, Jr. Krull-Schmidt and cancellation over local rings, Pac. J. Math. 46 (1973), 115-121.

10. P. Griffith, On the decomposition of modules and generalized left uniserial rings, Math. Ann. 184 (1970), 300-308.

11. J. P. Jans, On the indecomposable representations of algebras, Ann. Math. (2) 66 (1957), 418-429.

12. I. Kaplansky, Infinite Abelian Groups (revised edition), Ann Arbor, 1969.

13. C. Ringel and H. Tachikawa, QF-3 Rings, J. Reine Angew. Math. 272 (1975), 49-72.

14. R. B. Warfield, Jr., Decomposability of finitely presented modules, Proc. Amer. Math. Soc. 25 (1970), 167-172.

15. R. B. Warfield, Jr., Rings whose modules have nice decompositions, Math. Z. 125 (1972), 187-192.

16. R. B. Warfield, Jr., Countably generated modules over commutative Artinian rings, Pac. J. Math. 60 (1975), 289-302.

17. R. B. Warfield, Jr., Notes on cancellation, stable range, and related topics, University of Washington, 1975.